Methods in Enzymology

Volume 271
HIGH RESOLUTION SEPARATION
AND ANALYSIS OF
BIOLOGICAL MACROMOLECULES
Part B
Applications

METHODS IN ENZYMOLOGY

EDITORS-IN-CHIEF

John N. Abelson   Melvin I. Simon

DIVISION OF BIOLOGY
CALIFORNIA INSTITUTE OF TECHNOLOGY
PASADENA, CALIFORNIA

FOUNDING EDITORS

Sidney P. Colowick and Nathan O. Kaplan

Methods in Enzymology

Volume 271

High Resolution Separation and Analysis of Biological Macromolecules

Part B
Applications

EDITED BY

Barry L. Karger

DEPARTMENT OF CHEMISTRY
BARNETT INSTITUTE
NORTHEASTERN UNIVERSITY
BOSTON, MASSACHUSETTS

William S. Hancock

HEWLETT-PACKARD
PALO ALTO, CALIFORNIA

ACADEMIC PRESS

San Diego New York Boston London Sydney Tokyo Toronto

Academic Press, Inc.
A Division of Harcourt Brace & Company
525 B Street, Suite 1900, San Diego, California 92101-4495

United Kingdom Edition published by
Academic Press Limited
24-28 Oval Road, London NW1 7DX

International Standard Serial Number: 0076-6879

International Standard Book Number: 0-12-182172-2

PRINTED IN THE UNITED STATES OF AMERICA
96   97   98   99   00   01   MM   9   8   7   6   5   4   3   2   1

Table of Contents

Section I. Liquid Chromatography

Section II. Electrophoresis
A. Slab Gel Electrophoresis: High Resolution

B. Capillary Electrophoresis

Section III. Mass Spectrometry

Contributors to Volume 271

Article numbers are in parentheses following the names of contributors.
Affiliations listed are current.

SAMY ABDEL-BAKY (21), *BASF Corporation, Agricultural Products Center, Research Triangle Park, North Carolina 27709*

KARIMAN ALLAM (21), *Webb Technical Group, Raleigh, North Carolina 27612*

A. APFFEL (17), *Hewlett Packard Laboratories, Palo Alto, California 94304*

NEBOJSA AVDALOVIC (7), *Dionex Corporation, Sunnyvale, California 94088*

LOUISETTE J. BASA (6), *Genentech, Inc., South San Francisco, California 94080*

JAN BERKA (13), *Barnett Institute, Northeastern University, Boston, Massachusetts 02115*

ROBERT L. BRUMLEY, JR. (10), *GeneSys, Inc., Mazomanie, Wisconsin 53560*

ERIC C. BUXTON (10), *Department of Chemistry, University of Wisconsin, Madison, Wisconsin 53706*

J. CHAKEL (17), *Hewlett Packard Laboratories, Palo Alto, California 94304*

STEPHEN CHAN (16), *Mass Spectrometry Resource, Boston University Medical Center, Boston, Massachusetts 02118*

ROSANNE C. CHLOUPEK (2), *Genentech, Inc., South San Francisco, California 94080*

JOSEPH M. CORBETT (8), *Department of Cardiothoracic Surgery, National Heart and Lung Institute, Imperial College, Heart Science Centre, Harefield Hospital, Harefield, Middlesex UB9 6JH, United Kingdom*

MICHAEL J. DUNN (8), *Department of Cardiothoracic Surgery, National Heart and Lung Institute, Imperial College, Heart Science Centre, Harefield Hospital, Harefield, Middlesex UB9 6JH, United Kingdom*

FRANTISEK FORET (13), *Barnett Institute, Northeastern University, Boston, Massachusetts 02115*

JOHN FRENZ (20), *Department of Manufacturing Sciences, Genentech, Inc., South San Francisco, California 94080*

MICHAEL GIDDINGS (10), *Department of Chemistry, University of Wisconsin, Madison, Wisconsin 53706*

ROGER W. GIESE (21), *Barnett Institute and Bouve College, Northeastern University, Boston, Massachusetts 02115*

BETH L. GILLECE-CASTRO (18), *Protein Chemistry Department, Genentech, Inc., South San Francisco, California 94080*

DAVID R. GOODLETT (19), *Chemical Methods and Separations Group, Chemical Sciences Department, Pacific Northwest Laboratory, Richland, Washington 99352*

A. W. GUZZETTA (17), *Scios Nova, Inc., Mountain View, California 94043*

WILLIAM S. HANCOCK (17), *Hewlett Packard Laboratories, Palo Alto, California 94304*

ROBERT S. HODGES (1), *Department of Biochemistry and the Medical Research Council, Group in Protein Structure and Function, University of Alberta, Edmonton, Alberta T6G 2H7, Canada*

EDWARD R. HOFF (2), *Genentech, Inc., South San Francisco, California 94080*

STEVEN A. HOFSTADLER (19), *Chemical Methods and Separations Group, Chemical Sciences Department, Pacific Northwest Laboratory, Richland, Washington 99352*

L. J. JANIS (4), *Lilly Research Laboratories, Eli Lilly and Company, Lilly Corporate Center, Indianapolis, Indiana 46285*

DJURO JOSIC (5), *Octapharma Pharmazeutika Produktionsges.m.b.H, Research and Development Department, A-1100 Wien, Austria*

BARRY L. KARGER (13), *Department of Chemistry, Barnett Institute, Northeastern University, Boston, Massachusetts 02115*

KEN-ICHI KASAI (9), *Department of Biological Chemistry, Faculty of Pharmaceutical Sciences, Teikyo University, Sagamiko, Kanagawa 199-01 Japan*

P. M. KOVACH (4), *Lilly Research Laboratories, Eli Lilly and Company, Lilly Corporate Center, Indianapolis, Indiana 46285*

TERRY D. LEE (3), *City of Hope, Division of Immunology, Beckman Research Institute, Duarte, California 91010*

COLIN T. MANT (1), *Department of Biochemistry and the Medical Research Council, Group in Protein Structure and Function, University of Alberta, Edmonton, Alberta T6G 2H7 Canada*

MICHAEL MARCHBANKS (10), *Hazleton Wisconsin, Inc., Madison, Wisconsin 53704*

RANDY M. MCCORMICK (7), *Seurat Analytical Systems, Sunnyvale, California 94089*

T. M'TIMKULU (17), *Berlex Biosciences, Brisbane, California 94005*

MILOS V. NOVOTNY (14), *Department of Chemistry, Indiana University, Bloomington, Indiana 47405*

E. PUNGOR, JR. (17), *Berlex Biosciences, Brisbane, California 94005*

BRUCE B. REINHOLD (16), *Mass Spectrometry Resource, Boston University Medical Center, Boston, Massachusetts 02118*

VERNON N. REINHOLD (16), *Mass Spectrometry Resource, Boston University Medical Center, Boston, Massachusetts 02118*

EUGENE C. RICKARD (11), *Lilly Research Laboratories, Eli Lilly and Company, Lilly Corporate Center, Indianapolis, Indiana 46285*

R. M. RIGGIN (4), *Lilly Research Laboratories, Eli Lilly and Company, Lilly Corporate Center, Indianapolis, Indiana 46285*

MANASI SAHA (21), *BASF Corporation, Agricultural Products Center, Research Triangle Park, North Carolina 27709*

KIYOHITO SHIMURA (9), *Department of Biological Chemistry, Faculty of Pharmaceutical Sciences, Teikyo University, Sagamiko, Kanagawa 199-01 Japan*

JOHN E. SHIVELY (3), *City of Hope, Division of Immunology, Beckman Research Institute, Duarte, California 91010*

LLOYD M. SMITH (10), *Department of Chemistry, University of Wisconsin, Madison, Wisconsin 53706*

RICHARD D. SMITH (19), *Environmental Molecular Sciences Laboratory, Pacific Northwest National Laboratory, Richland, Washington 99352*

C. SOUDERS (17), *Berlex Biosciences, Brisbane, California 94005*

MICHAEL W. SPELLMAN (6), *Genentech, Inc., South San Francisco, California 94080*

JOHN T. STULTS (18), *Protein Chemistry Department, Genentech, Inc., South San Francisco, California 94080*

KRISTINE M. SWIDEREK (3), *City of Hope, Division of Immunology, Beckman Research Institute, Duarte, California 91010*

GLEN TESHIMA (12), *Department of Analytical Chemistry, Genentech, Inc., South San Francisco, California 94080*

JAMES R. THAYER (7), *Dionex Corporation, Sunnyvale, California 94088*

XINCHUN TONG (10), *Department of Chemistry, University of Wisconsin, Madison, Wisconsin 53706*

JOHN K. TOWNS (4, 11), *Lilly Research Laboratories, Eli Lilly and Company, Lilly Corporate Center, Indianapolis, Indiana 46285*

R. REID TOWNSEND (6), *Department of Pharmaceutical Chemistry, University of California at San Francisco, San Francisco, California 94143*

HAROLD R. UDSETH (19), *Chemical Methods and Separations Group, Chemical Sciences Department, Pacific Northwest Laboratory, Richland, Washington 99352*

JON H. WAHL (19), *Chemical Methods and Separations Group, Chemical Sciences Department, Pacific Northwest Laboratory, Richland, Washington 99352*

SHIAW-LIN WU (12), *Department of Analytical Chemistry, Genentech, Inc., South San Francisco, California 94080*

JOHN R. YATES (15), *Department of Molecular Biotechnology, School of Medicine, University of Washington, Seattle, Washington 98195*

KATRIN ZEILINGER (5), *Virchow-Klinikum der Humbold Universität, Experimentelle Chirurgie, 13353 Berlin, Germany*

Preface

All areas of the biological sciences have become increasingly molecular in the past decade, and this has led to ever greater demands on analytical methodology. Revolutionary changes in quantitative and structure analysis have resulted, with changes continuing to this day. Nowhere has this been seen to a greater extent than in the advances in macromolecular structure elucidation. This advancement toward the exact chemical structure of macromolecules has been essential in our understanding of biological processes. This trend has fueled demands for increased ability to handle vanishingly small quantities of material such as from tissue extracts or single cells. Methods with a high degree of automation and throughput are also being developed.

In the past, the analysis of macromolecules in biological fluids relied on methods that used specific probes to detect small regions of the molecule, often in only partially purified samples. For example, proteins were labeled with radioactivity by *in vivo* incorporation. Another approach has been the detection of a sample separated in a gel electrophoresis by means of blotting with an antibody or with a tagged oligonucleotide probe. Such procedures have the advantages of sensitivity and specificity. The disadvantages of such approaches, however, are many, and range from handling problems of radioactivity, as well as the inability to perform a variety of *in vivo* experiments, to the invisibility of residues out of the contact domain of the tagged region, e.g., epitope regions in antibody-based recognition reactions.

Beyond basic biological research, the advent of biotechnology has also created a need for a higher level of detail in the analysis of macromolecules, which has resulted in protocols that can detect the transformation of a single functional group in a protein of 50,000–100,000 daltons or the presence of a single or modified base change in an oligonucleotide of several hundred or several thousand residues. The discovery of a variety of posttranslational modifications in proteins has further increased the demand for a high degree of specificity in structure analysis. With the arrival of the human genome and other sequencing initiatives, the requirement for a much more rapid method for DNA sequencing has stimulated the need for methods with a high degree of throughput and low degree of error.

The bioanalytical chemist has responded to these challenges in biological measurements with the introduction of new, high resolution separation and detection methods that allow for the rapid analysis and characterization of macromolecules. Also, methods that can determine small differences in many thousands of atoms have been developed. The separation techniques include affinity chromatography, reversed phase liquid chromatography (LC), and capillary electrophoresis. We include mass spectrometry as a high resolution separation method, both given the fact that the method is fundamentally a procedure for separating gaseous ions and because separation–mass spectrometry (LC/MS, CE/MS) is an integral part of modern bioanalysis of macromolecules.

The characterization of complex biopolymers typically involves cleavage of the macromolecule with specific reagents, such as proteases, restriction enzymes, or chemical cleavage substances. The resulting mixture of fragments is then separated to produce a map (e.g., peptide map) that can be related to the original macromolecule from knowledge of the specificity of the reagent used for the cleavage. Such fingerprinting approaches reduce the characterization problem from a single complex substance to a number of smaller and thus simpler units that can be more easily analyzed once separation has been achieved.

Recent advances in mass spectrometry have been invaluable in determining the structure of these smaller units. In addition, differences in the macromolecule relative to a reference molecule can be related to an observable difference in the map. The results of mass spectrometric measurements are frequently complemented by more traditional approaches, e.g., N-terminal sequencing of proteins or the Sanger method for the sequencing of oligonucleotides. Furthermore, a recent trend is to follow kinetically the enzymatic degradation of a macromolecule (e.g., carboxypeptidase). By measuring the molecular weight differences of the degraded molecule as a function of time using mass spectrometry [e.g., matrix-assisted laser desorption ionization–time of flight (MALDI–TOF)], individual residues that have been cleaved (e.g., amino acids) can be determined.

As well as producing detailed chemical information on the macromolecule, many of these methods also have the advantage of a high degree of mass sensitivity since new instrumentation, such as MALDI–TOF or capillary electrophoresis with laser-based fluorescence detection, can handle vanishingly small amounts of material. The low femtomole to attomole sensitivity achieved with many of these systems permits detection more sensitive than that achieved with tritium or ^{14}C isotopes and often equals that achieved with the use of ^{32}P or ^{125}I radioactivity. A trend in mass spectrometry has been the extension of the technology to ever greater mass ranges so that now proteins of molecular weights greater than 200,000 and oligonucleotides of more than 100 residues can be transferred into the gas phase and then measured in a mass analyzer.

The purpose of Volumes 270 and 271 of *Methods in Enzymology* is to provide in one source an overview of the exciting recent advances in the analytical sciences that are of importance in contemporary biology. While core laboratories have greatly expanded the access of many scientists to expensive and sophisticated instruments, a decided trend is the introduction of less expensive, dedicated systems that are installed on a widespread basis, especially as individual workstations. The advancement of technology and chemistry has been such that measurements unheard of a few years ago are now routine, e.g., carbohydrate sequencing of glycoproteins. Such developments require scientists working in biological fields to have a greater understanding and utilization of analytical methodology. The chapters provide an update in recent advances of modern analytical methods that allow the practitioner to extract maximum information from an analysis. Where possible, the chapters also have a practical focus and concentrate on methodological details which are key to a particular method.

The contributions appear in two volumes: Volume 270, High Resolution Separation of Biological Macromolecules, Part A: Fundamentals and Volume 271, High Resolution Separation of Biological Macromolecules, Part B: Applications. Each volume is subdivided into three main areas: liquid chromatography, slab gel and capillary electrophoresis, and mass spectrometry. One important emphasis has been the integration of methods, in particular LC/MS and CE/MS. In many methods, chemical operations are integrated at the front end of the separation and may also be significant in detection. Often in an analysis, a battery of methods are combined to develop a complete picture of the system and to cross-validate the information.

The focus of the LC section is on updating the most significant new approaches to biomolecular analysis. LC has been covered in recent volumes of this series, therefore these volumes concentrate on relevant applications that allow for automation, greater speed of analysis, or higher separation efficiency. In the electrophoresis section, recent work with slab gels which focuses on high resolution analysis is covered. Many applications are being converted from the slab gel into a column format to combine the advantages of electrophoresis with those of chromatography. The field of capillary electrophoresis, which is a recent, significant high resolution method for biopolymers, is fully covered.

The third section contains important methods for the ionization of macromolecules into the gas phase as well as new methods for mass measurements which are currently in use or have great future potential. The integrated or hybrid systems are demonstrated with important applications.

We welcome readers from the biological sciences and feel confident that they will find these volumes of value, particularly those working at the interfaces between analytical/biochemical and molecular biology, as well as the immunological sciences. While new developments constantly occur, we believe these two volumes provide a solid foundation on which researchers can assess the most recent advances. We feel that biologists are working during a truly revolutionary period in which information available for the analysis of biomacromolecular structure and quantitation will provide new insights into fundamental processes. We hope these volumes aid readers in advancing significantly their research capabilities.

WILLIAM S. HANCOCK
BARRY L. KARGER

METHODS IN ENZYMOLOGY

VOLUME 91. Enzyme Structure (Part I)
Edited by C. H. W. HIRS AND SERGE N. TIMASHEFF

VOLUME 92. Immunochemical Techniques (Part E: Monoclonal Antibodies and General Immunoassay Methods)
Edited by JOHN J. LANGONE AND HELEN VAN VUNAKIS

VOLUME 93. Immunochemical Techniques (Part F: Conventional Antibodies, Fc Receptors, and Cytotoxicity)
Edited by JOHN J. LANGONE AND HELEN VAN VUNAKIS

VOLUME 94. Polyamines
Edited by HERBERT TABOR AND CELIA WHITE TABOR

VOLUME 95. Cumulative Subject Index Volumes 61–74, 76–80
Edited by EDWARD A. DENNIS AND MARTHA G. DENNIS

VOLUME 96. Biomembranes [Part J: Membrane Biogenesis: Assembly and Targeting (General Methods; Eukaryotes)]
Edited by SIDNEY FLEISCHER AND BECCA FLEISCHER

VOLUME 97. Biomembranes [Part K: Membrane Biogenesis: Assembly and Targeting (Prokaryotes, Mitochondria, and Chloroplasts)]
Edited by SIDNEY FLEISCHER AND BECCA FLEISCHER

VOLUME 98. Biomembranes (Part L: Membrane Biogenesis: Processing and Recycling)
Edited by SIDNEY FLEISCHER AND BECCA FLEISCHER

VOLUME 99. Hormone Action (Part F: Protein Kinases)
Edited by JACKIE D. CORBIN AND JOEL G. HARDMAN

VOLUME 100. Recombinant DNA (Part B)
Edited by RAY WU, LAWRENCE GROSSMAN, AND KIVIE MOLDAVE

VOLUME 101. Recombinant DNA (Part C)
Edited by RAY WU, LAWRENCE GROSSMAN, AND KIVIE MOLDAVE

VOLUME 102. Hormone Action (Part G: Calmodulin and Calcium-Binding Proteins)
Edited by ANTHONY R. MEANS AND BERT W. O'MALLEY

VOLUME 103. Hormone Action (Part H: Neuroendocrine Peptides)
Edited by P. MICHAEL CONN

VOLUME 104. Enzyme Purification and Related Techniques (Part C)
Edited by WILLIAM B. JAKOBY

VOLUME 105. Oxygen Radicals in Biological Systems
Edited by LESTER PACKER

VOLUME 106. Posttranslational Modifications (Part A)
Edited by FINN WOLD AND KIVIE MOLDAVE

VOLUME 107. Posttranslational Modifications (Part B)
Edited by FINN WOLD AND KIVIE MOLDAVE

Volume 143. Sulfur and Sulfur Amino Acids
Edited by WILLIAM B. JAKOBY AND OWEN GRIFFITH

Volume 144. Structural and Contractile Proteins (Part D: Extracellular Matrix)
Edited by LEON W. CUNNINGHAM

Volume 145. Structural and Contractile Proteins (Part E: Extracellular Matrix)
Edited by LEON W. CUNNINGHAM

Volume 146. Peptide Growth Factors (Part A)
Edited by DAVID BARNES AND DAVID A. SIRBASKU

Volume 147. Peptide Growth Factors (Part B)
Edited by DAVID BARNES AND DAVID A. SIRBASKU

Volume 148. Plant Cell Membranes
Edited by LESTER PACKER AND ROLAND DOUCE

Volume 149. Drug and Enzyme Targeting (Part B)
Edited by RALPH GREEN AND KENNETH J. WIDDER

Volume 150. Immunochemical Techniques (Part K: *In Vitro* Models of B and T Cell Functions and Lymphoid Cell Receptors)
Edited by GIOVANNI DI SABATO

Volume 151. Molecular Genetics of Mammalian Cells
Edited by MICHAEL M. GOTTESMAN

Volume 152. Guide to Molecular Cloning Techniques
Edited by SHELBY L. BERGER AND ALAN R. KIMMEL

Volume 153. Recombinant DNA (Part D)
Edited by RAY WU AND LAWRENCE GROSSMAN

Volume 154. Recombinant DNA (Part E)
Edited by RAY WU AND LAWRENCE GROSSMAN

Volume 155. Recombinant DNA (Part F)
Edited by RAY WU

Volume 156. Biomembranes (Part P: ATP-Driven Pumps and Related Transport: The Na,K-Pump)
Edited by SIDNEY FLEISCHER AND BECCA FLEISCHER

Volume 157. Biomembranes (Part Q: ATP-Driven Pumps and Related Transport: Calcium, Proton, and Potassium Pumps)
Edited by SIDNEY FLEISCHER AND BECCA FLEISCHER

Volume 158. Metalloproteins (Part A)
Edited by JAMES F. RIORDAN AND BERT L. VALLEE

Volume 159. Initiation and Termination of Cyclic Nucleotide Action
Edited by JACKIE D. CORBIN AND ROGER A. JOHNSON

Volume 160. Biomass (Part A: Cellulose and Hemicellulose)
Edited by WILLIS A. WOOD AND SCOTT T. KELLOGG

Section I

Liquid Chromatography

[1] Analysis of Peptides by High-Performance Liquid Chromatography

By COLIN T. MANT and ROBERT S. HODGES

I. Introduction

A. Focus

Even the most superficial perusal of the literature for the purpose of reviewing high-performance liquid chromatography (HPLC) separations of peptides quickly reveals that shortage of relevant material is certainly not a problem. This is due primarily to the tremendous development of high-performance chromatographic techniques in the past few years, in terms of scale, instrumentation, and column packings. In addition, there is an almost bewildering variety of mobile phases employed by various researchers for specific applications in all major modes of HPLC employed for peptide separations.

This chapter is aimed at laboratory-based researchers, both beginners and more experienced chromatographers, who wish to learn about peptide applications in HPLC. Thus, standard analytical applications in HPLC of peptides will be stressed, as opposed to micro- or preparative-scale chromatography. Only nonspecialized columns, mobile phases, and instrumentation readily available and easily employed by the researcher are described in detail. In addition, through the use of peptide standards specifically designed for HPLC, the researcher is introduced to standard operating conditions that should first be attempted in the separation of a peptide mixture.

B. Characterization of Peptides

The distinction between a peptide, polypeptide, or protein, in terms of the number of peptide residues they contain, is somewhat arbitrary. However, peptides are usually defined as containing 50 amino acid residues or less. Although molecules containing more than 50 residues usually have a stable 3-dimensional structure in solution, and are referred to as proteins, conformation can be an important factor in peptides as well as proteins. Secondary structure, e.g., α helix, is generally absent even under benign conditions for small peptides (up to ~15 residues); however, the potential for a defined secondary or tertiary structure increases with increasing peptide length and, for peptides containing more than 20–35 residues, folding of the peptide chain to internalize nonpolar residues is likely to become an increasingly important conformational feature. In addition, the presence of disulfide bridge(s) would be expected to affect peptide conformation and, thus, the retention behavior of a peptide in HPLC may differ from that in the fully reduced state.[1] Thus, conformation should always be a consideration when choosing the conditions for chromatography.

C. Peptide Detection

Peptide bonds absorb light strongly in the far ultraviolet (<220 nm), providing a convenient means of detection (usually between 210 and 220 nm). In addition, the aromatic side chains of tyrosine, phenylalanine, and tryptophan absorb light in the 250 to 290-nm ultraviolet range. The development of multiwavelength detectors, enabling the simultaneous detection of peptide bond and aromatic side-chain absorbance, has proved of immense value for the rapid separation and identification of peptides and proteins.

D. Major Modes of HPLC Used in Peptide Separations

Because amino acids are the fundamental units of peptides, the chromatographic behavior of a particular peptide will be determined by the number and properties (polarity, charge potential) of the residue side chains it contains. Thus, the major modes of HPLC employed in peptide separations take advantage of differences in peptide size (size-exclusion HPLC, or SEC), net charge (ion-exchange HPLC, or IEC), or hydrophobicity (reversed-phase HPLC, or RP-HPLC; and, to a lesser extent, hydrophobic interaction chromatography, or HIC). Within these modes, mobile-phase conditions may be manipulated to maximize the separation potential of a particular HPLC column.

E. Peptide Sources and Separation Goals

HPLC has proved versatile in the isolation of peptides from a wide variety of sources. The complexity of peptide mixtures will vary widely depending on the source, because peptides derived from various sources differ widely in size, net charge, and polarity. In addition, the quantity of peptides to be isolated will depend on their origin, e.g., peptides obtained from biological tissues are often found only in small quantities, whereas quantities of peptides obtained from protein cleavage or solid phase synthesis may be considerably larger.

[1] K. K. Lee, J. A. Black, and R. S. Hodges, in "HPLC of Peptides and Proteins: Separation, Analysis and Conformation" (C. T. Mant and R. S. Hodges, eds.), p. 389. CRC Press, Boca Raton, FL, 1991.

As a general rule, the approach to separation must be tailored to the separation goals, i.e., purification of a single peptide from a complex mixture (e.g., the purification of a synthetic peptide from synthetic impurities following solid phase peptide synthesis) will require a different approach to that necessary for separating all components of a complex mixture (e.g., peptide fragments resulting from tryptic cleavage of a protein). The former approach may only require the application of a single HPLC technique, i.e., taking advantage of only one property (size, charge, or polarity) of the peptide of interest. In contrast, the latter approach will generally require a combination of separation techniques (SEC, IEC, and RP-HPLC) (multidimensional or multistep HPLC) for efficient separation of all desired peptides. The reader is directed to Refs. 2–8 for selected practical examples of approaches to multidimensional HPLC of peptides, a brief review of which can be found in Ref. 9.

F. Peptide HPLC Standards

Common to all peptide applications in HPLC is the need to choose the correct column(s) and the most suitable mobile phase. The logical approach to this is the employment of standards, specifically peptide standards, to monitor the suitability of HPLC columns and conditions. Peptide standards are best suited for monitoring peptide retention behavior in HPLC, because it is preferable to use standards that are structurally similar to the sample of interest. Among other things, peptide standards allow the researcher to monitor column performance (efficiency, selectivity, and resolution), run-to-run reproducibility, column aging, instrumentation variations, and the effect of varying run parameters (e.g., the flow rate in SEC, IEC, and RP-HPLC or the gradient rate in IEC and RP-HPLC) and temperature. In addition, and importantly, peptide standards allow the researcher to identify nonideal peptide retention behavior on a particular HPLC column, as well as to develop approaches for manipulation or suppression of such nonideal behavior through changes in the mobile phase.

The value of peptide standards (or other types of standards, depending on the compounds to be separated) in monitoring peptide retention behav-

[2] H. Mabuchi and H. Nakahashi, *J. Chromatogr.* **213,** 275 (1981).

[3] N. Takahashi, Y. Takahashi, and F. W. Putnam, *J. Chromatogr.* **266,** 511 (1983).

[4] C. T. Mant and R. S. Hodges, *J. Chromatogr.* **326,** 349 (1985).

[5] J. Eng., C.-G. Huang, Y.-C. Pan, J. D. Hulmes, and R. S. Yalow, *Peptides* **8,** 165 (1987).

[6] P. Young, T. Wheat, J. Grant, and T. Kearney, *LC-GC* **9,** 726 (1991).

[7] K. Matsuoka, M. Taoka, T. Isobe, T. Okuyama, and Y. Kato, *J. Chromatogr.* **515,** 313 (1990).

[8] N. Lundell and K. Markides, *Chromatographia* **34,** 369 (1992).

[9] C. T. Mant and R. S. Hodges, *J. Liq. Chromatogr.* **12,** 139 (1989).

ior in HPLC, as well as HPLC column and instrument performance, cannot be overestimated. Indeed, routine monitoring of columns and instruments should be obligatory to ensure maximum separation efficiency, the frequency of monitoring dependent, of course, on how much a particular column or instrument is employed.

Peptide standards designed specifically to monitor the peptide-resolving capability of SEC, IEC, and RP-HPLC are commercially available. These standards are described in detail in the relevant sections of this chapter, where they serve to demonstrate clearly standard approaches to peptide separations in the above-described three major HPLC modes.

G. Further Reading

Several useful articles and reviews on HPLC of peptides can be found in Refs. 10–15. References 16 and 17 represent resource books in this area. For an extensive source of information on the early development of HPLC of peptides, the reader is directed to Ref. 18. A more current and comprehensive practical overview on this topic is supplied by Ref. 19.

Readily available journals that frequently contain up-to-date research articles on HPLC of peptides include the *Journal of Chromatography* (published by Elsevier, Amsterdam, The Netherlands), *Chromatographia* (Vieweg & Sohn, Wiesbaden, Germany), the *Journal of Liquid Chromatography* (Marcel Dekker, New York), *Bioseparation* (Kluwer, Dordrecht, The Netherlands), *Analytical Chemistry* (American Chemical Society, Washington, D.C.), *Analytical Biochemistry* (Academic Press, San Diego, CA), and *Pep-

[10] T. E. Hugli (ed.), "Techniques in Protein Chemistry." Academic Press, New York, 1989.
[11] P. T. Matsudaira (ed.), "A Practical Guide to Protein and Peptide Purification for Microsequencing." Academic Press, San Diego, CA, 1989.
[12] C. Fini, A. Floridi, V. N. Finelli, and B. Wittman-Liebold (eds.), "Laboratory Methodology in Biochemistry." CRC Press, Boca Raton, FL, 1990.
[13] J. G. Dorsey, J. P. Foley, W. T. Cooper, R. A. Barford, and H. G. Barth, *Anal. Chem.* **64,** 353R (1992).
[14] C. T. Mant, N. E. Zhou, and R. S. Hodges, *in* "Chromatography" (E. Heftmann, ed.), 5th Ed., p. B75. Elsevier, Amsterdam, The Netherlands, 1992.
[15] C. Schöneich, S. Karina Kwok, G. S. Wilson, S. R. Rabel, J. F. Stobaugh, T. D. Williams, and D. G. Vander Velde, *Anal. Chem.* **65,** 67R (1993).
[16] K. M. Gooding and F. E. Regnier (eds.), "HPLC of Biological Macromolecules: Methods and Applications." Marcel Dekker, New York, 1990.
[17] M. T. W. Hearn (ed.), "HPLC of Proteins, Peptides and Polynucleotides: Contemporary Topics and Applications." VCH Publishers, New York, 1991.
[18] W. S. Hancock (ed.), "Handbook of HPLC for the Separation of Amino Acids, Peptides and Proteins," Vols. I and II. CRC Press, Boca Raton, FL 1984.
[19] C. T. Mant and R. S. Hodges (eds.), "HPLC of Peptides and Proteins: Separation, Analysis and Conformation." CRC Press, Boca Raton, FL, 1991.

tide Research (Eaton Publishing, Natick, MA). The symposium volumes from the International Symposium on the separation and analysis of Proteins, Peptides, and Polynucleotides (ISPPP) and International Symposium on Column Liquid Chromatography (HPLC '92, HPLC '93, etc.) meetings are a particularly good source of information on peptide HPLC. For a continuing source of practical articles on peptide analysis by HPLC (plus a wealth of other practical aspects of HPLC), we recommend *LC-GC,* a monthly magazine on separation science (published by Aster Publishing Corporation, Eugene, OR).

II. Materials

A. HPLC Packings

Silica-based packings are still the most widely used for all major modes of HPLC, the rigidity of microparticulate silica enabling the use of high linear flow velocities of mobile phases. In addition, favorable mass transfer characteristics allow rapid analyses to be performed. However, most silica columns are limited to a pH range of pH 2.0–8.0, because the silica support is rapidly dissolved in the presence of basic eluents. Column packings based on organic polymers having a broad pH tolerance (often pH 0–14) are becoming increasingly used in all modes of HPLC. The most commonly employed of these materials are formed from cross-linked polystyrene-divinylbenzene[20]; other polymeric supports that show some promise include those based on polymethacrylate or vinyl alcohol copolymers. Useful overviews of silica- and polymer-based supports may be found in Refs. 16 and 21, with Ref. 22 representing a key publication in this area.

Non-silica-based supports have been successfully employed in SEC[14,23–25] and IEC[5,7,8,14,26–29] of peptides. Although the number of reported

[20] L. L. Lloyd, *J. Chromatogr.* **544,** 201 (1991).

[21] R. E. Majors, *LC-GC* **11,** 188 (1993).

[22] K. K. Unger, "Packings and Stationary Phases in Chromatographic Techniques." Marcel Dekker, New York, 1989.

[23] G. D. Swergold and C. S. Rubin, *Anal. Biochem.* **131,** 295 (1983).

[24] H. Sasaki, T. Matsuda, O. Ishikawa, T. Takamatsu, K. Tanaka, Y. Kato, and T. Hashimoto, *Sci. Rep. Toyo Soda,* **29,** 37 (1985).

[25] C. T. Mant, J. M. R. Parker, and R. S. Hodges, *J. Chromatogr.* **397,** 99 (1987).

[26] T. Isobe, T. Takayasu, N. Takai, and T. Okuyama, *Anal. Biochem.* **122,** 417 (1982).

[27] U.-H. Stenman, T. Laatikainen, K. Salminen, M.-L. Huhtala, and J. Leppäluoto, *J. Chromatogr.* **297,** 399 (1984).

[28] S. Burman, E. Breslow, B. T. Chait, and T. Chaudhary, *J. Chromatogr.* **443,** 285 (1988).

[29] T. W. L. Burke, C. T. Mant, J. A. Black, and R. S. Hodges, *J. Chromatogr.* **476,** 377 (1989).

applications of such supports to RP-HPLC of peptides is growing,[7,14,30–34] their value remains comparatively untested in this HPLC mode when one considers its extensive employment in the peptide field (much greater than that of SEC or IEC).[35] A comparison of RP-HPLC of peptides on a silica-based versus a polystyrene-based column is illustrated in Section III,D,4,c.

1. Size-Exclusion HPLC

The most useful size-exclusion columns currently available to researchers for peptide applications generally contain packings of 5- to 10-μm particle size and 60- to 150-Å pore size.[14,36] These columns are, in fact, designed mainly for protein separations. Thus, the range of required fractionation for peptides (~100–6000 Da) tends to be at the low end of the fractionation ability of most current columns. However, such columns are still of great potential value in the early stages of a peptide purification protocol or for peptide/protein separations.[4,9,14] A size-exclusion column designed specifically for separation of peptides in the molecular weight range 100–7000 (Superdex Peptide from Pharmacia) has been introduced and has shown much promise in the authors' laboratory.

Column dimensions for analytical size-exclusion columns are generally in the range of 25–60 cm in length and 4.5- to 8.0-mm internal diameter (i.d.). It should be noted that although columns at the upper end of this internal diameter range are often referred to as semipreparative, this is something of a misnomer in SEC considering the small capacity of these columns for preparative applications. A useful review of high-performance size-exclusion columns may be found in Ref. 37.

2. Ion-Exchange HPLC

High-performance ion-exchange chromatography (IEC), with separations based on solute charge, is becoming increasingly popular for the

[30] T. Sasagawa, L. H. Ericsson, D. C. Teller, K. Titani, and K. A. Walsh, *J. Chromatogr.* **307,** 29 (1984).
[31] T. Uchida, T. Ohtani, M. Kasai, Y. Yanagihara, K. Noguchi, H. Izu, and S. Hara, *J. Chromatogr.* **506,** 327 (1990).
[32] J. K. Swadesh, *J. Chromatogr.* **512,** 315 (1990).
[33] B. S. Welinder and H. H. Sørensen, *J. Chromatogr.* **537,** 181 (1991).
[34] T. J. Sereda, C. T. Mant, A. M. Quinn, and R. S. Hodges, *J. Chromatogr.* **646,** 17 (1993).
[35] R. E. Majors, *LC-GC* **7,** 468 (1989) and **9,** 686 (1991).
[36] F. Ahmed and B. Modrek, *J. Chromatogr.* **599,** 25 (1992).
[37] H. G. Barth and B. E. Boyes, *Anal. Chem.* **64,** 428R (1992).

analysis of peptides.[14,16,19] Both anion-exchange (AEC)[7,14,28,38–42] and cat-ion-exchange[3–6,8,9,14,28,29,38,42–46] (CEC) packings have proved useful for pep-tide applications. Common anion-exchange packings consist of primary, secondary, and tertiary (weak AEC) or quaternary amine (strong AEC) functionalities adsorbed or covalently bound to a support. These positively charged packings will interact with acidic (negatively charged) peptide residues (aspartic and glutamic acid above pH ~ 4.0), as well as the nega-tively charged C-terminal α-carboxyl group. Strong anion-exchange col-umns are the most useful mode of AEC for peptides, because the quater-nized supports, carrying an obligatory positive charge, yield essentially unchanged peptide elution times over the acidic to neutral pH range. In contrast, weak anion-exchange sorbents become increasingly protonated (i.e., increasingly positively charged) with a decrease in pH, leading to unpredictable peptide elution behavior. Common cation-exchange packings consist of carboxyl (weak CEC) or sulfonate (strong CEC) functionalities bound to a support matrix. These negatively charged packings will interact with the basic (positively charged) residues, histidine (pH < 6.0) and argi-nine and lysine (pH < 10.0), as well as the positively charged N-terminal α-amino group. The weakly acidic nature of the carboxyl group (pK_a ~ 4.0) means that the employment of such weak cation-exchange sorbents is limited to conditions of neutral pH; lowering the pH will result in progres-sive protonation of the negatively charged carboxyl group, making it unable to retain positively charged species. In contrast, a strong cation-exchange moiety such as a sulfonate group retains its negatively charged character over a wide pH range.

If a choice must be made concerning the type of ion-exchange column for general peptide applications, the researcher should purchase a strong cation-exchange column.[14,16,19] As indicated previously, the utility of such a column lies in its ability to retain its negatively charged character in the acidic to neutral pH range. Most peptides are soluble at low pH, where the side-chain carboxyl groups of acidic residues (glutamic acid and aspartic

[38] C. T. Mant and R. S. Hodges, *J. Chromatogr.* **327**, 147 (1985).

[39] M. Dizdaroglu, *J. Chromatogr.* **334**, 49 (1985).

[40] M. A. Jimenez, M. Rico, J. L. Nieto, and A. M. Gutierrez, *J. Chromatogr.* **360**, 288 (1986).

[41] Y. Sakanoue, E. Hashimoto, K. Mizuta, H. Kondo, and H. Yamamura, *J. Biochem.* **168**, 669 (1987).

[42] P. C. Andrews, *Peptide Res.* **1**, 93 (1988).

[43] P. J. Cachia, J. Van Eyk, P. C. S. Chong, A. Taneja, and R. S. Hodges, *J. Chromatogr.* **266**, 651 (1983).

[44] D. L. Crimmins, J. Gorka, R. S. Thoma, and B. D. Schwartz, *J. Chromatogr.* **443**, 63 (1988).

[45] A. J. Alpert and P. C. Andrews, *J. Chromatogr.* **443**, 85 (1988).

[46] R. S. Hodges, J. M. R. Parker, C. T. Mant, and R. R. Sharma, *J. Chromatogr.* **458**, 147 (1988).

acid) and the free C-terminal α-carboxyl group are protonated (i.e., uncharged), thereby emphasizing any basic, positively charged character of the peptides. In addition, the use of acidic eluents helps to extend the lifetime of silica-based packings, still the most widely used material for peptide separations by IEC. However, simply by changing the pH of the mobile phase from pH 3 to pH 6 (pK_a of glutamate and aspartate residues, ~4.0), the net charge on the peptides in a peptide mixture may be significantly altered, resulting in dramatic selectivity changes.

Features common to most ion-exchange columns best suited for analytical peptide applications are the particle size of the packings (5–10 μm; 6.5 μm is particularly common) and a pore size of 300 Å, the latter being a good compromise for IEC of both peptides and proteins.[38] Typical analytical column dimensions include a length of 15–25 cm and a 3- to 5-mm i.d. Further details can be found in [3] in volume 270.[46a]

3. Reversed-Phase HPLC

The excellent resolving power of RP-HPLC has resulted in its becoming the predominant HPLC technique for peptide separations.[14,16,19,35] Favored RP-HPLC sorbents for the vast majority of peptide separations continue to be silica-based supports containing covalently bound octyl (C_8) or octadecyl (C_{18}) functionalities.[14,19] In a fashion similar to IEC, features common to analytical RP-HPLC packings used in peptide separations include the particle size (5–10 μm) and a pore size of 300 Å. It is important for a solute to have easy access to the pores of a support, i.e., pore diffusion (solute transfer into and out of the porous structure of a packing material) should be unrestricted as much as possible. It has been calculated[47] that when the ratio of molecular diameter to pore diameter exceeds 0.2, the pore diffusion becomes restricted, leading to band spreading and a reduction in solute resolution. A pore size of 100 Å, a common size for small molecule separation by RP-HPLC, is suitable for peptides of up to only approximately 15–20 amino acids in length.[20,48] Hence, 300 Å represents an excellent compromise for separation of small peptides to medium molecular weight proteins. Analytical RP-HPLC column dimensions typically include a length of 10–25 cm and a 4- to 4.6-mm i.d. A general discussion of RP-HPLC can be found in [1] in volume 270.[48a]

[46a] G. Choudhary and C. Horváth, *Methods Enzymol.* **270,** Chap. 3, 1996.
[47] K. K. Unger, R. Janzen, and G. Jilge, *Chromatographia* **24,** 144 (1987).
[48] M. Hermodson and W. C. Mahoney, *Methods Enzymol.* **91,** 352 (1983).
[48a] M.-I. Aguilar and M. T. W. Hearn, *Methods Enzymol.* **270,** Chap. 1, 1996.

Details on all columns described in the present chapter may be found in the captions to Figs. 1–12.

B. HPLC Solvents

1. Water

Almost without exception, mobile phases employed for peptide separations in HPLC are aqueous based. Clean, pure water is an obvious requirement for HPLC applications, and HPLC-grade water can either be purchased (usually in 4-liter bottles) or purified on site, depending on demand.[49] For the applications described in the present chapter, both purchased HPLC-grade water and water treated by a water purifier (model HP 661A; Hewlett-Packard, Avondale, PA) are employed.

2. Other Solvents

Other solvents almost invariably refers to nonpolar solvents (e.g., acetonitrile, methanol, 2-propanol) employed either as the organic modifier in RP-HPLC, or as a means of suppressing nonideal hydrophobic interactions in SEC[25] or IEC.[29] All nonaqueous solvents commonly employed in HPLC are available in a highly pure form (spectroscopic or HPLC grade). The HPLC-grade acetonitrile employed in the present chapter was purchased from companies such as Fisher Scientific (Pittsburgh, PA) and J. T. Baker (Phillipsburg, NJ).

C. Mobile-Phase Additives

As a general guideline, any additives to a (pure) mobile-phase solvent should be HPLC-grade, if available, or of the highest quality that can be obtained.[49–51] The suitability of a particular additive can be quickly determined by an ultraviolet (UV) scan of an aqueous solution of the additive at the highest concentration at which it is likely to be employed [e.g., 0.1% (v/v) trifluoroacetic acid (TFA) for RP-HPLC, 1 M KCl for IEC]. If there are obvious UV-absorbing contaminants in the reagent, it can either be discarded in favor of a cleaner preparation or, when possible, it can be purified.

---

[49] C. T. Mant and R. S. Hodges, *in* "HPLC of Peptides and Proteins: Separation, Analysis and Conformation" (C. T. Mant and R. S. Hodges, eds.), p. 37. CRC Press, Boca Raton, FL, 1991.

[50] J. W. Dolan, *in* "HPLC of Peptides and Proteins: Separation, Analysis and Conformation" (C. T. Mant and R. S. Hodges, eds.), p. 31. CRC Press, Boca Raton, FL, 1991.

[51] J. W. Dolan, *LC-GC* **11,** 498 (1993).

What follows in this section is a description of mobile-phase additives used to obtain the various chromatograms illustrated in this chapter, together with approaches to their purification, where necessary.

1. Trifluoroacetic Acid

This ion-pairing reagent is by far the most extensively used mobile-phase additive in RP-HPLC.[14,19] The quality of this reagent is especially important, since impurities originally in the TFA or resulting from aging can cause excessive baseline noise.[49-51] Highly pure TFA is available from many commercial sources, e.g., Pierce (Rockford, IL), Aldrich (Milwaukee, WI), and Sigma (St. Louis, MO) (HPLC grade, spectrophotometric grade, and sequenator grade are all suitable for HPLC use).

2. Buffer Components

Triethylamine is a reagent frequently employed in both volatile and nonvolatile buffer systems over a wide pH range.[19,52] In its capacity as a major buffer component, this reagent is often used at concentrations of 0.20–0.25 M, making purity of this reagent an important consideration. Highly pure triethylamine can be readily purchased, although we have regularly used lesser-grade reagent, following distillation over ninhydrin, with no discernible problems.

The phosphate component of the triethylamine phosphate (TEAP) buffer employed in this chapter was an analytical-grade reagent obtained from J. T. Baker, the purity of which has invariably proved adequate.

Without exception, we have found that analytical-grade phosphate-based buffer systems, frequently employed in the authors' laboratory in SEC and IEC (and occasionally in RP-HPLC), require some form of purification prior to their use.[49] This is particularly apparent when they are used for gradient elution in IEC or RPC. If no effort is made to clean them up prior to gradient elution, contaminants in the analytical reagents produce unwanted peaks and drifting baselines. This is easily avoided by a simple cleanup procedure involving the extraction of contaminants by a chelating resin. We routinely prepare a stock solution [e.g., 1 liter of 1–2 M aqueous KH_2PO_4 (SEC, IEC) or $(NH_4)_2HPO_4$ (RP-HPLC)], add the chelating resin (Chelex 100; Bio-Rad, Richmond, CA) (~10 g/liter of solution), and stir for 1 hr. The phosphate solution is then aliquoted, diluted as desired, and filtered through a 0.22-μm pore size filter. It is then ready for use. The remainder of the phosphate solution is stored at 4° over the resin until further use.

[52] J. E. Rivier, J. Liq. Chromatogr. **1,** 343 (1978).

3. Salts

This laboratory frequently employs NaCl or KCl, either as the displacing salts for IEC or as a means to suppress nonideal adsorptive behavior in SEC (Section III,B,1). The authors have never found any particular need to use other than analytical-grade salts for these purposes following a quick spectral check of a solution of the reagents when they are first purchased.

For the occasional RP-HPLC applications carried out at neutral pH in the authors' laboratory, 0.1 M sodium perchlorate ($NaClO_4$) is routinely added to the mobile phase buffer to suppress any nonideal behavior (i.e., ionic interactions) with free silanols on silica-based columns (Section III,D,2). In the authors' experience, chelexing of analytical-grade $NaClO_4$ has never successfully removed all UV-absorbing contaminants from this reagent. Instead, it has proved necessary to pass a stock solution of the perchlorate through a preparative C_{18} column (following filtration through a 0.22-μm pore size filter) to remove any impurities. The purified perchlorate solution is then diluted as required. Because of the high solubility of $NaClO_4$ in aqueous media containing organic solvents, it is an ideal additive for RP-HPLC (and, frequently, IEC), for which the presence of an organic modifier such as acetonitrile is required.

4. Urea

For any studies requiring complete denaturation of peptides or proteins (e.g., to ensure predictable retention behavior during SEC), urea concentrations ranging from 6 to 8 M are quite typical. At these high concentration levels, reagent purity is obviously vital. Although highly pure urea (e.g., ultrapure grade from ICN Biochemicals, Cleveland, OH) is commercially available, it tends to be substantially more expensive than the analytical-grade reagent, which can be purified to a level suitable for HPLC applications by a straightforward procedure. Thus, following preparation of the concentrated aqueous urea solution (usually requiring some heat, because dissolving urea in water is a highly endothermic process), the solution is stirred over a mixed-bed resin [Bio-Rad AG 501-X8 (20–50 mesh) in our case] (~10 g/liter of solution) for about 30–60 min to remove dissolved impurities. The resin is removed by filtering through a sintered glass funnel and the supernatant is subsequently filtered through a 0.22-μm pore size filter. The urea solution is now ready for use.

One point to consider when using solvents containing urea is the possible presence of ammonium cyanate resulting from urea breakdown. The cyanate ion reacts with primary amine groups, resulting in the formation of blocked lysine side chains and N-terminal amino groups.[53,54] This potential problem can be eliminated for all practical purposes by using freshly prepared urea-containing solutions.

On a final note, the researcher should be wary of unintentional additives to the mobile phase and/or sample solvent. An example of this is contamination by oxidizing agents such as chromate left over from cleanup steps. Glassware subjected to soaking in chromic acid should be washed thoroughly with water (preferably distilled or HPLC-grade water) to avoid subsequent oxidation of amino acid side chains such as methionine.

D. Instrumentation

1. Liquid Chromatograph

Complete descriptions of the pumping systems employed for producing the chromatograms shown in this chapter are given in the captions to Figs. 1–12. A major point to note here is that the liquid chromatograph must have a gradient-making capability to be of any use for IEC or RP-HPLC of peptides.

2. Detection Systems

Although both fluorescence and electrochemical techniques have been employed for specific peptide analyses, UV detectors are, by far, the most commonly used detectors for peptide applications. A single-wavelength detector (set at 210 nm for peptide bond absorbance) may be adequate for many applications; however, a capability of simultaneous detection of two wavelengths at least (dual-wavelength detectors are available) is recommended. Advantage may then be taken of the presence of any aromatic residues (particularly tryptophan and tyrosine) in the peptide(s), i.e., absorbance detection at 210 and 280 nm. All of the chromatograms shown in the present chapter were produced by diode-array detectors (DADs), capable of scanning multiple wavelengths.

E. Peptide Standards

The structures and source of the peptide standards utilized for the present article are described in Section III under the individual modes of HPLC.

[53] G. R. Stark, *Methods Enzymol.* **25,** 579 (1972).
[54] R. L. Lundblad and C. M. Noyes, "Chemical Reagents for Protein Modification," Vol. I, p. 127. CRC Press, Boca Raton, FL, 1984.

III. Methods

A. Prerun Concerns

1. Equilibration of Columns

There are certain precautions the researcher should take prior to using a previously stored column, particularly when salts are to be employed in the mobile phase. Columns are frequently stored in aqueous–organic solvents (see Section III,G,2), and the introduction of salts into such an aqueous organic solution may cause salt precipitation in the lines and on the column if the salt and/or organic solvent concentration is high enough. Thus, prior to equilibration with mobile phases containing salts, the column should first be flushed with several column volumes of HPLC-grade water. The column may then be equilibrated with the initial mobile-phase eluent.

2. Blank Runs

During gradient elution, the chromatographic run generally begins with a weak eluent, followed by an increase in eluent strength (i.e., increasing ionic strength in IEC, increasing nonpolarity of mobile phase in RP-HPLC) over a period of time. If the weaker eluent contains UV-active or fluorescent impurities that are strongly retained by the stationary phase, they are concentrated at the top of the column during the equilibration period or during the initial stages of the gradient run.[49,55] As the eluent strength is increased, these impurities are eluted from the column and detected, resulting in spurious peaks and/or rapidly rising baselines. If the stronger eluent contains the impurities, similar results may be obtained, although of reduced magnitude (some baseline drift is typical), because less of it is generally passed through the column and the impurities in the stronger eluent are more quickly eluted as the solvent strength increases.

Prior to the first run performed on a previously stored column, a gradient run in the absence of sample (i.e., a "blank" run) should be carried out.[49] In fact, even before the blank run, we suggest that the column should be subjected to a rapid gradient wash, e.g., 100% eluent A (the weak eluent) to 100% eluent B (the strong eluent) in 15 min, followed by rapid reequilibration back to the desired starting conditions. This will serve to remove any impurities from the column that may have accumulated during storage. The subsequent blank run should always be run out to 100% eluent B, in case any strongly adsorbed impurities are present in the mobile phase. Another advantage of carrying out these initial runs without sample is that

[55] E. Johnson and B. Stevenson, "Basic Liquid Chromatography," p. 320. Varian Associates, Palo Alto, CA, 1978.

subsequent runs with the sample(s) will be more reproducible, because the column is then thoroughly conditioned.[56]

3. Sample Preparation

Proper sample preparation is important both for achieving the desired separation and as a means of preventing column problems. Sample preparation techniques include (1) simplification of the sample by removing unwanted materials that could harm the column, (2) putting the sample in the correct form for injection by proper solubilization, (3) optimizing the sample concentration for proper column loading, and (4) removal of particulate materials (usually down to a particle diameter of 0.2 μm) to prevent blockage of columns. Overviews of these sample preparation techniques can be found in Refs. 57 and 58.

In the present chapter, the peptide standards were obtained as lyophilized powders and only required dissolving in HPLC-grade water or in the starting eluent of the particular mode of HPLC. The peptide solutions were then filtered by syringe through disposable 0.2-μm pore size filters (Gelman Sciences, Ann Arbor, MI). It should be noted, however, that peptides containing a large proportion of nonpolar side chains and/or few charged residues may be somewhat insoluble in 100% aqueous solution. Under these circumstances, addition of a certain proportion of the employed organic modifier (generally acetonitrile or, for more hydrophobic peptide samples, 2-propanol) to the sample aqueous phase is usually effective. The level of organic modifier in such a mixed aqueous–organic solvent should be minimized to prevent premature elution of the sample from the column. Examples of this approach to the separation of extremely hydrophobic peptides may be found in Ref. 59.

B. Size-Exclusion HPLC

1. Mobile-Phase Selection

Aqueous size-exclusion chromatography is used for two purposes, namely separation and/or molecular weight determinations of solutes. Separation of peptides by a mechanism based solely on molecular size (ideal SEC) occurs only when there is no interaction between the solutes

[56] R. C. Chloupek, J. E. Battersby, and W. S. Hancock, *in* "HPLC of Peptides and Proteins: Separation, Analysis and Conformation" (C. T. Mant and R. S. Hodges, eds.), p. 825. CRC Press, Boca Raton, FL, 1991.

[57] C. T. Wehr and R. E. Majors, *LC-GC* **5,** 548 (1987).

[58] R. E. Majors, *LC-GC* **10,** 356 (1992).

[59] A. K. Taneja, S. Y. M. Lau, and R. S. Hodges, *J. Chromatogr.* **317,** 1 (1984).

and the column matrix. Although high-performance size-exclusion columns are designed to minimize nonspecific interactions, most modern SEC columns are weakly anionic (negatively charged) and slightly hydrophobic, resulting in solute–packing interactions and, hence, giving rise to deviations from ideal size-exclusion behavior (nonideal SEC).[14,25] Such nonspecific interactions must be suppressed if predictable solute elution behavior is required. Electrostatic effects between solutes and the column matrix may be minimized by the addition of salts to the mobile phase.[25] Thus, electrostatic (or ion-exchange) effects are minimized above an eluent ionic strength of about 0.1 M and aqueous phosphate or Tris buffers, often containing 0.1–0.4 M salts, are commonly employed as the mobile phase for SEC of peptides and proteins.[14,25] Hydrophobic interactions may be suppressed by the addition of low levels (e.g., 5–10%, v/v) of organic modifier, e.g., methanol or acetonitrile, to the mobile phase.[14,25] In the authors' experience, however, electrostatic effects are generally dominant during SEC of peptides, thus almost invariably requiring the presence of salts in the mobile phase to suppress these interactions. It is also worth mentioning that, again in the authors' experience, both hydrophobic and electrostatic effects are minimized by the inclusion of salts in the mobile phase.[25]

It should be noted that the multimodal characteristics of a particular size-exclusion column [i.e., when the separation mechanism (pure size exclusion versus ion exchange versus a mixture of the two) depends on mobile-phase composition[14,17,25]] may be extremely advantageous for specific peptide and protein applications. On a similar note, albeit with the separation mechanism based purely on a size-exclusion mechanism, one report[60] describes manipulation of the separation range for polypeptides of a size-exclusion column by varying the mobile-phase conditions. This manipulation is effected by shrinkage or swelling of a cross-linked hydrophilic matrix bound to a silica support.

2. Size-Exclusion HPLC Peptide Standards

The synthetic peptide standards shown in Table I were designed specifically for monitoring both ideal and nonideal behavior on size-exclusion columns.[25,36] The increasing size of the peptide standards enables the accurate molecular weight calibration of a column during ideal SEC; the increasingly basic (positively charged) character of the standards makes them sensitive to the anionic (negatively charged) character of a size-exclusion

---

[60] P. C. Andrews and A. J. Alpert, presented at the Tenth International Symposium on HPLC of Proteins, Peptides and Polynucleotides, Wiesbaden, Germany, October 29–31, 1990, Abstract 110.

TABLE I
Synthetic Size-Exclusion HPLC Peptide Standards[a]

Peptide standard	Sequence	Number of residues	Molecular weight	Net charge
1	Ac-Gly-Leu-Gly-Ala-Lys-Gly-Ala-Gly-Val-Gly-amide	10	826	+1
2	Ac-(Gly-Leu-Gly-Ala-Lys-Gly-Ala-Gly-Val-Gly)$_2$-amide	20	1593	+2
3	Ac-(Gly-Leu-Gly-Ala-Lys-Gly-Ala-Gly-Val-Gly)$_3$-amide	30	2360	+3
4	Ac-(Gly-Leu-Gly-Ala-Lys-Gly-Ala-Gly-Val-Gly)$_4$-amide	40	3127	+4
5	Ac-(Gly-Leu-Gly-Ala-Lys-Gly-Ala-Gly-Val-Gly)$_5$-amide	50	3894	+5

[a] Ac, N^α-Acetyl; amide, C^α-amide. These standards were purchased from Alberta Peptide Institute, University of Alberta, Edmonton, Alberta, Canada T6G 252.

column; the increasing hydrophobicity of the standards enables a determination of column hydrophobicity. In addition, the high glycine content of the standards minimizes or eliminates any tendency toward secondary structure.

In Fig. 1, the standards have been employed to illustrate nonideal and ideal SEC behavior on a silica-based column. Figure 1 shows elution profiles of peptide standards 1, 2, and 5 (10, 20, and 50 residues, respectively) using aqueous mobile-phase buffers of 50 mM KH$_2$PO$_4$–100 mM KCl, pH 6.5 (top, Fig. 1) or 5 mM KH$_2$PO$_4$–50 mM KCl, pH 6.5 (bottom, Fig. 1). In Fig. 1 (top), the peptides are eluted in order of decreasing size, as would be expected under ideal SEC conditions; in addition, the three peptides exhibited a linear log molecular weight versus elution time relationship. In contrast, in Fig. 1 (bottom), the column is exhibiting nonspecific interactions between the peptides and column matrix at this lower phosphate and KCl concentration, with the smallest peptide (peptide standard 1, 10 residues) being eluted first and the largest peptide (peptide standard 5, 50 residues) being eluted last. In addition, all three peptides are being retained longer than the total permeation volume of the column (denoted by arrow). By definition, under ideal SEC conditions, no molecule will be retained beyond the total permeation volume of the column. The column is, in fact, behaving like a cation-exchange column, the peptides being eluted in order of increasing positive charge (peptide standards 1, 2, and 5 possess a +1, +2, and +5 net charge, respectively) instead of decreasing size. With an increase in the ionic strength of the mobile phase, these electrostatic effects are suppressed to produce the peptide separation profile based on an ideal size-exclusion mechanism.

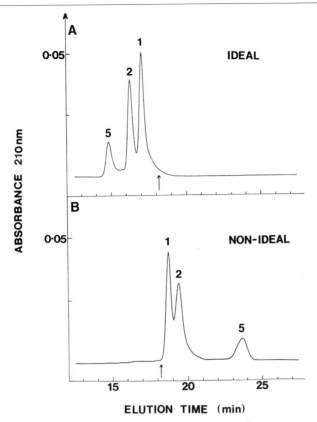

FIG. 1. Use of synthetic peptide standards to monitor nonideal and ideal size-exclusion behavior. Column: SynChropak GPC60 (300 × 7.8 mm i.d., 10-μm particle size, 60-Å pore size; SynChrom, Lafayette, IN). Instrumentation: The HPLC instrument consisted of a Varian Vista Series 5000 liquid chromatograph (Varian, Walnut Creek, CA) coupled to a Hewlett-Packard (Avondale, PA) HP1040A diode array detection system, HP85B computer, HP9121 disk drive, HP2225A Thinkjet printer, and HP7470A plotter. (A) Ideal SEC of peptide standards: mobile phase, 50 mM aqueous KH_2PO_4–100 mM KCl, pH 6.5; flow rate, 0.5 ml/min; room temperature. (B) Nonideal SEC of peptide standards: mobile phase, 5 mM aqueous KH_2PO_4–50 mM KCl, pH 6.5; flow rate, 0.5 ml/min; room temperature. The sequences of standards 1, 2, and 5 (10, 20, and 50 residues and +1, +2, and +5 net charge, respectively) are shown in Table I. The arrows denote the elution time for the total permeation volume of the column.

3. Standard Chromatographic Conditions for Size-Exclusion HPLC of Peptides

Optimum flow rates for analytical size-exclusion columns are generally in the range of 0.2–1.0 ml/min.[14,46] In addition, sample volume must be kept as small as possible.

a. NONDENATURING CONDITIONS. Suggested standard run conditions are as follows:

Isocratic elution: 50 mM aqueous KH$_2$PO$_4$, pH 6.5, containing 0.1
                     M KCl
Temperature: Room temperature
Flow rate: 0.5 ml/min

Figure 2 compares the elution profiles of peptide standards 1, 2, and 5 (10, 20, and 50 residues, respectively) obtained with two different silica-based size-exclusion columns (Fig. 2A and B) and one column containing a nonsilica, agarose-based packing (Fig. 2C) using the above-cited run conditions. The mobile-phase buffer is typical of standard nondenaturing run conditions applicable to both peptide and protein separations, and is a good place to start.[61] All three columns in Fig. 2 exhibited similar peptide elution profiles under ideal size-exclusion conditions, i.e., a linear log molecular weight versus elution time relationship was obtained. The longer retention times and somewhat broader peaks obtained with the agarose-based packing (Fig. 2C) are probably due to its significantly larger column volume compared to the two silica-based columns. All of these columns are readily resolving a 10-residue peptide from a 20-residue peptide.

b. DENATURING CONDITIONS. Many proteins and large peptides may deviate from ideal size-exclusion behavior, owing to conformational effects. In addition, the tendency of peptides or protein fragments to maintain or reform a particular conformation as opposed to a random coil configuration in nondenaturing media will complicate retention time prediction.[14,25] Agents such as 6 M guanidine hydrochloride, 8 M urea, and 0.1% sodium dodecyl sulfate (SDS) are frequently employed, both for suppressing nonspecific interactions and to denature solutes for molecular weight estimation.[2,25,62-65]

Under circumstances in which predictable elution behavior is required and the conformational character of a peptide–protein mixture in a particular mobile phase is uncertain the following mobile phase is recommended as a starting point:

50 mM aqueous KH$_2$PO$_4$, pH 6.5, containing 0.5 M KCl and 8 M urea

[61] C. T. Mant and R. S. Hodges, in "HPLC of Peptides and Proteins: Separation, Analysis and Conformation" (C. T. Mant and R. S. Hodges, eds.), p. 11. CRC Press, Boca Raton, FL, 1991.
[62] F. E. Regnier, Methods Enzymol. 91, 137 (1983).
[63] H. Mabuchi and J. Nakahashi, J. Chromatogr. 228, 292 (1982).
[64] Y. Shioya, H. Yoshida, and T. Nakajimi, J. Chromatogr. 240, 341 (1982).
[65] S. Y. M. Lau, A. K. Taneja, and R. S. Hodges, J. Biol. Chem. 259, 13253 (1984).

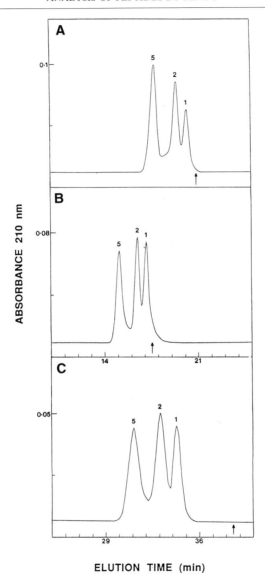

FIG. 2. Ideal SEC of synthetic peptide standards under recommended standard nondenaturing conditions on different SEC columns. Columns: (A) Altex Spherogel TSK G2000SW (300 × 7.5 mm i.d., 10-μm particle size, 130-Å pore size; Beckman Instruments, Berkeley, CA), (B) SynChropak GPC60 (300 × 7.8 mm i.d., 10 μm, 60 Å; SynChrom, Lafayette, IN), and (C) Pharmacia Superose 12 (300 × 10 mm i.d., 10 μm; Pharmacia, Dorval, Canada). Instrumentation: Same as in Fig. 1. Conditions: Mobile phase, 50 mM aqueous KH_2PO_4–0.1 M KCl, pH 6.5; flow rate, 0.5 ml/min; room temperature. The sequences of peptide standards 1, 2, and 5 (10, 20, and 50 residues, respectively) are shown in Table I. The arrows denote the elution time for the total permeation volume of the column. (From Ref. 25, with permission.)

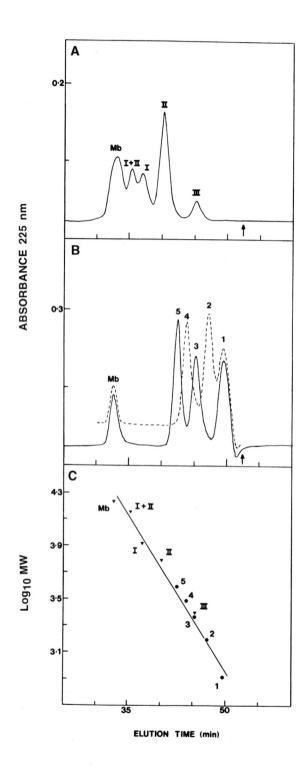

This mobile phase was employed to produce the elution profiles shown in Fig. 3. Figure 3A shows the elution profile of horse heart myoglobin (Mb) and its cyanogen bromide fragments (2500–17,000 Da); Fig. 3B shows the elution profile of myoglobin and peptide standards 1–5. The linear log molecular weight versus elution time relationship shown in Fig. 3C could be obtained only under the above-described highly denaturing conditions, owing to the tendency of the myoglobin fragments to reassociate under nondenaturing conditions.[25] The slower flow rate (0.2 ml/min) employed, compared to 0.5 ml/min in Fig. 2, was required to separate more fully the larger myoglobin fragments (I, I + II) from each other and from the whole protein molecule.

C. Ion-Exchange HPLC

1. Mobile-Phase Selection

Peptides may be removed from an ion-exchange column by either gradient (step or linear) or isocratic elution. The retention time of a peptide in AEC or CEC will depend on a number of factors, including buffer pH and the nature and ionic strength of the anion (e.g., chloride ion) or cation (e.g., sodium or potassium ions) employed for displacement of acidic or basic peptides, respectively.[14,46a] Linear gradient elution is generally the elution mode of choice when attempting to resolve mixtures of peptides with a wide range of net charges. Gradient elution of peptides is usually performed with salt gradients of either sodium or potassium chloride in phosphate, Tris or citrate mobile-phase buffers. A linear NaCl or KCl gradient in 5–50 mM KH$_2$PO$_4$ buffer, pH 6.5 (AEC or CEC) or pH 3.0 (strong CEC only), constitutes suitable elution conditions for most peptide separations and is certainly a good place to start.[61] Care should be taken in the choice of ionic strength of the starting buffer. If it is too high, weakly acidic or basic peptides that may otherwise be retained by anion- or cation-exchange columns, respectively, may be eluted with unretained com-

FIG. 3. Ideal SEC of myoglobin fragments and a mixture of synthetic peptide standards under recommended standard denaturing conditions. Column: Altex Spherogel TSK G2000SW (300 × 7.5 mm i.d.). Instrumentation: Same as in Fig. 1. Conditions: Mobile phase, 50 mM KH$_2$PO$_4$–0.5 M KCl–8 M urea, pH 6.5; flow rate, 0.2 ml/min; room temperature. (A) Elution profile of horse heart myoglobin (Mb) and its cyanogen bromide cleavage fragments (I, II, I + II, III). (B) Elution profile of horse heart Mb and synthetic peptide standards 1–5. (C) Plot of log MW versus elution time of Mb, cyanogen bromide fragments of Mb, and the five synthetic peptide standards. Peptide standards 1, 2, 3, 4, and 5 contain 10, 20, 30, 40, and 50 residues, respectively, and their sequences are shown in Table I. The arrows denote the elution time for the total permeation volume of the column (A and B). (From Ref. 25, with permission.)

pounds.[14] As a general rule, the optimum separation of two peptides by gradient elution on cation- or anion-exchange columns will be obtained when there is a net charge difference of at least 1 unit between the peptides.[14] However, it should be noted that the relative polypeptide chain lengths of two peptides of the same net charge can also have a profound effect on their separation.[29,46] Thus, Burke et al.[29] demonstrated that the critical factor when attempting to equate peptide elution behavior in CEC with peptide structure was the peptide net charge-to-mass ratio (expressed by the authors as net charge/ln N, where N is the number of residues in the peptide).

An ion-exchange separation is frequently useful as the penultimate step of a multistep (or multidimensional) peptide purification protocol, prior to a final purification (and desalting) RP-HPLC separation.[4,6-9]

Although, as the name implies, the separation mechanism of IEC is electrostatic in nature, ion-exchange packings may also often exhibit significant hydrophobic characteristics, giving rise to mixed-mode contributions to solute separations.[29,66-68] Although mixed-mode effects can enhance peptide or protein separations,[29,68] it has been shown that removal of nonspecific hydrophobic interactions may be necessary just to elute peptides from the ion-exchange matrix.[29] Thus, when only the predominant, i.e., ionic, stationary phase–peptide interaction is required (ideal ion-exchange behavior), the mobile phase must be manipulated so as to minimize nonspecific interactions, e.g., by the addition of an organic solvent such as 2-propanol or acetonitrile to the mobile-phase buffers to suppress hydrophobic interactions between the solute and the ion-exchange packing.

2. Ion-Exchange HPLC Peptide Standards

Figure 4 shows the separation of four synthetic cation-exchange peptide standards (peptides C1–C4) (Table II) on a silica-based strong cation-exchange column at pH 6.5 in the presence of 5% (Fig. 4, top) or 10% (Fig. 4, bottom) acetonitrile (v/v). Peptides C1–C4 contain, respectively, one to four basic residues (lysine residues) with no acidic residues present. Thus, over the pH range used for the majority of cation-exchange separations [pH 3.0 to 7.0 (strong CEC) or pH $>$ 4.5 to 7.0 (weak CEC)], the net charges of +1 to +4 for peptides C1–C4, respectively, do not change.[29] Any variation in retention behavior of the peptide standards would, therefore, be indicative of undesirable pH effects on the cation-exchange matrix. The

[66] W. Kopaciewicz, M. A. Rounds, J. Fausnaugh, and F. E. Regnier, *J. Chromatogr.* **266,** 3 (1983).

[67] L. A. Kennedy, W. Kopaciewicz, and F. E. Regnier, *J. Chromatogr.* **359,** 73 (1986).

[68] M. A. Rounds, W. D. Rounds, and F. E. Regnier, *J. Chromatogr.* **397,** 25 (1987).

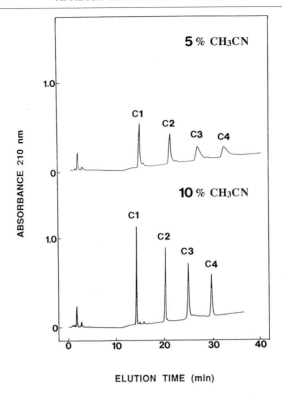

FIG. 4. Monitoring nonideal (hydrophobic) behavior in cation-exchange HPLC with synthetic peptide standards. Column: PolySULFOETHYL Aspartamide strong cation-exchange column (250 × 4.6 mm i.d., 5-μm particle size, 300-Å pore size; PolyLC, Columbia, MD). Instrumentation: Same as in Fig. 1, except for an HP9000 Series 300 computer. Conditions: Linear AB gradient (20 mM salt/min, following 5-min isocratic elution with buffer A), where buffer A is 5 mM aqueous KH_2PO_4, pH 6.5, and buffer B is buffer A plus 1 M NaCl, both buffers containing 5% (top) or 10% (bottom) acetonitrile (v/v); flow rate, 1 ml/min; room temperature. The sequences of peptide standards C1, C2, C3, and C4 (+1, +2, +3, and +4 net charge, respectively) are shown in Table II.

hydrophobicity of the standards increases from peptide C1 to peptide C4, with a concomitant increase in peptide sensitivity to potential hydrophobic solute–packing interactions. The presence of a tyrosine residue in peptides C2 and C4 enables their detection at 280 nm. The inclusion of acetonitrile in the mobile phase was found to be necessary to elute the four peptides from the column within a reasonable time and with reasonable peak shape. In the absence of acetonitrile, neither peptide C3 nor C4 (+3 and +4 net charge, respectively) were eluted from the column, while C2 (+2 net charge) was eluted as a broad, skewed peak. The improvement in the peptide

TABLE II
SYNTHETIC CATION-EXCHANGE HPLC PEPTIDE STANDARDS[a]

Peptide standard	Sequence	Net charge
1	Ac-Gly-Gly-Gly-Leu-Gly-Gly-Ala-Gly-Gly-Leu-Lys-amide	+1
2	Ac-Lys-Tyr-Gly-Leu-Gly-Gly-Ala-Gly-Gly-Leu-Lys-amide	+2
3	Ac-Gly-Gly-Ala-Leu-Lys-Ala-Leu-Lys-Gly-Leu-Lys-amide	+3
4	Ac-Lys-Tyr-Ala-Leu-Lys-Ala-Leu-Lys-Gly-Leu-Lys-amide	+4

[a] Ac, N^α-Acetyl; amide, C^α-amide. These standards were purchased from Alberta Peptide Institute.

elution profile of the four standards on addition of 5% acetonitrile (Fig. 4, top) and the further improvement at a level of 10% acetonitrile (Fig. 4, bottom) indicated that, in its absence, the column packing was exhibiting hydrophobic, in addition to ionic, characteristics. Synthetic anion-exchange peptide standards (-1, -2, -3, and -4 net charge) are also available from the same source as the cation-exchange standards.

3. Standard Chromatographic Conditions for Ion-Exchange HPLC of Peptides

Flow rates of 0.5–2.0 ml/min are favored for analytical ion-exchange separations.[14,46] The choice of gradient rate (increasing concentration of counterion per unit time) will be dictated by the complexity and charge distribution of the peptide mixture to be resolved, but an increase of 5–20 mM salt/min is suitable for most analytical purposes.[14]

Recommended standard run conditions for strong CEC of peptides are as follows:

Linear AB gradient: Buffer A: 5 mM aqueous KH$_2$PO$_4$, pH 6.5 or 3.0

Buffer B: 5 mM aqueous KH$_2$PO$_4$, pH 6.5 or 3.0, containing 0.5 M KCl

Note: Only the pH 6.5 conditions are applicable for AEC and weak CEC.

If acetonitrile is required, add 20% (v/v) acetonitrile to both buffers during preparation so as not to dilute the salt concentrations.

Gradient rate: 10-min isocratic hold of 100% buffer A, followed by a linear increasing salt gradient of 20 mM KCl/min

For example,

Time (min)	Mobile-phase composition
0	100% A
10	100% A
35	100% B

Flow rate: 1 ml/min for standard analytical columns (4- to 4.6-mm i.d.)

Temperature: Room temperature

Figure 5 shows the separation of the four peptide standards (C1–C4) on a silica-based strong cation-exchange column at pH 3.0 and pH 6.5

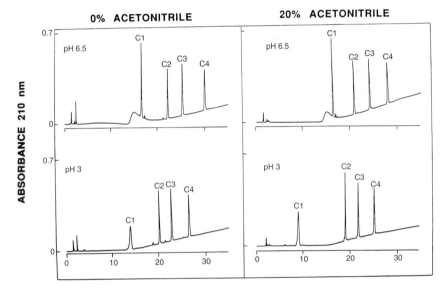

FIG. 5. Strong cation-exchange HPLC of a mixture of synthetic peptide standards at pH 6.5 (top) and pH 3.0 (bottom) under recommended standard conditions. Column: PolySUL-FOETHYL Aspartamide (200 × 4.6 mm i.d., 5-μm particle size, 200-Å pore size; PolyLC, Columbia, MD). Instrumentation: Same as in Fig. 4. Conditions: Linear AB gradient (20 mM KCl/min, following 10-min isocratic elution with buffer A), where buffer A is 5 mM aqueous KH_2PO_4, pH 3.0 or 6.5, and buffer B is buffer A plus 0.5 M KCl; for the runs in the presence of acetonitrile (right-hand profiles), both buffers contained 20% (v/v) acetonitrile; flow rate, 1 ml/min; room temperature. The sequences of peptide standards C1, C2, C3, and C4 (+1, +2, +3, and +4 net charge, respectively) are shown in Table II. (Reprinted with permission from C. T. Mant and R. S. Hodges (eds.), "HPLC of Peptides and Proteins: Separation, Analysis and Conformation." CRC Press, Boca Raton, FL, © 1991.)

following employment of the above-described standard conditions, both in the presence and absence of 20% acetonitrile.[61] From Fig. 5, all four peptide standards (C1, C2, C3, and C4; +1, +2, +3, and +4 net charge, respectively) were eluted by the gradient at pH 6.5. At pH 3.0, there was a slight decrease in peptide retention times and peptide C1 (+1 net charge) was now eluted during the initial 10-min isocratic hold of buffer A, instead of by the gradient (as seen at pH 6.5). The presence of 20% acetonitrile also served to reduce peptide retention times at both pH values, albeit not significantly, indicating that some small degree of hydrophobic peptide/packing interaction was present in the absence of the solvent. The sharpness of the peptide peaks (i.e., narrow bandwidths) eluted by the gradient indicates an efficient column. In fact, this is a good strong cation-exchange column, with only minor pH sensitivity and hydrophobic characteristics. In addition, its extremely advantageous ability to retain a weakly basic (+1 net charge) peptide is not, in the authors' experience, a widespread characteristic of commercially available cation-exchange columns.

D. Reversed-Phase HPLC

1. Mobile-Phase Selection

Reversed-phase silica-based columns may contain surface silanols which act as weak acids and are ionized above pH 3.5–4.0.[62,69] Chapter [1] in volume 270[48a] reviews the overall field of RP-HPLC and [2] in this volume[69a] focuses on peptide mapping. The weak acids may interact with the basic residues of peptides chromatographed on reversed-phase columns and have an adverse effect on resolution, characteristically producing long retention times and peak broadening. Although excellent separation of peptides and proteins may be obtained at acidic or neutral pH, the majority of RP-HPLC separations are carried out at pH values <3.0. Apart from the suppression of silanol ionization under these conditions, the silica matrix is more stable at low pH. Whatever the pH eventually employed by the researcher, the suppression of nonideal (ionic) interactions with the hydrophobic stationary phase is always an important concern.

Solutes are displaced from the hydrophobic stationary phase of a reversed-phase column by the introduction of an organic modifier into an aqueous mobile phase. Acetonitrile is the favored organic solvent for most peptide applications,[14,70,71] although a more nonpolar solvent (e.g., 2-propa-

[69] C. T. Mant and R. S. Hodges, *Chromatographia* **24**, 805 (1987).

[69a] E. R. Hoff and R. C. Chloupek, *Methods Enzymol.* **271**, Chap. 2, 1996 (this volume).

[70] C. T. Mant and R. S. Hodges, *LC-GC* **4**, 250 (1986).

[71] D. Guo, C. T. Mant, A. K. Taneja, J. M. R. Parker, and R. S. Hodges, *J. Chromatogr.* **359**, 499 (1986).

nol)[3,59,72-74] or a more polar solvent (e.g., methanol[72,74-77]) may be required for more hydrophobic or hydrophilic peptides, respectively. A useful list of properties of chromatographic solvents (e.g., chemistry, viscosity, and UV cutoff) can be found in Ref. 78.

Peptides are charged molecules at most pH values, and the presence of different counterions in the mobile phase will influence their chromatographic behavior. Differences in the polarities of peptides can be maximized through careful choice of ion-pairing reagent.[14,79,80] Thus, a cationic ion-pairing reagent [e.g., triethylammonium ion, tetrabutylammonium ion (usually employed as the phosphate salt)], whose use is limited to pH values >4.0, the pK_a of acidic side-chain groups, will show affinity for ionized carboxyl groups[14,77,81-85]; an anionic ion-pairing reagent (e.g., trifluoroacetic acid, heptafluorobutyric acid, hexanesulfonic acid) will interact with the protonated basic residues of a peptide.[14,79-84,86-89] The actual effect on peptide or protein retention time will depend strongly on the hydrophobicity of the ion-pairing reagent and the number of oppositely charged groups on the solute,[79,80] a principle illustrated in Section III,D,4,a.

Apart from the powerful peptide-resolving capability of RPC, the availability of volatile mobile phases (e.g., aqueous TFA–acetonitrile systems) makes it ideal as the final step in a multidimensional purification protocol.[4,9]

Although isocratic elution of peptides from reversed-phase columns is

[72] K. J. Wilson, A. Honegger, R. P. Stötzel, and G. J. Hughes, *Biochem. J.* **199**, 31 (1981).

[73] E. M. Domitas, K. R. Hamid, R. C. Hider, and U. Ragnarsson, *Biochim. Biophys. Acta* **911**, 285 (1987).

[74] K. Hermann, R. E. Lang, T. Unger, C. Bayer, and D. Ganten, *J. Chromatogr.* **312**, 273 (1984).

[75] M. Fenger and A. H. Johnsen, *J. Endocrinol.* **118**, 329 (1988).

[76] L. Lozzi, M. Rustici, A. Santucci, L. Bracci, S. Petreni, P. Soldani, and P. Neri, *J. Liq. Chromatogr.* **11**, 1651 (1988).

[77] G. R. Rhodes, M. J. Rubenfield, and C. T. Garvie, *J. Chromatogr.* **488**, 456 (1989).

[78] *LC-GC* **10**, 515 (1992).

[79] D. Guo, C. T. Mant, and R. S. Hodges, *J. Chromatogr.* **386**, 205 (1987).

[80] C. T. Mant and R. S. Hodges, *in* "HPLC of Peptides and Proteins: Separation, Analysis and Conformation" (C. T. Mant and R. S. Hodges, eds.), p. 327. CRC Press, Boca Raton, FL, 1991.

[81] H. P. J. Bennett, *J. Chromatogr.* **266**, 501 (1983).

[82] Z. Iskandarini, R. L. Smith, and D. J. Pietrzyk, *J. Liq. Chromatogr.* **7**, 111 (1984).

[83] G. Winkler, P. Briza, and C. Kunz, *J. Chromatogr.* **361**, 191 (1986).

[84] H. P. J. Bennett, *in* "HPLC of Peptides and Proteins: Separation, Analysis and Conformation" (C. T. Mant and R. S. Hodges, eds.), p. 319. CRC Press, Boca Raton, FL, 1991.

[85] A. Calderan, P. Ruzza, O. Marin, M. Secchieri, G. Borin, and F. Marchiori, *J. Chromatogr.* **548**, 329 (1991).

[86] K. Kalghatgi and C. Horváth, *J. Chromatogr.* **443**, 343 (1988).

[87] J. Ishida, M. Kai, and Y. Ohkura, *J. Chromatogr.* **356**, 171 (1991).

[88] H. P. J. Bennett, C. A. Browne, and S. Solomon, *J. Liq. Chromatogr.* **3**, 1353 (1980).

[89] W. M. M. Schaaper, D. Voskamp, and C. Olieman, *J. Chromatogr.* **195**, 181 (1980).

occasionally useful, e.g., for separation of closely related peptides such as diastereomers,[90–93] the organic modifier (usually acetonitrile) concentration range for practical isocratic separations of peptides is narrow compared to smaller organic molecules.[94] Peptides interact with a reversed-phase packing mainly by an adsorption/desorption ("on/off") mechanism and partition only to a limited extent,[71,94–97] unlike small organic molecules such as alkylphenones, which exhibit considerable partitioning as they pass down a reversed-phase column.[94,96,97] Thus, for the separation of a mixture of peptides of varying hydrophobicities, gradient (usually linear) elution is the elution mode of choice.

2. Reversed-Phase HPLC Peptide Standards

a. STANDARDS FOR MONITORING NONIDEAL BEHAVIOR. The cation-exchange standards shown in Table II and employed for the CEC separations shown in Figs. 4 and 5 were also developed to monitor the extent of silanol activity (i.e., the potential for nonspecific, ionic solute–packing interactions) on silica-based reversed-phase packings.[69] Figure 6 demonstrates the application of these standards to monitoring silanol activity on two different silica-based columns at pH values of 2.0, 4.5, and 7.0. The contrast in the performances of the two columns is striking. Column A (Fig. 6) exhibited significant ionic interactions with the basic peptide standards (apparent mainly from the retention behavior of peptides 3 and 4) over the entire pH range (pH 2.0–7.0) available to chromatographers on silica-based reversed-phase columns, even under conditions designed to suppress or block such interactions, i.e., low pH or the presence of amines or salts in the mobile phase.[14] Peptides with charges of +3 and +4 are not uncommon and, thus, this column is clearly unsuitable for peptide applications. In contrast, the elution profiles obtained on column B (Fig. 6) were excellent at all three pH values. However, some ionic interaction was apparent at pH 7.0 even on this column, stressing once again the advantage of employing mobile phases at pH values <4.0–4.5, if at all possible.

[90] W. L. Cody, B. C. Wilkes, and V. J. Hruby, *J. Chromatogr.* **314,** 313 (1984).

[91] L. Gozzini and P. C. Montecucchi, *J. Chromatogr.* **362,** 138 (1986).

[92] S. Görög, B. Herényi, O. Nyéki, I. Schön, and L. Kistaludy, *J. Chromatogr.* **452,** 317 (1988).

[93] V. J. Hruby, A. Kuwasaki, and G. Toth, *in* "HPLC of Peptides and Proteins: Separation, Analysis and Conformation" (C. T. Mant and R. S. Hodges, eds.), p. 379. CRC Press, Boca Raton, FL, 1991.

[94] C. T. Mant, T. W. L. Burke, and R. S. Hodges, *Chromatographia* **24,** 565 (1987).

[95] W. C. Mahoney, *Biochim. Biophys. Acta* **704,** 284 (1982).

[96] D. Guo, C. T. Mant, A. K. Taneja, and R. S. Hodges, *J. Chromatogr.* **359,** 519 (1986).

[97] R. S. Hodges and C. T. Mant, *in* "HPLC of Peptides and Proteins: Separation, Analysis and Conformation" (C. T. Mant and R. S. Hodges, eds.), p. 3. CRC Press, Boca Raton, FL, 1991.

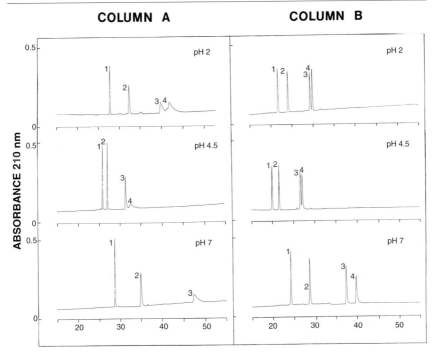

FIG. 6. Monitoring of nonideal (ionic) behavior in RP-HPLC with synthetic peptide standards. Column A: C$_{18}$ column (250 × 4.6 mm i.d., 5-μm particle size, 300-Å pore size). Column B: Aquapore RP-300 C$_8$ (200 × 4.6 mm i.d., 7 μm, 300 Å; Brownlee Laboratories, Santa Clara, CA). Instrumentation: The HPLC instrument consisted of a Spectra-Physics (Autolab Division, San Jose, CA) SP8700 solvent delivery system and SP8750 organizer module coupled to a Hewlett-Packard HP1040A detection system, HP3390A integrator, HP85 computer, HP9121 disk drive, and HP7470A plotter. Conditions: pH 2.0, linear AB gradient (1% B/min), where eluent A is 0.1% aqueous TFA and eluent B is 0.1% TFA in acetonitrile; pH 4.5, linear AB gradient (2% B/min, equivalent to 1% acetonitrile/min and 1 mM TEAP/min), where eluent A is 10 mM aqueous triethylammonium phosphate (TEAP), pH 4.5, and eluent B is 50% acetonitrile containing 60 mM TEAP (eluent B was prepared by mixing equal volumes of acetonitrile and an aqueous 120 mM solution of TEAP, pH 4.5); pH 7.0, linear AB gradient (1.67% B/min, equivalent to 1% acetonitrile/min and 1.67 mM NaClO$_4$/min), where eluent A is 10 M aqueous (NH$_4$)$_2$HPO$_4$, pH 7.0, and eluent B is 60% aqueous acetonitrile containing 100 mM NaClO$_4$. Flow rate, 1 ml/min; room temperature. The sequences of peptide standards 1–4 are shown in Table II (peptides C1–C4 in Table II and Figs. 4 and 5 denote peptides 1–4 in Fig. 6).

TABLE III
SYNTHETIC REVERSED-PHASE HPLC PEPTIDE STANDARDS[a]

Peptide standard	Sequence
I1	Ac-Arg-Gly-Gly-Gly-Gly-<u>Ile</u>-Gly-<u>Ile</u>-Gly-Lys-amide
I2	Ac-Arg-Gly-Gly-Gly-Gly-<u>Ile</u>-Gly-Leu-Gly-Lys-amide
S2	Ac-Arg-Gly-Gly-Gly-Gly-Leu-Gly-Leu-Gly-Lys-amide
S3	Ac-Arg-Gly-<u>Ala</u>-Gly-Gly-Leu-Gly-Leu-Gly-Lys-amide
S4	Ac-Arg-Gly-<u>Val</u>-Gly-Gly-Leu-Gly-Leu-Gly-Lys-amide
S5	Ac-Arg-Gly-<u>Val</u>-<u>Val</u>-Gly-Leu-Gly-Leu-Gly-Lys-amide

[a] Ac, N^α-Acetyl; amide, C^α-amide. The underlines denote amino acid substitutions in the peptide S2 sequence. These standards were purchased from Alberta Peptide Institute.

b. STANDARDS FOR MONITORING COLUMN PERFORMANCE. Table III shows the sequences of six synthetic peptide standards (I1, I2, S2, S3, S4, and S5), designed to monitor RP-HPLC column performance.[14,98] The hydrophobicity of the peptide analogs increases only slightly between S2 and S4—between S2 and S3 there is a change from an α-H to a β-CH$_3$ group, between S3 and S4 there is a change from a β-CH$_3$ group to two methyl groups attached to the β-CH group, between S4 and S5 there is a change from an α-H to an isopropyl group attached to the α-CH group. The hydrophobicity variations between I1, I2, and S2 are even more subtle. There is a change of only an isoleucine to a leucine residue between I1 and I2 and between I2 and S2. Guo et al.[71] demonstrated that leucine is slightly more hydrophobic than isoleucine, although these residues contain the same number of carbon atoms. Because isoleucine is β branched, the β carbon is close to the peptide backbone and not as available to interact with the hydrophobic stationary phase compared to the conformation of the leucine side chain. Thus, this peptide mixture enables precise determination of the resolving power of a reversed-phase column (demonstrated in Fig. 7 for an analytical C$_{18}$ column).

3. Standard Chromatographic Conditions for Reversed-Phase HPLC of Peptides

The best approach to most analytical peptide separations is to employ aqueous trifluoroacetic (TFA) to TFA–acetonitrile linear gradients (pH

[98] T. W. L. Burke, C. T. Mant, and R. S. Hodges, in "HPLC of Peptides and Proteins: Separation, Analysis and Conformation" (C. T. Mant and R. S. Hodges, eds.), p. 307. CRC Press, Boca Raton, FL, 1991.

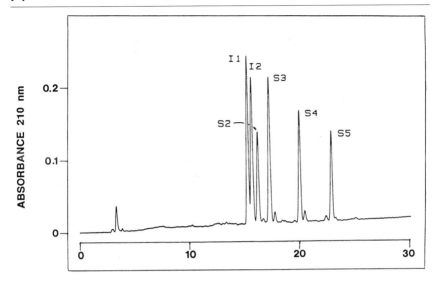

ELUTION TIME (min)

Fig. 7. Monitoring of reversed-phase column performance with synthetic peptide standards under recommended standard conditions. Column: SynChropak RP-P C_{18} (250 × 4.6 mm i.d., 6.5-μm particle size, 300-Å pore size; SynChrom, Lafayette, IN). Instrumentation: Same as in Fig. 1, except for a Hewlett-Packard HP 1090 liquid chromatograph (including autosampler) and an HP9000 Series 300 computer. Conditions: Linear AB gradient (1% B/min), where eluent A is 0.1% aqueous TFA and eluent B is 0.1% TFA in acetonitrile; flow rate, 1 ml/min; room temperature. The sequences of the peptide standards are shown in Table III.

2.0) at a flow rate of 0.5–2.0 ml/min.[9,14,46] A gradient steepness of 0.5–2.0% acetonitrile/min is suitable for most purposes.[9,14,46]

Recommended standard run conditions for RPC of peptides are as follows:

Linear AB gradient: Eluent A: 0.1% aqueous TFA
                        Eluent B: 0.1% TFA in acetonitrile
Gradient rate: 1% B/min

For example,

Time (min)	Mobile-phase composition
0	100% A
50	50% A, 50% B

Flow rate: 1 ml/min for standard analytical columns (4- to 4.6-
                  mm i.d.)
Temperature: Room temperature

These conditions were employed to produce the excellent elution profile of the peptide standards shown in Fig. 7.[61]

TFA absorbs at 210 nm, a wavelength used frequently for peptide bond detection (see Section III,F). Hence, a less than 0.1% concentration (e.g., 0.05–0.08% TFA) is sometimes useful for trace analysis, owing to a lower background signal. In addition, it should be noted that a lower TFA concentration may be required in eluent B than eluent A to counter baseline drift due to absorbance by the organic modifier in eluent B.[99] Again, this only tends to be a concern for trace analysis (and/or very sensitive detection wavelength).

4. Optimization of Peptide Elution Profiles in Reversed-Phase HPLC

Out of the three major modes of HPLC, RP-HPLC offers the widest scope for manipulation of mobile and stationary phase characteristics to improve peptide separations. The simplest approach to improving peptide elution profiles once an initial run has been carried out is to vary the gradient rate, where peptide resolution is improved with a decreasing gradient rate (this also applies to IEC). The value of varying the flow rate under gradient-elution conditions is somewhat limited; however, the tendency for peptides to diffuse decreases with increasing flow rate, producing smaller bandwidths and, hence, improved resolution.[98]

There is a limit to what can be achieved through gradient rate and/or flow rate variations on a specific column and under specific mobile-phase conditions. Other options to optimize peptide separations include varying the mobile-phase conditions and/or the type of reversed-phase column employed. Several of these alternatives are illustrated as follows (Figs. 8–11).

a. EFFECT OF ION-PAIRING REAGENT. A classic demonstration of the value of manipulating counterion (in this example, anionic counterion) hydrophobicity for optimizing peptide separations is shown in Fig. 8. A mixture of seven basic peptides (10–14 residues) with varying numbers of positively charged residues (a peptide number also denotes its net positive charge) was subjected to RP-HPLC on an analytical C_{18} column. The peptides demonstrated increasing retention times with increasing hydropho-

[99] C. T. Mant and R. S. Hodges, in "HPLC of Peptides and Proteins: Separation, Analysis and Conformation" (C. T. Mant and R. S. Hodges, eds.), p. 69. CRC Press, Boca Raton, FL, 1991.

bicity of the counterion: $HFBA^-$ (Fig. 8C) > TFA^- (Fig. 8B) > $H_2PO_4^-$ (Fig. 8A). In addition, the greater the net charge on a peptide, the greater the effect on its retention time on increasing counterion hydrophobicity. Thus, the elution order of the peptides changed from one counterion to another. For instance, the elution order of peptides 1, 3, and 6 (containing one, three, and six positively charged residues, respectively) was reversed as the counterion changed from $H_2PO_4^-$ (Fig. 8A) to $HFBA^-$ (Fig. 8C). It should be noted that, although TFA is the counterion of choice for most purposes, it produced the least effective resolution of this particular peptide mixture, highlighting the value of counterion variations in optimizing peptide separations.

Anionic ion-pairing reagents are generally used only at low concentrations (0.05–0.5%, v/v) in the mobile phase. Counterion concentration may have a marked effect on the retention times of peptides,[79,80] as demonstrated in Fig. 8B (insets). As the concentration of TFA in the elution solvents was increased from 0.01% (left inset, Fig. 8B) to 0.1% (middle, Fig. 8B) to 0.4% (right inset, Fig. 8B), the retention times increased and the elution order of peptides 1, 3, and 6 (+1, +3, and +6 net charge, respectively) changed accordingly. The greater the net charge of a peptide, the more its retention time increased with increasing TFA concentration. Thus, the elution order of these three peptides changed from 3, 6, and 1 in 0.01% TFA (left inset, Fig. 8B) to 1, 3, and 6 in 0.4% TFA. It is interesting to note that this change in peptide elution order is different from that observed when increasing the hydrophobicity of the counterion, i.e., 6, 3, and 1 with $H_2PO_4^-$ (Fig. 8A) to 1, 3, and 6 with $HFBA^-$ (Fig. 8C). It is not a good idea to use consistently high concentrations of acidic ion-pairing reagents to separate peptide mixtures, because these can have a deleterious effect on silica-based RP-HPLC columns by gradually cleaving the n-alkyl functionalities from the reversed-phase support. However, Guo et al.[79] demonstrated that the greatest effect of variations in TFA (and HFBA) concentration on peptide retention was apparent over the 0.01 to 0.2% range of reagent concentration, with only a limited effect at higher acid levels. These results clearly demonstrate the importance of consistency in the concentration of ion-pairing reagent in the mobile phase for accurate run-to-run comparisons of peptide separations. In addition, and in a manner similar to counterion variations, the effect of counterion concentration on a contaminating peptides in a synthetic peptide mixture can also be a test of homogeneity of the peptide of interest.

b. EFFECT OF VARYING FUNCTIONAL GROUPS ON STATIONARY PHASE. Although RP-HPLC on stationary phases containing alkyl chains (e.g., C_8, C_{18}) as the functional ligand is still the method of choice for most peptide separations,[14,19] less hydrophobic cyanopropyl (CN) packings also have

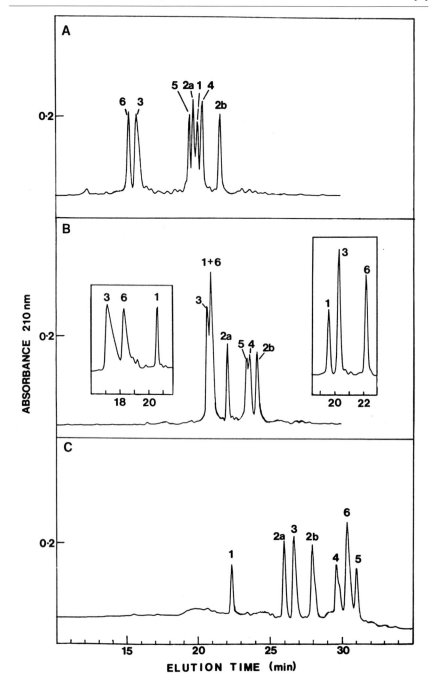

certain advantages for such applications. Kirkland and co-workers[100] reported the development of stable silica-based bonded phases (the cyanopropyl group is somewhat labile in acidic mobile phases), based on protecting
the siloxane bond between the silica and the functional group with bulky
side chains, e.g., two isopropyl groups in place of the usual two methyl
groups.

Figure 9 compares the separation of two peptide analogs on a C_8 column
(Fig. 9A) with that obtained on a sterically protected CN column (Fig. 9B).
Peptides 8W and 8F differ by only two residues: Two phenylalanine residues
in 8F are replaced by two tryptophan residues in 8W. From a poor separation on the C_8 column ($\Delta t = 0.3$ min only) (Fig. 9A), the peptides were
separated by 4.7 min on the CN column (Fig. 9B). This dramatic change
in the separation of the two peptides between the C_8 and CN columns
implies significant selectivity differences between the dipolar cyanopropyl
and the octyl functionalities. Other examples of the utility of complementary
selectivities of alkyl versus CN functionalities may be found in Ref. 101.

c. EFFECTS OF VARYING THE PACKING SUPPORT. It has been noted[102]
that although RP-HPLC on porous PSDV (polystyrene-divinylbenzene)
matrices is generally equivalent to that on alkylsilane-derivatized silica,
there are exceptions.[103] For instance, $\pi-\pi$ interactions may play a role in
the retention of some solutes (e.g., aromatic molecules, aromatic side chains
in peptides) on PSDV matrices; also, because PSDV matrices are hydrocarbons, they may be expected to give different selectivities to a silica-
based packing.

Figure 10 compares the elution profiles of a mixture of 18-residue peptide analogs run on a silica-based C_8 column (Fig. 9, top) and PSDV (Fig.
9, bottom). The letters above the individual peaks denote the single amino
acid substitutions (at position 9) made to produce each analog. These

[100] J. J. Kirkland, J. L. Glajch, and R. D. Farlee, *Anal. Chem.* **61,** 2 (1988).
[101] N. E. Zhou, C. T. Mant, J. J. Kirkland, and R. S. Hodges, *J. Chromatogr.* **548,** 179 (1991).
[102] K. A. Tweeten and T. N. Tweeten, *J. Chromatogr.* **359,** 111 (1986).
[103] N. B. Afeyan, S. P. Fulton, and F. E. Regnier, *J. Chromatogr.* **544,** 267 (1991).

FIG. 8. Effect of anionic ion-pairing reagent on the separation of a mixture of basic peptides
in RP-HPLC. Column: SynChropak RP-P C_{18} (250 × 4.1 mm i.d., 6.5-μm particle size, 300-Å
pore size; SynChrom, Lafayette, IN). Instrumentation: Same as in Fig. 1. Conditions: Linear
AB gradient (1% B/min), where eluent A is water and eluent B is acetonitrile, both eluents
containing 0.1% H_3PO_4 (A), TFA (B), or HFBA (C); flow rate, 1 ml/min; room temperature.
(B) Insets: left, 0.01% TFA in eluents A and B; right, 0.4% TFA in eluents A and B. The
peptides contain between 5 and 14 residues; the numbers above the peptide peaks denote
the number of positive charges they contain. (From Ref. 79, with permission.)

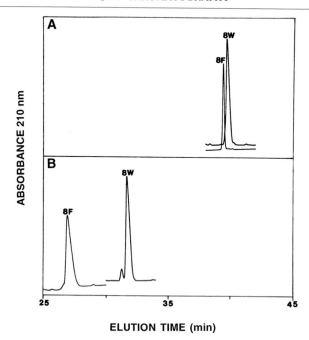

FIG. 9. Effect of stationary phase group of silica-based column on selectivity of peptide separation. Columns: (A) Zorbax SB-C$_8$ (150 × 4.6 mm i.d., 5-μm particle size, 94-Å pore size); (B) Zorbax SB-300CN (150 × 4.6 mm i.d., 6 μm, 300 Å) (Rockland Technologies, Newport, DE). Instrumentation: Same as in Fig. 7. Conditions: Linear AB gradient (1% B/min), where eluent A is 0.05% aqueous TFA and eluent B is 0.05% TFA in acetonitrile; flow rate, 1 ml/min; room temperature. The sequence of the peptide analogs is Ac-Gly-X-X-(Leu)$_3$-(Lys)$_2$-amide, where position X is occupied by Trp (8W) or Phe (8F). Ac, N^α-Acetyl; amide, C^α-amide. (From Ref. 101, with permission.)

synthetic peptide analogs were designed to have high potential for forming amphipathic α helices in a hydrophobic environment such as those characteristic of RP-HPLC. The hydrophobic face of the helix (the preferred binding domain) would be expected to interact with the hydrophobic stationary phase.[104] The amino acid substitutions were always made in the middle of this face, ensuring intimate interaction of the residue at this position with the reversed-phase packing. The PSDV sorbent was packed in the authors' laboratory. Although the subsequent peptide elution profiles were satisfactory, commercially packed columns of this material produce even better column efficiencies and elution profiles.

From Fig. 10, despite a generally similar peptide elution pattern on the

[104] N. E. Zhou, C. T. Mant, and R. S. Hodges, Peptide Res. **3,** 8 (1990).

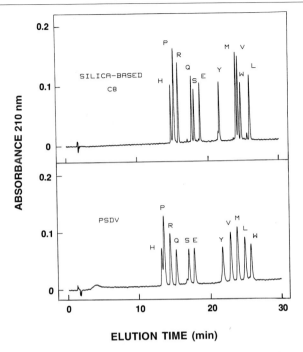

FIG. 10. Effect of column support on selectivity of peptide separations in RP-HPLC. Columns: (*upper*) silica-based Zorbax SB-300 C_8 (150 × 4.6 mm i.d., 5-μm particle size, 250-Å pore size; Rockland Technologies, Newport, DE); (*lower*) PLRP-S polystyrene-divinyl-benzene (PSDV) sorbent (5 μm, 100 Å) laboratory-packed into a column of 150 × 4.6 mm i.d. (Polymer Laboratories, Church Stretton, Shropshire, UK). Instrumentation: Same as in Fig. 7. Conditions: Linear AB gradient (1% B/min, starting from 2% acetonitrile), where eluent A is 0.05% aqueous TFA and eluent B is 0.05% TFA in acetonitrile; flow rate, 1 ml/min; room temperature. The sequence of the peptide analogs is Ac-Glu-Ala-Glu-Lys-Ala-Ala-Lys-Glu-X-Glu-Lys-Ala-Ala-Lys-Glu-Ala-Glu-Lys-amide, where X is substituted by various amino acid residues; the letters above the peptide peaks (one-letter amino acid code) denote the amino acid residues occupying position X.

two columns, certain significant column selectivity differences are apparent. The simplest way to visualize these differences is to compare the retention behavior of the serine (S)- and glutamic acid (E)-substituted analogs with the other peptides. The effect of the two different packings on the elution behavior of these two peptides relative to each other and to most of the other peptides appears to be negligible, making them useful internal standards. The major differences in elution behavior between the two columns involve the glutamine (Q)-, tyrosine (Y)-, methionine (M)-, and tryptophan (W)-substituted analogs. On the C_8 column, the Q analog is eluted 1.2 min prior to the E analog; the Y, M, and W analogs are eluted, respectively,

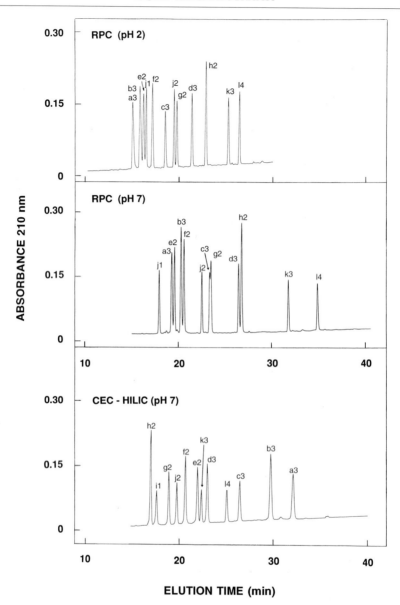

Fig. 11. Effect of pH on peptide elution profiles in RP-HPLC and comparison of RP-HPLC with mixed-mode HILIC-CEC for peptide separations. Columns: (*Upper* and *middle*) Zorbax SB-300 C$_8$ (150 × 4.6 mm i.d., 5-μm particle size, 250-Å pore size; Rockland Technologies, Newport, DE); (*lower*) polysulfoethyl aspartamide (PolySULFOETHYL Aspartamide) strong cation-exchange column (200 × 4.6 mm i.d., 5 μm, 300 Å; PolyLC, Columbia, MD).

2.7, 5.2, and 5.9 min after the E analog. In contrast, on the PSDV column, the Q analog is now eluted 2.7 min prior to the E analog; while the Y, M, and W analogs are eluted, respectively, 4.0, 6.1, and 8.1 min after the E analog. The overall relative differences in retention time of these four analogs (relative to analog E) on the two columns is 1.5 min (Q), 1.3 min (Y), 0.9 min (M), and 2.2 min (W). These shifts in relative retention times of these analogs relative to the other peptides has produced an elution order change of two peptide pairs between the columns. Thus, the M and V analogs and L and W analogs have a reversed order of elution between the two columns. The aromatic nature of the side chains of tyrosine (Y) and tryptophan (W) (through potential π–π interactions with the PSDV matrix) may offer some explanation for the observed selectivity differences of the two columns. If so, it is interesting that a single aromatic side chain in an 18-residue peptide would have such an effect. One could speculate that the presence of the tyrosine or tryptophan residue in the center of the hydrophobic face of the preferred peptide-binding domain may help to amplify differences between the packings. The unexpected effect of the glutamine and methionine side chains on the retention behavior of the analogs between the two packings only serves to emphasize the potentially useful, if unpredictable (to date), qualitative nature of testing stationary phases for peptide separations.

d. EFFECT OF pH. Although the resolving power of the acidic aqueous TFA–acetonitrile mobile phase makes it particularly suitable for peptide separations, manipulation of elution profiles through pH variations of the mobile phase may frequently be advantageous.

Figure 11 (top and middle) compares the elution profiles of a mixture of basic (positively charged) peptides on a silica-based C_8 column at pH 2 (top, Fig. 11) and pH 7 (middle, Fig. 11). The presence of sodium perchlorate in the pH 7 system is to ensure suppression of any undesirable ionic interactions between negatively charged free silanols on the silica surface and

---

Instrumentation: (*Upper* and *middle*) same as in Fig. 7; (*lower*) same as in Fig. 6. Conditions: (*upper*) linear AB gradient (1% B/min), where eluent A is 0.1% aqueous TFA and eluent B is 0.1% TFA in acetonitrile; (*middle*) linear AB gradient (2% B/min, equivalent to 1% acetonitrile/min), where buffer A is 10 mM aqueous $(NH_4)_2HPO_4$, pH 7, and buffer B is 50% aqueous acetonitrile containing 10 mM $(NH_4)_2HPO_4$, pH 7, both buffers also containing 200 mM $NaClO_4$; (*lower*) linear AB gradient (2% B/min, equivalent to a linear increasing salt gradient of 5 mM $NaClO_4$/min and a linear decreasing acetonitrile gradient of 0.8% acetonitrile/min), where buffer A is 5 mM aqueous triethylammonium phosphate (TEAP), pH 7, containing 90% (v/v) acetonitrile and buffer B is 5 mM aqueous TEAP, pH 7, containing 250 mM $NaClO_4$ and 50% (v/v) acetonitrile. Flow rate, 1 ml/min; room temperature. The peptides contain either 10 or 11 residues; the numbers above the peptide peaks denote the number of potentially positive charges they contain.

positively charged residues in the peptides. The difference in elution patterns between the pH 2 and pH 7 runs is quite clear, with some peptides being better resolved at pH 2 (e.g., b3 and f2; d3 and h2; c3 and g2), whereas others show an improved separation at the higher pH (e.g., i1, j2, k3). The variation in elution order at pH 7, plus the general increase in peptide retention times at this higher pH, are probably due to a combination of effects, e.g., ion-pair formation of the negatively charged hydrophilic perchlorate counterion with positively charged residues coupled with a possible promotion of hydrophobic interactions of the peptides with the stationary phase by the high perchlorate concentration. In addition, more recent work[34] has demonstrated that the pK_a value of a positively charged α-amino group (generally about pH 9.0–9.5 in the free amino acid) is reduced to about pH 6.0–6.8 by the hydrophobic environment characteristic of RP-HPLC, depending on the N-terminal residue. Peptides a3, b3, c3, and d3 all contain a free N-terminal arginine (pK_a of α-amino group on an N-terminal arginine was measured as pH 6.1 by Sereda et al.[34]). Thus, at pH 7, this N-terminal α-amino group would be expected to be essentially deprotonated, i.e., neutral. From the pH 7 run, peptides having identical sequence are eluted in close pairs (a–e, b–f, c–g, and d–h) at pH 7; this is in contrast to their wide separation at pH 2, where peptides e, f, g, and h are the peptides with their N terminals acetylated, evidence that the free α-amino groups of peptides a, b, c, and d are not charged (or are only negligibly charged) at pH 7. This combination of possible effects on peptide elution patterns as the pH is raised makes the effect of pH a difficult thing to predict. The equation is complicated even further if acidic residues (potentially negatively charged at pH values above 4.0–5.0) are present in peptides. Mobile-phase systems above pH 5.0, in fact, are sometimes necessary for highly acidic peptides, where solubility is a problem at acidic pH.

The use of alkaline mobile phases (i.e., above pH 7.0) will, of course, require the use of a non-silica-based packing, e.g., PSDV. In fact, volatile 0.1 M ammonium bicarbonate mobile phases have proved useful in the pH range 7.0 to 11.0.[14,30,105,106] However, in the authors' opinion, very high pH values have only a limited use for peptide (and protein) purification by RP-HPLC.

Overviews and examples of peptide separations at pH values in the range of 4.0–7.0 (and above) can be found in Refs. 14 and 19.

[105] D. R. Knighton, D. R. K. Harding, J. R. Napier, and W. S. Hancock, *J. Chromatogr.* **249**, 193 (1982).
[106] G. E. D. Jackson and N. M. Young, *Anal. Biochem.* **162**, 251 (1987).

E. Novel Methods of Peptide Analysis by HPLC

Apart from the three major modes of HPLC suitable for peptide separations, it is worth mentioning three other chromatographic modes available to the researcher. Two of these modes, affinity chromatography (AFC) and hydrophobic interaction chromatography (HIC), are almost invariably employed for protein separations (indeed, the latter technique was developed for this purpose) and are mentioned only briefly. A more recent chromatographic technique, hydrophilic interaction chromatography (HILIC), promises to be the most useful novel technique for general peptide separations.

1. High-Performance Affinity Chromatography

Out of a somewhat limited source of published material on high-performance affinity chromatography (AFC) of peptides, the following two articles represent a useful source of information on the topic.

Van Eyk et al.,[107] in an article reviewing AFC of peptides and proteins, gave a general overview of the general principles of AFC, coupled with a discussion of the advantages of AFC over conventional affinity chromatography. In addition, approaches to choice of ligand, ligand attachment, buffers, and sample recovery techniques were reviewed. A specific example of AFC of peptides offered was that of examining the binding affinities of multiple glycine-substituted troponin I peptide analogs on a troponin C AFC column. Troponins I and C are part of a multiprotein complex found in the thin filaments of muscle. A comparison of step-gradient elution versus linear gradient elution of such peptides from the silica-based AFC column was described, as well as flow rate effects. This article represents a useful starting point for the researcher when considering AFC for peptide applications.

An excellent review of AFC of a specific kind of peptide was published by Chaiken.[108] This paper summarized observations of selective chromatographic separation of native sense peptides using immobilized antisense peptides, where antisense peptides are sequences of amino acids encoded in the antisense strand of DNA.

[107] J. E. Van Eyk, C. T. Mant, and R. S. Hodges, in "HPLC of Peptides and Proteins: Separation, Analysis and Conformation" (C. T. Mant and R. S. Hodges, eds.), p. 479. CRC Press, Boca Raton, FL, 1991.
[108] I. Chaiken, J. Chromatogr. **597**, 29 (1992).

2. Hydrophobic Interaction Chromatography

Hydrophobic interaction chromatography (HIC) was developed as an alternative to RP-HPLC for general purification, owing to the denaturing conditions of RP-HPLC.[109] Thus, HIC conditions (stationary and mobile phase) are selected with the view of maintaining tertiary and quaternary native protein structure. Generally, such columns and conditions have no advantage for peptides over that of RP-HPLC stationary and mobile phases. Alpert,[109a] for instance, subjected a number of peptides to RP-HPLC and HIC and compared their chromatographic behavior. Although, in most cases, selectivities as evaluated by elution order were similar or identical for the two methods, better efficiency was exhibited with RP-HPLC. However, for some peptides, HIC may be a better choice. Thus, HIC performed better than RP-HPLC for purification of an extremely hydrophobic peptide, a synthetic analog of a lipoprotein fragment.[109a] In addition, HIC may also give better resolution for peptides large enough to possess secondary or tertiary structure, e.g., synthetic calcitonin analogs.[110] In addition, HIC may occasionally supplement RP-HPLC for purification of very hydrophilic peptides when such peptides are not retained under reversed-phase conditions.[109a]

3. Hydrophilic Interaction Chromatography

Hydrophilic interaction chromatography (HILIC) has been promoted as a novel chromatographic mode for application to the separation of a wide range of solutes.[111] Separation by HILIC, in a manner similar to normal phase chromatography (to which it is related), depends on hydrophilic interactions between the solutes and the hydrophilic stationary phase, i.e., solutes are eluted from an HILIC column in order of increasing hydrophilicity (decreasing hydrophobicity). HILIC is characterized by separations being effected by a linear gradient of decreasing organic modifier concentration, i.e., starting from a high concentration of organic modifier (typically, 70–90% aqueous acetonitrile).

In their evaluation of the potential HILIC for peptide separations, Zhu et al.[112] noted that the HILIC packing (polyhydroxyethylaspartamide) exhibited some cation-exchange characteristics, i.e., the packing contained a slight negative charge. These researchers then reasoned that, because

[109] S.-L. Wu and B. L. Karger, Methods Enzymol. **270,** Chap. 2, 1996.

[109a] A. J. Alpert, J. Chromatogr. **444,** 269 (1988).

[110] M. L. Heinitz, E. Flanigan, R. C. Orlowski, and F. E. Regnier, J. Chromatogr. **443,** 229 (1988).

[111] A. J. Alpert, J. Chromatogr. **499,** 177 (1990).

[112] B.-Y. Zhu, C. T. Mant, and R. S. Hodges, J. Chromatogr. **548,** 13 (1991).

the charged character of a cation-exchange sorbent would confer on it considerable hydrophilic character, such a column should be examined for possible peptide separations by HILIC. Classic examples of this approach may be found in Ref. 113, with Fig. 11 (bottom) representing a summary of what may be achieved when manipulating cation-exchange column selectivity to separate peptides based on a mixed-mode hydrophilic and ionic interaction mechanism.

It has been noted previously that all ion-exchange sorbents exhibit some hydrophobic character that may be suppressed by the addition of low levels of organic modifier, e.g., acetonitrile, to the mobile-phase buffers. Zhu *et al.*[113] noted that as the level of acetonitrile was raised from 10 to 90% (in 10% steps), the separation process (linearly increasing gradients of sodium perchlorate with the same level of acetonitrile in both buffers) became increasingly mixed-mode (hydrophilic and ionic interactions), with HILIC interactions becoming dominant at high acetonitrile concentrations. The hydrophilic interactions were apparent from observing that peptides with a particular net positive charge were eluted prior to other peptides with lower net charges (all things being equal, the opposite should be true on CEC). These observations are summarized in Fig. 11 (bottom). The same positively charged peptides separated by RP-HPLC at pH 2 (top, Fig. 11) and pH 7 (middle, Fig. 11) were now separated by a mixed HILIC–CEC mechanism on a polysulfoethylaspartamide strong cation-exchange column. Elution conditions were manipulated to elute all peptides within times similar to those observed for the RP-HPLC runs. Thus, a level of 90% acetonitrile was present in buffer A, while only 50% was present in buffer B; hence, elution was carried out by a decreasing acetonitrile gradient in addition to an increasing gradient of sodium perchlorate concentration. The excellent peptide resolution achieved under these conditions (Fig. 11, bottom) compares favorably with the RP-HPLC separations. In fact, Zhu *et al.*[113] suggest that mixed-mode HILIC–CEC may rival RP-HPLC for peptide separations. HILIC effects are apparent from the elution of peptide h2 (+2 net charge) before i1 (+1) and l4 (+4) before c3, b3, and a3 (+3); h2 has greater overall hydrophilicity than i1, as is also true of l4 over that of c3, b3, and a3. In addition, the order of elution within peptide groups of the same net charge (e.g., a3–d3) also reflects the hydrophilic character of these peptides, i.e., in order of increasing hydrophilicity (decreasing hydrophobicity: d3 < c3 < b3 < a3). Note the order reversal when comparing HILIC effects (Fig. 11, bottom) and RP-HPLC hydrophobic effects (Fig. 11, top and middle); thus, the peptide elution order of a3, b3, c3 then d3 in RP-HPLC is reversed to d3, c3, b3 then a3 during HILIC–CEC.

[113] B.-Y. Zhu, C. T. Mant, and R. S. Hodges, *J. Chromatogr.* **594,** 75 (1992).

The ideal hydrophilic interaction sorbent, i.e., one lacking ionic characteristics and only exhibiting an HILIC mechanism, has yet to be developed.
Although the development of such a sorbent is certainly worthwhile, the
potential for excellent mixed-mode HILIC and ionic separations of peptides
on an ion-exchange column should not be overlooked or underestimated.

F. Detection Wavelength

The RP-HPLC elution profiles shown in Fig. 12 demonstrate the effect
of varying detection wavelength (200 to 230 nm, in 10-nm steps) on the
sensitivity of detection of the six peptide standards shown in Table III.[98]

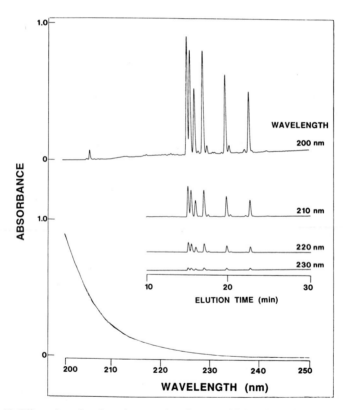

FIG. 12. Effect of varying detection wavelength on sensitivity of UV detection of peptides
during RP-HPLC. Column and instrumentation: Same as in Fig. 7. Conditions: Same as in
Fig. 7. *Top* and *middle:* Elution profiles of synthetic peptide standards (sequences shown in
Table III) at wavelengths of 200, 210, 220, and 230 nm. *Bottom:* Spectrum (200–250 nm) of
peptide standard I1. (Reprinted with permission from C. T. Mant and R. S. Hodges (eds.),
"HPLC of Peptides and Proteins: Separation, Analysis and Conformation." CRC Press, Boca
Raton, FL, © 1991.)

wavelength increases from 200 nm (~850 mAU for peptide I1, the first eluted peak) to 230 nm (~15 mAU for I1). On each 10-nm increase in detection wavelength, there is an approximate fourfold decrease in sensitivity.

The decrease in absorbance of the peptide bond with increasing detection wavelength is clearly apparent from the UV spectrum (200–250 nm) of peptide standard I1 shown at the bottom of Fig. 12. Although not shown, the absorption maxima for the peptide bond is actually ~187 nm. However, several practical considerations preclude the routine use of wavelengths below 200 nm for solute detection; indeed, the use of wavelengths below 210 nm tends to be the exception rather than the rule. Detection below 210 nm can suffer from interference due to impurities present in buffers, solvents, or even the sample, which may obscure the peak(s) of interest.[51] HPLC-grade solvents are the norm for most analytical purposes, especially where detectors are being used at high sensitivity. However, chemical groups in the pure solvents themselves may also contribute significantly to UV absorbance at wavelengths below 210 nm. The "UV cutoff" values— the wavelength in nm at which the absorbance of a 1-cm-long cell filled with the solvent has an absorbance of 1.0, measured against water as reference—of the popular RP-HPLC solvents methanol, acetonitrile, and 2-propanol are 205, 188, and 205 nm, respectively. Thus, the excellent sensitivity, with little detection interference, of the 200-nm wavelength peptide elution profile shown in Fig. 12 is due partly to the low UV cutoff value of acetonitrile as well as the cleanliness of the sample. The quality of the detection system is, of course, also an important factor. The common use of 210 nm as the preferred detection wavelength for most analytical reversed-phase applications (and, indeed, for those of other HPLC modes) is a good compromise between detection sensitivity and potential detection interference.

G. Postrun Concerns

1. Column Washing

The extended use of any column will result in deterioration of resolution, often due to the accumulation of noneluted solutes, e.g., hydrophobic peptides/proteins. The most general technique for removing retained solutes from all types of HPLC columns is gradient elution from 0.1% aqueous TFA to 0.1% TFA in 2-propanol.[49] Repeated gradient elutions may be necessary for tightly bound solutes. If this general technique does not produce entirely satisfactory results, cleaning procedures specific to a particular type of HPLC column (and usually included in the manufacturer instructions accompanying a particular column) may be required.

2. Column Storage

Both silica- and polymer-based SEC columns are best stored in 10–20% aqueous alcohol (usually methanol) to prevent bacterial growth.

IEC columns are generally stored in 10–20 or 100% alcohol (usually methanol). These columns should not be stored in ionic solutions.

RP-HPLC columns should be stored in a high concentration of organic solvent. The columns are then cleaned as well as stored after each day's use. The concentration of organic solvent should be significantly higher than the concentrations used in the mobile phase and the authors recommend >50% aqueous organic solvent (e.g., acetonitrile) for most purposes. The storage solvent should not contain acidic mobile-phase additives such as TFA.

When employing mobile phases containing salts for peptide applications, it is important to ensure that the column is salt free prior to storage in a solution containing organic solvents. This is particularly important when high salt and/or high organic solvent concentrations are employed in the mobile phase and storage solution, respectively. Thus, the column should be flushed with several column volumes of water prior to storage. In the case of RP-HPLC, it is also important to remove efficiently all traces of acids prior to storage.

A review of column maintenance techniques can be found in Ref. 49.

H. Microbore and Preparative Reversed-Phase HPLC of Peptides

Although the present chapter does not deal with these topics in detail, the researcher should be aware that efficient small-scale (microbore, narrow-bore) and preparative chromatography (of substantial levels of peptide samples) may be achieved on analytical LC instrumentation.

The advantages of employing HPLC columns with internal diameters smaller than the conventional 4- to 4.6-mm i.d. include increased detection sensitivity, higher concentrations of eluted solutes, and decreased solvent consumption. In comparing peptide resolution on conventional, narrow-bore, and microbore reversed-phase columns, Burke et al.[114] supported the observation of Schlabach et al.[115] that by simply replacing a conventional column with a narrow-bore column, a substantial improvement can be made in terms of both increased detection sensitivity and decreased solvent consumption.

[114] T. W. L. Burke, C. T. Mant, and R. S. Hodges, in "HPLC of Peptides and Proteins: Separation, Analysis and Conformation" (C. T. Mant and R. S. Hodges, eds.), p. 679. CRC Press, Boca Raton, FL, 1991.
[115] T. D. Schlabach, L. R. Zieske, and K. J. Wilson, in "HPLC of Peptides and Proteins: Separation, Analysis and Conformation" (C. T. Mant and R. S. Hodges, eds.), p. 661. CRC Press, Boca Raton, FL, 1991.

Several articles[94,116–118] have demonstrated the effectiveness of slow gradient rates (0.1% organic modifier/min) for reversed-phase purification of up to ~20 mg of a sample mixture containing peptides closely related in hydrophobicity, on standard analytical columns. In addition, a novel preparative technique, sample displacement chromatography (SDC), has been reported to resolve up to 60–80 mg of a peptide mixture on standard analytical reversed-phase columns.[119–123] This technique is notable for the major separation process taking place in the absence of an organic modifier.

IV. Computer Simulation of Peptide Elution Profiles

Once an initial chromatographic run has been carried out, preferably with the recommended starting conditions described in Sections III,B,3 (SEC), III,C,3 (IEC), and III,D,3 (RP-HPLC), the researcher then has the option of deciding how to optimize the separation. The selection of optimal conditions may be critical in establishing an analytical method, with changes in flow rate (SEC, IEC, RPC) and gradient rate (IEC, RP-HPLC) often proving effective in optimizing peptide separations.[9,96,98]

A computer program, called ProDigest-LC (available from Alberta Peptide Institute, University of Alberta, Edmonton, Alberta, Canada), has been developed that assists scientists in devising methods of SEC, IEC, and RP-HPLC for the analytical separation and purification of peptides.[46,124,125] The program simulates peptide elution profiles at varying flow rates (SEC, IEC, RP-HPLC) and gradient rates (IEC, RP-HPLC), with no

[116] J. M. R. Parker, C. T. Mant, and R. S. Hodges, *Chromatographia* **24**, 832 (1987).

[117] R. S. Hodges, T. W. L. Burke, C. T. Mant, and S. M. Ngai, *in* "HPLC of Peptides and Proteins: Separation, Analysis and Conformation" (C. T. Mant and R. S. Hodges, eds.), p. 773. CRC Press, Boca Raton, FL, 1991.

[118] T. W. L. Burke, J. A. Black, C. T. Mant, and R. S. Hodges, *in* "HPLC of Peptides and Proteins: Separation, Analysis and Conformation" (C. T. Mant and R. S. Hodges, eds.), p. 783. CRC Press, Boca Raton, FL, 1991.

[119] T. W. L. Burke, C. T. Mant, and R. S. Hodges, *J. Liq. Chromatogr.* **11,** 1229 (1988).

[120] R. S. Hodges, T. W. L. Burke, and C. T. Mant, *J. Chromatogr.* **444,** 349 (1988).

[121] R. S. Hodges, T. W. L. Burke, and C. T. Mant, *J. Chromatogr.* **548,** 267 (1991).

[122] C. T. Mant, T. W. L. Burke, A. J. Mendonca, and R. S. Hodges, *in* "Proceedings of the 9th International Symposium on Preparative and Industrial Chromatography, p. 241, April 6–8, Nancy, France, 1992.

[123] R. S. Hodges, T. W. L. Burke, A. J. Mendonca, and C. T. Mant, *in* "Chromatography in Biotechnology" (C. Horváth and L. S. Ettre, eds.), p. 59. American Chemical Society, Washington, D.C., 1993.

[124] C. T. Mant, T. W. L. Burke, N. E. Zhou, J. M. R. Parker, and R. S. Hodges, *J. Chromatogr.* **485,** 365 (1989).

[125] C. T. Mant and R. S. Hodges, *in* "HPLC of Peptides and Proteins: Separation, Analysis and Conformation" (C. T. Mant and R. S. Hodges, eds.), p. 705. CRC Press, Boca Raton, FL, 1991.

prior information about the peptides except their amino acid composition. The experiments simulated on the computer by the program eliminate the time-consuming trial-and-error approach to peptide purification that often results in considerable loss of sample.

Of great value to the beginner is the teaching aid feature of ProDigest-LC. The program is designed to help the student or researcher to select the correct conditions for chromatography (HPLC mode, column, and mobile phase) and the option of examining the effect of varying flow rate, gradient rate, sample size, and sample volume on the separation. Thus, ProDigest-LC simulates peptide elution profiles without even having to carry out an actual chromatographic run. Having manipulated the program until the desired separation has been simulated, the researcher may then carry out the run.

Also, commercially available to the researcher is an optimization computer program, DryLab G (LC Resources, Inc., Orinda, CA).[126-128] This program requires a minimum of two experimental runs, following which the effects on the peptide elution pattern of manipulating parameters, such as gradient rate, gradient shape (multisequential gradients), and flow rate may be simulated until the desired resolution is achieved (see [7] in volume 270[128a] for details).

For a review of optimization programs, in addition to other aspects of computer-assisted method development in chromatography, the reader is directed to Ref. 129.

Acknowledgments

This work was supported by the Medical Research Council of Canada. We thank Dawn Lockwood for typing the manuscript.

[126] J. W. Dolan, D. C. Lommen, and L. R. Snyder, *J. Chromatogr.* **485,** 9 (1989).

[127] L. R. Snyder, J. W. Dolan, and D. C. Lommen, *J. Chromatogr.* **485,** 65 (1989).

[128] L. R. Snyder, J. W. Dolan, and D. C. Lommen, *in* "HPLC of Peptides and Proteins: Separation, Analysis and Conformation" (C. T. Mant and R. S. Hodges, eds.), p. 725. CRC Press, Boca Raton, FL, 1991.

[128a] L. R. Snyder, *Methods Enzymol.* **270,** Chap. 7, 1996.

[129] R. W. Giese, J. K. Haken, K. Macek, and L. R. Snyder (eds.), "Computer Assisted Method Development in Chromatography." *J. Chromatogr.* **485.** Elsevier, Amsterdam, The Netherlands, 1989.

[2] Analytical Peptide Mapping of Recombinant DNA-Derived Proteins by Reversed-Phase High-Performance Liquid Chromatography

By Edward R. Hoff and Rosanne C. Chloupek

Peptide mapping by reversed-phase high-performance liquid chromatography (RP-HPLC) is a key method for the characterization and quality control of recombinant DNA-derived proteins. High resolution separations are routinely achieved in run times of several minutes to a few hours. Peptide mapping has been used extensively as a technique to generate purified protein fragments for amino acid sequence analysis and amino acid composition analysis for primary structure determination.

The advent of recombinant DNA techniques enabled the production of large quantities of highly purified proteins as commercial products. This resulted in a need for high-resolution analytical "fingerprinting" to confirm the primary sequence of protein products by comparison to a reference material and to quantify mutations or posttranslational variants that could copurify with the protein of interest. Although DNA sequencing has been used for mutation detection, the limit of sensitivity for mutation detection was 10 to 15%.[1] Peptide mapping has detected mutations from 1 to 5%,[1,2] and posttranslational modifications at levels of 5 to 10% of the total protein.[3]

Figure 1 shows a comparison of high-resolution tryptic maps of recombinant tissue-type plasminogen activator (rtPA) and rtPA with a single-site mutation at residue 275, where glutamic acid has been substituted for arginine.[4] The mutation is clearly detected in this example at a 4% level, in spite of the number and complexity of the peptide mixture. Therefore, it is important that optimum separation conditions be found so that amino acid substitutions are detected with high probability.[5] Also, the method must be rugged and reproducible so that comparison of sample and a reference material is possible over time. In this chapter we focus on methods

[1] R. J. Harris, A. A. Murnane, S. L. Utter, K. L. Wagner, E. T. Cox, G. D. Polastri, J. C. Helder, and M. B. Sliwkowski, *Bio/Technology* **11**, 1293–1297 (1993).

[2] R. L. Garnick, *Dev. Biol. Stand.* **76**, 117–130 (1992).

[3] C. L. Clogston, Y. Hsu, T. C. Boone, and H. S. Lu, *Anal. Biochem.* **202**, 375–383 (1992).

[4] Committee on Process Development and Manufacturing, Biotechnology Division, Pharmaceutical Manufacturers Association, *BioPharm* **4**(2), 22–27 (1991).

[5] W. S. Hancock, E. Canova-Davis, J. Battersby, and R. Chloupek, *in* "Biotechnologically Derived Medical Agents: The Scientific Basis of Their Regulation" (J. L. Gueriguian, V. Fattorusso, and D. Poggiolini, eds.), pp. 29–49. Raven Press, New York, 1988.

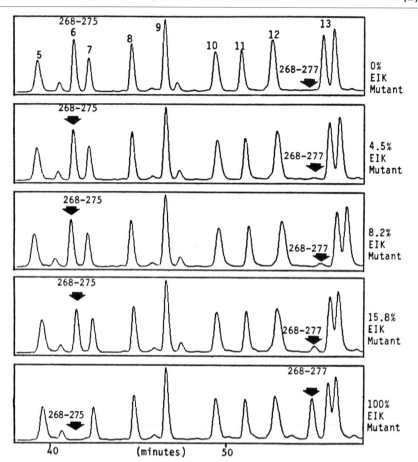

FIG. 1. Recombinant tissue type-plasminogen activator (rtPA) spiked with a single-site mutant. These chromatograms show a detection limit of about 5% for a substitution of Arg-275 with glutamic acid. The rtPA was alkylated and digested with trypsin in 1% ammonium bicarbonate. Chromatographic conditions were as follows: column, Waters (Milford, MA) NovaPak C_{18}, 3.9 × 150 mm (5-μm particle diameter); solvent A, 50 mM sodium phosphate, pH 2.8; solvent B, acetonitrile; gradient, 0–30% solvent B in 90 min starting at injection, then 30–60% in 30 min; flow rate, 1.0 ml/min; detection, ultraviolet (UV) at 214 nm; column temperature, 35°; sample, 80–100 μg of trypsin-digested RCM rtPA. [Reprinted with permission from Committee on Process Development and Manufacturing, Biotechnology Division, Pharmaceutical Manufacturers Association. Assessment of genetic stability for biotechnology products. *BioPharm* **4**(2), 22–27 (1991).]

using trypsin as the cleavage agent, although other specific and reproducible peptide bond-cleavage reagents are briefly discussed.

Sample Preparation

General Considerations

Analytical peptide mapping consists of chemically or enzymatically cleaving the peptide bonds of 0.5–5 mg of a protein of interest. The resulting peptide fragments are then separated by reversed-phase HPLC. The ideal enzyme-catalyzed cleavage of a protein would be highly site specific and quantitative in order to minimize artifacts such as nontheoretical or incomplete cleavages that complicate the map. The enzyme digestion would also produce fragments of a size convenient for both chromatographic separation and for subsequent analysis, such as amino-terminal sequencing or mass spectrometry. Optimal digest conditions are enzyme–substrate dependent, thus the effects of pH, temperature, enzyme-to-substrate ratio, and incubation time are factors for investigation. Chemical cleavage by cyanogen bromide at methionine residues has at times been employed, but it is not a generally useful approach because of the insolubility of the large protein fragments that are produced due to the typically low number of methionine residues in proteins.

Once peptide fragments are obtained, chromatographic conditions are investigated for optimal resolution in a minimum run time.[6–8] Reproducibility is paramount to the successful application of this method and can be affected by enzyme quality, sample stability, separation conditions, column performance, and instrument performance.[9]

Digest Buffer Selection

The first step in developing a peptide map is to identify a suitable digest buffer. Digest buffers are selected on the basis of the optimal pH for enzymatic activity (Table I), compatibility with subsequent analytical techniques, and solubility of the protein substrate. In the case of trypsin with

[6] C. T. Mant and R. S. Hodges (eds.), "High Performance Chromatography of Peptides and Proteins: Separation, Analysis and Conformation." CRC Press, Boca Raton, FL, 1991.

[7] M. T. W. Hearn (ed.), "HPLC of Proteins, Peptides and Polynucleotides." VCH Publications, New York, 1991.

[8] R. C. Chloupek, W. S. Hancock, and L. Snyder, *J. Chromatogr.* **594**, 65–73 (1992).

[9] R. C. Chloupek, J. E. Battersby, and W. S. Hancock, *in* "HPLC of Peptides and Proteins: Separation, Analysis and Conformation" (C. T. Mant and R. S. Hodges, eds.), pp. 825–833. CRC Press, Boca Raton, FL, 1991.

TABLE I
ENZYMES USED AS CLEAVAGE AGENTS

Enzyme	Cleavage sites	Notes
Trypsin	Carboxy side Arg, Lys	Active at pH 7.5 to 9; difficult to cleave Arg-Pro, Lys-Pro, Arg or Lys with Asp, Glu, or cysteic acid; some chymotryptic-like activity
V-8 (*Staphylococcus aureus* protease, Glu-C)	Carboxy side Glu, Asp	Active at pH 4 to 9; ammonium bicarbonate (pH 8) or ammonium acetate (pH 4) for Glu-X cleavage; phosphate (pH 7.8) for Glu-X and Asp-X cleavage; difficult to cleave Glu-Pro and Asp-Pro bonds; active in up to 0.1% SDS
Lys-C (endoproteinase Lys-C)	Carboxy side Lys	Active at pH 7.5 to 9 in ammonium bicarbonate or Tris-HCl; difficult to cleave Lys-Pro
Arg-C (endoproteinase Arg-C)	Carboxy side Arg	Active at pH 7.5 to 9 in ammonium bicarbonate or Tris-HCl
Asp-N (endoproteinase Arg-C, aspartyl protease)	Amino side Asp, cysteic acid	Active at pH 7.5 to 8 in ammonium bicarbonate or Tris-HCl; inhibited by EDTA
Clostripain	Carboxy side Arg	Active at pH 7 in 50 mM sodium phosphate, 10 mM DTT
Chymotrypsin	Carboxy side Tyr, Phe, Trp, Leu	Active pH 7.5 to 9; slight cleavage at Met, Gln, Asn, His, Thr; very slight cleavage at Ser, Gly, Val, Ile; adjacent C-terminal side basic amino acids can enhance cleavage; adjacent C-terminal side Pro inhibits cleavage

optimal activity near pH 8, the most common buffer is ammonium bicarbonate. Ammonium bicarbonate is volatile and can be removed by lyophilization, resulting in an essentially salt-free peptide mixture that can be advantageous for subsequent analyses. It is important to note that differences in digest buffer components can greatly affect the elution profile of a peptide map. Figure 2 illustrates the significant differences between a tryptic digest profile of recombinant human γ-interferon (rIFN-γ) when digested in a sodium acetate buffer versus an ammonium bicarbonate buffer digest at the same ionic strength and pH. In this example, differences were found to be due to nonspecific or incomplete cleavage of rIFN-γ. In a tryptic digest of recombinant human growth hormone (rhGH) the use of a sodium acetate digest buffer reduced both nonspecific cleavages and deamidation of T12 compared to an ammonium bicarbonate digest buffer.

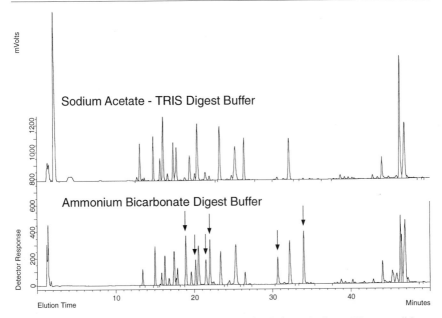

FIG. 2. The effect of digest buffers on γ-interferon (IFN-γ) tryptic digest. Digest conditions: trypsin (10 μl of 1-mg/ml trypsin solution in water) was added to 1.0 ml of 1-mg/ml IFN-γ solution in 1% ammonium bicarbonate buffer and then incubated at 30° for 6 hr; a second addition of trypsin (10 μl) was added and the digest incubated for a total of 24 hr. A solution consisting of 100 mM sodium acetate, 10 mM Tris, and 2 mM calcium chloride adjusted to pH 8.3 with acetic acid was used under the same conditions, except the digest was quenched by adding 50 μl of phosphoric acid. Chromatographic conditions: column, Alltech (Deerfield, IL) Nucleosil C$_{18}$, 4.6 × 150 mm (5-μm particle diameter); solvent A, 50 mM sodium phosphate, pH 2.8; solvent B, acetonitrile; gradient, 0–20% solvent B over 30 min starting at injection, then 20–40% over 20 min; flow rate, 1.0 ml/min; detection, UV at 214 nm; column temperature, 40°; sample, 80–100 μg of trypsin-digested IFN-γ. The arrows indicate the significant peaks that have been incompletely digested when compared to the digest with sodium acetate-TRIS digest buffer.

Small amounts of calcium can be a valuable additive that increases trypsin stability. Of course, a protein stabilizer works only if it stabilizes the enzyme more than the substrate. In the case of recombinant human DNase, calcium confers resistance to trypsin digestion, but DNase can be fully digested if pretreated by a calcium-chelating agent (Fig. 3).

The need for denaturants must also be considered when selecting a suitable digest buffer. Peptide bonds buried in a protein matrix can be resistant to hydrolysis by proteases. Denaturing conditions that disrupt the secondary and tertiary structure of the polypeptide chain often improve quantitative cleavage of sensitive bonds. Conditions are sought that will

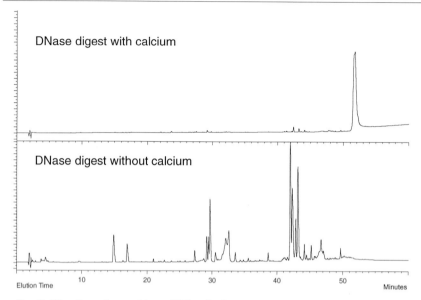

FIG. 3. Digestion of recombinant DNase in the presence or absence of calcium. DNase was pretreated with 40 mM Bis–Tris, 10 mM EGTA at 37° for 10 min, then the buffer was exchanged in two Sephadex G-25 columns. Both columns were equilibrated with 100 mM Tris at pH 8; one column also contained 1 mM calcium chloride. Trypsin (10 μl of a 1-mg/ml trypsin solution in 0.1 mM HCl) was added to 1.0 ml of a 1-mg/ml DNase solution and incubated at 30° for 6 hr; a second addition of trypsin (10 μl) was made and the digest was incubated for a total of 24 hr. The digest was quenched by adding 50 μl of phosphoric acid. The calcium-minus conditions were identical, except that DNase was also pretreated with EDTA. Chromatographic conditions: column, Mac-Med Analytical (Chadds Ford, PA), Zorbax SB-C8, 4.6 × 150 mm (5-μm particle diameter); solvent A, 0.1% TFA in water; solvent B, 0.1% TFA in acetonitrile; gradient, 0–60% solvent B over 60 min starting at injection; flow rate, 1.0 ml/min; detection, UV at 214 nm; column temperature, 35°; sample, 80–100 μg of trypsin-digested DNase.

destabilize the protein structure yet allow full activity of the protease. Destabilization can be achieved by denaturing solvent conditions or by chemical modification of the disulfide bridges. Ubiquitin, for example, is remarkably stable to the action of proteases, but was successfully digested by trypsin (5%, w/w) in the presence of 6.5 M urea.[10] High-sensitivity tryptic mapping of subnanomolar quantities of protein has been shown to be successful with the routine use of 8 M urea and 4% trypsin (w/w).[11] A range of useful proteolytic cleavages of ribonuclease has also been demon-

[10] M. J. Cox, R. Shapira, and K. Wilkinson, *Anal. Biochem.* **154,** 345–352 (1986).
[11] K. L. Stone and K. R. Williams, *in* "Macromolecular Sequencing and Synthesis: Selected Methods and Applications" (D. H. Schlesinger, ed.), pp. 7–24. A. R. Liss, Inc., New York, 1988.

strated in the presence of 20% (v/v) methanol, ethanol, 2-propanol, or aceto-nitrile.[12]

Sodium dodecyl sulfate (SDS) is a strong denaturant, but it is difficult to remove from a peptide mixture once it has been introduced. Sodium dodecyl sulfate can interfere with digestion, as well as subsequent chroma-tography. As little as 0.001% SDS adversely affected the digestion of trans-ferrin in a subnanomole digestion example.[13] In the presence of SDS a reversed-phase column can act as an anion exchanger for cationic peptides containing basic side chains or can lead to increased peptide retention by forming hydrophobic ion pairs.[14]

Chemical Modification

When the native structure of a protein is stabilized by disulfide bridges, chemical modification of the cystines is often advantageous because of the uniformity, stability, and solubility of the resulting derivatives. Standard methods based on the work of Crestfield *et al.*[15] involve treatment of the protein with either 2-mercaptoethanol or dithiothreitol (DTT), in the pres-ence of 8 M urea or 6 M guanidine hydrochloride and EDTA for reduction of disulfide bonds. Iodoacetic acid or iodoacetamide is then used to convert free sulfhydryl groups to an anionic carboxymethylcysteine residue. The chelating agent EDTA minimizes contamination of heavy metal ions, which can catalyze reoxidation of the free sulfhydryl groups.

Buffer Exchange Techniques

Buffer exchange of protein solutions has traditionally been accom-plished by placing a sample in prepared dialysis membrane tubing and soaking it in a bath containing at least a 10-fold higher volume of the desired buffer. Dialysis by this method is a time-consuming procedure, typically taking a day with at least two changes of the dialysis bath with fresh buffer. The method can be an advantage when handling large volumes of sample, but a distinct disadvantage for handling multiple small-volume samples. A more contemporary approach for buffer exchange of samples is the use of prepacked, disposable size-exclusion columns such as the

[12] K. G. Weinder, *Anal. Biochem.* **174,** 54–64 (1988).

[13] K. L. Stone, M. B. LoPresti, and K. R. Williams, Enzymatic digestion of proteins and HPLC peptide isolation in the sub-nanomole range. *In* "Laboratory Methodology in Biochemistry: Amino Acid Analysis and Protein Sequencing" (C. Fini, A. Floridi, V. N. Finelli, and B. Wittman-Liebold, eds.), pp. 181–205. CRC Press, Boca Raton, FL, 1990.

[14] W. S. Hancock and J. T. Sparrow, "HPLC Analysis of Biological Compounds: A Laboratory Guide," Vol. 26 of Chromatographic Science Series, p. 90. Marcel Dekker, New York, 1984.

[15] A. M. Crestfield, S. Moore, and W. H. Stein, *J. Biol. Chem.* **238,** 622–627 (1963).

PD-10 by Pharmacia (Piscataway, NJ). Column equilibration and protein elution can be accomplished in less than 1 hr, and recoveries of protein are 70 to 80%. Recovery of low nanomolar and subnanomolar samples can be more problematic owing to significant losses that can occur by sample transfer or buffer exchange. One solution in this case is to use dilution with the desired buffer and eliminate exchanges.[11] Commercially available membrane filtration devices (Amicon, Danvers, MA) with 3- to 30-kDa pore sizes for volumes as little as 50 μl can also be useful.

Trypsin Digestion

While the ideal enzyme does not exist, trypsin enjoys wide use because of its reproducible specificity for the carboxyl groups of lysine and arginine, high specific activity, availability at a high quality, and low cost per analysis. The relative rates of cleavage of lysine and arginine are influenced by the chemistry of the side chains in the immediate vicinity.[16] Trypsin rarely cleaves at lysine–proline or arginine–proline bonds, although cleavage of arginine–proline bonds has been reported.[17] Residues with negatively charged side chains adjacent to an arginine or a lysine can also reduce or even completely prevent cleavage by trypsin (i.e., aspartic acid, glutamic acid, cysteic acid and S-carboxymethylcysteine). Amino-terminal arginine, amino-terminal lysine, or sequential residues of arginine and lysine are also cleaved at reduced rates, leading to peptide heterogeneity as observed with T16–17–18 in human growth hormone (Fig. 4).

A comparison of the digestion of the oxidized insulin B chain using trypsin from four sources yielded variable levels of nontheoretical cleavages, illustrating the importance of high-quality trypsin preparations.[11] TPCK-treated trypsin is often used for analytical mapping because it reduces any residual chymotryptic activity common to even highly purified trypsin preparations. TPCK (L-1-tosylamide-2-phenylethyl chloromethyl ketone) is an inhibitor of chymotryptic activity. Even with TPCK treatment, significant chymotryptic-like cleavages can occur (Fig. 5, T10 in rhGH).

Finally, quenching of the trypsin activity terminates the digest and reduces chymotryptic-like or other nonspecific cleavages. The addition of soybean trypsin inhibitor or acidification to a pH <4 are common quenching methods. Acidification is most convenient because it adjusts the sample pH closer to that of the mobile phase of reversed-phase chromatography and often has the added benefit of increasing peptide solubility.

[16] C. B. Kasper, in "Molecular Biology, Biochemistry and Biophysics—Protein Sequence Determination" (S. B. Needleman ed.), p. 132. Springer-Verlag, New York, 1975.
[17] J. E. Shively and R. J. Paxton, in "Practical Spectroscopy Series" (C. N. McEwen and B. S. Larsen, eds.), Vol. 8, p. 25. Marcel Dekker, New York, 1990.

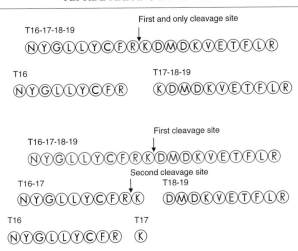

FIG. 4. A bead diagram illustrating the effect on digestion results of multiple basic residue (arginine, lysine) heterogeneity. Above, the first cleavage leads to an amino-terminal basic residue that is not further cleaved. Below, the first cleavage leads to a C-terminal base that can be further cleaved.

Standard Digest Protocol for Soluble Protein

Reagents

Trypsin (Worthington, Freehold, NJ), TPCK treated or equivalent
HPLC-grade water
Ammonium hydroxide, 100 mM
Hydrochloric acid, 0.10 mM

Equipment

PD-10 column, Pharmacia (Sephadex G-25 or equivalent)
Oven or water bath for sample temperature control
Ultraviolet (UV) spectrophotometer for reading samples at 280 nm

Sample Preparation

1. Equilibrate a PD-10 column at ambient temperature with 25 ml of 100 mM ammonium bicarbonate and load up to 2.5 ml of a protein sample per column (1–5 mg, typically).
2. Collect 1-ml fractions and read at 280 nm.
3. Pool peak fractions.

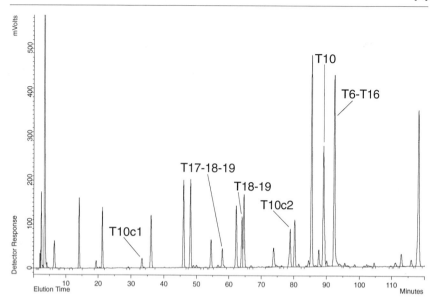

FIG. 5. An rhGH tryptic map illustrating peptide digest heterogeneity. Digest conditions: trypsin (50 μl of a 1-mg/ml trypsin solution in 0.1 mM HCl) was added to 2-mg/ml hGH in 100 mM sodium acetate–10 mM Tris–2 mM calcium chloride adjusted to pH 8.3 with acetic acid/sodium hydroxide and incubated at 37° for 2 hr; a second addition of trypsin was added and the digest was incubated for a total of 4 hr; the digest was quenched by adding 250 μl of phosphoric acid. Chromatographic conditions were as follows: column, Alltech (Deerfield, IL) Nucleosil C_{18}, 4.6 × 150 mm (5-μm particle diameter); solvent A, 50 mM sodium phosphate, pH 2.8; solvent B, acetonitrile; gradient, 0–40% solvent B over 120 min starting at injection; flow rate, 1.0 ml/min; detection, UV at 214 nm; column temperature, 40°; sample, 160–200 μg of trypsin-digested rhGH.

Digested Method

1. Prepare fresh, 1-mg/ml TPCK-treated trypsin in 0.01 mM HCl.
2. Add 10 μl of trypsin solution per milligram of protein in the sample solution, mix, and cap the tube.
3. Incubate at 37° for 2 hr, then add another 10 μl of trypsin solution per milligram of protein in the sample solution, mix, and cap the tube.
4. Incubate at 37° for another 2 hr and then quench the trypsin activity by acidifying the sample to a pH < 4 with hydrochloric acid.

For optimal conditions, digest buffers and a digestion time course should be investigated. Digestion is considered complete when the intact protein peak is absent and the major peptide peaks are invariable. Overdigestion promotes nonspecific cleavages usually seen initially as minor peaks near the baseline, followed by a gradual increase in area as the major peptide peaks lose area.

Standard Reduction and S-Carboxymethylation for Proteins Difficult to Digest in Native Form and Containing Disulfide Bridges (Cystines)

Reagents

HPLC-grade water
RCM buffer (buffer for standard reduction and S-carboxymethylation):
  Urea (USP grade), 8 M
  Tris base (molecular weight 121), 0.35 M
  EDTA (disodium, molecular weight 372), 0.25 M
Dithiothreitol (DTT), 1.0 M
Sodium hydroxide, 1 N
Iodoacetic acid (1.0 M), prepared fresh in the 1 N sodium hydroxide
  Caution: Wear gloves and eye protection. Use a fume hood for preparation of the solution. Store the solution in the dark.
Ammonium bicarbonate, 100 mM

Equipment

PD-10 column, Pharmacia (Sephadex G-25 or equivalent)
Oven or water bath for sample temperature control
UV spectrophotometer for reading samples at 280 nm

RCM Sample Preparation

1. Equilibrate a PD-10 column at ambient temperature with 25 ml of RCM buffer and load up to 2.5 ml of a protein sample (1–5 mg, typically). The viscosity of this buffer will cause the flow rate to be slow. Overnight dialysis with an appropriate membrane can be substituted for this buffer exchange.
2. Collect 1-ml fractions and read at 280 nm.
3. Pool peak fractions.

Standard Reduction and S-Carboxymethylation Method

1. Add 10 μl of 1.0 M dithiothreitol per milliliter of protein sample. Mix, cap the tube, and incubate at 37° for 1 hr. The final concentration of dithiothreitol is 10 mM.
2. Add 50 μl of 1 M iodoacetic acid in 1 N sodium hydroxide, mix, cap the tube, and incubate at ambient temperature in the dark for 30 min.
3. Equilibrate a PD-10 column at ambient temperature with 25 ml of 100 mM ammonium bicarbonate and load up to 2.5 ml of a protein sample per column (1–5 mg, typically).

4. Collect 1-ml fractions and read at 280 nm.
5. Pool peak fractions.
6. Digest as suggested above.

Reversed-Phase High-Performance Liquid Chromatography Analysis

Columns and Mobile Phases

The most common separation technique for peptide digests utilizes 5- or 10-μm silica-based column packings with covalently bonded alkane stationary phases (e.g., C_{18}) operated with aqueous trifluoroacetic acid (TFA) and acetonitrile gradients.[6,14] Kirkland et al.[18,19] have developed a stationary phase that is a new class of silane-modified silica reported to increase stability at elevated temperatures and low pH, resulting in longer term reproducibility and stability of peptide separations. The problem of chemical stability of the siloxane bonds, if neutral or basic separation conditions are desired, can be circumvented by using poly(styrene-divinylbenzene)-based packings that are stable over the entire pH range.[20]

As stated before, the volatile nature of the aqueous TFA and acetonitrile solvent system is of benefit when postcolumn techniques are to be performed on collected peptides such as sequencing, amino acid analysis, capillary electrophoresis, or mass spectrometry. Approximately equal levels of TFA should be added to organic and aqueous phases to obtain a stable baseline when monitored at 214 nm, due to the significant absorbance of 0.1% TFA in water (approximately 0.25 au). Other substitutes for TFA include sodium phosphate, phosphoric acid, hydrochloric acid, and perchlorate. Sodium phosphate is the most useful owing to its broad range of buffering for modifying peptide selectivity. Separations are typically performed at temperatures in the 25–50° range and require 80–120 min to resolve most components. Computer simulation of peptide gradient separations by programs such as DryLab G/plus (LC Resources, Walnut Creek, CA) can significantly reduce the development time of optimal separation conditions with a practical run time.[8]

Equipment

The separation is usually monitored in the ultraviolet range of light from 190 to 280 nm, depending on the degree of sensitivity desired and

[18] J. J. Kirkland, J. L. Glajch, and R. D. Farlee, Anal. Chem. **61,** 2–11 (1989).
[19] J. J. Kirkland and J. L. Glajch, LCGC **8,** 140–150 (1990).
[20] U. Esser and K. K. Unger, in "High Performance Chromatography of Peptides and Proteins: Separation, Analysis and Conformation" (C. T. Mant and R. S. Hodges, eds.), pp. 273–278. CRC Press, Boca Raton, FL, 1991.

the chromophore to be detected. The absorbance of the peptide bonds is detected in the 214-nm region, whereas the chromophores tyrosine and tryptophan are detected in the 280-nm region. Work utilizing a diode array detector and a computer workstation to construct a peptide spectral library for peak identification by UV spectral matching from 220 to 350 nm has shown utility for identifying peptide peaks and determining peptide peak purity for quality control analyses.[21]

Analytical peptide mapping is demanding on the solvent delivery system of a liquid chromatograph. Typical gradients have a 1%/min rate of change, with reported gradients as shallow as 0.33%/min (flow, 1 ml/min, 4.6-mm i.d. column). A demanding test of gradient performance is to run a peptide map at a 0.33%/min gradient rate and determine the standard deviation of peptide retention times. Systems with standard deviations of peptide retention times that are less than 0.1 min are acceptable for most analytical peptide maps using standard 4.6-mm i.d. analytical columns.[22] Systems designed or optimized for microbore columns offer new possibilities in performing analytical peptide mapping on even smaller sample sizes. Commercially packed capillary columns typically have a 320-μm i.d. and 15-cm length and are used with a flow rate of 5 μl/min. LC Packings produces a unique fused silica capillary Z cell for the Applied Biosystems, Inc. (ABI; Foster City, CA) 270 UV detector with a path length of 3 mm and an approximate volume of 35 nl. Particularly important parameters for the use of microbore systems are the system dwell volume (dead volume between the pumps and the top of the column) and UV detector flow cell volume.

Standard High-Performance Liquid Chromatography Conditions

Reagents

UV-grade acetonitrile [J. T. Baker (Phillipsburg, NJ) or equivalent]
HPLC-grade trifluoroacetic acid [Pierce (Rockford, IL) or equivalent]
HPLC-grade water

Equipment

HPLC instrument capable of programmed gradient elution, UV detection at 210–220 nm, acquisition of the chromatographic data, and integration of peak areas
Reversed-phase column, 100-Å pore, 5- to 10-μm particle size 25 × 0.46-cm i.d. or 25 × 0.2-cm i.d.

[21] H. J. Sievert, S. L. Wu, R. Chloupek, and W. S. Hancock, *J. Chromatogr.* **499,** 221–234 (1990).
[22] E. R. Hoff, *LCGC* **7**(4), 320–326 (1989).

Method

Solvent A: 0.1% Trifluoroacetic acid (1 ml of TFA/liter purified water)
Solvent B: 0.08% trifluoroacetic acid in acetonitrile (0.8 ml of TFA/liter acetonitrile)
Flow rate: 1 ml/min, 4.6-mm i.d. (0.2 ml/min, 2.0-mm i.d.)
Temperature: 40°
Gradient: 0–60% solvent B in 60 min
Sample: 100 μl, 1-mg/ml digest (10–100K molecular weight). *Note:* 20 μl, 1-mg/ml digest for the 2.0-mm i.d. column

Peptide Naming Convention

Our convention for naming peptides that result from the proteolytic digestion of a known protein sequence is first to identify the theoretical peptides that would result from a given protease, and use the first letter of the enzyme name (e.g., T for trypsin, C for chymotrypsin, V for V-8) followed by the theoretical number beginning with the N-terminal peptide as number one (e.g., T1, C1, V1).

Peptides resulting from nonspecific (nontheoretical) cleavages are indicated with a lower-case letter of the enzyme that would be characteristic of that type of cleavage. For example, T10 in hGH has two chymotryptic-like peptides that result from trypsin digestion that are designated T10c1 and T10c2).

Data Analysis

Peptide maps are often evaluated as an identity test with no criteria for peak area measurements. A reference is compared to a sample in a manner similar to an infrared scan identity test, where the presence of the major bands yields a positive identification. The specification for positive identity can be: "The chromatographic profile of the test specimen should be consistent with the profile of the Reference Standard."[23]

Area analysis of the peptide peaks can detect the loss of peak area due to mutation or posttranslational modification. Depending on the reproducibility of the peptide map, a specific variant can be detected at the 5 to 20% level by loss of peak area. New variants as low as 1% of the total protein can be detected if the elution of the variant peptide peak does not coelute with another peptide and is adequately resolved.

Area analysis requires normalization of the area responses to account for protein concentration differences between a sample and reference.

[23] Somatropin monograph. *In* "USP Pharmacopeial Previews," November–December, p. 1255. The United States Pharmacopeial Convention, Inc., Rockville, MD (1990).

There are two ways to normalize the peak areas: area percent normalization or internal reference normalization. The area normalization method divides each peak (i) by the total protein-related area, as in Eq. (1).

$$\text{Peptide area}_i \% = (A_i/A_t) \times 100\% \tag{1}$$

where A_i is the peak area of individual peak i and A_t is the total area of all protein-related peaks. The normalized peak response of each peak in the sample is then compared to the normalized peak response of a reference material, as described in Eq. (2).

$$\text{Percent difference}_i = [(PA_{i_s} - PA_{i_r})/PA_{i_r}] \times 100\% \tag{2}$$

where PA_{i_s} is the peptide area percent for peak i in the sample and PA_{i_r} is the peptide area percent for peak i in the reference.

Usually the percent difference from the reference material is calculated. The pitfalls of using total protein area is that not all peak areas are reproducible. Often hydrophobic peptides either precipitate in the digest solution or chromatograph unpredictably. The inclusion of this variable area in the equation will add variability to the result.

The internal reference normalization method selects a single reproducible peak to normalize the peak responses. The peak used as a standard should not elute too near the void volume, which would result in integration errors due to void volume peak interference, or too late in the gradient, thus indicating a hydrophobic peak that may be quite variable due to precipitation. Peptides containing sites prone to degradation, such as deamidation or oxidation, should also be avoided because the potential for multiple peaks due to the products of these reactions. Equation (3) shows the single peak normalization calculation.

$$NPR_i = A_i/A_{ISTD} \tag{3}$$

where NPR_i is the normalized peptide ratio for an individual peak i, A_i is the area of an individual peak i, and A_{ISTD} is the area of the internal reference peak in the same chromatogram as peak i. Equation (4) shows the calculation of the percent difference from the reference material.

$$\text{Percent difference}_i = (NPR_{i_s} - NPR_{i_r})/NPR_{i_r} \times 100\% \tag{4}$$

where NPR_{i_s} is the normalized peptide ratio [Eq. (3)] of peak i from a sample chromatogram, and NPR_{i_r} is the normalized peptide ratio [Eq. (3)] of peak i from a reference chromatogram.

When nonspecific cleavages or variants due to artifacts of the assay occur, summation of the peak area percents or normalized peptide ratio should be considered (e.g., T10 and T14 in rhGH).

Specific Calculations of Variant

In addition to the overall specification, the level of a single variant can be calculated. The most simple case is when the parent peptide(s) and the related variant peptide(s) are resolved from the other peptide peaks in the chromatogram. When the peptides are resolved, then the calculation can be performed solely on the basis of the sample chromatogram, with no dependence on a reference chromatogram. In that case Eq. (5) can be applied.

$$\text{Percent variant} = A_v/(A_v + A_p) \times 100\% \qquad (5)$$

where A_v is the peak area of the juvenile peptide(s) and A_p is the peak area of the parent peptide(s).

The assumption is made that the extinction coefficients of the peptide variants at the detection wavelength are not significantly changed for the variants. If the extinction coefficient does change, then a correction factor can be used to adjust the area of the variant.

For the situation in which the variant peptides of interest are merged with other peptide peaks, then calculations based on a reference chromatogram are required, as shown in Eq. (6).

$$\text{Percent variant peptide} = \left(1 - \frac{A_s/A_{istds}}{A_r/A_{istdr}}\right) \times 100\% \qquad (6)$$

where A_s is the peak area of the sample parent peptide, A_{istds} is the peak area of the sample internal reference peak, A_r is the peak area of the reference parent peptide, and A_{istdr} is the peak area of the reference internal reference peak

In Eq. (6) an assumption is made that there is only a single variant.

Finally, it should be pointed out that merely calculating a number does not mean that the number is correct. Through the process of assay validation a determination can be made as to which peaks can be quantitated. Well-resolved peptide peaks are often quantifiable to 1%. An assay validation should include linearity, accuracy, precision, ruggedness, and peak characterization elements.[9] Accuracy and precision are particularly important for quantitation. Assay precision will tell if a peak gives a reproducible response; some peaks do not. For assay accuracy, add-backs of variant proteins prior to digestion can assess the quantitative power of the method and the limit of detection of a particular variant.

The Future

Although peptide maps have long been used for characterization of the primary structure of unknown proteins, in conjunction with amino acid analysis and amino-terminal sequencing, comparative peptide mapping has become an essential tool for meeting pharmaceutical licensing requirements and quality control of recombinant DNA-derived protein products.[24] Use of an optimized and reproducible peptide-mapping methodology, as outlined in this chapter, enables routine detection of small changes in primary structure of a protein. This includes oxidation states of cysteine or methionine residues,[25-27] proteolysis,[25] deamidation,[28] mistranslation,[29] or other possible events leading to the alteration of the primary structure of a protein.

Electrospray ionization mass spectrometry has emerged as a key new technology for peptide map characterization[30] (and see [17] in this volume[30a]). The accessibility of on-line mass information for previously characterized peptide maps enables rapid identification of theoretical and variant sequences, especially useful in the cases of coelution of minor amounts. As a peptide map development tool, on-line mass detection allows rapid identification of peaks during testing of various separation conditions. Earlier methods development strategies relying on peptide identification by amino acid analysis or amino-terminal sequencing required at least days and milligram quantities of material to generate the necessary peptide characterization data.

Rapid identification of carbohydrate peptides and identification of glycosylation mutants based on known carbohydrate structures or consensus glycosylation sites is now possible using on-line mass spectrometry.[31] Glyco-

[24] R. L. Garnick, N. J. Soli, and P. A. Papa, *Anal. Chem.* **60**, 2546–2557 (1988).

[25] E. Canova-Davis, R. C. Chloupek, I. P. Baldonado, J. E. Battersby, M. W. Spellman, L. J. Basa, B. O'Connor, R. Pearlman, C. Quan, J. T. Chakel and W. S. Hancock, Analysis by FAB-MS and LC of proteins produced by either biosynthetic or chemical techniques. *Am. Biotechnol. Lab.* **6**(4), May, 8–17 (1988).

[26] M. Kunitani, P. Hirtzer, D. Johnson, R. Haleenbech, A. Boosman, and K. Koths, *J. Chromatogr.* **359**, 391–402 (1986).

[27] W. F. Bennett, R. C. Chloupek, R. Harris, E. Canova-Davis, R. Keck, J. Chakel, W. S. Hancock, P. Gellerfors, and B. Paulu, Characterization of natural-sequence recombinant human growth hormone. *In* "Proceedings of the International Congress on Advances in Growth Hormone and Growth Factor Research, September 28–30, Milan, Italy" (E. E. Muller, D. Cocchi, and V. Locatelli, eds.), pp. 29–48. Pythagora Press, Rome, and Springer-Verlag, Berlin, 1989.

[28] B. A. Johnson, J. M. Shirokawa, W. S. Hancock, M. W. Spellman, L. J. Basa, and D. W. Aswad, *J. Biol. Chem.* **264**, 14262–14271 (1989).

[29] G. Bogosian, B. N. Violand, E. J. Dorward-King, W. E. Workman, P. E. Jung, and J. F. Kanee, *J. Biol. Chem.* **264**, 531–539 (1989).

[30] V. Ling, A. W. Guzzetta, E. Conova-Davis, J. T. Stults, W. S. Hancock, T. R. Covey, and B. I. Shushan, *Anal. Chem.* **63**, 2909–2915 (1991).

[30a] W. S. Hancock, A. Apffel, D. Chakel, C. Sunders, T. M'Timkulu, E. Pungor, Jr., and A. W. Guzzetta, *Methods Enzymol.* **271**, Chap. 17, 1996 (this volume).

[31] A. W. Guzzetta, L. J. Basa, W. S. Hancock, B. A. Keyt, and W. F. Bennett, *Anal. Chem.* **65**, 2953–2962 (1993).

peptides have been shown to be detectable at signal intensities similar to nonglycosylated peptides[32] and in characteristic patterns related to carbohydrate heterogeneity.[30] The ability to derive useful structural information on microgram amounts of material represents a significant advance for comparative studies of the structure of glycoproteins.

Finally, the availability of routine on-line mass detection of peptides or proteins is changing the design of tryptic mapping methods because coelutions are frequently resolved by differences in mass. The additional dimension of information afforded by measuring the mass-to-charge ratio reduces the need for baseline resolution of peptides as long as a reproducible peptide map is achieved.

Conclusion

The extensive experience gained from biotechnology applications of RP-HPLC peptide mapping, plus the concomitant evolution of the related hardware, software, and reagents needed for this technique, has benefited the entire field of high-resolution structural analysis of biological macromolecules. In this chapter we have focused on analytical tryptic peptide mapping to illustrate the principles of the technique and the quantitative power of peak analysis for accurate variant and mutation analysis. Although tryptic maps were selected as examples, any specific and reproducible peptide bond-cleavage reagent could be employed.

Acknowledgments

The authors acknowledge Genentech, Inc., researchers that have contributed to the state of the art of analytical peptide mapping. In particular, we acknowledge the guidance and support for this work by William S. Hancock, John V. O'Connor, and Robert L. Garnick.

[32] M. E. Hernling, G. D. Roberts, W. Johnson, S. A. Carr, and T. R. Covey, *Biomed. Environ. Mass Spectrom.* **19,** 677–691 (1990).

[3] Trace Structural Analysis of Proteins

By KRISTINE M. SWIDEREK, TERRY D. LEE, and JOHN E. SHIVELY

Introduction

Structural analysis of proteins on limited amounts of sample is important. In the past, the conventional method of determining structure was to purify the protein, to fragment it chemically or enzymatically, and to separate the resulting fragments by one or more different high-performance liquid chromatography (HPLC) steps. Fragments were then analyzed by amino acid composition analysis and automated Edman chemistry degradation. Structural analysis of posttranslational modifications such as glycosylation or the identification of unusual amino acids was sometimes accomplished but often failed because of inability to determine the nature of the modification. In most cases, the starting quantity of sample needed to perform more comprehensive structural analysis was in the 1- to 10-nmol range. The introduction of micro-HPLC systems using capillary columns with inner diameters smaller than 500 μm made it possible to purify proteins and peptides at the low picomole level, but currently it is still difficult to perform the structural analysis of low quantities of sample by automated sequence analysis. Over the past decade the level of sensitivity for automated sequencing has been improved but it presently seems to be close to the limits with the existing instrumentation.

The improvement of various mass spectrometric methods has given the protein chemist the opportunity to implement mass spectrometry into structural protein analysis. With the introduction of laser desorption time of flight (LD-TOF) and electrospray mass spectrometry (ES-MS) mass determination of proteins can be carried out at the low picomole level.[1-3] The connection of micro-HPLC with ES-MS (LC-MS) makes available the on-line analysis of picomole amounts of protein and peptide mixtures, saving the time-consuming step of off-line purification of individual components, and reducing sample loss inherent in multiple manipulations. Tandem mass spectrometry (MS/MS) carried out by ES-MS and secondary ion mass spectrometry (SIMS) instruments allows analysis in the subpicomole range.[4] Furthermore, the combination of MS/MS with LC-MS presents the possibility of performing structural analysis on picomole amounts of sample. However, one should realize that we have not reached the point yet at which we can completely analyze the structure of only a few picomoles of protein in a short period of time. Sometimes the structural analysis of 90% of the primary structure of a medium sized protein is performed in less than a month, but the analysis of the remaining 10% will take a long time. Even though some techniques available for the structural analysis of proteins and peptides have a detection limit in the subpicomole range, 10–100 times

[1] M. Karas and F. Hillenkamp, *Methods Enzymol.* **193,** 280 (1990).

[2] J. B. Fenn, M. Mann, C. K. Meng, S. F. Weng, and C. M. Whitehouse, *Science* **246,** 64 (1989).

[3] C. G. Edmonds and R. D. Smith, *Methods Enzymol.* **193,** 412 (1990).

[4] D. F. Hunt, J. Shabanowitz, M. A. Moseley, A. L. McCormack, H. Michel, P. A. Martino, K. B. Tomer, and J. W. Jorgenson, *in* "Methods in Protein Sequence Analysis" (H. Jörnvall, J.-O. Höög, and A.-M. Gustavsson, eds.). Birkhäuser Verlag, Basel, Switzerland, 1991.

the amount of sample are often required to determine the remaining 10% of the primary structure of the protein.

In this chapter we present examples for a variety of structural analyses of low amounts of proteins and peptides utilizing different mass spectrometric techniques in combination with micro-HPLC. For the demonstration of the different techniques standard proteins as well as "real-life" samples are used. The standard protein used is horse cytochrome c digested with the endoproteinase Lys-C. This chapter is intended to show possible strategies to solve some often encountered structural problems.

Material and Sources

Fused silica capillaries (PT Polymicro Technologies, Phoenix, AZ)
Fittings, ferrules, and zero dead volume connectors (Valco, Houston, TX)
2-Methoxyethanol (Aldrich, Milwaukee, WI)
Trifluoroacetic acid (TFA), sequencer grade (Pierce, Rockford, IL)
Acetonitrile, HPLC grade (J. T. Baker, Phillipsburg, NJ)
Endoproteinase Lys-C (*Lysobacter enzymogenes*) (Boehringer Mannheim, Indianapolis, IN)
Trypsin, talylsulfonyl phenylalanyl chloromethyl ketone (TPCK) treated (Sigma, St. Louis, MO)
Chymotrypsin (Sigma)
Horse heart cytochrome c (Sigma)
Glycopeptides: Glycopeptides are isolated from carcinoembryonic antigen (CEA) as described[5]

Solutions and Preparation

Micro-HPLC, solvent A: 1 ml of trifluoroacetic acid in 999 ml of water
Micro-HPLC, solvent B: 0.7 ml of trifluoroacetic acid in 99.3 ml of water, 900 ml of acetonitrile

The concentration of trifluoroacetic acid in buffer B is different for conventional HPLC as noted in the text or figure captions.
Water is obtained from a Milli-Q system (Millipore, Bedford, MA).

Mass Spectrometry

All mass spectrometric analyses are performed on triple quadrupole mass spectrometers (TSQ-700) from Finnigan-MAT (San Jose, CA)

---

[5] K. M. Swiderek, C. S. Pearson, and J. E. Shively, *in* "Techniques in Protein Chemistry IV" (Hogue Angeletti, ed.), p. 127. Academic Press, San Diego, CA, 1993.

equipped with an electrospray ion source (ESI) operating at atmospheric pressure or a cesium ion gun (Phrasor Scientific, Inc., Duarte, CA) for SIMS. Mass spectra are recorded in the positive ion mode.

Electrospray Ionization

The inner needle for the electrospray is made from fused silica and its inner diameter varies from 10 to 50 μm, depending on the required flow rate. For standard operation using a flow rate of 2 μl/min, a needle with an inner diameter of 50 μm provides a stable spray. The electrospray needle is operated at a voltage differential of 3–4 kV, the conversion dynode is set to -15 kV. The drying gas is nitrogen, with a temperature setting at the instrument of around 200°. A sheath flow of 2-methoxyethanol for all liquid chromatography mass spectrometry (LC-MS) experiments is used. The sheath flow rate is adjusted to the same flow rate in the LC system and is typically 2 μl/min. The nitrogen sheath gas is delivered into the source at 30 psi. Samples are introduced directly from an on-line micro-HPLC system described as follows. Scans are continually acquired every 3 sec in a mass range from 500 to 2000. Below the mass of 500, solvent cluster ions derived from the sheath liquid 2-methoxyethanol are dominant and are filtered out by setting the scan range above 500. The data collection is monitored using the base peak and reconstructed ion current (RIC) profile. The base peak represents the highest intensity per scan and the RIC profile is the continuous collection of the total ion current per scan. Spectra are generated by averaging the scans containing the peak, and mass assignments are made by using the Finnigan-MAT BIOMASS data reduction software.

Secondary Ion Mass Spectrometry

The ion source acceleration potential is 8 kV, the conversion dynode is set to -12 kV. Scans are acquired every 7 sec in a scan range from 400 to 4000. After HPLC separation the sample is concentrated to 10–20 μl and 1 μl of sample is applied directly onto the sample stage, which has been previously covered with 3-mercapto-1,2-propanediol (thioglycerol) as matrix. Dry samples are dissolved in solvent A before loading.

Liquid Chromatography

Liquid chromatography is carried out using fused silica columns and solvent delivery systems developed and built in our laboratory. Variation in column design and the dimensions of the fused silica tubing used allow for the construction of both preparative scale columns and microcolumns suitable for direct interfacing with a mass spectrometer (LC-MS). The preparation of these columns, their solvent delivery systems, and a discussion of the options available for ultraviolet (UV) detection are described in the following sections. It should be mentioned at this point as well that all solvents used for the liquid chromatography should be at least of HPLC-purity grade.

Column Preparation

Columns can be inexpensively and easily packed in any laboratory equipped with an HPLC system or other source of high-pressure solvent delivery. Any HPLC pump capable of running at its maximum pressure setting without shutting off is preferred. High-pressure syringe pumps (100DM; Isco, Lincoln, NE) and reciprocating pumps (100A; Beckman, Fullerton, CA) have been used with equal facility in our laboratory. Columns intended for LC-MS have an inner diameter between 100 and 250 μm, flow rates between 0.5 to 2 μl/min, and are packed with Vydac (The Separations Group, Hesperia, CA) 5-μm C_{18} RP support.[6,7] Fresh bulk material as well as reclaimed packings from previously used 4.6-mm i.d. columns (after thorough washes of the resin) have been used with equal results. The construction of these columns is as follows.

A piece of fused silica tubing (column) with an outer diameter of 360 μm and a length of about 200 mm is connected to a short piece of smaller outer diameter (typically 150 μm) fused silica tubing. This line will be inserted into the column to serve as a transfer line and to maintain the position of the column frit. The inner diameter of all transfer lines, both postcolumn and postflow cell, should be 50 μm or less to maintain the resolution gained from the microcolumn. In any case all transfer lines should be as short as possible to minimize column back pressure. The frit may be cut from a variety of membranes such as glass fiber (GF/A), Teflon (Zytex), or hydrophilic polyvinylidene difluoride (PVDF), using the column as a cookie cutter. Once cut, the frit remains at the tip of the column and is positioned about 10 mm into the column, using the transfer line as a ramrod. The glass fiber will serve as a frit at the end of the column. The final column assembly is then glued into place with a standard two-part epoxy resin (available in any hardware store) and allowed to cure overnight. Prior to packing the column is fitted to a Valco zero dead volume (ZDV) reducing union (1/16 to 1/32 in.), using a 0.4-mm i.d. graphite ferrule, and filled with 100% acetonitrile from the HPLC pump. A reservoir such as an empty 2.1-mm i.d. HPLC column (frits removed) is filled with a slurry of

---

[6] T. D. Lee and M. T. Davis, *Protein Sci.* **1,** 935 (1992).
[7] M. T. Davis, D. C. Stahl, K. M. Swiderek, and T. D. Lee, *METHODS: A Companion to Methods in Enzymology,* **6,** 304–314 (1994).

the desired column resin in acetonitrile, using a disposable syringe fitted with a blunt 14-gauge needle and secured with a finger-tight fitting (Upchurch, Oak Harbor, WA). The column is attached to the reservoir through a short section of 0.04-in. i.d. stainless steel HPLC tubing. Once the empty column is connected the syringe is removed and the reservoir then attached to the HPLC pump through a stainless steel high-pressure two-way valve. Pressure is built up against the closed valve. The pressure required varies with the dimensions of the column to be packed. It is typically set to 6000 psi for columns with 250-μm i.d. and smaller and reduced to 4000 psi for larger inner diameter columns. When this pressure is reached, the valve should be opened quickly to force the resin into the column by the sudden release of the pressure. Sometimes it is necessary to tab gently at the reservoir to keep the resin moving and settling into the column. Before disconnecting the column the pressure is released through an upstream three-way valve to waste. A sudden depressurization should be avoided at all times. The column is disconnected from the reservoir, where the majority of the packing remains, and set aside. The excess packing is retrieved by closing the three-way valve and flushing the reservoir. Following the removal of the reservoir, the column is reattached to the pump and allowed to pack for an additional 10 min. Extending the time beyond 10 min is of little benefit. This step should be performed with a pressure of around 6000 psi for any size of column. It should be examined if empty tubing at the top of the column (dead volume) must be cut off before the column can be used.

With this technique fused silica columns of any inner diameter can be prepared. We routinely use 0.53-mm i.d. columns (0.74-mm o.d.) to isolate proteins and peptides on the 0.1- to 1-nmol level. If columns with larger inner diameters are made, the dimensions of the transfer line from the column end to the UV detector can be increased. Typically the inner diameter is then 50 μm and the flow rate applied is 20 μl/min. Fractions eluting from those columns can be collected and further processed. Proteins can be used for further digestion or peptides can be analyzed by automated sequence degradation and/or mass spectrometry.

Ultraviolet Detection

If the columns are prepared for the on-line connection of liquid chromatography to a mass spectrometer it can be helpful to have a UV detector in-line. Because the sample will be analyzed by mass spectrometry, UV detection is not really necessary but it will ensure the proper delivery of the sample from the column into the mass spectrometer. A variety of options are available if UV detection is desired. Capillary flow cells are commercially available for many different detectors. They are quite sensitive and gaining in popularity despite their high costs (LC Packings). However, suitable on-column flow cells that give up little in sensitivity, but have an improved baseline, can be constructed from materials on hand.

A flow cell can be glued to the transfer line at the end of the column (on-column flow cell). To prepare such a flow cell, the polyimide coating in the middle (10 mm) of a piece of fused silica tubing (70 mm) is burned off. The inner diameter of the tubing should be 100 to 200 μm. The fused silica transfer line from the column exit is inserted on one end and a fused silica transfer line long enough to reach the mass spectrometer is inserted into the other end of the flow cell. Both lines should meet in the clear part of the flow cell and leave only a small space of 1 mm in between them. This will ensure that the volume of the flow cell is as small as possible, preventing again the loss of resolution. The transfer lines should then be glued in place as described previously. A flow cell prepared in such a way can be used with the 759A UV detector (Applied Biosystems, Foster City, CA) and the HT optical cell, distributed by this manufacturer as well. Some simple modifications may be necessary for use in other detectors using a ball lens-capillary holder (Spectrophysics, Linear, etc.), but the basic design is the same.

Columns prepared for the preparative isolation of proteins and peptides have larger dimensions and faster flow rates so that the dimensions of the flow cell can be larger. For this type of column, UV detectors with small enough flow cells are commercially available so that the attachment of an on-column flow cell can be omitted. The transfer line at the end of the column is directly attached to the flow cell of the UV detector. The detectors used in our laboratory are the SPD-6A and SPD-10A from Shimadzu (Tokyo, Japan). Before using those detectors for microcapillary HPLC, the standard flow cell must be exchanged against the micro- and semimicroflow cell for the 590 and 590A version, respectively. Both flow cells must be modified as well, because they are equipped with large inner diameter tubing from the manufacturer. This tubing must be exchanged against 100-μm i.d. fused silica tubing to maintain the resolution for the collection of fractions.

High-Performance Liquid Chromatography Equipment

There are several options for a microcapillary HPLC system. The most common and simple way to get started with HPLC equipment available in many laboratories is by stream splitting. In our laboratory the ABI 140A/B delivery system or the older adequate version of this syringe pump from Brownlee is used. These pumps cannot be operated at a flow rate of only 2 μl/min for the smaller columns and 20 μl/min for the larger columns.

To achieve an accurate delivery a stainless steel T-valve splits the flow between the HPLC and the rheodyne valve before injection. One side of the T-valve goes to waste, the other through the Rheodyne (Cotati, CA) valve to the microcapillary column. To actually split the flow some back pressure on the waste side must be generated with, e.g., a previously used HPLC column or fused silica tubing with a very narrow inner diameter. The back pressure can be adjusted by changing the length of the tubing or by putting more than one previously used HPLC column in-line so that the desired flow rate is forced through the microcapillary column. The disadvantages of using such a system for microcapillary HPLC are large dead volumes, system refilling, and repressurization steps, which extend the time of a duty cycle considerably.

A more complex approach is to build an HPLC delivery system suitable for the correct delivery of low flow rates. On the system used in our laboratory for LC-MS the flow rate of the HPLC system is established with an Isco high-pressure syringe pump operating in the constant pressure mode and delivering 0.1% aqueous trifluoroacetic acid. For a column with an inner diameter of 250 μm and a length of 200 mm, the flow rate during sample injection is typically about 20 μl/min (ca. 4000 psi) and 2 μl/min (400 psi) during sample elution. The pressure settings generating the desired flow rate must be adjusted with each individual column. They might also vary according to other parameters creating back pressure, such as the length and inner diameter of the transfer line from the HPLC to the mass spectrometer or the dimensions of the inner needle. We have found that it is easiest to run the transfer line from the UV detector all the way to the needle tip, using the fused silica transfer line as the inner needle itself. This avoids unnecessary junctions generating dead volumes or possible leakage. Samples are loaded into a sample loop and injected onto the column using a Rheodyne injection valve. The sample injector is placed upstream relative to the gradient valve to minimize extra column dead volume. During the time of sample injection, a gradient is preformed and loaded into a stainless steel capillary loop with a volume of 200 μl, using two programmable low-pressure syringe pumps (model 4400-001; Harvard Apparatus, South Natick, MA) delivering appropriate amounts of solvent A and solvent B. The gradient loop is backfilled in reverse order to place the start of the gradient at the front of the loop. The time of injection varies with the volume of the sample and depends also on the anticipated size of the injection artifact. After the sample is injected, the gradient filled loop is switched on-line, and the gradient is pushed onto the column with a flow rate of 2 μl/min delivered from the Isco pump. A typical gradient runs from 2 to 72% solvent B over 35 min. A volume of 72% solvent B is placed at the end of the gradient to prevent erosion of the end of the gradient due to diffusion with solvent A, used to push the gradient through the system. After the elution of the sample, the flow rate is reset to 20 μl/min to reequilibrate the column as quickly as possible.

The setup described above with a flow rate of 2 μl/min during sample analysis is used to analyze about 5–30 pmol of sample. For the analysis of even smaller amounts, columns with an inner diameter of 160 or 100 μm are used, reducing the flow rate to 1 μl/min or less. The pressure values of the Isco pump are changed according to the desired flow rate. The flow rate should be checked periodically at the end of the needle and the pressure settings readjusted if needed.

The advantage of such a system is a constantly controlled system pressure, time and dead volume compression through pressure programming, and automatic column reequilibration without refilling or repressurization. The duty cycle is at least 50% shorter compared to the stream-splitting system, described above.

Sample Preparation

The problem inherent to the handling of low picomole amounts of samples is the loss of sample before any analysis can be done. Proteins and peptides can stick to tubes, especially if the concentration of sample is low and the volume high. Sometimes the risk of losing the sample can be reduced by silanizing the tubes. But at all times it is best to keep the concentration as high as possible and to reduce the volume to a minimum.

Enzymatic Fragmentation

All enzymatic fragmentations are carried out as described by Lee and Shively.[8]

Mass Analysis of Proteins

The mass analysis of small amounts of proteins within a possible mass accuracy of 0.01% is one of the most striking features of an electrospray mass spectrometer. Sample preparation is of critical importance to successful electrospray analysis. Electrospray MS is sensitive to contaminants such as salts and detergents. By coupling an HPLC directly to the mass spectrometer, problems of sample contamination are virtually eliminated. At the same time, all components in a mixture can be analyzed during the course of a single run.

The analysis of 12 pmol of mitochondrial NADH dehydrogenase serves to illustrate the salient features of the method (Fig. 1). NADH dehydrogenase is a membranal enzyme consisting of three subunits (I, II, and III)

[8] T. D. Lee and J. E. Shively, *Methods Enzymol.* **193,** 361 (1990).

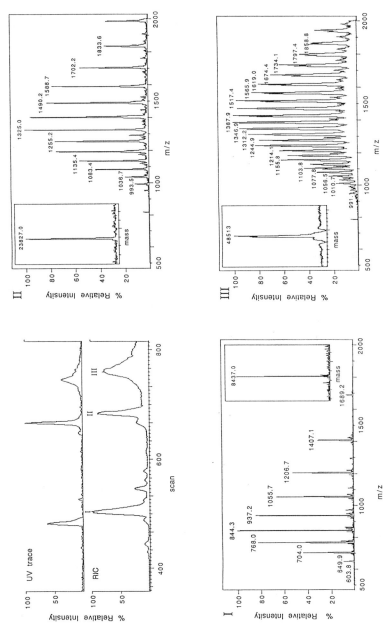

Fig. 1. LC-MS analysis of 12 pmol of mitochondrial NADH dehydrogenase. The separation of the three subunits in the UV trace, the reconstructed ion current (RIC), the spectra derived from peaks I, II, and III, and their deconvolution to the corresponding masses are shown.

and has a total molecular mass of 80,000 Da as determined by sodium dodecyl sulfate–polyacrylamide gel electrophoresis (SDS–PAGE). The three subunits separate during HPLC analysis (Fig. 1). The shift in time between peaks in the UV and reconstructed ion current (RIC) profiles are the result of sequential passage of the HPLC effluent through the UV detector and then into the mass spectrometer. Spectra are presented both in the normal fashion, in which the mass-to-charge ratio (x axis = m/z) is plotted with respect to relative ion abundance, and after computer processing (deconvolution) to yield directly the molecular weight (x axis = mass). The observed masses of subunits I and II correlate well with those calculated from the known amino acid sequence, 8438.4 and 23,815.5 Da for subunits I and II, respectively. For subunit III, only the cDNA sequence is known and the calculated mass derived from this sequence is 50,466.8 Da. However, the observed mass of the protein is smaller than expected. The primary sequence of the protein would have to be determined to explain the difference.

Structural Analysis of Peptides

For the analysis of peptides sample purity is of concern as well. In addition, peptides often exist as complex mixtures either as a result of isolation from a natural system or from an intentional enzymatic degradation of a larger protein structure. Off-line chromatographic separation and subsequent analysis of peptide components are always tedious and prone to failure when sample amounts are low. Many peptides and also proteins will stick to collection tubes after the removal of buffer and detergents by purification procedures such as reversed-phase HPLC (see also [2] in this volume[9]). Direct LC-MS analysis provides an effective alternative. All components of a mixture can be analyzed during the course of one LC-MS run and sample handling between the HPLC separation, and mass spectral analysis is eliminated. Cytochrome c (20 pmol), digested with Lys-C (CCKCD), is applied to LC-MS analysis as illustrated in Fig. 2. The elution of the peptides from the column is detected at a wavelength of 200 nm (top, Fig. 2). The mass spectral analysis of the peptides is monitored as a base peak as well as a reconstructed ion current profile. The base peak represents the largest ion intensity per scan; the reconstructed ion current is the total ion current per scan. Figure 3 shows the sequence of horse cytochrome c and the Lys-C fragments identified by LC-MS analysis. The application of only a few picomoles of sample to LC-MS analysis generates mass information on all peptides above the mass of 500 Da, covering 88%

---

[9] E. R. Hoff and R. C. Chloupek, *Methods Enzymol.* **271**, Chap. 2, 1996 (this volume).

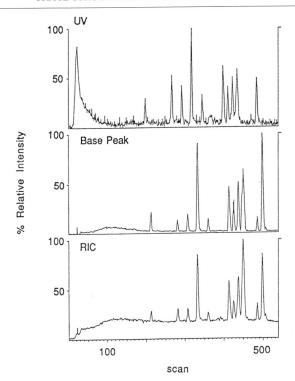

FIG. 2. LC-MS analysis of 20 pmol of cytochrome *c*, Lys-C digested (CCKCD). The upper trace shows the UV absorbance profile at 200 nm. The middle and lower traces display the base beak and reconstructed ion profile, respectively.

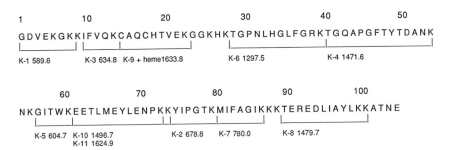

FIG. 3. Sequence of horse cytochrome *c* and location of the fragments identified by LC-MS analysis. The N-terminal glycine is acetylated.

of the complete sequence. Even more information could be obtained by lowering the scan range for monitoring ions other than solvent ions. For analysis of smaller amounts (<3 pmol), columns with inner diameters less than 250 μm are used in order to reduce the flow rate and thus maintain adequate sample concentration for the ES-MS analysis. The lower flow rates have the disadvantage that the electrospray is less stable. To overcome this problem and to retain the chromatographic resolution, the inner diameter of the electrospray needle is reduced. In Fig. 4, the LC-MS analysis of 500 fmol of Lys-C-digested cytochrome c is shown. The inner diameter of the electrospray needle used in this analysis is 10 μm. The HPLC is performed on a fused silica capillary column with an inner diameter of 100 μm packed with C_{18} reversed-phase adsorbent to reduce the flow rate to 1 μl/min. The flow rate of the sheath liquid is adjusted to the same value as well.

Tandem Mass Spectrometry Analysis of Peptides

Tandem mass spectrometry (MS/MS) is, for small amounts of peptides, a sensitive technique to determine information on the primary structure of the material. If the peptide of interest is already purified, the easiest and fastest method of analysis is by SIMS/MS.

To reduce the risk of losing peptides through unnecessary purification steps, it is sometimes advisable to omit the last reversed-phase HPLC step and to analyze a peptide mixture directly by on-line LC-MS/MS. This

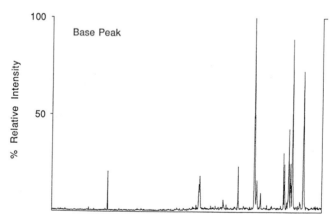

FIG. 4. LC-MS analysis of 500 fmol of horse cytochrome c, Lys-C digested. The base peak profile is displayed. The profile should be compared to the base peak profile of the analysis of 20 pmol of cytochrome c in Fig. 2. All peptides found in this application can be identified.

technique has the advantage of separating peptides from a mixture into single components and obtaining structural data at the same time. The Lys-C digest of cytochrome c, as discussed previously, is used to demonstrate the capability of existing mass spectrometric instrumentation to perform LC-MS/MS analysis. To perform LC-MS/MS analysis on as many peptides as possible in one run it is necessary to automate the process. For automation, a program using the Finnigan-MAT instrument controlled language (ICL) is written to switch the instrument into MS/MS mode whenever an ion of interest is observed. The base peak intensity of this ion must be high enough to trigger the switch. After four scans or when the ion intensity falls below a preset value, the instrument switches back to normal scan mode. By limiting the number of scans taken in the MS/MS mode, fragmentation data on even closely eluting peptides can be generated. At the same time the m/z values are stored in a user list. If a second run is desired, the instrument can be programmed to ignore ions in the user list, and MS/MS analysis will be carried out only on ions that were not analyzed in the first run. MS/MS data on all peptides underlined in the sequence of cytochrome c, which is shown in Fig. 3, are obtained in a single run. The MS/MS data are generated on the singly or doubly charged ions of the peptides depending on which of the signals is more intense. However, MS/MS spectra generated from multiply charged ion species are in general more complex and difficult to interpret.

Structural Analysis of Glycopeptides

The isolation of glycopeptides can be difficult, especially when the species are derived from highly glycosylated proteins, which often resist enzymatic digestion. For the structural analysis of glycopeptides by mass spectrometry it is also important to remove any contaminating peptides from the glycopeptide fraction. Figure 5 demonstrates the purification scheme of two CEA glycopeptides, using different HPLC techniques. The isolated glycopeptides contain glycosylation sites 1 and 26 of the 28 existing sites on CEA. To perform a successful digestion of CEA we use a double digest with trypsin and chymotrypsin, resulting in the peptide map shown in Fig. 5A. The most successful approach to further purification of glycopeptides is to use hydrophilic HPLC. Even after several chromatography steps on reversed-phase columns, glycopeptide-containing fractions that show poor resolution on a C_{18} column (Fig. 5B) can be separated into several peaks on a polyhydroxyethyl aspartamide column (PolyLC, Columbia, MD), using a linear gradient of 80% acetonitrile to H_2O in 10 mM triethylamine phosphate, pH 3.0, over 1 hr, as shown in Fig. 5C and D. Owing to heterogeneity

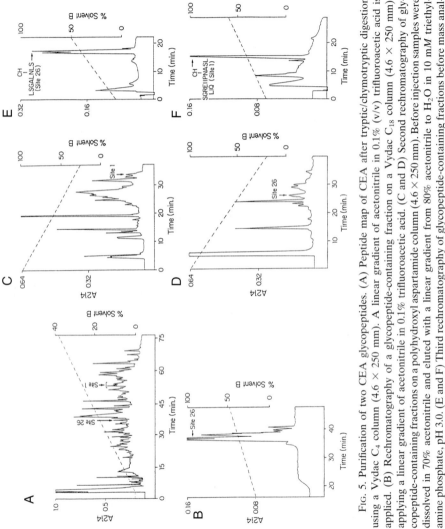

Fig. 5. Purification of two CEA glycopeptides. (A) Peptide map of CEA after tryptic/chymotryptic digestion using a Vydac C_4 column (4.6 × 250 mm). A linear gradient of acetonitrile in 0.1% (v/v) trifluoroacetic acid is applied. (B) Rechromatography of a glycopeptide-containing fraction on a Vydac C_{18} column (4.6 × 250 mm), applying a linear gradient of acetonitrile in 0.1% trifluoroacetic acid. (C and D) Second rechromatography of glycopeptide-containing fractions on a polyhydroxyl aspartamide column (4.6 × 250 mm). Before injection samples were dissolved in 70% acetonitrile and eluted with a linear gradient from 80% acetonitrile to H_2O in 10 mM triethylamine phosphate, pH 3.0. (E and F) Third rechromatography of glycopeptide-containing fractions before mass analysis on a Vydac C_{18} column (2.1 × 250 mm). A linear gradient of acetonitrile in 0.1% trifluoroacetic acid is applied.

in the carbohydrate structure, glycopeptides often elute as broad peaks from this column. Also, the retention times of glycopeptides are longer compared to those of unmodified peptides.

Prior to mass spectral analysis of these fractions, triethylamine phosphate must be removed by a reversed-phase HPLC step using a Vydac C_{18} column with a fast gradient of acetonitrile in 0.1% trifluoroacetic acid (Fig. 5E and F). At this step it is helpful to verify the purity of samples by N-terminal sequencing before subjecting them to mass analysis. After sequence analysis the expected mass for the peptide portion can be calculated. The mass of the peptide with the complete carbohydrate structure attached (parent mass) is sometimes difficult to determine due to heterogeneity of the carbohydrate portion or additional fragmentation that occurs during mass analysis. Both factors account for the lower signal intensity of glycopeptides in mass spectra compared to the analysis of an unglycosylated peptide. The difference of the observed parent mass of the glycopeptide in the SIMS analysis and the mass of the peptide portion reveals the mass of the carbohydrate portion.

As an example, the interpretation of the carbohydrate portion of two glycopeptides isolated from CEA is shown in Fig. 6. The primary structure of the peptides is listed in Table I. Knowledge of the mass of the peptide portion assists in assigning structures to the fragment ions related to the carbohydrate.

Although electrospray mass spectrometry allows the analysis of large proteins by displaying an envelope of mass to charge states, it leads to difficulties in the structure determination of the carbohydrate portion of glycopeptides because the sample is heterogeneous with respect to mass. The large number of signals due to both sample heterogeneity and fragmentation in the mass analysis results from the fact that different fragments exhibit multiple charge states. This makes it difficult to interpret the resulting spectra. In most cases it is much easier to start from the simpler structural analysis of glycopeptides by SIMS and then proceed to ES-MS. Another problem of the structural analysis of glycopeptides is the relatively large amount of sample required just to purify the peptides over several chromatographic steps. Here the proposed structure of the two CEA glycopeptides could be further verified by carbohydrate composition analysis.

Conclusion

With the introduction of mass spectrometry into protein chemistry a new, powerful tool has been implemented to perform different structural analyses on small amounts of samples. The analysis of macromolecules with electrospray mass spectrometry in combination with liquid chromatography

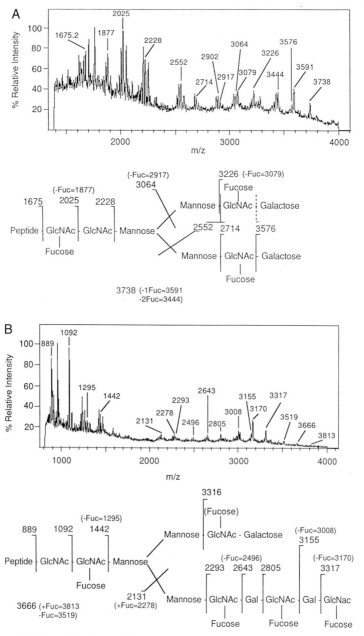

FIG. 6. SIMS analysis of two glycopeptides isolated from CEA. The peptide sequences are shown in Table I. (A) SIMS analysis of the glycopeptide carrying glycosylation site 1. (B) SIMS analysis of the glycopeptide carrying glycosylation site 26.

TABLE I

EXPERIMENTAL AND CALCULATED MASS VALUES FOR PROTONATED MOLECULAR IONS IN
MASS SPECTRA OF TWO CARCINOEMBRYONIC ANTIGEN GLYCOPEPTIDES

Peptide	Sequence	Observed m/z	Calculated m/z (peptide)	Carbohydrate portion (Da)
A (site 1)	SGREIIYPNASLLIQ	3738	1674.9	2063.1
B (site 26)	LSGANLNLS	3666	889.0	2777

within an error of ±0.01% is a reliable technique for mass determination. Accurate mass determination of a protein with known primary structure can indicate possible posttranslational modifications of the protein as well as errors in protein sequence analysis or cDNA-derived sequences. The success of a variety of experiments such as site-directed mutagenesis or covalent modification studies can be verified quickly without employing large amounts of sample.

The mass analysis of proteins by matrix-assisted laser desorption time of flight (MALDI-TOF) has become a quick, reliable, and less expensive alternative to electrospray mass spectrometry. However, there are some disadvantages to this type of mass spectrometer such as the little understood sample preparation. So far, many matrices are known to be suitable for the analysis of different components such as proteins, peptides, carbohydrates, lipids, oligonucleotides, and so on. But little is known about which matrix works best and how to prepare the samples to obtain the best results. However, only 1 pmol (or even less) of sample need be applied to the sample stage so that one can try a variety of conditions without wasting too much sample. MALDI-TOF can be used for the analysis of peptides as well.

Lately, more and more laboratories have been trying to implement this technique in the analysis of whole-peptide mixtures without prior separation of the peptides. However, the further structural analysis of peptides by MS/MS is not possible yet. At this point SIMS/MS and LC-MS/MS still seem to be the better choices for structural analysis, whereas LC-MS/MS is preferred for the analysis of complex protein and peptide mixtures in the low picomole range. LC-MS/MS can provide helpful information on the primary structure of individual components in mixtures without the need for complete purification necessary for SIMS/MS.

Mass analysis of peptide maps derived from different endoproteolytic fragmentations by MALDI-TOF or LC-MS alone will be helpful. The comparison of the obtained masses from these maps with the corresponding theoretical maps of proteins listed in protein databases will reveal if the analyzed protein is a known or an unknown. If the structure of the sample is indeed known, the information on this structure will be available after only one LC-MS or MALDI-TOF analysis without the need for further sequence analysis.

The structural analysis of posttranslational modifications such as glycosylation is still a challenge because of the need for large amounts of sample for purification and analysis of the peptides carrying the modification. For the characterization of glycopeptides SIMS will provide the least complex spectra and requires at the same time the least amount of sample. The analysis of glycopeptides by electrospray mass spectrometry is still complicated and needs more software development to help deconvolute spectra deriving from glycopeptide mass analysis. Another improvement would be to integrate the on-line separation and mass analysis of glycopeptides, keeping the number of chromatographic steps to a minimum and thus reducing the amount of starting material needed for the purification of glycopeptides.

It is unlikely that mass spectrometry will replace existing techniques such as microsequence analysis completely. Rather it should be viewed as a complementary technique taking advantage of higher sensitivity and the ability to analyze mixtures without complete separation.

Acknowledgments

We thank Dr. S. Chen for generous provision of mitochondrial NADH dehydrogenase, D. C. Stahl for contributing to the work on the LC-MS/MS analysis, and Mike Davis for technical input on the liquid chromatography section. This work was supported in part by NIH Grants CA 37808 and CA 33572.

[4] Protein Liquid Chromatographic Analysis in Biotechnology

By L. J. JANIS, P. M. KOVACH, R. M. RIGGIN, and J. K. TOWNS

High-performance liquid chromatography (HPLC) has been an essential tool for the characterization of proteins since the pioneering work of Regnier and others[1] in the 1970s. Given the wide versatility, relative ease of use, and high resolution of the technique; it is not surprising that HPLC is considered to be perhaps the most valuable tool for characterization of virtually any biosynthetic protein. However, the successful, or at least optimal, application of HPLC for the characterization of a biosynthetic protein

[1] F. E. Regnier, Methods Enzymol. 91, 137 (1982).

requires a thorough evaluation of several factors, including (1) the specific objectives for the assay (e.g., product identification, purity quantitation, and process monitoring), (2) the chemical and physical stability and other physicochemical properties of the parent protein, (3) the types of impurities or "related substances" that are likely to be present, and (4) the various HPLC separation modes (e.g., size exclusion, reversed phase) that are potentially capable of meeting the assay goals for the specific protein of interest.

This discussion explores the specific challenges and experimental approaches that are applicable to the characterization of biosynthetic proteins, focusing primarily on selected proteins that are mass produced in the pharmaceutical and agricultural industries. Fundamental aspects of protein HPLC have been described in earlier chapters and are not presented here. However, considerable emphasis is given to those practical aspects of protein HPLC that are critical to successful characterization of biosynthetic protein products. Representative examples are presented to illustrate the various HPLC approaches that have been successful in the biotechnology industry. Additional examples from the literature are presented in less detail in the introductory section of this chapter. However, no attempt has been made to conduct an exhaustive literature review, because such reviews necessarily must be updated periodically and are readily available elsewhere.[2] While the focus of this chapter is protein and peptide analysis, some relevant examples of preparative separations are included in the literature review as appropriate to illustrate certain discussion points. Throughout this chapter the term *related substances* refers to proteins that are structurally related to the parent biosynthetic protein of interest. The term *impurities* refers to other protein impurities (e.g., antibodies arising from affinity purification media and processing enzymes). This chapter does not discuss the use of HPLC for determination of nonprotein impurities (e.g., surfactants) in biosynthetic proteins, although such applications are widespread throughout the biotechnology industry.

General Considerations

Method Objectives

The first consideration in defining an HPLC method development strategy is to establish the objectives for the method. Protein separations are conducted in the biotechnology industry for a variety of analytical purposes,

[2] J. G. Dorsey, J. P. Foley, W. T. Cooper, R. A. Barford, and H. G. Barth, *Anal. Chem.* (Fundamental Reviews) **64,** 353R (1992); and subsequent biannual issues.

including (1) fermentation/cell culture process monitoring, (2) purification process monitoring, (3) purified bulk material characterization, (4) final product characterization, and (5) product stability evaluation. For purified bulk material and final product characterization, the method objectives may further be segregated into product identity, purity, and potency (or content) assays.[3] Each analytical application may place different requirements on the assay. The objective of fermentation/cell culture process monitoring, for example, is to track the yield of product during the fermentation cycle. Hence determination of specific impurities may be of little interest in this case; however, assay time must be minimized to allow timely feedback of information to the fermentation operation. However, purified bulk material characterization requires detailed knowledge of specific impurity levels, but can be conducted over a longer time interval. Hence, for process monitoring one may choose to employ a single, rapid HPLC technique that provides adequate resolution of the product from impurities, whereas for final product characterization a battery of less rapid, high-resolution HPLC techniques and other techniques may be needed in order to determine individual impurities. Likewise product identity, purity, and potency/content methods have differing objectives that must be considered in developing appropriate HPLC assays.

Physicochemical Properties of Proteins

Proteins vary widely in their physicochemical properties and these properties can have a great impact on the conditions necessary for HPLC analysis. Solution stability can differ widely from one protein to another, and for a given protein solution stability can be highly dependent on pH, temperature, presence of denaturants, and other factors. Such factors must be explored as part of the assay development process in order to select appropriate conditions under which artifactual degradation is minimized. Certain proteins, most notably glycoproteins,[4] exist in a variety of biologically active forms (the term *microheterogeneity* is frequently used to describe this feature). In developing purity assays for such proteins one must consider where the various biologically active forms elute in the assay and how they are to be accounted for in the definition of "purity." Similarly, certain proteins readily self-associate into noncovalent multimers.[5] In some instances one may wish to determine only covalent related substances and, hence, it is appropriate to conduct the HPLC assay using conditions that dissociate

[3] R. M. Riggin and N. A. Farid, in "Analytical Biotechnology, ACS Symposium Series" (C. Horváth and J. G. Nikelly, eds.), Vol. 434, p. 113. American Chemical Society, Washington, D.C., 1990.
[4] P. Knight, *Biotechnology* **7,** 37 (1989).
[5] S. Formisano, M. L. Johnson, and H. Edelhoch, *Biochemistry* **17,** 1468 (1978).

the noncovalent multimers, whereas in other instances measurement of the extent of self-association may be an objective of the assay. Other physicochemical properties to be considered include dry state stability (e.g., on exposure to humidity, heat, and light), presence or absence of free thiol groups, effect of denaturants [e.g., solvents used in reversed-phase (RP)-HPLC] on secondary or tertiary structure, and presence or absence of bound metals.

Types of Related Substances (or Impurities)

The specific related substances, or impurities, to be separated will differ from one protein to another. Furthermore, the quantity of such related substances will depend to some extent on the production process employed, because many of the related substances arise during specific process steps and the extent of subsequent removal will depend on the purification process in place at that time. Usually, one cannot predict all of the related substances that will be observed (i.e., experimentation is required to elucidate the spectrum of related substances present). However, knowledge of the protein sequence, processing reagents, and other processing conditions often allows one to predict some of the related substances that are likely to be observed. Such predictions provide a useful starting point for the development of appropriate HPLC conditions, which can then be further refined as experimental data are obtained. Table I summarizes some of the classes of related substances commonly observed in biosynthetic proteins. Some of the related substances arise from the fermentation/cell culture process (e.g., glycosylation variants) whereas other related substances, such

TABLE I
RELATED SUBSTANCES COMMONLY ENCOUNTERED IN BIOSYNTHETIC PROTEINS

Related substance	Site affected	Useful HPLC modes[a]
Deamido forms	Asparagine or glutamine	IEC; RP-HPLC
Sulfoxides	Methionine	RP-HPLC; SEC
Proteolytically clipped forms	Diverse	HIC; SEC
N-Terminal variants (less frequently C-terminal variants)	N terminus or C terminus	RP-HPLC; HIC
Disulfide isomers	Cystine groups	RP-HPLC
Dimers/aggregates	Diverse	HIC; IEC; RP-HPLC
Glycosylation variants	Diverse (asparagine for N-linked and threonine, serine, or tyrosine for O-linked carbohydrates)	HIC; IEC; RP-HPLC

[a] IEC, Ion-exchange HPLC; RP-HPLC, reversed-phase HPLC; HIC, hydrophobic interaction chromatography; SEC, size-exclusion HPLC.

as methionine sulfoxide derivatives and deamidated forms, arise primarily as a result of degradation during processing or storage of the protein. Knowledge of the origin of the various related substances is relevant to optimization of the purification process and to the selection of appropriate HPLC methods to monitor product purity and stability.

High-Performance Liquid Chromatography Separation Modes

As described in earlier chapters and in Table I, a wide variety of HPLC separation modes is available. Owing to the myriad of potential related substances arising from the production process (e.g., in-process intermediates) and from degradation of the biosynthetic protein, no single technique is likely to be sufficient for complete characterization of a biosynthetic protein product. One must therefore select a battery of complementary techniques that will ensure that the necessary information is obtained, without placing an unnecessary analytical burden on the production or quality control laboratories. Considerations for selection of an appropriate battery of techniques are described in the section Application Examples.

Multidimensional Approach

Hence, both the methods development and quality control processes must rely on the use of a multidimensional approach. For example, a single peak observed for RP-HPLC separation of a protein (see [2] in this volume[5a]) may not be an indication that the protein is highly pure, because it is quite common for related substances to coelute with the parent protein. The possibility of a coeluting related substance must be examined by one or more of the following approaches. The main peak can be isolated and reinjected onto an alternate separation system such as a sodium dodecyl sulfate–polyacrylamide gel electrophoresis (SDS–PAGE), isoelectric focusing (IEF), capillary electrophoresis (CE), or dissimilar HPLC system, taking appropriate measures to avoid decomposition of the protein after collection from the HPLC system. If the HPLC solvent system is not compatible with the second separation system, or if suitable stability of the isolated main peak is not attainable, then the nonfractionated protein product can be subjected to one of the alternate separation procedures. On-line spectroscopic monitoring using diode array detection can be of value in establishing peak purity in favorable cases,[6] but usually protein spectra are too similar to provide much discrimination. Advances in electrospray mass spectrometry (ESMS) as both an off-line technique and on-line HPLC/ ESMS technique allow for much more definitive proof of peak purity,[7]

[5a] E. R. Hoff and R. C. Chloupek, *Methods Enzymol.* **271,** Chap. 2, 1996 (this volume).

[6] H. J. Sievert, S. L. Wu, R. Chloupek, and W. S. Hancock, *J. Chromatogr.* **499,** 221 (1990).

[7] S. A. Carr, M. E. Hemling, M. F. Bean, and G. D. Roberts, *Anal. Chem.* **63,** 2802 (1991).

although this technique is not suitable for direct analysis of many glycoproteins and one must operate the HPLC/ESMS system using only volatile
buffers in the mobile phase.

In most quality control schemes, both size-exclusion and RP-HPLC
methods will be employed to provide high-resolution, quantitative data, in
addition to various spectroscopic, electrophoretic, and immunochemical
procedures for more specialized purposes. Conventional electrophoretic
techniques often do not directly add to the information obtained, because
HPLC frequently offers better resolution of related substances. However,
a key advantage of conventional electrophoretic techniques is the ability
to detect virtually any protein, thereby ensuring that any unanticipated
protein impurities would be detected. In developing the HPLC methods
to be used in the battery of tests, one must also consider the information
obtained from the other techniques and seek to understand the relationship
between those sets of information. The objective of such an assessment is
to obtain the maximum amount of relevant information consistent with an
efficient use of analytical resources.

Experimental Considerations

High-Performance Liquid Chromatography Hardware

In this chapter the discussion is limited to a description of a few considerations specifically relevant to biosynthetic protein analysis. HPLC separation of closely related substances requires extremely precise control of
the solvent composition, because RP-HPLC elution time is much more
dependent on organic solvent composition for proteins than for small organic compounds. Continuous sparging of solvents (e.g., with helium for
degassing purposes) should generally be avoided due to the resulting change
in organic solvent composition. In some cases, presaturation of the helium
with organic vapor by passing the sparge gas through a solvent reservoir
(containing mobile phase) ahead of the mobile-phase reservoir can minimize this problem. Single-pump, low-pressure mixing HPLC gradient systems are widely used for protein analysis. However, when using solvents
that are difficult to mix (e.g., *n*-propanol), we have noted that dual-pump,
high-pressure mixing is advantageous because the two solvent streams are
continuously delivered, rather than delivered in a segmented fashion.

Another hardware consideration is the choice of detector, and more
specifically detector wavelength. Generally either 280- or 214-nm wavelengths are employed for protein analysis. Detection at 214 nm is more
sensitive and more generally applicable because absorption at this wavelength is due to the peptide backbone. However, some HPLC mobile-phase
constituents may interfere at this wavelength (e.g., acetate salts). Detection

at 280 nm is somewhat less sensitive, but can be used to detect most proteins (absorption at 280 nm is primarily due to the aromatic amino acid side chains of tryptophan, phenylalanine, and tyrosine). In some cases, dual-wavelength or diode array detectors may be useful in order to provide more detailed spectral information for each peak observed. However, in most cases the UV spectra of related proteins are similar.

The use of microbore HPLC and packed capillary HPLC for characterizing biosynthetic proteins is becoming more widespread. However, such techniques have not been widely adopted for routine quality control or methods development within the biotechnology industry owing to limitations in hardware, and the relatively large amounts of assay material available. As hardware advances occur, this situation may change because the advantage of lower solvent consumption is clearly attractive. Furthermore, such techniques are more compatible with HPLC/MS systems, which are now widely available and offer profound advantages for methods development.

Sample Preparation

Samples must be introduced into the HPLC system in a suitable matrix and must have sufficient stability to allow routine use of the procedure. In general the sample solution should be similar in composition (e.g., pH) to the HPLC mobile phase so as to minimize the effect of the sample injection on separation efficiency. One exception to this rule is RP-HPLC, wherein the organic solvent composition of the sample should be less than that of the initial mobile phase in order to focus the injected proteins at the head of the column (in most cases little or no organic solvent needs to be added to the protein solution prior to injection). Clearly, the protein must have adequate stability in the selected solvent and in many cases stability must be enhanced by use of a refrigerated autoinjector. When conducting size-exclusion HPLC for determining noncovalent multimers, one must avoid the use of denaturants (e.g., organic solvents) or extreme pH conditions that might disturb the equilibrium between monomer and multimer forms (the appropriate solution content must be selected on the basis of the ultimate usage conditions for the protein product). In some instances inert ingredients or preservatives present in the formulated drug product may interfere in the assay, requiring the use of alternative conditions, or careful removal of the ingredients prior to the assay.

Column Manufacturing Issues

The heart of the HPLC separation system is the column. Unfortunately, this is the one component over which the analyst has the least direct control. Consequently it is important to work closely with the column manufacturer to obtain several independent column manufacturing batches, and to investigate whether a particular method performs consistently on a number of column manufacturing batches. In extreme cases, it may be necessary to adopt a method-specific column performance specification for use by the manufacturer or the customer. Whenever possible, an HPLC method to be used on a routine basis should be validated using two alternative sources (brands) of columns, in order to minimize the impact of column manufacturing problems.

Assay Development and Validation Strategy

Optimal assay development and thorough assay validation are requirements for the successful use of any technique for product characterization and quality control. During the development phase one must evaluate a range of possible separation modes and, wherever possible, separation effectiveness should be verified using authentic related substance standards. Purity determinations using a single HPLC method should not be regarded as valid unless the purity has been corroborated by several other criteria (e.g., demonstrating separation of several potential related substance standards under the same assay conditions, reassay of the isolated peak on a dissimilar separation system, HPLC/MS verification). In cases where established methods exist (e.g., bioassay), one must determine the qualitative and quantitative relationship between the proposed HPLC method and the existing method. Method validation should include assessment of the following criteria: accuracy, precision, linearly, range, limit of detection (for trace level assays), and ruggedness.[8]

Application Examples

Literature Overview

Representative applications demonstrating the usefulness of HPLC for characterization of biosynthetic proteins are summarized in Table II.[9-36]

---

[8] Validation of compendial methods. *In* "U.S. Pharmacopeia," XXIInd Ed., p. 1710. United States Pharmacopeial Convention, Inc., Rockville, MD, 1990.

[8a] Deleted in proof.

[9] E. P. Kroeff and R. E. Chance, *in* "Hormone Drugs," p. 148. U.S. Pharmacopeial Convention, Inc., Rockville, MD, 1982.

[10] H. W. Smith, L. M. Atkins, D. A. Binkley, W. G. Richardson, and D. J. Miner, *J. Liquid Chromatogr.* **8,** 419 (1985).

[11] E. P. Kroeff, R. A. Owens, E. L. Campbell, R. D. Johnson, and H. I. Marks, *J. Chromatogr.* **461,** 45 (1989).

RP-HPLC, hydrophobic interaction chromatography (HIC), size-exclusion chromatography (SEC), and ion-exchange HPLC (IEC) are widely used for biosynthetic protein characterization. On the other hand, the last technique listed in Table II, immobilized-metal affinity chromatography (IMAC), is frequently used for preparative separations, owing to its very high loading capacity, but is less commonly used for analytical purposes. The application examples in the following sections highlight the most important points to be considered in developing and using an HPLC method for characterization of biosynthetic proteins.

[12] B. S. Welinder and F. H. Andresen, *in* "Hormone Drugs," p. 163. U.S. Pharmacopeial Convention, Inc., Rockville, MD, 1982.

[13] E. P. Kroeff, R. A. Owens, E. L. Campbell, R. D. Johnson, and H. I. Marks, *J. Chromatogr.* **461**, 45 (1989).

[14] G. W. Becker, P. M. Tackett, W. W. Bromer, D. S. Lefeber, and R. M. Riggin, *Biotechnol. Appl. Biochem.* **10**, 326 (1988).

[15] R. M. Riggin, G. K. Dorulla, and D. J. Miner, *Anal. Biochem.* **167**, 199 (1987).

[16] P. Gellerfors, B. Pavlu, K. Axelsson, C. Nyhlen, and S. Johansson, *Acta Paediatr. Scand. Suppl.* **370**, 93 (1990).

[17] B. S. Welinder, H. H. Sorensen, and B. Hansen, *J. Chromatogr.* **398**, 309 (1987).

[18] P. Oroszlan, S. Wicar, G. Teshima, S.-L. Wu, W. S. Hancock, and B. L. Karger, *Anal. Chem.* **64**, 1623 (1992).

[19] D. C. Wood, W. J. Salsgiver, T. R. Kasser, G. W. Lange, E. Rowold, B. N. Violand, A. Johnson, R. R. Leimgruber, G. R. Parr, N. R. Siegel, N. M. Kimack, C. E. Smith, J. F. Zobel, S. M. Ganguli, J. R. Garbow, G. Bild, and G. G. Krivi, *J. Biol. Chem.* **264**, 14741 (1989).

[20] R. Bischoff, D. Clesse, O. Whitechurch, P. Lepage, and C. Roitsch, *J. Chromatogr.* **476**, 245 (1989).

[21] P. A. Hartman and J. D. Stodola, *J. Chromatogr.* **444**, 177 (1988).

[22] K. Benedek, B. Hughes, M. B. Seaman, and J. K. Swadesh, *J. Chromatogr.* **444**, 191 (1988).

[23] K. T. Shitanishi and S. W. Herring, *J. Chromatogr.* **444**, 107 (1988).

[24] H. Ming, B. D. Burleigh, and G. M. Kelly, *J. Chromatogr.* **443**, 183 (1988).

[25] M. G. Kunitani, R. L. Cunico, and S. J. Staats, *J. Chromatogr.* **443**, 205 (1988).

[26] S.-L. Wu, W. S. Hancock, B. Pavlu, and P. Gellerfors, *J. Chromatogr.* **500**, 595 (1990).

[27] R. M. Riggin, C. J. Haviland, D. K. Clodfelter, and P. M. Kovach, *Anal. Chim. Acta* **249**, 201 (1991).

[28] E. Canova-Davis, I. P. Baldonado, and G. M. Teshima, *J. Chromatogr.* **508**, 81 (1990).

[29] G. W. Becker, R. R. Bowsher, W. C. MacKellar, M. L. Poor, P. M. Tackitt, and R. M. Riggin, *Biotechnol. Appl. Biochem.* **9**, 478 (1987).

[30] R. M. Riggin, C. J. Shaar, G. K. Dorulla, D. S. Lefeber, and D. J. Miner, *J. Chromatogr.* **435**, 307 (1988).

[31] E. Canova-Davis, G. M. Teshima, T. J. Kessler, P. J. Lee, A. W. Guzzetta, and W. S. Hancock, *in* "Analytical Biotechnology, ACS Symposium Series" (C. Horváth and J. G. Nikelly, eds.), Vol. 434, p. 90. American Chemical Society, Washington, D.C., 1990.

[32] W. Schroder, M. L. Dumas, and U. Klein, *J. Chromatogr.* **512**, 213 (1990).

[33] M. Kunitani, G. Dollinger, D. Johnson, and L. Kresin, *J. Chromatogr.* **588**, 125 (1991).

[34] F.-M. Chen, G. S. Naeve, and A. L. Epstein, *J. Chromatogr.* **444**, 153 (1988).

[35] J. S. Patrick and A. L. Lagu, *Anal. Chem.* **64**, 597 (1992).

[36] J. Porath, *J. Chromatogr.* **443**, 3 (1988).

TABLE II
HPLC METHODS REPORTED FOR CHARACTERIZATION OF BIOSYNTHETIC PROTEINS

Biosynthetic protein	Analysis objective	Comments	Refs.
RP-HPLC			
Human insulin	Determination of product potency		9, 10
Human insulin	Determination of related substances (e.g., A21 desamido)		9, 11, 12
Human insulin	Determination of related substances	In-process monitoring through purification steps	13
Human growth hormone (hGH)	Separation of desamido and sulfoxide derivatives from hGH. Separation of N-methionyl-hGH from hGH	Neutral pH mobile phase; C_4 column; isocratic elution	14, 15
hGH	Separation of extensively oxidized hGH from hGH	Acidic pH mobile phase C_{18} column: gradient elution.	16
hGH	Separation of several hGH derivatives	Various assay conditions described.	17
hGH	Evaluation of conformational effects on hGH separation	Various conditions evaluated.	18
Bovine somatotropin (BST)	Process monitoring for principal component/purity evaluation	Acidic pH mobile phase; C_8 column; gradient elution.	19
Hirudin	Process monitoring	Acidic pH mobile phase; C_{18} column; gradient elution.	20
Fertirelin (analog of luteinizing hormone-releasing factor)	Purity evaluation	Various conditions described	21
Malaria antigen	Purity evaluation	Acidic pH mobile phase; C_4 column; gradient elution.	22

(*continued*)

Biosynthetic Human Insulin

Human insulin is a small protein (5808 Da, 51 amino acid residues, p*I* 5.5). It is composed of two peptide chains (A and B) linked by two interchain disulfide bonds and has another intrapeptide disulfide bond on the A-chain (Fig. 1, page 98). Human insulin is closely related in structural, chemical, and physical properties to pork and to beef insulin (Table III). These are the primary therapeutic forms of insulin although the advent of recombinant

TABLE II (*continued*)

Biosynthetic protein	Analysis objective	Comments	Refs.
Human factor X	Separation of factor X from factor Xa	Acidic pH mobile phase; Diphenyl column; gradient elution.	23
Insulin-like growth factor-I (IGF-I)	Separation of various reduced and oxidized forms	Acidic pH mobile phase; C_8 column; gradient elution.	24
Hydrophobic interaction chromatography			
hGH	Purity evaluation; separation of single/two-chain forms	Method optimization study	26
hGH	Same as above	Sodium sulfate reverse gradient; neutral pH; Phenyl-5PW column	16
Tumor necrosis factor (TNF)	Evaluation of noncovalently associated forms of TNF	Ammonium sulfate reverse gradient; ethylene glycol gradient; neutral pH; Phenyl 5-PW column	25
Plasminogen activator (PA)	Separation of glycoforms and single/two-chain forms	Ammonium phosphate reverse gradient; neutral pH; Phenyl 5-PW column	27
Human relaxin	Purity evaluation	Ammonium sulfate reverse gradient; neutral pH; Phenyl 5-PW column; RP-HPLC found to be more discriminating	28
Size-exclusion HPLC			
hGH	Determination of noncovalent dimer and aggregates	Neutral pH mobile phase; nondissociating conditions	29
hGH	Product potency	Same as above	30

DNA techniques has made the large-scale production of human insulin the primary source of therapeutic insulin. Insulin (Ins) may be crystallized as zinc–insulin crystals and in solution, in the presence of zinc, will form hexameric insulin–zinc (Ins_6Zn_2) species. Most insulin formulations contain some amount of zinc, as well as preservatives and agents to maintain isotonicity. This discussion focuses on HPLC methods that were developed for the characterization of bulk human insulin material and formulations and for product stability evaluation. This includes methods for the evaluation of product identity, purity, and content/potency.

TABLE II (*continued*)

Biosynthetic protein	Analysis objective	Comments	Refs.
Tissue plasminogen activator (t-PA)	Separation of single/two-chain forms	Disulfides reduced prior to injection	31
Factor VIII	Determination of related substances	90% formic acid mobile phase used to dissociate aggregated forms	32
Polyethylene glycol (PEG)-modified proteins	Evaluation of extent of PEG incorporation	Various naturally derived proteins derivatized with PEG	33
Ion-exchange HPLC			
hGH	Comparison to RP-HPLC for purity evaluation	RP-HPLC found to be more discriminating	14
hGH	Determination of deamidated forms	NaCl gradient; neutral pH mobile phase; DEAE-5PW column	16
Hirudin	Purity evaluation; purification	NaCl gradient; neutral pH mobile phase; Mono Q column	20
IgM monoclonal antibody	Purification; purity evaluation	Mono Q column; fractions reinjected onto mixed-mode ABx column for purity assessment	34
Human proinsulin fusion protein	Quantitation of fermentor yield	In-process monitoring of crude fermentation broth using 2-dimensional HPLC (SEC and IEC)	35
Immobilized-metal affinity chromatography (IMAC)			
Various proteins and peptides	Evaluation of technique	Good example of IMAC characteristics	36

The objective of an identity method is to verify that the correct chemical species is present in the product. One must take into consideration other similar products from which human insulin must be differentiated, such as other insulins. RP-HPLC readily separates human insulin from pork and beef insulin, which differ from human by one and three amino acids, respectively (Table III), and from an analog of human insulin in which the B28 and B29 residues are transposed (Fig. 2) (A. M. Korbas, unpublished data, 1994).

The purity of the human insulin drug product is determined through the use of several HPLC assays. These assays were developed on the basis of the type of related substances that are expected to form during the

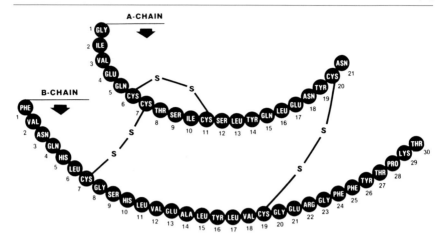

Fig. 1. Structure of human insulin.

manufacture and storage of insulin. On the basis of the insulin structure, there are several possible deamidation sites and the possibility of formation of covalent high molecular weight proteins (mostly dimeric species). Deamidation was determined to occur at the A21 position under acidic pH conditions and at the B3 position under neutral or basic pH conditions. Formation of other related substances is possible depending on the conditions.

A high-performance size-exclusion chromatography (SEC) method was developed to monitor the formation of covalent high molecular weight proteins (HMWPs). Covalent HMWPs have little bioactivity and may also be immunoreactive. To determine the covalent HMWPs, the sample preparation and HPLC mobile-phase conditions were developed to dissociate the insulin–zinc complex. Sample treatment is critical to the separation, particularly in the case of insulin formulations that may contain high levels of zinc or the highly basic modifying protein–protamine. In these cases,

TABLE III
CHEMICAL PROPERTIES OF VARIOUS INSULINS

Insulin species	Formula (atomic)	Mass (Da)	Amino acid at position:		
			A8	A10	B30
Human	$C_{257}H_{383}N_{65}O_{77}S_6$	5808	Thr	Ile	Thr
Porcine	$C_{256}H_{381}N_{65}O_{76}S_6$	5778	Thr	Ile	Ala
Bovine	$C_{254}H_{377}N_{65}O_{75}S_6$	5734	Ala	Val	Ala

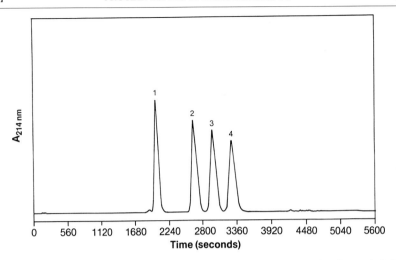

FIG. 2. RP-HPLC identity chromatogram of insulins. Peak 1, beef insulin; peak 2, Lys (B28) Pro (B29) biosynthetic human insulin; peak 3, biosynthetic human insulin; peak 4, pork insulin. Column: Vydac Protein & Peptide C_{18}, 25×0.46 cm. Chromatographic conditions: A, 18% Acetonitrile–82% 0.20 M sodium sulfate, pH 2.3 (adjusted with phosphoric acid); B, 50% acetonitrile–50% 0.20 M sodium sulfate, pH 2.3 (adjusted with phosphoric acid); flow rate, 1.0 ml/min; gradient, elute 19% B isocratically for 3600 sec, increase percent B linearly from 19 to 49% over 1380 sec; injection volume, 20 μl; detection, UV at 214 nm; column temperature, 40°.

EDTA or heparin, respectively, must be added to the sample prior to chromatography. Acetonitrile modifier was used in the SEC mobile phase to prevent noncovalent association. This is a demanding SEC separation because the relatively low molecular weights of the monomeric and dimeric insulin species place them at the lower limit of the molecular weight separation ranges of most SEC columns. Column selection and mobile phase development proved to be critical in attaining the necessary resolution. The SEC of a stressed human insulin sample demonstrates good resolution of monomeric and multimeric insulin species (Fig. 3).

An RP-HPLC purity method was developed to separate human insulin and its related substances, such as the major impurity, A21 desamidoinsulin (Fig. 4A). A21 desamidoinsulin elutes after the parent human insulin peak under isocratic conditions followed by more hydrophobic derivatives, including dimeric species, that are eluted in the gradient section of the chromatogram. The high molecular weight protein species are efficiently eluted in a group by using a fast, steep gradient because it is not necessary to separate individual covalent polymers for the purposes of this assay. If, however, one is interested in observing individual covalent polymeric species, such

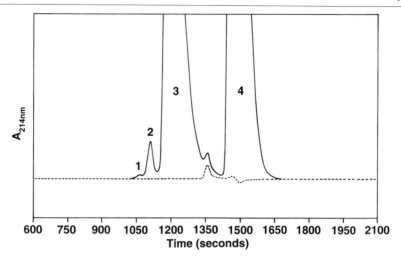

Fig. 3. SEC chromatogram of a formulation of biosynthetic human insulin after temperature stress (ca. 1% HMWP content). Peak 1, tetramer; peak 2, dimer; peak 3, monomer; peak 4, formulation preservative. Column: DuPont Zorbax GF250 S, 25 × 0.94 cm. Chromatographic conditions: 31% acetonitrile–69% 0.1 *M* ammonium phosphate, pH 7.5; flow rate, 0.5 ml/min; injection volume, 20 μl; detection, UV at 214 nm; column temperature, ambient. (——) Blank injection.

as for accelerated product stability investigations, the slope of the gradient may be decreased, thus allowing the peaks to be resolved, as shown in Fig. 4B (R. E. Heiney, unpublished data, 1992). The tradeoff is a significant increase in the assay time.

Another RP-HPLC purity method was developed for product stability evaluation. Under the neutral pH conditions of human insulin formulations, the asparagine at position 3 of the B-chain may be deamidated. The formation of an isoaspartyl derivative at this position is also possible. Under the acidic mobile-phase conditions described previously, the B3 desamidoinsulin species are not resolved from the parent human insulin (Fig. 5A). Selection of a neutral pH mobile phase based on ammonium phosphate permits good resolution of the B3 and isoaspartyl B3 desamidoinsulins from human insulin in a stressed human insulin product (Fig. 5B). This demonstrates how the selectivity changes drastically as the pH is changed about the p*I* of insulin (pI ≅ 5.8). Furthermore, the selectivity may be affected by the choice of buffer salt and mobile-phase modifier.

The insulin content of bulk drug product and formulations may be readily determined by RP-HPLC. In the case of biosynthetic human insulin, the purity of the material is so great that it may be considered as a highly purified chemical entity. An HPLC assay was developed that

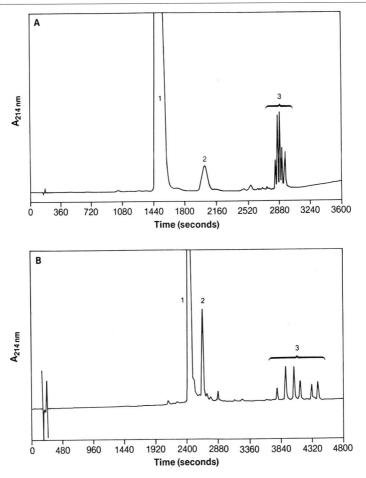

FIG. 4. RP-HPLC purity chromatograms of biosynthetic human insulin as a function of gradient conditions. Peak 1, biosynthetic human insulin; peak 2, human A21 desamido insulin; peak 3, human insulin high molecular weight proteins. (A) Column: Vydac Protein & Peptide C_{18}, 25 × 0.46 cm. Chromatographic conditions: A, 18% acetonitrile–82% 0.20 M sodium sulfate, pH 2.3 (adjusted with phosphoric acid); B, 50% acetonitrile–50% 0.20 M sodium sulfate, pH 2.3 (adjusted with phosphoric acid); flow rate, 1.0 ml/min; gradient, elute 22% B isocratically for 2160 sec, increase percent B linearly from 22 to 67% over 1500 sec, hold at 67% B for 360 sec; injection volume, 20 μl; detection; UV at 214 nm; column temperature, 40°. (B) Column: DuPont Zorbax C_8, 150 Å, 25 × 0.46 cm. Chromatographic conditions: A, 20% 1.125 M ammonium sulfate, pH 2.0–80% water; B, 40% acetonitrile–20% 1.125 M ammonium sulfate, pH 2.0–40% water; flow rate, 1.0 ml/min; gradient, increase percent B linearly from 60 to 76% over 4800 sec, increase percent B linearly from 76 to 100% over 360 sec, hold at 100% B for 720 sec; injection volume, 20 μl; detection; UV at 214 nm; column temperature: 40°.

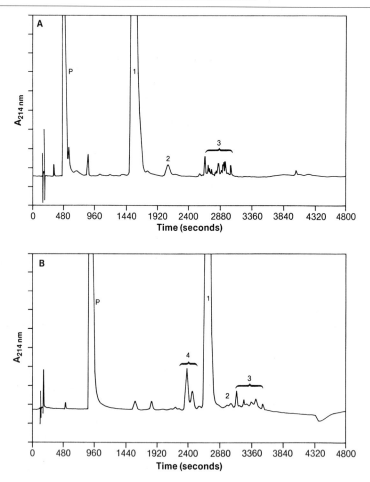

Fig. 5. RP-HPLC purity chromatograms of biosynthetic human insulin as a function of mobile-phase pH. Peak 1, biosynthetic human insulin; peak 2, human A21 desamido insulin; peak 3, human insulin high molecular weight proteins; peak 4, human B3 desamido insulins. P = preservatives. (A) Chromatographic conditions same as Fig. 4A. (B) Column: Beckman Ultrasphere ODS, 25 × 0.46 cm. Chromatographic conditions: A, 10% acetonitrile–50% 0.2 M ammonium phosphate dibasic, pH 7.0 (adjusted with phosphoric acid)–40% water; B, 50% acetonitrile–50% 0.2 M ammonium phosphate, dibasic, pH 7.0 (adjusted with phosphoric acid); flow rate, 1.0 ml/min; gradient, elute isocratically at 37% B for 900 sec, step to 40% B and elute isocratically for 900 sec, step to 43% B and elute isocratically for 900 sec, increase percent B linearly from 43 to 69% over 1200 sec, hold at 69% B for 300 sec; injection volume, 20 μl; detection, UV at 214 nm; column temperature, 40°.

TABLE IV
DETERMINATION OF HUMAN INSULIN POTENCY BY RABBIT
BIOASSAY AND HPLC

Zinc–insulin crystal lot summary	Bioassay (U/mg)	HPLC assay (U/mg)
Mean[a]	28.94	28.65
Range	3.5	1.2
Standard deviation	0.9	0.27
% RSD	3.1	0.9
Lots (n)	42	42

[a] Determined on a dried basis.

allowed the accurate and precise determination of the amount of parent insulin and its major related substance, A21 desamidoinsulin (90% relative bioactivity). The other related substances with significant bioactivity are present in insignificant amounts and higher molecular weight peptides of insulin have essentially no biological activity.[3] The content results (milligrams of insulin per milligram of protein) by HPLC were transformed to HPLC potency values (units of insulin activity per milligram of protein) using the maximum specific activity of human insulin as determined from bioassay of highly purified materials (168×10^6 IU/mol insulin, equivalent to 184 U/mg N or 28.85 U/mg[37]). These values were compared to results obtained by rabbit bioassay.[38] A good comparison is observed between the two assay systems (Table IV). HPLC has the further advantages of the separation of insulin species, greater precision, lower cost, and higher throughput compared to the bioassay. The use of RP-HPLC to control the content and potency of recombinant human insulin products permits tighter quality control and provides a more consistent product for the diabetic patient.

Biosynthetic Human Growth Hormone

Biosynthetic human growth hormone (hGH) has a rather large spectrum of potential degradation products as it contains 9 asparagine and 13 gluta-

[37] M. Pingel, A. Volund, E. Sorensen, and A. R. Sorensen, Assessment of insulin potency by chemical and biological methods. In "Hormone Drugs" (J. L. Gueriguian, E. D. Bansome, Jr., and A. S. Outschoorn, eds.), pp. 200–207.

[38] "U.S. Pharmacopeia," XXIInd Ed. United States Pharmacopeial Convention, Inc., Rockville, MD, 1990.

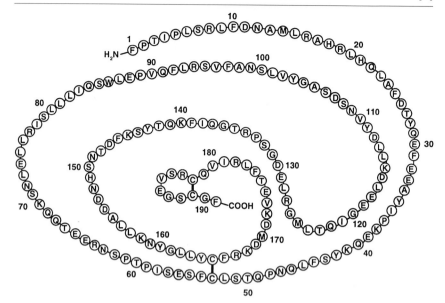

FIG. 6. Structure of human growth hormone (hGH).

mine residues, which could be deamidated, and 3 methionine residues, which could be oxidized (Fig. 6). In addition, this protein is known to form noncovalent dimeric and oligomeric species. Chemical derivatives of hGH differ from the parent molecule in either charge (e.g., desamido derivatives) or hydrophobicity (e.g., methionine sulfoxide derivatives). Consequently, charge separation techniques such as IEF and ion-exchange chromatography or hydrophobic chromatographic techniques such as RP-HPLC have proven to be useful tools for determining levels of such chemically modified forms.[15] The chromatograms of purified sulfoxide (in which the methionine at position 14 has been oxidized) and desamido (in which the asparagine at position 149 has been converted to an aspartic acid) derivatives along with undegraded hGH and an isolated hGH-related substances fraction on a reversed-phase HPLC column and an anion-exchange column are shown in Fig. 7. Figure 7A shows the sulfoxide and the desamido derivatives elute before undegraded hGH standard and are well resolved from it. The isolated related substances fraction is a mixture of approximately five components. The two major components of this mixture have the same retention times as the sulfoxide and the desamido derivatives, thereby indicating that the major degradation products of hGH are probably these derivatives. However, it is important to recognize that retention time alone is not sufficient to prove identity. As discussed in other chapters, liquid chromatography/

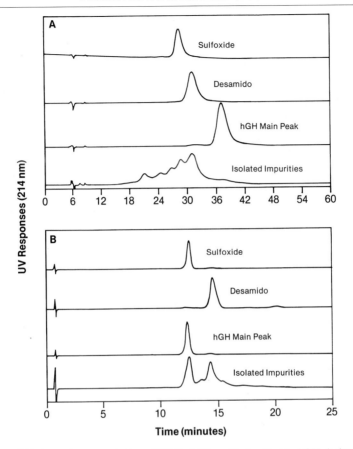

FIG. 7. (A) Isocratic reversed-phase HPLC of (1) purified sulfoxide hGH derivative, (2) purified desamido-hGH derivative, (3) undegraded hGH, and (4) isolated hGH impurities. (B) High-performance anion-exchange chromatography of the same four samples. (Reprinted from Ref. 14, with permission.) (A) Column: Vydac C$_4$, 25 × 0.46 cm. Chromatographic conditions: 29% n-propanol–71% 0.05 M Tris-HCl, pH 7.5; flow rate: 0.5 ml/min; injection volume, 20 μl; detection, at 214 nm; column temperature; 45°. (B) Column: Mono Q HR 5/5 column. Chromatographic conditions: A, 30% acetonitrile–70% 0.05 M Tris-HCl, pH 7.5; B, same as buffer A with the addition of 0.3 M NaCl; flow rate, 1.0 ml/min; gradient, hold 100% A for 3 min, increase percent B linearly from 0 to 100% over 25 min; injection volume, 50 μl; detection, UV at 214 nm; column temperature, 45°.

mass spectrometry (LC/MS) techniques are now available that can be used to confirm identity based on both molecular mass and chromatographic retention time.[7] In this particular work off-line MS analysis (using fast-atom bombardment ionization) was used to confirm the identities of the various related substances.[14] The same four samples were examined using anion-exchange chromatography (Fig. 7B). The desamido derivative has a later retention time than the undegraded hGH standard, demonstrating that this derivative is much more acidic than unmodified hGH. In contrast, the sulfoxide derivative has a retention time that nearly matches that of the hGH standard, i.e., the net charge on this derivative is no different than the standard. The isolated related substances fraction contains major components coeluting with the sulfoxide and desamido derivatives on this column, as was observed on the RP-HPLC column (Fig. 7A). These two complementary modes of separation strengthen the conclusion that sulfoxide and desamido derivatives of hGH are the principal degradation products in rhGH formulations.[14]

Determination of the best mode (isocratic vs gradient) for the RP-HPLC separation of hGH is a good example of what is required of an analytical method in assessing the purity of a protein and further illustrates the power of the isocratic mode of separation. The chromatographic profiles obtained for a typical biosynthetic hGH production lot under isocratic and gradient operating conditions are shown in Fig. 8A and B, respectively. As shown by the bottom chromatogram in Fig. 8A, no significant related substance peaks are evident in biosynthetic hGH when the main peak is shown on scale. However, when the main component is expanded off scale (as illustrated by the upper chromatogram in Fig. 8A), related substances identified as mono- and didesamido derivatives of hGH are observed at a total concentration of approximately 3%.[15] The lot of biosynthetic hGH shown in Fig. 8A, when assayed using gradient conditions, failed to reveal the presence of any related substances (Fig. 8B). An alternate gradient RP-HPLC procedure reported in the literature[39] was found to give results comparable to those in Fig. 8B. These results illustrate the fundamental rule that the assessment of purity for a protein product is significantly affected by the selectivity of the analytical method.

In addition to the chemically related substance associated with hGH, dimers and higher molecular weight oligomers are observed.[29] These "size variants" can be either covalent (e.g., disulfide or amide linkages) or noncovalent. Noncovalent variants can exist through an association of hydrophobic regions of two or more proteins molecules or by ionic attraction. It is therefore necessary to evaluate both dissociating and nondissociating conditions when analyzing such variants in order to assess their biological properties relative to monomeric hGH. Figure 9 illustrates the effectiveness of using SEC-HPLC under nondissociating conditions in the isolation and characterization of the dimeric and oligomeric forms of hGH in a typical production lot.[30]

[39] W. J. Kohr, R. Keck, and R. N. Harkins, *Anal. Biochem.* **122,** 348 (1982).

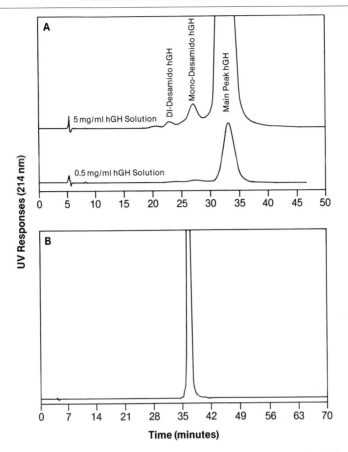

FIG. 8. (A) Isocratic and (B) gradient RP-HPLC of a typical biosynthetic hGH lot. (Reprinted from Ref. 15, with permission.) (A) Column: Vydac Protein Pak C_4. Chromatographic conditions: 71% 0.05 M Tris, pH 7.5–29% n-propanol; flow rate, 0.5 ml/min; injection volume, 20 μl; detection, UV at 220 nm; column temperature, 45°. (B) Column: Vydac Protein Pak C_4. Chromatographic conditions: A, 0.05 M Tris, pH 7.5; B, n-propanol; flow rate, 0.6 ml/min; gradient, start with 15% B, increase percent B linearly from 15 to 50% over 35 min; injection volume, 20 μl; detection, UV at 220 nm; column temperature, 45°.

In addition to the chromatographic assays employed in the determination of purity of biosynthetic hGH, the potency assay for hGH is also based on HPLC methodology. The hGH potency assay provides a good example of the care that must be taken in choosing the proper chromatographic mode for analysis. RP-HPLC was found not to be an appropriate technique for establishing the potency of hGH, because hGH preparations contain biologically inactive, noncovalent dimer that is dissociated to monomeric

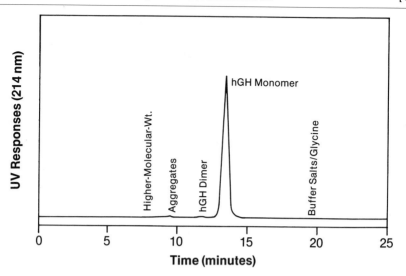

Fɪɢ. 9. SEC chromatogram of biosynthetic human growth hormone. Column: TSK G3000SW, 60 × 0.76 cm. Chromatographic conditions: 0.025 M ammonium bicarbonate, pH 7.6; flow rate, 1.0 ml/min; injection volume, 20 μl; detection, UV at 214 nm; column temperature, ambient.

hGH under RP-HPLC conditions.[29] On the other hand, size-exclusion HPLC was found to be an appropriate technique because both covalent and noncovalent dimer are separated from monomeric hGH using this technique (Fig. 9). The size-exclusion HPLC profile of the purified hGH dimer obtained from an hGH process sidestream is shown in the upper chromatogram as compared to the monomeric hGH profile in the lower chromatogram of Fig. 10A. Chemical derivatives of hGH (e.g., sulfoxide and desamido) are not separated from hGH monomer by this SEC-HPLC method. However, such derivatives were shown to have biopotencies equivalent to unmodified hGH and therefore are included in the potency calculation[30] (the low level of chemically related substances is determined by a separate RP-HPLC purity assay). The hGH dimer was also examined by sodium dodecyl sulfate-polyacrylamide gel electrophoresis (SDS–PAGE), a traditional assay technique for evaluating purity. As shown in Fig. 10B the dimer was indistinguishable from monomeric hGH, and only a small portion of material eluted at the position expected for dimeric hGH (M_r approximately 44,000). A similar result was obtained using native PAGE, indicating that the dimer is dissociated to monomeric hGH under these conditions, as well. These data indicate that size-exclusion HPLC data are critical to the assessment of the quality of commercial preparations of

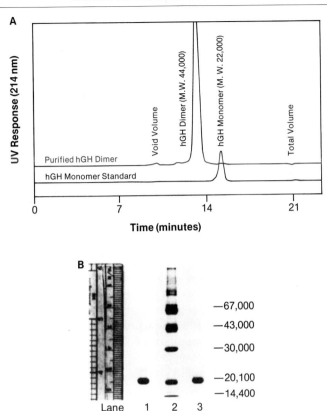

FIG. 10. (A) HP-SEC chromatogram of purified human growth hormone dimer fraction under nondissociating conditions. (B) SDS–PAGE of purified human growth hormone dimer. (Reprinted from Ref. 29, with permission.) (A) Same conditions as in Fig. 9. (B) Lane 1, purified dimer; lane 2, molecular weight calibration standards; lane 3, monomeric hGH reference standard. Electrophoresis system: A discontinuous buffer system using a vertical slab gel unit. Electrophoretic conditions: Run under nonreducing conditions; sample, 2.0 mg/ml in 0.38 *M* Tris-HCl, pH 8.8, containing 20% glycerol and 4% SDS. A 250-μl aliquot of this solution was added to 200 μl of reagent water and 50 μl of a 0.01% bromphenol blue tracking dye solution. This solution was heated for 90 sec in a boiling water bath and then a 3-μl aliquot was assayed.

hGH, because the commonly employed electrophoretic techniques do not distinguish between noncovalent dimeric forms of hGH and the desired monomeric hGH product. In addition to the selectivity, the SEC-HPLC potency assay was found to correlate well with the conventional bioassays while offering much greater precision than bioassays or immunochemical assays. Other techniques based on light scattering (e.g., low-angle laser light scattering, or LALLS) coupled with HPLC can also provide useful information concerning the molecular size of such noncovalent dimers and aggregates.[40]

Purity/Microheterogeneity of Modified Tissue Plasminogen Activator

In this example the protein to be characterized was a 358-residue glycoprotein containing 2 potential sites for glycosylation (positions 12 and 279) and 9 disulfide bridges (structure shown in Ref. 27). On the basis of analogy to the full-length tissue plasminogen activator, the Arg^{106}-Ile^{107} bond was expected to be susceptible to proteolytic cleavage (autolysis and/or by plasmin), resulting in a "two-chain" form of the molecule. To develop an appropriate purity assay for this complex molecule, a detailed assessment of the chemical structures present in the product was made, and several potential assay systems were evaluated. Because the two-chain form of the molecule has different biological properties than the single-chain form, development of a method to establish quantitatively the level of two-chain plasminogen activator (PA) was given high priority. Ultimately a hydrophobic interaction chromatography (HIC) method was found to be suitable for this purpose. Representative chromatograms, as well as the experimental conditions, are shown in Fig. 11. Typical production lots of PA contained two major peaks, with a trace amount of two earlier eluting peaks (the first peak is barely visible in most cases). Milligram quantities of each of the four components were isolated and further characterized. Reinjection of the isolated fractions demonstrated each peak to be a unique component (not an artifact of the separation process). Amino acid compositions for all four peaks were identical, thereby demonstrating that each peak was PA related. N-Terminal sequence analysis and SDS–PAGE, under reducing conditions, demonstrated that peaks 1 and 2 were two-chain forms of PA (cleaved between positions 106 and 107) whereas peaks 3 and 4 were single-chain forms. Treatment of peak 4 with plasmin resulted in the formation of peak 2, and treatment of peak 3 with plasmin resulted in the formation of peak 1. These results demonstrate that peak 1 is a two-chain form of peak 3, with peaks 2 and 4 having a similar relationship. Monosaccharide analysis of peaks 3 and 4 confirmed that peak 3 contained approximately twice the monosaccharide composition of peak 4. N-Terminal sequence analysis confirmed the presence of asparagine at position 12 (a potential glycosylation site) in peak 4, but not in peak 3, indicating that the former is not glycosylated at this site. Subsequent peptide-mapping studies confirmed that peak 3 was glycosylated at both potential glycosylation sites, whereas peak 4 was glycosylated only at one site (position 279).

[40] H. H. Stuting and I. S. Krull, *J. Chromatogr.* **539,** 91 (1991).

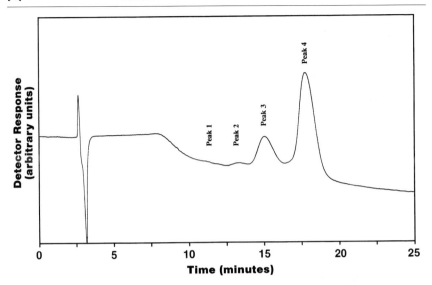

FIG. 11. HIC chromatogram of a typical PA lot. Separation performed on a TSK Phenyl 5PW column. Mobile phase A, 1 M potassium phosphate–2 M urea–0.02% sodium azide (pH 8.5). Mobile phase B same as A, except potassium phosphate concentration was 0.02 M. A 200-μl volume of a 1-mg/ml solution of PA in mobile phase A was injected. The gradient was held at 20% B for 5 min, followed by an increase to 50% B over 1 min and then a linear increase from 50 to 100% B over the next 15 min. Detection was by UV at 280 nm. (Reprinted from Ref. 27, with permission.)

This study demonstrates that extreme care must be taken in developing and validating methods for determining purity, or microheterogeneity, or complex protein products. One must also guard against inaccurate interpretation of the data. For example, in the PA product described earlier, each "peak" is actually a family of components having a variety of oligosaccharide structures. Hence the "purity" value that one obtains by this assay would not be expected to correlate with "purity" determined by another technique in which the various glycoforms are separated (e.g., anion-exchange chromatography). The goal of the assay and the meaning of the results must be clearly communicated in order to avoid misinterpretation.

Discussion

HPLC is clearly a powerful technique for characterization and routine quality control of biosynthetic proteins. Its merits include high resolution, wide applicability, ease of automation, and quantitative capability. In most cases HPLC offers better quantitation and higher resolution than conventional protein separation techniques, such as electrophoresis. HPLC is much more widely applied to biosynthetic protein characterization than is CE, although the latter technique offers high separation efficiency. The key advantages of HPLC, relative to CE, include the availability of a wider range of separation modes, more stable operating performance, and better dynamic range (due to sample capacity limitations for CE). However, CE serves a key role in protein assay development as an "orthogonal" separation technique that can be used to validate RP-HPLC methods.

One essential point to remember relative to HPLC analysis of biosynthetic proteins is the requirement for a multidimensional (or multitechnique) approach to both methods development and quality control. Biosynthetic proteins are usually of high purity. HPLC methods are generally used to separate and quantify relatively small quantities of proteins (related substances) that are structurally similar to the parent biosynthetic protein; hence high resolution and high specificity are required. Validation that sufficient resolution is achieved by a particular HPLC method requires corroborating evidence from other techniques, such as HPLC/MS, capillary zone electrophoresis (CZE), and/or dissimilar HPLC separation modes. Whenever possible, resolution should be verified through the use of authentic samples of the related substances of interest.

Future Developments

Advances in HPLC methodology should continue at a rapid pace and will enhance the use of this technology for protein characterization. Several developments are expected to result from the need to increase assay speed; currently some separations require 2 hr per assay. Greater assay speed has been demonstrated using nonporous particles,[41] small porous particles (e.g., 3 μm), and standard diameter particles having large conduit pores.[42] As more experience is gained with these systems, and a wider range of separation media becomes available, such high-speed separations should become more widely used for biosynthetic protein analysis. Advances in automated methods development should also improve analytical efficiency and allow for more comprehensive development of fully optimized methods.

[41] Y. Kato, S. Nakatani, T. Kitamura, Y. Yamasaki, and T. Hashimoto, *J. Chromatogr.* **502**, 416 (1990).
[42] N. B. Afeyan, N. F. Gordon, I. Mazsaroff, L. Varady, S. P. Fulton, Y. B. Yang, and F. E. Regnier, *J. Chromatogr.* **519**, 1 (1990).
[43] Deleted in proof.

The rapid advances in combined HPLC/MS, especially the advent of electrospray mass spectrometry (ESMS), have already impacted HPLC analysis of biosynthetic proteins.[7] This trend is expected to continue, and probably accelerate, as refinements are made and lower cost systems become available. Most companies now developing biosynthetic proteins for commercial purposes have access to, and are employing, HPLC/MS (either directly coupled, or off-line) for protein characterization as part of the overall research and development process.

[5] Membrane Proteins

By DJURO JOSIC and KATRIN ZEILINGER

Introduction

Depending on the type of interaction, membrane proteins are either embedded into a lipid bilayer or associated with membrane structures, in the latter case usually by ionic interaction or by hydrogen bonds. To isolate and analyze these proteins, they must be extracted from the membrane structures (i.e., solubilized). Solubilization is mostly carried out with detergents. The guidelines for membrane solubilization using detergents, as described by Helenius and Simons,[1] are still in principle valid. Chaotropic reagents such as urea and guanidine hydrochloride are now used less frequently. Diluted sodium hydroxide and sodium bicarbonate (between pH 10 and pH 12), concentrated salt solutions, and complex-forming substances, such as EDTA and ethylene glycol-bis(β-aminoethylether)-N,N,N',N'-tetraacetic acid (EGTA), are used for solubilization of extrinsic membrane proteins. Other methods, e.g., repeated freezing and thawing, can be applied to dissolve the structures mechanically, thereby allowing most membrane-associated proteins to be removed.[2–4]

The effort necessary for solubilization of the proteins increases with growing complexity of the membrane structure. Through the use of various reagents in different steps, the membrane proteins can be preseparated

[1] A. Helenius and K. Simons, *Biochim. Biophys. Acta* **415,** 29 (1975).
[2] L. M. Hjemeland and A. Crambach, *Methods Enzymol.* **104,** 305 (1984).
[3] J. van Renswoude and C. Kempf, *Methods Enzymol.* **104,** 329 (1984).
[4] D. Josic, W. Schütt, R. Neumeier, and W. Reutter, *FEBS Lett.* **185,** 182 (1985).

METHODS IN ENZYMOLOGY, VOL. 271

according to their respective solubility and hydrophobic characteristics. This in turn allows further separation by use of different, mainly chromatographic and electrophoretic, methods.

A complete solubilization scheme for plasma membranes of liver and Morris hepatomas has been developed. The scheme includes a combination of several reagents such as detergents, diluted sodium hydroxide, and highly concentrated salt solutions, in addition to freezing and thawing. This kind of treatment is the best possible preparation of membrane proteins from the sample for subsequent chromatographic and electrophoretic separation.[4,5] If possible, the proteins should also retain their biological activity.

By selective extraction, a preseparation of membrane proteins according to their hydrophobic characteristics is achieved. The pretreatment facilitates subsequent separation and provides a first guideline for the choice of detergent. When choosing the detergent for the running buffers, one should use the one with which the protein was solubilized, if possible. It is also advisable to use less denaturing detergents such as 3-[(3-cholamidopropyl)-dimethyl-ammonio]-1-propanesulfonate (CHAPS) or octylglucoside for solubilization and likewise as an additive to the running buffers. Thereby the chance of retaining the biological activity is increased.

There are further difficulties that must be taken into consideration, especially when membrane proteins are separated, namely a tendency toward association and aggregation and the possibility of nonspecific interactions with the support.

In this chapter, to illustrate the strategy for high-resolution separation of membrane proteins, the chromatographic and electrophoretic separation of proteins from the plasma membranes of liver and Morris hepatomas is shown. Three groups of proteins were chosen with different degrees of hydrophobicity. Membrane-associated annexins [calcium-binding protein (CBP) 65/67, CBP 35, and CBP 33] are less hydrophobic and can be solubilized without using detergents. These proteins can be phosphorylated, but they are not glycosylated.[6] Dipeptidyl-peptidase IV (DPP IV; EC 3.4.14.5) and the cell–cell adhesion protein cell-CAM (cell adhesion molecule, or gp110) are intrinsic membrane proteins, and they are also glycosylated.[7] Other membrane proteins, e.g., from eukaryotic cells or of bacterial and viral origin, can be assigned to one of the above-mentioned groups in accordance with their behavior in chromatographic as well as electrophoretic separations.

[5] D. Josic, H. Baumann, and W. Reutter, *Anal. Biochem.* **142,** 473 (1984).
[6] Y.-P. Lim, M. D. Thesis, Freie Universität Berlin, Berlin, Germany, 1991.
[7] A. Becker, Ph.D. Thesis, Freie Universität Berlin, Berlin, Germany, 1989.

Strategies Having General Validity

Size-Exclusion High-Performance Liquid Chromatography

With size-exclusion high-performance liquid chromatography (SE-HPLC), a satisfactory resolution in membrane protein separation can hardly be achieved under nondenaturing conditions. As the basis of this method of separation is the size of the molecules, nonspecific interactions between sample and support and among different sample components must be suppressed. This requires the addition of denaturing agents. The interactions are reduced considerably by chaotropic reagents or sodium dodecyl sulfate (SDS). Under denaturing conditions, e.g., when SDS is added, resolution and yield are optimized in SE-HPLC, along with good reproducibility of results.[5]

Reversed-Phase High-Performance Liquid Chromatography

Reversed-phase HPLC offers the best resolution of all chromatographic methods. However, it is rarely used for separating membrane proteins. Under the separation conditions usually required in reversed-phase HPLC, i.e., the use of an acidic application buffer/phosphate buffer or 0.1% (v/v) trifluoroacetic acid and elution with an acetonitrile gradient or 2-propanol gradient, the hydrophobic proteins can be recovered only partly, sometimes not at all. The addition of highly concentrated, organic acids, above all formic or acetic acid, to both buffers, was first proposed for the separation of hydrophobic proteins.[8] Heukeshoven and Dernik have developed and used this method for the separation of hydrophobic proteins from poliovirus.[9] With regard to their hydrophobic characteristics and their behavior in chromatographic separation, these proteins are similar to membrane proteins. When both the application buffer and the elution buffer (2-propanol) contained 60% (v/v) formic acid, even the proteins from different strains of poliovirus could be selectively separated. Welinder *et al.*[10] have shown that the addition of acidic acid improved resolution and yield considerably, in the separation of membrane proteins that had been isolated from erythrocyte ghosts. Columns with synthetic supports showed better performance than did columns with supports based on silica gel. A separation of membrane proteins from erythrocyte ghosts is shown in Fig. 1.

---

[8] W. Schwarz, J. Born, H. Tiedemann, and I. Molnar, Separation of proteins by SE- and RP-HPLC. *In* "Practical Aspects of Modern HPLC" (I. Molnar, ed.). Walter de Gruyte, Berlin, 1982.

[9] J. Heukeshoven and R. Dernick, *Chromatographia* **19,** 95 (1985).

[10] B. S. Welinder, H. H. Sörensen, and B. Hansen, *J. Chromatogr.* **462,** 255 (1989).

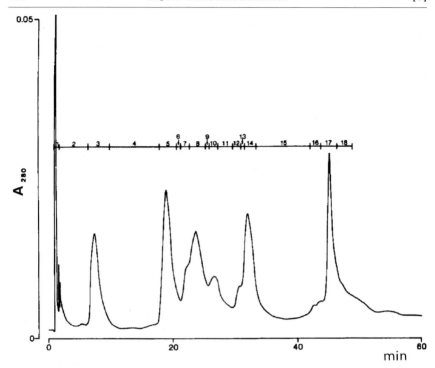

FIG. 1. Separation of SDS-solubilized erythrocyte membrane proteins by reversed-phase HPLC. Column, TSK Phenyl 5 PW RP (75 × 4.6-mm i.d.); buffer A, 20% acetic acid; buffer B, acetic acid–acetonitrile (40:60, v/v); gradient, 80% A (10 min), 80 to 50% A (35 min), 50 to 0% A (15 min), and 0% A (15 min); flow rate, 0.5 ml/min. The membrane proteins with higher molecular weights are eluted with a high concentration of acetonitrile (peak 3 and following). (Reprinted from *J. Chromatogr.*, **462**, B. S. Welinder, H. H. Sörensen, and B. Hanson. High performance liquid chromatographic separation of membrane proteins isolated from erythrocyte ghosts, pp. 255–268. Copyright 1989 with kind permission of Elsevier Science–NL, Sara Burgerhartstraat 25, 1055 KV Amsterdam, The Netherlands.)

The application of organic solvents and acids usually causes the proteins to lose their biological activity. However, RP-HPLC is a useful method with good resolution for the isolation of those membrane proteins, whose primary structure is to be investigated subsequently. The method is also used for analysis of membrane proteins for peptide mapping after proteolytic digestion.[11]

Ion-Exchange High-Performance Liquid Chromatography

Ion-exchange HPLC is a chromatographic method, the degree of resolution of which is surpassed only by RP-HPLC. Ion-exchange HPLC became

[11] U. Willems, "Membrane Proteins." Pharmacia monograph, pp. 1–18. Freiburg, Germany, 1990.

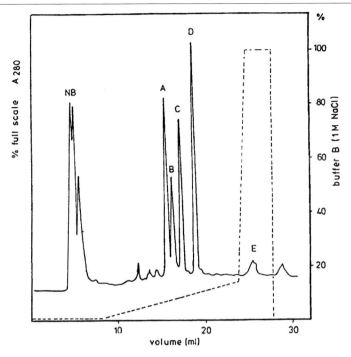

Fig. 2. Anion-exchange HPLC of the crude cell wall elicitor preparation (CEP), isolated from germ tube walls of the phytopathogenic fungus *Puccinia graminis* ureidospores. Six milligrams of CEP in 10 ml of Tris-HCl buffer, pH 8.5, containing 4% betaine, was applied to a Mono Q HR 5/5 column (Pharmacia-LKB). Elution was performed at a flow rate of 1 ml/min with a linear gradient. Peak C contains the highly enriched glycoprotein of interest. (Reprinted from *J. Chromatogr.*, **521**, B. Beissmann and H. J. Reisener. Isolation and purity determination of a glycoprotein elicitor from wheat stem rust by medium-pressure liquid chromatography, pp. 187–197. Copyright 1990 with kind permission of Elsevier Science–NL, Sara Burgerhartstraat 25, 1055 KV Amsterdam, The Netherlands.)

particularly useful for the separation of membrane proteins after supports on a polymer basis were introduced and the quality of supports based on agarose was improved.[12] Because of their hydrophilic characteristics, these supports have a lower level of nonspecific interactions with the sample. Therefore high resolution is achieved in many cases. The separation of plant cell wall proteins by anion-exchange HPLC is shown in Fig. 2. In the case of other membrane proteins, which are even more hydrophobic, interaction with the support is still too strong. Therefore other methods must be chosen, or detergents must be added to the separation buffers (see page 119).

[12] B. Beissmann and H. J. Reisener, *J. Chromatogr.* **521**, 187 (1990).

Other Chromatographic Methods

Hydrophobic interaction HPLC can also be used for the isolation of membrane proteins. However, interaction with hydrophobic groups must be prevented from growing too strong. Otherwise the sample can no longer be recovered from the column, not even through the use of detergents or, in extreme cases, by organic solvents.[13]

Affinity chromatographic methods and similar methods are often used for separating membrane proteins. Apart from lectin affinity chromatography, immunoaffinity chromatography is the most widely used method. As some membrane proteins have a receptor function, they can be bound specifically and eluted selectively through affinity chromatography with immobilized ligands. Transferrin receptor from plasma membranes of various mamalian cells binds with high affinity to immobilized diferric transferrin. However, its affinity for apotransferrin is low. Consequently the complex of transferrin and transferrin receptor can be dissociated by chelation of ferric ions after the addition of chelating reagents. In this way transferrin receptor can be eluted from the column under mild conditions.[14] The affinity chromatographic methods are described in Figs. 3 and 5 in more detail.

Electrophoretic Methods and High-Performance
  Membrane Chromatography

Methods of high-performance liquid chromatography, in which membranes and compact disks are used, are called high-performance membrane chromatography (HPMC). They have considerable potential because of their high resolution, short running times, and low nonspecific interactions, especially in the case of membrane protein separation.[15]

The number of reports dealing with the separation of membrane proteins by high-performance capillary electrophoresis is still rather small. First results indicate that practically all methods used for the separation of water-soluble proteins can also be applied to the separation of hydrophobic membrane proteins.[16] Apart from the guidelines mentioned above, the know-how obtained in high-performance capillary electrophoresis and in

[13] S. C. Goheen, HPLC purification of detergent-solubilized membrane proteins. *In* "HPLC of Proteins, Peptides and Polynucleotides" (M. T. W. Hern, ed.). VCH, New York, 1991.
[14] D. Schell, "Membrane Proteins." Pharmacia monograph, pp. 141–144. Freiburg, Germany, 1990.
[15] D. Josic, J. Reusch, K. Löster, O. Baum, and W. Reutter, *J. Chromatogr.* **590,** 59 (1992).
[16] D. Josic, K. Zeilinger, W. Reutter, A. Böttcher, and G. Schmitz, *J. Chromatogr.* **516,** 89 (1990).

the preparative free-flow isotachophoresis of serum lipoproteins can be transferred to the separation of membrane proteins.[17]

Solubilization of Liver and Morris Hepatoma Plasma Membranes

Animals and Reagents

Male or female Wistar or Buffalo rats (Institut für Molekularbiologie und Biochemie, Berlin, Germany), weighing about 160–180 g each, are fed a commercial diet, containing 18–20% (w/w) protein (Altromin R, Altromin; Lage, Lippe, Germany). Chemicals of analytical reagent grade are purchased from Merck (Darmstadt, Germany) or Sigma (Munich, Germany). All detergents are purchased from Sigma.

Plasma Membranes

Plasma membranes are isolated by zonal centrifugation using the method of Pflegler et al.[18] A Kontron centrifuge with zonal rotor is used (Kontron, Munich, Germany). Membrane purity is routinely checked by assays for marker enzymes, as described by Tauber and Reutter.[19] The protein content is determined according to the procedure of Lowry et al.[20]

Buffers and Solutions

Tris-buffered saline (TBS), containing 0.01 M Tris-HCl, pH 7.4, and 0.155 M sodium chloride, is used in the experiments. Depending on the particular step of solubilization, 1% (w/w) of one of the following detergents is added to TBS: CHAPS, Nonidet P-40 (NP-40), Triton X-100, Triton X-100 reduced, and Triton X-114. In the case of alkaline extraction, a 2 mM solution of sodium hydroxide is used. For calcium complexing 0.01 to 0.05 M sodium EDTA or sodium EGTA, pH 7.4, with 1% (w/v) octylglucoside or 1% (w/v) CHAPS is used.

Stepwise Extraction

Scheme I shows the stepwise solubilization of plasma membranes from liver and Morris hepatomas. This kind of solubilization is a useful method of sample preparation for subsequent electrophoretic and chomatographic separation. In the first step, freezing and thawing, the plasma membranes

---

[17] D. Josic, A. Böttcher, and G. Schmitz, *Chromatographia* **30,** 703 (1990).

[18] R. C. Pflegler, N. G. Anderson, and F. Snyder, *Biochemistry* **7,** 2826 (1968).

[19] R. Tauber and W. Reutter, *Eur. J. Biochem.* **83,** 37 (1978).

[20] D. H. Lowry, N. J. Rosenbrough, N. J. Farr, and R. J. Randall, *J. Biol. Chem.* **193,** 265 (1951).

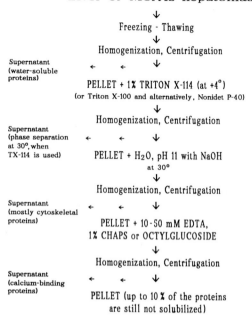

PLASMA MEMBRANES
Liver or Morris hepatomas

↓

Freezing - Thawing

↓

Homogenization, Centrifugation

Supernatant
(water-soluble
proteins)   ←   ←   ↓

PELLET + 1% TRITON X-114 (at +4°)
(or Triton X-100 and alternatively, Nonidet P-40)

↓

Homogenization, Centrifugation

Supernatant
(phase separation
at 30°, when   ←   ←   ↓
TX-114 is used)   PELLET + H_2O, pH 11 with NaOH
at 30°

↓

Homogenization, Centrifugation

Supernatant
(mostly cytoskeletal   ←   ←   ↓
proteins)   PELLET + 10-50 mM EDTA,
1% CHAPS or OCTYLGLUCOSIDE

↓

Homogenization, Centrifugation

Supernatant
(calcium-binding   ←   ←   ↓
proteins)   PELLET (up to 10 % of the proteins
are still not solubilized)

SCHEME I. Stepwise solubilization of plasma membranes.

are mechanically destroyed, and the membrane-associated proteins are solubilized. No detergents need be used in this step. However, as the proteins tend to aggregate, the addition before separation of a mild detergent such as CHAPS or octylglucoside is recommended.

In the second step, the hydrophobic membrane proteins are extracted by nonionic detergents such as NP-40, Triton X-100, or Triton X-114. Triton X-114 is water soluble at 4°, but forms water-insoluble micelles at 25°. These are detergent rich and can easily be separated by centrifugation from the aqueous, detergent-poor phase. This method of extraction, including the subsequent phase separation, was developed by Bordier[21] and has proven to be an efficient preparatory step for further separation of a large fraction of the membrane proteins. The proteins from both the detergent-poor and the detergent-rich phase can be subjected, after dilution with a cold buffer, to further chromatographic or electrophoretic separation, on a preparative as well as analytical scale.

The patterns of proteins extracted from plasma membranes of the liver

[21] C. Bordier, *J. Biol. Chem.* **256,** 1604 (1981).

differ clearly from those obtained from the strongly dedifferentiated, fast-growing Morris hepatoma 7777.[4] It must be pointed out that the protein concentration in the extract should not exceed 2 mg/ml, when phase separation is carried out with Triton X-114. With protein concentrations above this level, phase separation will be either incomplete or fail altogether. However, not all the membrane proteins follow this simple rule. Phase separation will cause a number of membrane proteins to be distributed in varying degrees between the detergent-rich and the detergent-poor phases.

The typical behavior of mixed distribution is shown by the intrinsic membrane protein DPP IV. Although this enzyme is an intrinsic membrane protein, it appears after phase separation with Triton X-114 in both phases, in the aqueous phase with almost one-third of its enzymatic activity and in the detergent phase with two-thirds of its enzymatic activity. In comparison, the hydrophobic, intrinsic membrane protein cell-CAM appears almost exclusively in the detergent-rich phase.[7] Although the method of phase separation does not indicate the localization of all proteins in the membrane proteins, it helps to separate the proteins according to their water solubility, before more complicated separation methods are applied.

After treatment with nonionic detergents, the pellet still contains non-solubilized proteins. In the case of liver plasma membranes, the amount of proteins that are still not solubilized is between 20 and 30%. A large part of these are membrane-associated proteins from the cytoskeleton. These proteins can be extracted by treatment with dilute sodium hydroxide at pH 11. The extraction should last between 10 and 30 min and should be carried out at room temperature. This step, i.e., alkaline extraction, is also run before the extraction by detergents, i.e., in other solubilization schemes, which were developed especially for the isolation and identification of membrane-associated, cytoskeletal proteins.[22] However, this kind of treatment can lead to partial denaturing and consequently to the loss of activity of some membrane proteins, extracted by Triton or NP-40. To avoid these consequences, especially in the case of several membrane-associated enzymes, Scheme I places the alkaline extraction after the extraction with nonionic detergent.

In the following step, the pellet is treated with a solution of 1% zwitterionic detergent CHAPS and with 0.01 to 0.05 M EDTA or EGTA. After pretreatment as described above, only a few proteins will remain, so that calcium-binding proteins can be highly enriched.[4,5] These calcium-binding proteins have been identified as membrane-associated annexins.[6]

[22] A. Hubbard and A. Ma, *J. Cell Biol.* **96**, 230 (1983).

Separation of Integral Membrane Proteins Dipeptidyl-Peptidase IV
and Cell-CAM

High-Performance Liquid Chromatography

The HPLC system consists of two pumps, a programmer, a spectrophotometer with a deuterium lamp, an RH 7125 loop injection valve (all from Knauer, Berlin, Germany), and a Frac-100 fraction collector (Pharmacia, Freiburg, Germany).

Protein recovery is determined by protein measurement with the method of Lowry *et al.*[20] Recovery of dipeptidyl peptidase IV is measured by determining the enzymatic activity.[23] Cell-CAM is determined in an enzyme-linked immunosorbent assay (ELISA), using monoclonal antibodies.[7]

Columns

The following columns are used: TSK 5PW DEAE for anion exchange, particle size 10 μm, pore size 100 nm, column dimensions 75 × 7.5 mm (Tosohaas, Stuttgart, Germany); crown ether, bound to Eupergit C30N, particle size 30 μm, pore size 50 nm, column dimensions 80 × 8.0 mm (Säulentechnik Knauer, Berlin); concanavalin A (ConA), wheat germ agglutinin (WGA), arginine, and antibodies are immobilized to tresyl-activated TSK-Toyopearl, 30-μm particle size, 100-nm pore size (Tosohaas), column dimensions 60 × 8.0 mm, unless otherwise stated in the figure captions. The immobilization procedure has been described earlier.[15]

Buffers

Buffers Used for Anion-Exchange HPLC and Arginine-High-Performance Affinity Chromatography (HPAC)
Buffer A: 0.01 *M* Tris-HCl, pH 7.4, containing 0.1% (v/v) Triton
X-100, reduced
Buffer B: 0.01 *M* Tris-HCl, pH 7.4, containing 1 *M* NaCl and 0.5%
(v/v) Triton X-100, reduced

Buffers Used for Crown Ether Affinity Chromatography
Buffer A: 0.01 Tris-HCl, pH 7.4, containing 0.02 *M* KCl and 0.1%
(v/v) Triton X-100, reduced
Buffer B: 0.01 *M* Tris-HCl, pH 7.4, containing 1 *M* NaCl and 1%
(v/v) Triton X-100, reduced

Before each run the column is saturated by a 5-ml injection of 0.5 *M*
KCl solution.

---

[23] W. Kreisel, R. Heussner, B. Volk, R. Büchsel, and W. Reutter, *FEBS Lett.* **147,** 85 (1982).

Buffers Used for ConA-HPAC and WGA-HPAC
    Buffer A: 0.01 M Tris-HCl, pH 7.4, containing a 1 mM concentration
        each of Ca^{2+} and Mg^{2+} and between 0.1 and 0.5% (v/v) Triton
        X-100, reduced
    Buffer B: Buffer A containing a 0.02 to 0.2 M concentration of
        α-methyl-D-glucopyranoside or α-methyl-D-mannopyranoside for
        ConA-HPAC or 0.2 M N-acetyl-D-glucosamine for WGA-HPAC

Buffers Used for Immunoaffinity HPLC
    Buffer A: TBS, pH 7.4, containing between 0.1 and 0.5% (v/v) Triton
        X-100, reduced
    Buffer B: 0.2 M sodium citrate, pH 2.4, containing 0.5% (v/v) Triton
        X-100, reduced

Electrophoretic Methods

    For sodium dodecyl sulfate–polyacrylamide gel electrophoresis (SDS–
PAGE), dialyzed and freeze-dried samples are dissolved in 62.5 mM Tris-
HCl buffer, pH 6.8, containing 3% (w/v) SDS, 5% (v/v) 2-mercaptoethanol,
10% (v/v) glycerol, and 0.001% (w/v) bromphenol blue. The experiments
with SDS–PAGE are carried out by the Laemmli method,[24] using equipment obtained from Bio-Rad (Munich, Germany). Between 100 and 150
μg of protein in the case of complex samples, or between 5 and 10 μg for
samples that contain less than five components, is applied to each track.
    Two dimensional electrophoresis is performed by the method of O'Far-
rel.[25] For Western blotting, a method close to the procedure of Towbin *et
al.*[26] has been chosen. A nitrocellulose membrane (Schleicher & Schüll,
Düsseldorf, Germany) is used. The glycoproteins are detected with ConA
peridoxase or WGA peridoxase (Sigma). Single proteins are detected after
incubation with specific monoclonal or polyclonal antibodies, in the subsequent step, with gold-coupled anti-mouse antibodies or anti-rabbit antibodies (BioTrend, Cologne, Germany).

High-Performance Capillary Isotachophoresis

    For all separations by high-performance capillary isotachophoresis
(HPCITP) a Beckman (Munich, Germany) system is used. The experiments
are also carried out with a system developed by Schmitz and co-workers.[17]
PTFE capillaries (20 cm $\times$ 200 to 400-μm i.d.; Labochrom, Sinsheim, Germany) are used for HPCITP. As the leading electrolyte in ITP, 5 mM H_3PO_4

[24] U. K. Laemmli, *Nature (London)* **227,** 680 (1970).
[25] P. H. O'Farrel, *J. Biol. Chem.* **250,** 4007 (1975).
[26] H. Towbin, T. Staehlin, and J. Gordon, *Proc. Natl. Acad. Sci. U.S.A.* **76,** 4350 (1979).

is used. The electrolyte contains 0.25% (w/v) hydroxypropylmethylcellulose (HPMC) in order to increase the viscosity and to suppress electroendoosmotic movement in the capillary. Ammediol (2-ammino-2-methyl-1,3-propanediol) is added as a counterion to reach pH 9.2. The terminating electrolyte contains 100 mM valine and is adjusted with ammediol to pH 9.4.

The HPCITP experiments with the system from Beckman is carried out at a voltage of 5 to 6 kV; the current is between 240 μA (beginning) and 50 μA (end of separation). With the system developed by Schmitz and coworkers,[17] the separation is started with a constant current of 150 μA, and during the 10-min run the current is reduced to 100 μA (the corresponding voltage is 7 kV) and subsequently to 50 μA (6 kV) before detection.

The proteins are monitored spectrophotometrically at 280 nm. The temperature during the separation is kept at a constant 20° (Beckman apparatus) or 10° (system by Schmitz and co-workers[17]).

Isolation Procedure

In Scheme II the isolation of DPP IV and of cell-CAM is shown. By stepwise extraction of plasma membranes of the liver, about 90% of the enzymatic activity of DPP IV is solubilized with the nonionic detergent Triton X-100 or Triton X-114. The use of these detergents also allows the almost complete solubilization of cell-CAM. In phase separation by Triton X-114, DPP IV is not quantitatively separated in the aqueous or the detergent-rich phase. As far as the adhesion protein cell-CAM is concerned,

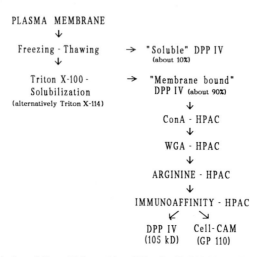

SCHEME II. Isolation of dipeptidyl peptidase IV and cell-CAM from the plasma membranes of rat liver.

more than 90% of the protein is found in the detergent-rich phase. However, the protein is still not separated from DPP IV. To increase the yield of DPP IV and at the same time achieve a separation of the cell-CAM, phase separation after Triton X-114 extraction is omitted from the isolation scheme. To allow the spectrophotometric detection of the proteins at 280 nm, reduced samples of Triton X-100 or Triton X-114 are used. They have low background absorbance at this wavelength.

If only the isolation of DPP IV is intended, anion-exchange HPLC with a DEAE column can be the first isolation step, with DPP IV appearing in the middle of the gradient as a rather broad peak. The much more hydrophobic, highly glycosylated cell-CAM binds to the column. However, it is recovered only at low yield (the chromatograms are not shown here). This is typical of the behavior of hydrophobic glycoproteins and therefore of the difficulties arising with their isolation. Despite improved hydrophilic supports, their nonspecific interaction with the matrix is difficult to control. Also, there is the problem of microheterogeneity in the case of highly glycosylated proteins. These two factors render the separation of some glycoproteins by ion-exchange HPLC, in this case cell-CAM, rather ineffective, if nondenaturing conditions are chosen for the process. Less hydrophobic proteins (e.g., proteins from plant cell walls) and DPP IV, which are all glycosylated to a lesser degree, could be separated much better (cf. Fig. 2).

Figure 3 shows ConA-HPAC of a Triton X-100 extract from plasma membranes of the liver. This step is effective, leading to considerable enrichment of both proteins. In the case of liver, further glycoproteins are observed, as shown in the ConA blot in Fig. 3b, part C. The glycoprotein cell-CAM is eluted together with DPP IV and can be detected by immunoblot in the region between 100 and 120 kDa (not shown here). Elution behavior of DPP IV from Morris hepatoma 7777 is different. Here the enzymatic activity can be eluted only with 0.2 M methyl-α-D-mannopyranoside.[27] This indicates a different glycosylation of the enzyme in malignant cells. Cell-CAM appears only in traces in hepatoma 7777, and here it cannot be detected. Further enrichment of the proteins is achieved by WGA-HPAC, arginine-HPAC or, as an alternative, crown ether HPAC. The chromatograms with their respective SDS–PAGE runs are shown in Figs. 4 and 5. However, the separation between DPP IV and cell-CAM is not possible with any of the methods used so far. This is shown in Fig. 5b. It is not clear whether the aggregation has a physiological background. Preliminary investigations indicate that this is not the case.[27] The cell-CAM can subsequently be separated by immunodepletion with immobilized antibodies. A

[27] D. Josic, A. Becker, and W. Reutter, *In* "HPLC of Proteins, Peptides and Polynucleotides" (M. T. W. Hearn, ed.), p. 469. VCH Publishers, New York, 1991.

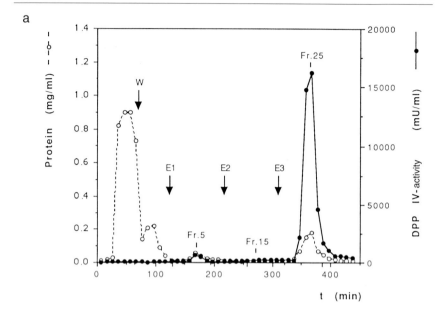

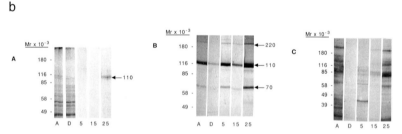

FIG. 3. Concanavalin A-HPAC of a Triton X-100 extract from plasma membranes of liver. (a) Chromatography: 40 ml of Triton X-100 extract (protein content, 48 mg; DPP IV activity, 600 mU) was applied to a ConA-HPAC column. The column was subsequently washed with 80 ml of 1% (v/v) Triton X-100 in the application buffer (W). The bound glycoproteins were eluted in three steps, with the following eluents used in succession: (1) 0.2 M α-methyl-D-glucopyranoside in the application buffer (E1); (2) 0.2 M α-methyl-D-mannopyranoside in the application buffer (E2); (3) 0.2 M α-methyl-D-mannopyranoside in the application buffer, to which 1% Triton X-100 had been added (E3). Single fractions of 10 ml each were collected. After dialysis against 0.05% Triton X-100 in TBS, pH 8.0, 25 μl was taken from each eluate for SDS–PAGE. Column, Toyopearl ConA (250 × 30 mm). Chromatographic conditions: flow, 1 ml/min; pressure, 1–2 atm; room temperature. (b) Electrophoretic detection: SDS–PAGE, Coomassie blue staining (A), immunoblot (B), and ConA blot (C) of eluates from ConA-HPAC of plasma membranes of the liver. Quantities of 25 μl of the loading sample A, of the flow-through D, and of each of the fractions 5, 15, and 25 (Fig. 3a) were applied. For the immunoblot, polyclonal anti-DPP IV antibodies were used. The arrows mark the monomer and the dimer DPP IV at 110 and at 220 kDa, respectively. A cross-reacting protein is marked in the immunoblot at 70 kDa.

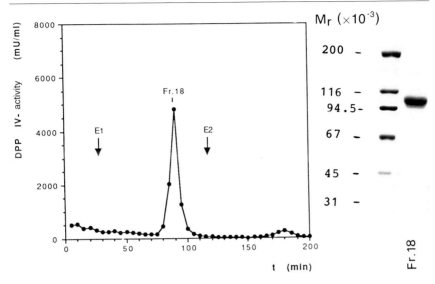

FIG. 4. Second purification step in the isolation process of DPP IV from plasma membranes of the liver under nondenaturing conditions, wheat germ agglutinin-HPAC. The eluate from the ConA-HPAC column with DPP IV activity (see Fig. 3) was applied after dialysis against the application buffer on a Toyopearl WGA-HPAC column (dimensions, 300 × 10.0 mm). The column was rinsed with 5 column volumes of application buffer, and the bound glycoproteins were eluted with 0.2 M N-acetylglucosamine in the application buffer. Other chromatographic conditions as in Fig. 3. The electrophoretic presentation of the eluted proteins is shown on the right-hand side.

high concentration of detergent in the buffer, 1% Triton X-100, is required to suppress aggregation. DPP IV does not bind to the anti-cell-CAM antibody column, and is obtained in a pure form in flowthrough (not shown here).

Figure 6 shows the HPCITP of a WGA eluate, after the second affinity chromatography step. In a single step the two hydrophobic membrane proteins were separated from one another by this kind of capillary electrophoresis, and a 95-kDa contamination was separated from both.

Membrane-Associated Proteins

Columns

In addition to the columns used for the isolation of intrinsic membrane proteins (see above), a hydroxylapatite column with a precolumn is used, dimensions 120 × 8.0 mm, particle size 10 μm, pore size 100 nm. The

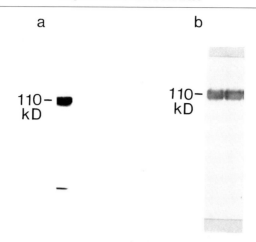

FIG. 5. Third purification step in the isolation process of DPP IV from plasma membranes of the liver under nondenaturing conditions using arginine HPAC. The eluate from WGA-HPAC was dialyzed against 5 mM Tris-HCl, pH 7.4, and applied to a Toyopearl-Arginine column. DPP IV was eluted with 0.2 M NaCl and 0.1% octylglucoside. (a) SDS–PAGE of the fraction containing the DPP IV activity. (b) Immunoblot with monoclonal anti-cell-CAM antibodies (antibody 9.2). The amount of eluted protein was about 8 mg, containing between 30 and 40% cell-CAM, after quantitative detection in an ELISA (cf. Ref. 7).

dimensions of the precolumn with the same material are 30 × 8.0 mm (Säulentechnik). For RP-HPLC a SynChropak RP4 column, dimensions 250 × 4.6 mm (Bischoff Analysentechnik, Leonberg, Germany) is used.

Buffers

For hydroxylapatite HPLC, the application buffer is used: buffer A with 1 to 10 mM sodium phosphate, pH 7.0, and elution buffer; buffer B with 0.5 M sodium-phosphate, pH 7.0, with or without addition of detergent. The other chromatographic conditions are listed in the figure captions. In RP-HPLC the application buffer is 60% (v/v) formic acid. Elution is carried out with a mixture of 60% (v/v) 2-propanol and 40% (v/v) formic acid.

Phase Separation with Triton X-114

The sample with CBP 65/67 is thoroughly dialyzed against TBS, pH 7.4, to remove octylglucoside. To the dialyzed sample 1% (v/v) Triton X-114 is added at 4°. Phase separation is carried out at 25°, as described by Bordier.[21]

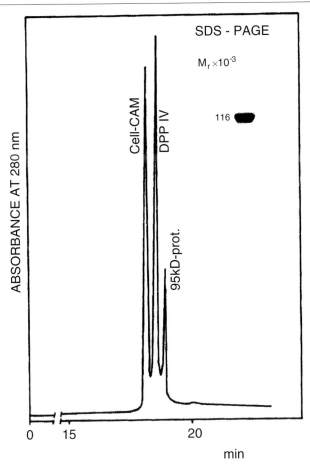

FIG. 6. Separation of intrinsic membrane proteins from liver by HPCITP. The three peaks belong to the three main components in the mixture: DPP IV, cell-CAM, and a still unidentified 95,000-D protein. The separation was carried out on a Beckman HPCE system. (Reprinted from *J. Chromatogr.*, **516**, D. Josic, K. Zeilinger, W. Reutter, A. Böttcher, and G. Schmitz, p. 89. Copyright 1990 with kind permission of Elsevier Science–NL, Sara Burgerhartstraat 25, 1055 KV Amsterdam, The Netherlands.)

Separation

Membrane-associated proteins are usually less hydrophobic than intrinsic membrane proteins. As can be seen in Scheme I, these proteins are solubilized by dilute sodium hydroxide at pH 11, using concentrated salt solutions or, in the case of calcium-binding proteins, complex-forming reagents.

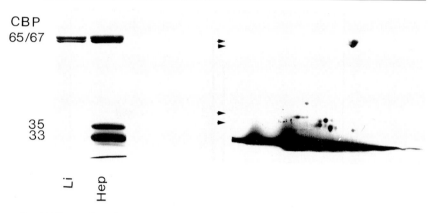

FIG. 7. Electrophoretic separation of membrane-associated, calcium-binding proteins from liver and Morris hepatoma 7777. *Left:* SDS–PAGE of EDTA extract (see also Scheme I). *Right:* Two-dimensional electrophoresis of EDTA extract from Morris hepatoma 7777. Isoelectric focusing was carried out in a rather broad range of pH, between 3.0 and 11.0. The separated polypeptides were detected by Coomassie blue staining.

This chapter presents the isolation and separation of annexins, i.e., calcium-binding proteins, from liver and Morris hepatoma 7777. Annexins can be solubilized easily by calcium complexing with EDTA or with EGTA.[4,6,28] However, it is difficult to solubilize the proteins in the presence of calcium. If the calcium concentration during solubilization is kept at a minimum level of about 0.5 mM, as was the basis in Scheme I, the proteins stay in the pellet. Highly enriched proteins can then be extracted with EDTA in the penultimate step of solubilization.

In plasma membranes of the liver, CBP 65/67, with an apparent molecular weight of 65,000 and 67,000 in SDS–PAGE, forms the major share of the proteins belonging to the group of annexins. In the plasma membranes of Morris hepatomas, annexins CBP 35 and 33, which have lower molecular weights, are more common. With rising malignancy of the Morris hepatomas, the fraction of annexins with a lower molecular weight, CBPs 33 and 35, increases in relation to the CBP 65/67.[4–6]

Figure 7 shows the two-dimensional (2D) electrophoresis of an EDTA extract (cf. also Scheme I) from plasma membranes of Morris hepatoma 7777. Despite their microheterogeneity, all three annexins, CBP 65/67 and especially CBPs 33 and 35, show a rather uniform behavior in the chromatographic separation. None of the separation methods, not even reversed-phase HPLC, results in more than four peaks, one for each CBP group. A

[28] C. E. Creutz, *Science* **258**, 924 (1992).

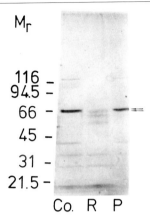

FIG. 8. Behavior of CBP 65/67 after phase separation in a 1% Triton X-114 solution at 25°; SDS–PAGE of the proteins before and after phase separation is shown. Co., Protein mixture before phase separation; R, detergent-rich phase; P, detergent-poor (aqueous) phase.

further isolation of single CBP groups, similar to 2D electrophoresis, can be achieved only by means of monoclonal antibodies or methods of high-performance capillary electrophoresis.

In phase separation with Triton X-114, the annexins appear in the aqueous phase in the presence of 1 mM EDTA. This is shown in Fig. 8 for CBP 65/67. However, when calcium is present, a change in the conformation of the annexins takes place, and the hydrophobic regions on their surfaces are exposed. This allows a convenient separation of annexins from other proteins by calcium-induced hydrophobic interaction chromatography (HIC).[29] Again further isolation of the different groups (CBP 65/67, CBP 35, and CBP 33) is not possible, not even using gradient elution (not shown here; cf. Ref. 6).

CBP 65/67, CBP 35, and CBP 33 can be separated from each other by RP-HPLC, hydroxylapatite (HA)-HPLC, crown ether-HPAC, or collagen-HPAC. The interaction of annexins with hydroxylapatite is probably due to their calcium-binding characteristics. As far as crown ether is concerned, the mechanism of retention is more difficult to interpret. A possible explanation would be interaction between phosphate groups, resulting from the phosphorylation of the annexins and the potassium of the crown ether. The separation by collagen-HPAC is based on the specific binding of the annexins to collagen (chromatographic separation not shown here). Hydroxylapatite-HPLC and crown ether-HPAC are to date the only two

[29] T. C. Südhof and D. Stone, *Methods Enzymol.* **139,** 30 (1987).

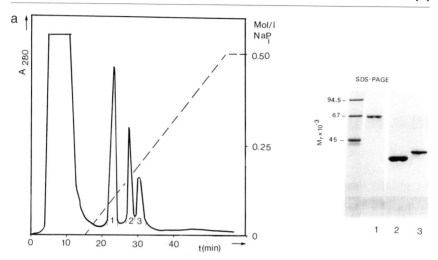

FIG. 9. Separation of calcium-binding proteins from the plasma membranes of Morris hepatoma 7777 under denaturing and nondenaturing conditions. Extracts obtained after calcium-complexing (see Scheme 1), each containing 5 mg of protein, were separated by hydroxylapatite-HPLC and crown ether-HPAC. (a) Hydroxylapatite-HPLC. The sample was extensively dialyzed against 5 mM Tris-HCl, pH 8.0, and applied to a hydroxylapatite column. The column was rinsed with 10 ml of 1 mM sodium phosphate buffer, pH 7.0, containing 0.1% octylglucoside. The proteins were eluted with a linear gradient up to 0.5 M sodium phosphate, pH 7.0, with 0.1% octylglucoside. In this way three fractions were obtained: fraction 1 with CBP 65/67; fraction 2 with CBP 33; and fraction 3 with CBP 35 (cf. the electrophoretic presentation). The first peak in flowthrough contains proteins that did not bind to the column, as well as remnants of Triton X-114 from previous extraction steps. Chromatographic conditions: hydroxylapatite column with precolumn (Pentax, Hamburg, Germany); flow, 1.0 ml/min; pressure, 15 atm; room temperature. The buffer gradient is shown by a dashed line. (b) Crown ether-HPAC. The sample was dialyzed against 10 mM Tris-HCl, pH 7.4 (buffer A), and applied to a crown ether-HPAC column. The column had previously been saturated with potassium ions by an injection of 5 ml of 0.5 M KCl. After rinsing with 10 ml of buffer A with 0.1% octylglucoside, the proteins were eluted with a gradient up to 1 M NaCl in buffer A with 0.1% octylglucoside. Chromatographic conditions: column, Eurocrown (Knauer Säulentechnik), with immobilized crown ether on a Eupergit-C30N support; flow, 1.0 ml/min; pressure, 4 atm (Eupergit column); room temperature. The buffer gradient is shown (dashed line). (c) Reversed-phase HPLC. The sample was dialyzed against 1 mM Tris-HCl, pH 7.5, and 0.5 ml was applied to the RP-HPLC column. The column was rinsed with buffer A and subsequently eluted. The gradient is shown (dashed line). Peak 1, CBP 33; peaks 2 and 3, CBP 35 (differences in phosphorylation); peak 4 (hatched), CBP 65/67.

chromatographic methods that allow complete separation of the three membrane associated groups of annexins from liver and Morris hepatomas under nondenaturing conditions. Separation by RP-HPLC takes place under strongly denaturing conditions. Similar to the experiments mentioned above

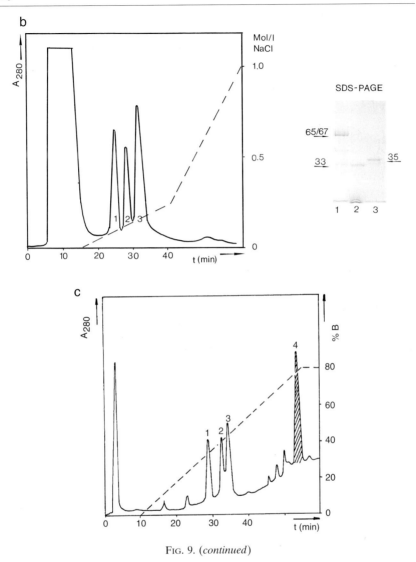

Fig. 9. (*continued*)

(cf. Refs. 7–9 and Fig. 2), the addition of 40–60% formic acid is also necessary for the separation of annexins, in order to achieve complete recovery and good resolution. Separation experiments with annexins involving different HPLC methods are shown in Fig. 9a–c.

The annexins isolated under nondenaturing conditions can be used for further investigations concerning, e.g., calcium binding, phospholipid binding, or phospholipase A_2 inhibition.[6] If annexins are isolated under denaturing conditions, e.g., by size-exclusion HPLC in the presence of SDS and 2-mercaptoethanol, or by reversed-phase HPLC (cf. Fig. 9a), they lose their binding and inhibition capacity, either completely or to a large extent. Denatured polypeptides can be used for determining amino acid sequence or for peptide mapping.

Conclusion

The separation experiments with annexins, which represent on the one hand the membrane-associated proteins and on the other hand the intrinsic membrane proteins DPP IV and cell-CAM, contain general guidelines to be followed in all experiments concerning the separation of hydrophobic proteins. Because of their hydrophobicity and possible changes in conformation and microheterogeneity, which are usually caused by their different degrees of glycosylation or phosphorylation, it is difficult to establish a general rule for this rather complex group of proteins. Appropriate sample preparation is as important as the right choice of separation method and detergent. The HPLC methods that are frequently used for the separation of water-soluble proteins, namely size exclusion, hydrophobic interaction, and reversed phase, cannot always be relied on for success. However, in most cases most specialized methods such as those described here must be chosen, above all affinity chromatography. Apart from the complexity of the samples, special treatment of these proteins is required because of nonspecific interactions with the support, which must be kept under control.

The new methods such as HPMC and capillary electrophoresis have not been used frequently so far. However, first results indicate that these methods, especially HPCE, can be applied successfully, with high resolution, to the separation of membrane proteins.

[6] Identification and Characterization of Glycopeptides in Tryptic Maps by High-pH Anion-Exchange Chromatography

By R. Reid Townsend, Louisette J. Basa, and Michael W. Spellman

Introduction

This chapter describes the application of high-pH anion-exchange chromatography with pulsed amperometric detection (HPAEC–PAD)[1,2] to the identification and partial characterization of glycopeptides within a reversed-phase high-performance liquid chromatography (HPLC) peptide map. HPAEC is a versatile chromatographic method that can be used for the detection and quantitation of monosaccharides released by acid hydrolysis[2] and for the analysis of intact oligosaccharides released enzymatically[3] or chemically.[4] A large number of other analytical methods, including mass spectrometry,[5,6] enzymatic microsequencing,[7] nuclear magnetic resonance (NMR),[8] and capillary electrophoresis,[9] can also be applied to the characterization of glycopeptides; in general, a combination of several complementary analytical methods will be needed to achieve a definitive carbohydrate structure determination. In this chapter we present the type of information that can be obtained using a single, comparatively simple, analytical method.

In the mid-1970s, HPLC methods were introduced for the analysis of mono- and oligosaccharides (for reviews, see Hicks[10] and Honda[11]). Most of these methods required derivatization both for detecting and increasing

[1] R. D. Rocklin and C. A. Pohl, *J. Liq. Chromatogr.* **6,** 1577 (1983).

[2] M. R. Hardy, R. R. Townsend, and Y. C. Lee, *Anal. Biochem.* **170,** 54 (1988).

[3] A. L. Tarentino, C. M. Gomez, and T. H. Plummer, Jr., *Biochemistry* **24,** 4665 (1985).

[4] T. Patel, J. Bruce, A. Merry, C. Bigge, M. Wormald, A. Jaques, and R. Parekh, *Biochemistry* **32,** 679 (1993).

[5] S. A. Carr, M. E. Hemling, M. F. Bean, and G. D. Roberts, *Anal. Chem.* **63,** 2802 (1991).

[6] A. W. Guzzetta, L. J. Basa, W. S. Hancock, B. A. Keyt, and W. F. Bennett, *Anal. Chem.* **65,** 2953 (1993).

[6a] Deleted in proof.

[7] C. J. Edge, T. W. Rademacher, M. R. Wormald, R. B. Parekh, T. D. Butters, D. R. Wing, and R. A. Dwek, *Proc. Natl. Acad. Sci. U.S.A.* **89,** 6338 (1992).

[8] H. van Halbeek, *Methods Enzymol.* **230,** 132–168 (1994).

[9] R. S. Rush, P. L. Derby, T. W. Strickland, and M. F. Rohde, *Anal. Chem.* **65,** 1834 (1993).

[9a] Deleted in proof.

[10] K. B. Hicks, *Adv. Carbohydr. Chem. Biochem.* **46,** 17 (1988).

[11] S. Honda, *Anal. Biochem.* **140,** 1 (1984).

analyte hydrophobicity for reversed-phase HPLC (RP-HPLC) (for review, see Townsend[12]). An HPLC method, HPAEC-PAD, has been introduced[1,2] that bypasses the derivatization steps by using pulsed electrochemical detection on gold electrodes.[13] HPAEC takes advantage of the fact that the hydroxyl groups of carbohydrates are weakly acidic and, at pH values greater than pH 12, can be resolved by anion-exchange chromatography. Separations are carried out on pellicular polymeric stationary phases that are stable under alkaline conditions. This method has been shown to be useful for resolving and quantitating the constituent monosaccharides released by acid hydrolysis[2] and for resolving N-linked oligosaccharides.[14,15] Because no derivatization is required, HPAEC-PAD is attractive for the analysis of a large numbers of fractions, such as those that result from RP-HPLC of proteolytic digests of glycoproteins. We detail, herein, the usefulness of this HPLC method for monosaccharide survey analysis of a tryptic map of CHO cell-expressed recombinant tissue plasminogen activator (tPA), a glycoprotein that has been the subject of other characterization studies.[16–18] We also demonstrate the applicability of HPAEC-PAD to the further characterization of the oligosaccharides released from the individual glycopeptides.

Materials and Methods

Equipment and Buffers

Glass-distilled water from a Corning (Corning, NY) MegaPure apparatus

NaOH solution (Fisher Scientific, Pittsburgh, PA), 50% (w/w)

Plastic pipettes (25-ml)

Glass hydrolysis vials (Chromacol, Trumbull, CT)

Autosampler vials (Sun Brokers, Wilmington, NC)

Disposable, limited-volume sample vials (12 × 32 mm)

[12] R. R. Townsend, in "HPLC of Glycoconjugates and Preparative/Process Chromatography of Biopolymers" (L. Ettre and C. Horváth, eds.). American Chemical Society, Washington, D.C., pp. 86–101, 1993.

[13] D. C. Johnson and T. Z. Polta, Chromatogr. Forum 1, 37 (1986).

[14] M. R. Hardy and R. R. Townsend, Proc. Natl. Acad. Sci. U.S.A. 85, 3289 (1988).

[15] L. J. Basa and M. W. Spellman, J. Chromatogr. 499, 205 (1990).

[16] R. C. Chloupek, R. J. Harris, C. K. Leonard, R. G. Keck, B. A. Keyt, M. W. Spellman, A. J. S. Jones, and W. S. Hancock, J. Chromatogr. 463, 375 (1989).

[17] M. W. Spellman, L. J. Basa, C. K. Leonard, J. A. Chakel, J. V. O'Connor, S. Wilson, and H. van Halbeek, J. Biol. Chem. 264, 14100 (1989).

[18] R. J. Harris, C. K. Leonard, A. W. Guzzetta, and M. W. Spellman, Biochemistry 30, 2311 (1991).

Caps for disposable 12 × 32 mm sample vials
Teflon/silicone septa
Concentrated (13 *N*) trifluoroacetic acid (TFA), in sealed 1-ml ampoules (Pierce, Rockford, IL)
MonoStandards (Dionex, Sunnyvale, CA)
Sodium acetate, anhydrous (Mallinckrodt AR, St. Louis, MO)
Disposable filter system, 1 liter, 0.22-μm pore size (Corning)
Peptide: *N*-glycosidase F (PNGase F), recombinant (Boehringer Mannheim, Indianapolis, IN)

Preparation of Eluents

1. Two- or 4-liter plastic bottles (Dionex Corporation) are rinsed with glass-distilled water.

2. Eluent 1 (E1) is installed as a 4-liter bottle of glass-distilled water to provide sufficient eluent for performing monosacchride analysis for 48 hr.

3. Eluent 2 (E2) (2 liters of 0.2 *M* NaOH) is prepared by pipetting 20.8 ml of 50% NaOH (19.3 *M*) into 1980 ml of freshly degassed glass-distilled water. The solution is swirled several times with the plastic pipette and transferred to a 2-liter plastic container on the instrument. The NaOH solution is drawn to avoid precipitated sodium carbonate on the bottom of the container. (The NaOH solution, rather than solid NaOH, is used to minimize contamination with carbonate, which has a high affinity for quaternary ammonium columns.)

4. E3 (2 liters of 0.1 *N* NaOH) is prepared by pipetting 10.4 ml of 50% NaOH (19.3 *M*) into 1990 ml of freshly degassed glass-distilled water.

5. E4 (2 liters of 0.1 *N* NaOH containing 0.5 *M* sodium acetate) is prepared by dissolving 82 g of sodium acetate in 1990 ml of glass-distilled water. This solution is filtered through a 0.22-μm pore size disposable filter and then degassed prior to the addition of 10.4 ml of 50% NaOH.

Between analyses the eluents are kept under helium (5–7 psi). If pressure is lost, the hydroxide-containing eluents should be remade.

Chromatographic Apparatus

The chromatographic apparatus is a Dionex GlycoStation, equipped with a gradient pump, a PAD-II detector, and an autosampler, all under software control. An automated system is needed for the practical analysis of large numbers of samples because variations in the time between injections (as will often occur with manual injections) can result in significant drift in retention times of monosaccharides (1–3 min). Furthermore, three or four chromatographic runs are usually required at system startup to achieve consistent retention times and peak areas.

The system is equipped with a CarboPac PA1 column (4 × 250 mm) and a guard column (Dionex Corporation). The pulse potentials and durations for the PAD-II detector are 0.05 V for 480 msec; 0.60 V for 120 msec; and −0.60 V for 60 msec. The time constant is set to 3 sec. The Dionex "basic" PAD cell, equipped with a thin gasket, is used.

Monosaccharide Separations

Monosaccharides are analyzed at a flow rate of 1 ml/min, using the following elution program: $t = 0$, E1 = 92%, E2 = 8%; $t = 25$ min, E1 = 92%, E2 = 8%; $t = 27$ min, E1 = 0%, E2 = 100%; $t = 37$ min, E1 = 0%, E2 = 100%; and $t = 39$ min, E1 = 92%, E2 = 8%.

Oligosaccharide Separations

Oligosaccharide separations are carried out using essentially the same chromatographic apparatus and column described above. The column is preequilibrated at a flow rate of 1 ml/min, with E3 = 96%, E4 = 4%. Oligosaccharides are eluted using the following gradient program: $t = 0$, E3 = 96%, E4 = 4%; $t = 5$, E3 = 96%, E4 = 4%; $t = 60$, E3 = 60%, E4 = 40%; $t = 61$, E3 = 40%, E4 = 60%; $t = 75$, E3 = 40%, E4 = 60%; $t = 76$, E3 = 96%, E4 = 4%; $t = 95$, E3 = 96%, E4 = 4%. Detection is facilitated by the postcolumn addition of 0.3 N NaOH at a flow rate of 0.8 ml/min.

Acid Hydrolysis for Neutral and Amino Sugar Analysis

A 10% aliquot of each of the fractions from the RP-HPLC separation is dried in a glass hydrolysis vial. To each sample vial is added 0.4 ml of 2 N TFA, and the samples are placed in an oven at 100° for 3 hr. After the vials are allowed to cool at room temperature for 15 min, they are spun 1–2 min in a microcentrifuge prior to opening, and then the acid is evaporated using a SpeedVac (Savant, Hicksville, NY) apparatus. We have previously reported that polypropylene vials can be used for hydrolysis.[19] Subsequently, we have found that some vial lots give intolerable amounts of D-glucose (Glc) and interfering peaks. The high-quality glass vials have proved satisfactory.

PNGase F Digestion of Glycopeptides

The conditions used for PNGase F digestion are essentially as reported previously.[19] Lyophilized aliquots (35% of total collected) of glycopeptides

---

[19] M. R. Hardy and R. R. Townsend, *Methods Enzymol.* **230**, 208–225 (1994).

are reconstituted in 0.06 ml of 12.5 mM sodium phosphate buffer, pH 7.6. One manufacturer unit of PNGase F is added to each sample, and the samples are incubated for 18 hr at 37°. After the samples have cooled to room temperature, 5 μl of 0.175 M acetic acid is added to each sample. After 2 hr at room temperature, the samples are brought to dryness in a SpeedVac.

Results and Discussion

Monosaccharide Survey Analysis

A sample of recombinant tPA was reduced, S-carboxymethylated, and treated with trypsin as described by Chloupek et al.[16] and then analyzed by RP-HPLC (Fig. 1). The peaks from the RP-HPLC separation were collected manually, and aliquots of all 62 fractions were analyzed for neutral and amino monosaccharides after acid hydrolysis. The hydrolysis conditions used here (2 N TFA for 3 hr at 100°) are minimal for complete de-N-

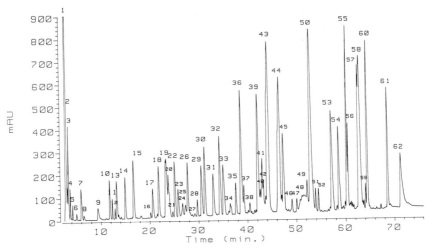

Fig. 1. RP-HPLC tryptic map of recombinant tPA. Chromatography was carried out at a flow rate of 1 ml/min on a Hewlett-Packard 1090M system equipped with a Vydac C$_{18}$ column (4.6 × 250 mm). Solvent A was 0.1% (v/v) aqueous TFA; solvent B was 0.08% (v/v) TFA in acetonitrile. The column was preequilibrated using 100% solvent A. Peptides were eluted using the following gradient program: $t = 0$, A = 100%, B = 0%; $t = 3$, A = 100%, B = 0%; $t = 53$, A = 75%, B = 25%; $t = 80$, A = 48%, B = 52%; $t = 81$, A = 40%, B = 60%; $t = 82$, A = 100%, B = 0%. Peaks were detected by absorbance at 214 nm. Injected sample amount was 0.54 mg/analysis. Peaks were collected manually from five replicate RP-HPLC runs.

acetylation of 2-deoxy-2-acetamido-D-glucose (GlcNAc) and 2-deoxy-2-acetamido-D-galactose (GalNAc).[2] Although the release of the amino sugars is low under these conditions, they are a practical compromise for survey monosaccharide analysis.

Fractions 1 through 22 were devoid of peaks with elution times of the external standards [L-fucose (Fuc), 2-deoxy-2-amino-D-galactose (GalN), 2-deoxy-2-amino-D-glucose (GlcN), D-galactose (Gal), Glc, and D-mannose (Man)]. Figure 2 shows the HPAE chromatograms of fractions 23–27 after acid hydrolysis. The quantities of individual monosaccharides in the fractions are compiled in Table I. D-Glucose was found in most fractions and was judged to be a contaminant; its variable amount (Table I) is likely due to the different volumes of the fractions. Significant amounts of Fuc, GlcN, Gal, and Man were found in fractions 24, 25, and 26, and the ratios of Man:GlcN:Gal:Fuc were 3:6.5:4.2:0.9, 3:5.3:3.0:0.9, and 3:4.3:2.0:0.9, respectively. Approximately 10-fold less carbohydrate was found in fraction 23 and likely represents contamination from peak 24. Peak fraction 27 was also found to contain less than 50 pmol of the above-

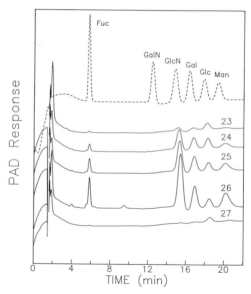

FIG. 2. Monosaccharide analysis of fractions 23–27 from the RP-HPLC separation of a tryptic digest of recombinant tPA. The dried hydrolysates were dissolved in water (200 μl) and 150 μl was analyzed using the Dionex GlycoStation, configured as described. Each chromatogram has a full-scale deflection of 600 mV. Elution positions of monosaccharide standards are indicated on the upper trace. See Materials and Methods for chromatographic conditions.

TABLE I

MONOSACCHARIDE ANALYSIS OF CARBOHYDRATE-CONTAINING
FRACTIONS FROM RP-HPLC OF TRYPTIC DIGEST OF
RECOMBINANT TISSUE PLASMINOGEN ACTIVATOR[a]

Fraction number	Fuc	GlcN	Gal	Glc	Man
22	—	—	—	0.89	—
23	0.040	0.27	0.19	0.74	—
24	0.24	1.7	1.1	1.3	0.79
25	0.47	2.8	1.6	1.2	1.6
26	1.0	5.0	2.3	1.4	3.5
27	—	—	—	0.53	—
38	—	—	—	0.33	—
39	0.038	0.23	0.24	0.26	0.15
40	0.17	1.2	0.69	0.29	0.64
41	0.57	2.7	1.3	0.25	2.1
42	0.27	1.6	0.69	0.17	1.1
43	0.55	2.9	0.25	0.56	6.8
44	0.31	1.7	0.084	0.37	4.1
45	—	—	—	0.91	—
54	—	—	—	—	—
55	0.64	—	—	0.84	—
56	0.092	—	—	0.23	—
57	—	—	—	0.078	—

[a] Content in nanomoles per sample. Samples correspond to
10% aliquots of collected peptides.

cited monosaccharides. These data indicated that fractions 24–26 contained
glycopeptides, and the ratios of constituent monosaccharides suggested
that their oligosaccharide structures were of the fucosylated N-acetyllactosamine-type oligosaccharides.

Monosaccharides were found in fractions 38–45, as shown in Fig. 3.
Two distinct profiles were observed. Fractions 40–42 contained
Man : GlcN : Gal : Fuc in ratios that were similar to those found in fractions
23–27 (fraction 40, 3 : 5.6 : 3.3 : 0.8; fraction 41, 3 : 3.8 : 1.9 : 0.82; fraction 42,
3 : 4.4 : 1.9 : 0.74). In fractions 43 and 44 the Fuc and Gal were markedly
decreased (Fig. 3) and the ratio of Man : GlcN was 2.3 : 1 in both. The data
suggested that these fractions represented two closely eluting glycopeptides,
one containing glycopeptides with N-acetyllactosamine-type chains and the
other containing oligomannosidic-type chains.

Fractions 55 and 56 contained peaks with the same retention times as
Fuc (Fig. 4). In addition to the contaminating Glc, a peak with a retention
time similar to that of Man was observed. The profiles of the standard
monosaccharides run before and after the analyses of fractions 54–56 show

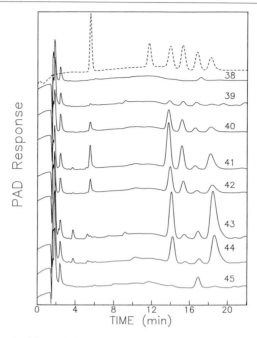

PAD Response

TIME (min)

FIG. 3. Monosaccharide analysis of fractions 38–45 from the RP-HPLC separation of a tryptic digest of recombinant tPA. Monosaccharide standards are identified in Fig. 2. All other conditions were as described in the caption to Fig. 2.

that this peak has a retention time greater than that of Man (Fig. 4, grid line). This was confirmed by analysis of a comix of a portion of the hydrolysate of fraction 55 and the monosaccharide standards. When this was done, two partially separated peaks with a retention time of ~17 min were seen (data not shown), indicating that the peak eluting after Glc in the hydrolysate in fraction 55 was not Man. We have intermittently observed this peak in samples from a variety of sources and do not know its identity.

Glycopeptides containing fucosyl-, oligomannosidic-, and N-acetyl-lactosamine-type modifications were identified by survey monosaccharide analysis of the recombinant tPA tryptic map (Figs. 2–4). It is not clear why the proportion of Gal and GlcN varied among the fractions containing the N-acetyllactosamine-type glycopeptides. Interestingly, the ratio of one Fuc to three Man residues varied only slightly (0.7 to 1.0) in the glycopeptide fractions containing the N-acetyllactosamine-type oligosaccharides; this result is consistent with the one fucose residue known to be on the chitobiose core of oligosaccharides from recombinant tPA.[17]

The recombinant tPA amino acid sequence is known to include potential

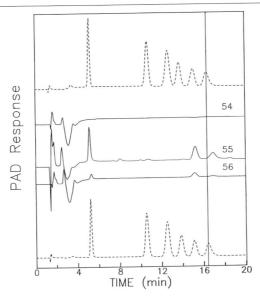

FIG. 4. Monosaccharide analysis of fractions 54–56 from the RP-HPLC separation of a tryptic digest of recombinant tPA. Each chromatogram has a full-scale deflection of 3000 mV. All other conditions were as described in the caption to Fig. 2.

sites of N-linked glycosylation at Asn-117, Asn-184, and Asn-448.[16] Fractions that were determined from the monosaccharide survey analysis to contain glycopeptides (Table I) were then subjected to quantitative amino acid analysis to identify and quantitate the peptide(s). From the amino acid analysis, RP-HPLC peaks 24 to 26 were found to contain peptide 441–449 (N-linked site at Asn-448), peaks 40–42 were found to contain peptide 163–189 (N-linked site at Asn-184), and peaks 43 and 44 were found to contain peptide 102–129 (N-linked site at Asn-117). Peaks 55 and 56, which were found in the monosaccharide survey analysis to contain only fucose (Table I), were confirmed to contain peptide 56–82, which has been shown[18] to have a residue of fucose glycosidically linked to Thr-61.

Oligosaccharide Profiling

The glycopeptides whose monosaccharide compositions were consistent with N-linked oligosaccharides (see above) were further subjected to digestion with PNGase F to release the oligosaccharides, which were analyzed by HPAEC (Figs. 5–7). The oligosaccharide separations were consistent with the interpretation of the monosaccharide survey analyses and permitted further conclusions to be drawn about the populations of oligosac-

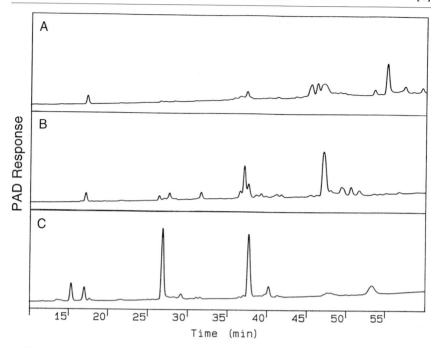

FIG. 5. HPAEC analysis of oligosaccharides released by PNGase F digestion of fractions 24 (A), 25 (B), and 26 (C) from the RP-HPLC tryptic map of recombinant tPA. Chromatographic conditions are described in Materials and Methods.

charides attached to each of the N-glycosylation sites. Under the conditions employed, HPAEC separations of oligosaccharides are dominated by the number of formal charges[14,15]; in this case, the separation reflects the number of residues of sialic acid attached to the oligosaccharides.

Peaks 24, 25, and 26 from the RP-HPLC tryptic map arise from glycopeptide 441–449. The HPAEC elution characteristics of the oligosaccharides released from this glycopeptide by PNGase F indicated that they were all of the *N*-acetyllactosamine type and, furthermore, indicated that RP-HPLC was capable of resolving different oligosaccharides attached to the same peptide. Thus, the oligosaccharides released from fraction 24 (Fig. 5A) eluted mostly in the region of the chromatogram characteristic of tetrasialyl structures (retention time of ~54 min); the oligosaccharides released from fraction 25 (Fig. 5B) were predominantly trisialyl (retention time of ~48 min); and the oligosaccharides released from fraction 26 (Fig. 5C) consisted of a mixture of disialyl (retention time of ~38 min) and monosialyl (retention time of ~27 min) structures. These results demonstrate that the RP-HPLC tryptic mapping conditions are capable of some resolution of differ-

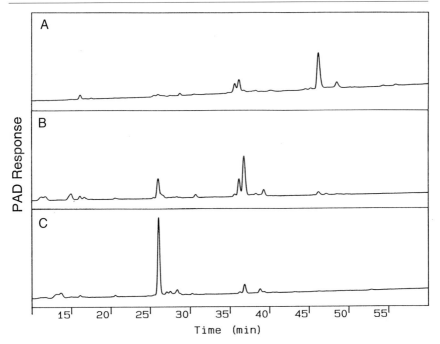

FIG. 6. HPAEC analysis of oligosaccharides released by PNGase F digestion of RP-HPLC fractions 40 (A), 41 (B), and 42 (C). Conditions are as described in Materials and Methods.

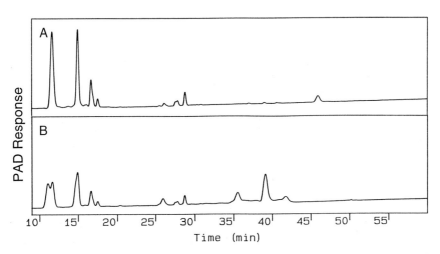

FIG. 7. HPAEC analysis of oligosaccharides released by PNGase F digestion of RP-HPLC fractions 43 (A) and 44 (B).

ent glycoforms of the same peptide, with larger (and therefore more hydrophilic) oligosaccharides eluting before smaller ones. The relationship between oligosaccharide structure and glycopeptide elution position in RP-HPLC has been studied in greater detail by liquid chromatography–mass spectrometry (LC–MS).[6]

RP-HPLC peaks 40–42 were found to contain glycopeptides in the monosaccharide survey analysis (Table I) and shown by amino acid analysis to contain peptide 163–189. The HPAEC profiles of the oligosaccharides released from these fractions (Fig. 6) confirmed that this glycopeptide carried N-acetyllactosamine-type structures, as had been indicated by the monosaccharide survey analysis. The partial resolution of glycoforms by RP-HPLC observed above with the 441–449 glycopeptide was also observed here. Thus, the oligosaccharides released from fraction 40 (Fig. 6A) consisted predominantly of trisialyl structures (retention time of ~48 min), those released from fraction 41 consisted of a mixture of disialyl (retention time of ~38 min) and monosialyl (retention time of ~27 min) structures, and those released from fraction 42 were almost exclusively monosialyl structures.

RP-HPLC fractions 43 and 44 were also found to be glycosylated by survey monosaccharide analysis, but had monosaccharide compositions suggestive of oligomannosidic structures (Table I). Amino acid analysis demonstrated that both of these fractions contained glycopeptide 102–129. The HPAE profiles of the oligosaccharides released from fractions 43 and 44 are shown in Fig. 7. Fraction 43 was found to contain almost exclusively neutral oligosaccharides (retention times between 10 and 20 min; Fig. 7A), with the major components having retention times characteristic of $Man_5GlcNAc_2$ (~11 min), $Man_6GlcNAc_2$ (~14.5 min), and $Man_7GlcNAc_2$ (~16 min). The minor peaks in the monosialyl region of the chromatogram (retention times between 26 and 30 min) likely arise from the hybrid-type oligosaccharides that have been reported in previous studies of recombinant.[6,17] Fraction 44 (Fig. 7B) gave a more heterogeneous pattern of N-linked oligosaccharides than did fraction 43; this pattern included all of the peaks found with fraction 43 and also included several more acidic peaks. It is not clear if these more acidic peaks arise from oligosaccharides or from peptide contaminants.

In summary, the results from oligosaccharide profiling of individual glycopeptide peaks were consistent with those of the monosaccharide survey analysis. These profiles confirmed the presence of N-acetyllactosamine-type oligosaccharides attached to Asn-184 and Asn-448 and oligomannosidic structures attached to Asn-117. The combined HPAEC results reported here are also consistent with those reported from previous structure analysis of recombinant tPA glycosylation.[6,17] The focus of the analyses reported here has been on the identification of glycopeptides and on determining the major classes of oligosaccharides attached to each. HPAEC can be applied to the further characterization of desialylated oligosaccharides (see, for example, Ref. 17). Under such conditions, HPAEC separations are influenced by anomeric configurations and linkage positions of the individual glycosidic residues.

Acknowledgment

The monosaccharide analyses were performed with the expert technical assistance of Lorri Reinders.

[7] High-Resolution Nucleic Acid Separations by High-Performance Liquid Chromatography

By JAMES R. THAYER, RANDY M. MCCORMICK, and NEBOJSA AVDALOVIC

Introduction

The central role of nucleic acids in genetic engineering has spawned the development of numerous techniques for isolating, characterizing, and separating these compounds. Workers in this field have enjoyed the tolerance of genetic techniques for minimal purity of the components they manipulate. However, the realization of the therapeutic, agricultural, and medical diagnostic utility of various forms of nucleic acids has created a need, sometimes imposed by regulatory agencies, for higher purity components. Examples of nucleic acids for which high purity is now expected or required include the following: plasmids and restriction fragments used for DNA sequencing of master cell banks and genetic mapping, oligonucleotide probes and polymerase chain reaction (PCR) products to be probed for forensic and diagnostic applications, primers for PCR reactions, modified oligonucleotides in therapeutic applications, and synthetic ribozymes. Industrial research and development has responded to these demands by providing methods that are faster and offer greater purity or capacity than traditional electrophoretic and ultracentrifugal methods for purifying nucleic acids. For some applications, the new methods must also be amenable to scaleup.

A comparison of high-performance liquid chromatography (HPLC) with traditional electrophoretic and ultracentrifugal techniques reveals that the weakest aspects of the traditional techniques are actually the strengths of HPLC. Automation of HPLC sampling, data reduction, fraction collection, and reporting functions enhances quantification and promotes greater throughput and recovery. HPLC also offers the opportunity for scaleup. In comparison, gel electrophoretic and ultracentrifugal methods are difficult to automate and quantify, present problems with recovery, are not readily scaleable, and for some applications, offer little practical advantage over current HPLC techniques (e.g., synthetic oligonucleotide separation[1]).

This chapter is organized in sections on applications and procedures for a wide variety of nucleic acids. Single-stranded (ss) nucleic acid separations include: (1) synthetic oligonucleotides (e.g., PCR and sequencing primers, oligonucleotide probes, and ribozymes), and (2) phosphorothioate (antisense) oligonucleotides. Double-stranded (ds) nucleic acid separations include: (1) restriction fragments (RFs), (2) PCR products, and (3) plasmids.

Numerous informative reviews on the use of HPLC for separations of nucleic acids have been published.[2–10] However, some new anion-exchange HPLC columns are now available that were not evaluated in prior reviews. Among these are columns packed with nonporous resins containing diethylaminoethyl (DEAE) or quaternary alkylamines as anion-exchange sites. Examples of these columns, which are capable of resolution far superior to their predecessors, are DEAE-NPR (Toyo Soda, Tokyo, Japan), Gen-Pak Fax (Millipore, Bedford, MA), and NucleoPac PA100 (Dionex, Sunnyvale, CA). An evaluation of three of these columns for fast restriction fragment separation (less than 20 min) indicated little practical advantage of one column over the others.[11] Figure 1 shows the separation of restriction fragments produced by *Hae*III digestion of pBR322 on these three columns. In Fig. 1, the DEAE-NPR and Gen-Pak Fax are compared with a NucleoPac

[1] P. J. Oefner, G. K. Bonn, C. G. Huber, and S. Nathakarnkitkool, *J. Chromatogr.* **625**, 331 (1992).

[2] J. A. Thompson, *BioChromatography* **1**(1), 16 (1986).

[3] G. Zon and J. A. Thompson, *BioChromatography* **1**(1), 22 (1986).

[4] J. A. Thompson, *BioChromatography* **1**(2), 68 (1986).

[5] J. A. Thompson, *BioChromatography* **2**(1), 4 (1987).

[6] J. A. Thompson, *BioChromatography* **2**(2), 68 (1987).

[7] J. A. Thompson, *BioChromatography* **2**(4), 166 (1987).

[8] G. Zon, Purification of synthetic oligonucleotides. *In* "High Performance Liquid Chromatography in Biotechnology" (W. S. Hancock, ed.), pp. 307–397. Wiley, New York, 1990.

[9] L. W. McLaughlin and R. Bischoff, *J. Chromatogr.* **418**, 51 (1987).

[10] D. Riesner, HPLC of plasmids, DNA restriction fragments and RNA transcripts. *In* "HPLC of Proteins, Peptides, and Polynucleotides" (M. T. W. Hearn, ed.), pp. 689–736. Verlag Chemie, Weinheim, New York, 1987.

[11] E. Katz, Comparison of Anion Exchange HPLC Columns for DNA Analysis. The Pittsburgh Conference, March 9–12, 1992, Book of Abstracts, Paper 1200.

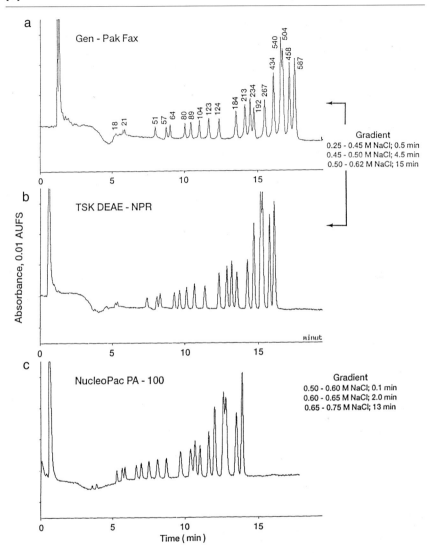

FIG. 1. Separation of the components of a *Hae*III digest of pBR322 on three different anion-exchange columns: (a) TSK DEAE-NPR (4.6-mm i.d. × 35 mm long, middle trace); (b) Millipore Gen-Pak Fax (4.6-mm i.d. × 100 mm long, top trace); (c) Dionex NucleoPac PA100 (4.0-mm i.d. × 50 mm long, bottom trace). Flow and gradient conditions as indicated. [Prepared by E. Katz (Perkin-Elmer Corporation, Norwalk, CT).]

PA100 *guard* column. The DEAE-NPR, Gen-Pak Fax, and similar nonporous products contain media with 2- to 3-μm diameter particles in stainless steel columns. These products contain weak anion-exchange functions (DEAE), precluding use at pH 12.4, where hydrogen bonding is abolished (use of high-pH eluents prohibits formation, during chromatography, of double-stranded structures by oligonucleotides having internally complementary sequences). These DEAE columns are short (3.5–10 cm × 0.46-cm diameter) and exhibit substantial back-pressure (1500–2250 psi at 0.5 ml/min, 20°, 4.6-mm i.d. × 10 cm[12]). We describe the use of the NucleoPac PA100 column (Dionex) that contains a strong anion-exchange function (quaternary amine, allowing the use of eluents in pH range from pH 0.7 to pH 12.5), and is available in columns of polyetheretherketone (PEEK) (inert to halide salts) and in a variety of column dimensions (4 × 50 mm to 22 × 250 mm). This column exhibits low back-pressure (800–1200 psi at 1.0 ml/min, 24°, 4.0-mm i.d. × 25 cm with a 4.0-mm i.d. × 5-cm long guard column), indicating its utility for scaleup.

Equipment

The Dionex DX-300 liquid chromatograph employed in this work consists of a gradient pump (AGP), an eluent degas module (EDM-II, used to degas the eluent before use and to apply a 5- to 10-psi helium overpressure to the eluent bottles during chromatography), a variable-wavelength absorbance detector (VDM-II), an AS3500 autosampler equipped with sample cooler and column oven, and a biocompatible injection system. These components are controlled by Dionex AI450 software running on a Dell 316SX computer. Fractions are collected with a Gilson (Middleton, WI) FC-80 fraction collector. Ethanol precipitates are sedimented in a Fisher (Pittsburgh, PA) 235C microcentrifuge placed in a 4° chromatography refrigerator (Revco, Asheville, NC) or by using an Eppendorf (Hamburg, Germany) 5402 refrigerated microcentrifuge. Samples are dried in a Virtis (Gardiner, NY) model 12SL lyophilizer or a Savant (Hicksville, NY) SpeedVac (SVC100 equipped with a VP100 vacuum pump and RT490 refrigerated trap).

Preparation for Anion-Exchange High-Performance Liquid
 Chromatography of DNA

Preparation of Eluents

The eluent system we use contains four components: E1, deionized distilled H_2O; E2, 0.2 *M* NaOH; E3, 0.25 *M* Tris-HCl, pH 8; and E4, 0.375

---

[12] D. J. Stowers, J. M. B. Keim, P. S. Paul, Y. S. Lyoo, M. Merion, and R. M. Benbow, *J. Chromatogr.* **444**, 47 (1988).

M NaClO$_4$. See the Appendix for specific gradient conditions used for each application.

E1: Deionized Distilled H$_2$O

Water is deionized to <5 μS, glass distilled, and stored in a glass carboy fitted with a vapor trap to minimize absorption of CO$_2$. This water is used for the preparation of all eluents.

E2: 0.2 M NaOH

Carbonate, at high pH, is a divalent anion that adversely affects resolution of nucleic acids eluted by monovalent anions (e.g., CH$_3$CO$^-$, Cl$^-$, ClO$_4^-$). Commercial NaOH pellets are coated with a layer of sodium carbonate produced by adsorption of CO$_2$ from the air. Hence, use of NaOH pellets for eluent preparation is counterproductive. Prepare NaOH eluents by diluting a 50% (w/w) solution of NaOH solution (e.g., Fisher) that is low in carbonate. Pipette 20.8 ml of 50% NaOH into 1980 ml of distilled, deionized H$_2$O in a 2-liter graduated cylinder, using a wide-bore 25-ml plastic pipette. Mix the solution for 2 to 3 min with a magnetic stirrer, then transfer the entire solution to a plastic 2-liter eluent reservoir connected to the HPLC system. Degas with helium sparging for ~5 min, seal the reservoir, and maintain 4- to 8-psi overpressure until the reservoir is empty or refilled.

E3: 0.25 M Tris-Cl–HCl

Dissolve 30.28 g of Tris–OH [e.g., Sigma (St. Louis, MO) Trizma base] in 800 ml of distilled, deionized H$_2$O while stirring with a magnetic stirrer, and bring to pH 8 with 2 N HCl (e.g., Fisher, ~60 to 65 ml) using a calibrated pH meter. Bring to 1 liter with distilled, deionized H$_2$O after removing the stirring bar, replace the stirring bar, and stir for 2 to 5 min. Transfer the entire solution to a glass 1-liter reservoir connected to the HPLC system. Degas with helium sparging for ~5 min, seal the reservoir, and maintain 4- to 8-psi overpressure until the reservoir is empty or refilled.

E4: 0.375 M NaClO$_4$

Under the conditions used, the E4 solution has exhibited no capacity for oxidation of nucleic acid samples or for the components of common HPLC instrumentation.

Both anhydrous NaClO$_4$ (Aldrich, Milwaukee, WI) and the monohydrate (Fluka, Ronkonkoma, NY) have yielded acceptable results. Dissolve 45.92 g (anhydrous) or 52.67 g (monohydrate) of NaClO$_4$ in 800 ml of distilled, deionized H_2O in a 1-liter graduated cylinder, using a magnetic stirrer. Remove the stir bar, bring to 1 liter with distilled, deionized H_2O, replace the stirrer, and mix for 2 to 5 min. Transfer the entire solution to a 1-liter reservoir connected to the HPLC. Degas with helium for ~5 min, seal the reservoir, and maintain 4- to 8-psi overpressure until the reservoir is empty, or refilled.

Single-Stranded Nucleic Acids

Synthetic Oligonucleotides

Oligonucleotides are used as probes for specific sequences in diagnostic and forensic tests, primers for DNA sequencing and PCR amplification, therapeutics that inhibit the retrovirus replication cycle when containing modified bases or backbone, catalytically active ribozymes, and a wide variety of related applications.

A detailed discussion of the synthesis of these compounds is outside the scope of this chapter. Useful discussions of this process are available.[8,13,14] Synthetic oligonucleotides are typically 10 to 30 bases long, and are usually constructed using automated DNA synthesizers. The products of these synthesizers contain several components that must be removed prior to use. The oligonucleotides are simultaneously released from the support, phosphate deprotected in the instrument, and delivered in ammonia. The delivered product typically includes truncated sequences, some partially depurinated oligomers, deprotected oligomers, and the target, full-length fully dimethoxytrityl (DMT)-protected oligonucleotide. For forensic and diagnostic probes, integrity and purity of the oligonucleotide are paramount.

Purification from "Machine-Grade" Single-Stranded DNA. Historically, purification has been accomplished by a combination of reversed-phase extraction (i.e., sample-preparation cartridges) and denaturing polyacrylamide gel electrophoresis (PAGE). PAGE is labor intensive and not amenable to scaleup. Reversed-phase HPLC is also useful,[7] but is usually also accompanied by PAGE as an orthogonal method to confirm the product purity. In this section, we discuss the use of strong anion-exchange HPLC on a nonporous resin for purification and analysis of oligonucleotides.

The ammonia solution delivered by the synthesizer is brought to 4%

[13] S. Agrawal and J.-Y. Tang, *Tetrahedron Lett.* **31**(52), 7541 (1990).

[14] S. Gryaznov and R. Letsinger, *Nucleic Acids Res.* **20**(8), 1979 (1992); see also **20**(13), 3403 (1992).

(v/v) triethylamine to keep the solution slightly basic during initial processing or storage at $-20°$ (prevents unintentional deprotection).

Removing the contaminants prior to anion-exchange chromatography can be advantageous, if a large fraction of the oligomers in the solution are truncated or depurinated, and especially when column capacity overload is approached. Prepurification can be accomplished with any of the numerous reversed-phase cartridges designed for that purpose. The following procedure was developed on a C_{18} Sep-Pak (Millipore).[15]

Reversed-Phase Extraction. The dried synthetic oligonucleotide mixture is resuspended in an appropriate volume (~ 2.0 ml for a 0.2-μmol synthesis) of distilled, deionized H_2O containing 30 mM triethylammonium bicarbonate (TEABC). The hydrophobic character of the dimethoxytrityl (DMT)-protecting group affords a simple method for separating the protected oligonucleotide from partially deprotected contaminants. The aqueous sample is applied to the Sep-Pak cartridge that has been preconditioned with 10 ml of acetonitrile (CH_3CN), 5 ml of 100 mM TEABC in 30% CH_3CN, and 5 ml of 25 mM TEABC. After application, the 5'-OH-terminated ("truncated") oligomers are eluted with 15 ml of 25 mM TEABC in 10% CH_3CN. The 5'-DMT-terminated oligonucleotides are then eluted with 10 ml of 0.1 M TEABC with 30% CH_3CN. The DMT-oligonucleotide fraction is evaporated to complete dryness (or to a gum) in a centrifugal concentrator [e.g., Savant (Hicksville, NY) SpeedVac], and stored at $-20°$. The capacity of the Sep-Pak is ~ 50 A_{260} (1 to 2 mg) of total oligonucleotide.

Deprotection of the "trityl-on" enriched oligonucleotides is accomplished by complete resuspension in 1 ml of 3% (v/v) acetic acid (for a 0.2-μmol synthesis) at $24°$, followed by incubation for 10 min. After incubation, the solution is neutralized with a few drops of concentrated NH_4OH. The released "trityl-on" residues may be removed by passing the neutralized solution through a Sep-Pak, as described above, or by extraction into an equal volume of ethyl acetate (vortex, allow to separate by spinning in a microcentrifuge, and discard the upper layer). A small aliquot (e.g., 10 μl) may be sampled and diluted for quantification of enriched oligonucleotide by measuring absorbance at 260 nm.

Anion-Exchange Separations of Oligonucleotides

Anion-exchange chromatography at different pH values affords somewhat orthogonal or complementary separations of oligonucleotides. Elution by $NaClO_4$ at pH 8 usually results in the target oligonucleotide being

[15] K. M. Lo, S. S. Jones, N. R. Hackett, and H. G. Khorana, *Proc. Natl. Acad. Sci. U.S.A.* **81,** 2285 (1984).

separated in a single peak that elutes after truncated or other undesirable components. However, oligonucleotides with palindromic or other forms of self-complementary sequences may form loops or hairpins that alter their chromatographic behavior. Also, truncated oligonucleotides containing residual DMT residues may coelute with full-length detritylated oligonucleotides.

Another condition that may interfere with the resolution of synthetic oligonucleotides is the presence of several consecutive guanine residues, which are capable of hydrogen-bonding associations atypical of Watson–Crick base pairing. Such oligonucleotides may exhibit broad peaks with low absorbance at 260 nm. Solutions with pH values above 12.4 (25 mM NaOH) are incapable of maintaining hydrogen bonding. Chromatography in 25 mM NaOH is thus recommended for oligonucleotides suspected of containing poly(G) runs or self-complementary sequences. This feature of high-pH chromatography and the potential problems accompanying pH 8 chromatography indicate that both purity assessment and purification should be done at pH 12.4.

Programs useful for the separation of oligonucleotides 8 to 30 bases long at pH 8 and pH 12.4 are listed in the Appendix as programs 1 and 2. These gradients will adequately separate standard phosphodiester oligonucleotides from 8 to 30 bases long. Each guanine and thymine base will contribute an extra negative charge at pH 12.4, thus oligonucleotides high in these bases will tend to elute later, relative to oligomers containing less G and T, at the higher pH.

Recovery of oligonucleotides (except phosphorothioates) after collection may employ ethanol precipitation, or by conversion of buffer to a volatile salt with a desalting column [e.g., Pharmacia (Piscataway, NJ) NAP converted to the ammonium acetate form], followed by vacuum evaporation of the volatile salt (e.g., SpeedVac). For ethanol precipitation, a 0.1 vol of 2.5 M ammonium acetate is added to the collected fractions, the solution gently mixed, 2.5 vol of ice-cold ethanol added, the solution mixed again, and the resulting suspension (precipitate not visible) centrifuged at 12,000 g for 7 min in a refrigerated microcentrifuge (Eppendorf 5402) or in a standard microcentrifuge (Fisher 235C) in a 4° chromatography refrigerator. Phosphorothioates should be concentrated by vacuum evaporation after buffer exchange on Pharmacia NAP columns (Sephadex G-25) using 10 mM ammonium acetate as eluent because of their low recovery from ethanol precipitates.

Sample Preparation

Ten to 500 μl of suspension buffer (TE: 10 mM Tris-HCl, 1 mM EDTA, pH8) as a control, or 0.1 to 1000 μg of sample in TE, are placed in autosampler vials [Sun Brokers (Wilmington, NC) vials, caps, and septa]. They are then capped and placed into chilled (8°) vial racks in the autosampler.

System Equilibration

The first run in a schedule should contain a blank run of Tris-EDTA (TE) or H_2O to elute any compounds left stored in or adsorbed onto the column since its last use. Following the blank run, samples containing 50 ng or more can be analyzed without further system equilibration.

Analytical Oligonucleotide Separations

Under the conditions described above, injections (~0.05–10 μg/peak) yield symmetric peaks. An example of a 2.7-μg injection of the Cetus PCR01 primer (base composition, G8 C5 T7 A5; Perkin-Elmer Cetus, Norwalk, CT) is shown in Fig. 2. These traces clearly show the presence of numerous truncated sequences typical of crude or "machine-grade" oligonucleotides at both pH 8 and pH 12.4.

Semipreparative Purifications. Semipreparative purifications (>100 μg of oligonucleotide per peak) can be performed by scaling up the column diameter (9 × 250 mm = 5-fold scaleup, 22 × 250 mm column = 30-fold scaleup), or by overloading the column. To scale up directly by increasing

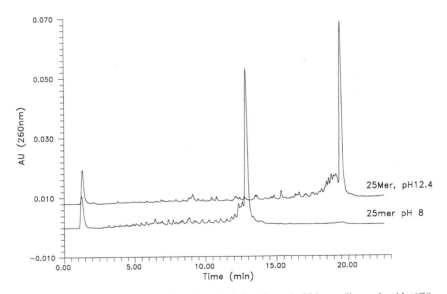

FIG. 2. Analytical-scale separation of a detritylated, crude 25-base oligonucleotide (G8, C5, T7, A5) from truncated contaminants on the NucleoPac PA100 column at 30°. Gradient conditions given in Appendix, gradients 1 and 2.

the column diameter, the flow rate should be raised by the same factor as the column capacity. Hence, a 50-μg purification on a 4 × 250 mm column at 1 ml/min could be scaled to 250 μg at 5 ml/min (9 × 250 mm column), or 1.5 mg at 30 ml/min (22 × 250 mm column). Under these conditions, the chromatograms would appear essentially identical.

Larger scale purifications can also be accomplished by overloading the analytical capacity of the NucleoPac column. This approximates the process of displacement chromatography. Under these conditions, the target analyte (i.e., the full-length oligonucleotide) can act as an eluent, displacing shorter length oligomers ahead of itself as it migrates through the column. The resulting trace does not indicate resolution of the full-length from smaller oligomers, because the latter are immediately replaced within the detector by the "pushing" full-length oligonucleotide. Figure 3 shows a trace typical of this chromatographic mode. A 1-mg sample of machine-grade Cetus PCR-01 primer was applied to the column, here in 0.15 M ammonium acetate. The smaller oligonucleotides were eluted with a step change to

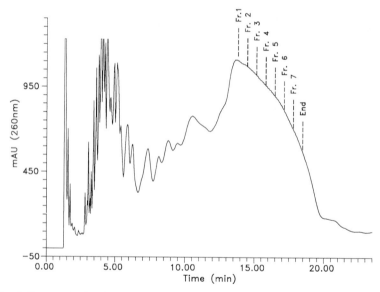

FIG. 3. Preparative loading of a 4 × 250 mm NucleoPac PA100, using a volatile eluent (ammonium acetate). One milligram of the same 25-base oligomer shown in Fig. 1 was purified by application in 0.15 M ammonium acetate. A step to 0.9 M in 1 min elutes most of the smaller truncated oligomers. The target detritylated 25-mer is eluted using quasidisplacement mode chromatography by a gradient from 0.9 to 1.1 M ammonium acetate in 20 min (1.5 ml/min). Each of the indicated fractions (1–7) exhibited the presence of a single 25-base oligonucleotide. The yield was 60% (24 of 40 A_{260}) at a final purity of >97%.

0.9 *M* ammonium acetate in 1 min. Elution of the longer oligonucleotides was accomplished with a shallow gradient (0.9 to 1.1 *M* ammonium acetate over 20 min). The resulting trace exhibits peaks of increasing width as the last (target) peak pushes the smaller oligomers ahead of it. Finally, the target oligomer elutes in a broad peak between 13.5 and 19.5 min. In this experiment, seven 1.0-ml fractions were collected between 13.5 and 18.5 min. Rechromatography of aliquots of the collected fractions revealed an essentially pure oligonucleotide component in each of these fractions. The collected fractions comprised 60% of the total A_{260}-absorbing material applied to the column [24 of 40 optical density units] at a purity greater than 97%.

Extended Length Oligonucleotides

The NaClO$_4$ gradient programs devised for normal-length oligonucleotides can, with minor modification, be applied to the separation of longer oligomers. Programs 3 and 4 in the Appendix describe the modified gradient programs.

Under these conditions, analytical-scale injections ($\sim$0.05–20 μg) of oligonucleotides of 30 to >70 bases yield sharp symmetric peaks. Examples of purity checks of gel-purified oligomers with 44 and 76 bases are shown in Fig. 4. Although less prominent in these gel-purified samples, the traces

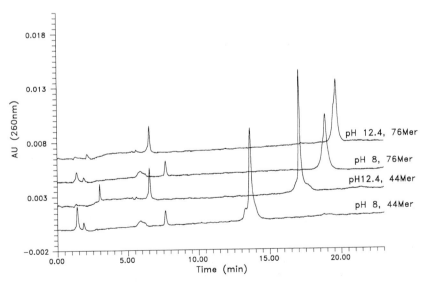

Fig. 4. Purity test of gel-purified 44- and 76-base oligonucleotides using the NucleoPac PA100 at 30°. Gradient methods 3 and 4 of the Appendix were used.

reveal the presence in contaminants of both oligonucleotide preparations at each pH. The pH 8 traces exhibit the presence of some "$n - 1$"-mer in both the 44-mer and 77-mer traces. The pH 12.4 traces demonstrate some "$n + 1$"-mer in the 44-mer, as well as some "$n - 1$"-mer in both oligonucleotide preparations. Injections represent 20 pmol (44-mer) or 8.5 pmol (76-mer).

Phosphorothioate (Antisense) DNA

Phosphorothioate DNA differs from normal (phosphodiester) DNA by a substitution of ionized sulfur for ionized oxygen on the phosphate backbone. These modified oligonucleotides can be synthesized as phosphodiesters using hydrogen phosphonate chemistry, and treated postsynthetically with sulfur in carbon disulfide. Alternatively, phosphorothioates can be synthesized by phosphoramidate chemistry, with the exception of the oxidation step. Where iodine is normally used to oxidize the phosphoramidate to phosphate, replacement of iodine with tetraethylthiuram disulfide, or other appropriate thiol reagent, results in a phosphorothioate product. The latter approach offers greater repetitive yield (efficiency) and allows for the synthesis of mixed phosphodiester/phosphorothioate oligonucleotides.

The sulfur contributes a subtantially greater hydrophobicity to the DNA, thus altering some of the characteristics affecting its manipulation. The sulfur also contributes to a much greater affinity for the NucleoPac anion-exchange media. Hence, the gradient methods that elute phosphodiester DNA will not adequately elute phosphorothioate DNA of similar length. Another alteration in the procedure for working with these modified oligonucleotides involves the conditions used to deprotect. These oligonucleotides are deprotected by incubation in 80% acetic acid for 3 hr at 24°. Ethyl acetate extraction of these modified oligonucleotides may yield poor recoveries. Gradient methods for elution of phosphorothioate oligonucleotides 10–30 bases in length are provided as methods 5 (pH 8) and 6 (pH 12.4) in the Appendix.

Chromatograms of phosphorothioate oligonucleotides also exhibit peaks that are fairly broad but often well separated, because conversion of phosphoramidate to phosphorothioate during the oxidation step, described above, results in chiral phosphorus atoms between each pair of bases. The phosphoramidate oxidation occasionally results in a residual phosphodiester bond. Even a single residual phosphodiester linkage in a 27-base phosphorothioate oligonucleotide will subtantially decrease the affinity of the oligonucleotide for the anion-exchange phase. This allows the separation of fully phosphorothioated oligonucleotides from those harboring residual phosphodiesters. For example, an oligonucleotide with 15 bases that contains 14 phosphorothioate linkages is separable from the same sequence 15-mer containing 13 phosphorothioate and 1 phosphodiester bond. If both truncated and incompletely phosphorothioated but full-length sequences are present, overlapping retention of the component oligonucleotides and their chiral counterparts will contribute to broad peaks and apparently incomplete resolution when compared to standard phosphodiester DNA.

Figure 5 compares the chromatography of a prepurified 15-mer phosphorothioate at pH 8 and pH 12.4. At pH 8, the full-length, fully phosphorothioated oligomer appears as a broad peak indistinguishable from sequences containing truncated sequences or residual phosphodiester linkages. At pH 12.4, the full-length, fully phosphorothioated oligomer is better resolved (blank traces are included to reveal "system" peaks that elute under these conditions).

The complexity of this sample matrix and the hydrophobicity of the phosphorothioates suggest that a reversed-phase sample cleanup step would be effective. Figure 6 compares the pH 8 and pH 12.4 chromatography of a phosphorothioate oligomer containing 15 thymidine residues. The sample was prepurified by the trityl-on method, using a polymer reversed-phase column (Hamilton, Reno, NV, PRP-1). The prepurification procedure should result in the presence of only full-length oligomers. However, anion-exchange chromatography on the NucleoPac column reveals the presence

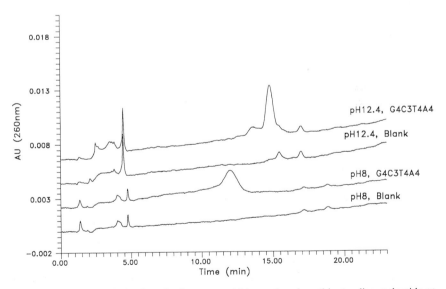

FIG. 5. Purity analysis of a mixed-sequence 15-base phosphorothioate oligonucleotide at pH 8 and 12.4 and at 30°. Gradient methods 5 and 6 of the Appendix were used. Traces from the blanks are included to identify "eluent system" peaks.

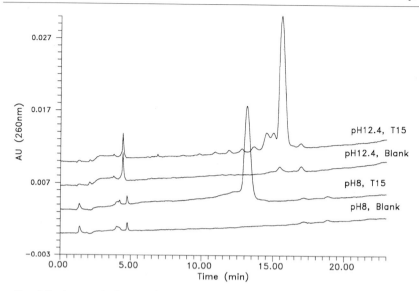

FIG. 6. Purity test of a "reversed-phase prepurified" T15 phosphorothioate oligonucleotide at pH 8 and 12.4 using the NucleoPac PA100 at 30°. Gradient methods 5 and 6 of the Appendix were used. Traces from the blanks are included to identify "eluent system" peaks.

of at least eight earlier-eluting oligomers. Bergot and Egan[16] and Bergot and Zon[17] described the separation of the full-length oligonucleotides containing differing numbers of residual phosphodiester linkages by anion-exchange chromatography. Subsequently, isolated oligomers were characterized by nuclear magnetic resonance (NMR) spectroscopy. The authors have now confirmed the separation of these species on the NucleoPac column (B. J. Bergot, personal communication, 1990).

Recovery by ethanol precipitation is substantially poorer for phosphorothioates than for phosphodiesters. When the target oligonucleotide contains more than 20 bases, ultrafilters with nominal molecular weight cutoff (NMWC) values of 5000 or less can be used to concentrate and desalt phosphorothioates. Alternatively, gel-permeation media (e.g., Sephadex G-25, as in Pharmacia NAP desalting columns) can be used to exchange salt from the purification step with a volatile buffer (e.g., 10 mM ammonium acetate) that is removed by vacuum evaporation (SpeedVac or lyophilizer). This desalting approach typically results in >80% recovery.

[16] B. J. Bergot and W. Egan, *J. Chromatogr.* **599**, 35 (1992).
[17] B. J. Bergot and G. Zon, *Ann. N.Y. Acad. Sci.* **30**, 310 (1991).

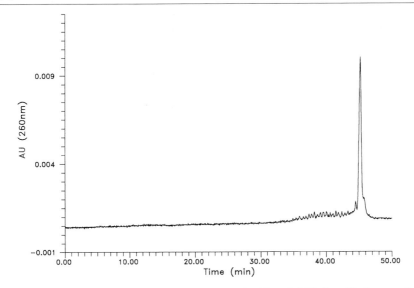

FIG. 7. Analysis of "purified" synthetic oligonucleotides. Gel-filled capillaries are also useful for determining the purity of synthetic oligonucleotides. This RP-HPLC-purified oligonucleotide was found to be only 70% pure, with the presence of at least 20 failure sequences detected. Separation conditions: capillary length, 50 cm total, 45 cm effective; capillary inner diameter, 75 μm; gel, polyacrylamide; injection, 20 sec at 0.5 kV, cathode electroinjection; separation field strength, 225 V/cm, 5 μA, 55 mW; capillary cooling, forced air cooling, ambient temperature; detection, 260-nm absorbance; sample, RP-HPLC-purified 32-mer, proprietary sequence.

High-Performance Liquid Chromatography versus Capillary Electrophoresis

For the purposes of analytical high-resolution separation, one can efficiently use gel-filled capillaries in 7.5 M urea and capillary electrophoresis equipment (e.g., Dionex CES1) to separate single-stranded nucleic acids. Figure 7 shows an example from the unpublished work of R. McCormick, H. Kumar, and N. Avdalovic (1993). It shows an HPLC (trityl on)-purified 32-mer profile run on a gel-filled capillary (6T/3.33C). This purified oligomer is found to be only 70% pure with at least 20 failure sequences detected.

The choice of the system (HPLC versus gel-filled capillary electrophoresis) to be used for the analysis of single-stranded DNA samples will depend on many factors, not the least of which is the availability of stable, gel-filled capillaries giving reproducible migration times. For oligonucleotides of 8–30 bases, HPLC has been characterized as superior, in speed of analysis, and practically equivalent in resolution, compared to gel-filled capillary electrophoresis.[1]

Double-Stranded DNA

Restriction Fragments and Polymerase Chain Reaction Products

Restriction enzymes are nucleases that reproducibly cut both strands of duplex DNA at sequences that are unique for each restriction enzyme. One use for these enzymes is to generate specific dsDNA sequences for insertion into and excision from plasmid or phage vectors during cloning. The size of cloned sequences varies widely and depends on the vectors used. Most of the plasmid and phage vectors used for cloning are substantially larger than the sequences inserted into them. Restriction nucleases recognize sequences from 4 to 14 bases long, and their activity results in cuts with or without overlapping (complementary) ends.

The polymerase chain reaction (PCR) process allows prodigious amplification of specific DNA sequences in a relatively short time. This is accomplished by polymerase-mediated extension of oligonucleotide primers complementary to, and bracketing, the desired sequence. With each cycle of extension, the concentration of the target sequence is doubled, until the primer, polymerase, or nucleoside triphosphate (NTP) concentration becomes limiting. The resulting dsDNA sequences are usually in the 50- to 2000-bp range and do not harbor "sticky" ends. For subsequent use, these sequences may require purification from the reaction matrix, which often contains high concentrations of glycerol, protein, and either oligonucleotide primers or NTPs.

Polyacrylamide and agarose gel electrophoresis are proven methods for separating and recovering such sequences, but they are labor-intensive techniques. Hence, a rapid method for automatic and unattended purification/recovery of restriction fragments and PCR products is valuable. The method should resolve nanogram to tens of microgram quantities of dsDNA in the 50- to 2000-bp size range, be relatively insensitive to sequence and base composition, and retain components primarily according to length. The method should also be fairly rapid, allowing high throughput, but be inexpensive enough to accommodate only occasional use.

The eluent system employed for the separation of ssDNA species can also be used to resolve many restriction DNA fragments. Gradient method 7 (see the Appendix) allows the separation of linear DNA species in the range of 72 to more than 1300 bp when restriction nuclease HaeIII is used

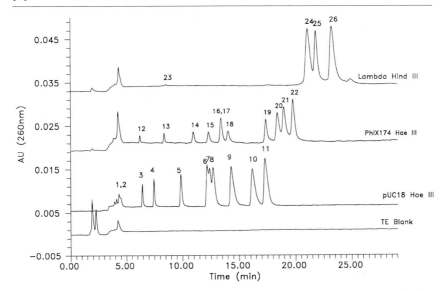

FIG. 8. Separation of DNA restriction fragments by NaClO$_4$ elution with a NucleoPac PA100 column. The bottom trace is a blank to show "system peaks." The middle traces show the separation of blunt-ended DNA fragments generated by treatment of pUC18 (lower middle trace) and ϕX174 (upper middle trace) with *Hae*III. Peaks 1–11 from pUC18: 1 (11 bp), 2 (18 bp), 3 (80 bp), 4 (102 bp), 5 (174 bp), 6 (267 bp), 7 (257 bp), 8 (298 bp), 9 (434 bp), 10 (458 bp), 11 (587 bp). Peaks 12–22 from ϕX 174: 12 (72 bp), 13 (118 bp), 14 (194 bp), 15 (234 bp), 16 and 17 (271 and 281 bp), 18 (310 bp), 19 (603 bp), 20 (872 bp), 21 (1078 bp), 22 (1353 bp). Peaks 23–26 from phage λ: 23 (125 bp), 24 (564 + 2027 + 2322 + 4361 + 6557 bp), 25 (9416 bp), 26 (23100 bp). Top trace: Fragments from the digestion of λ DNA by *Hind*III. This nuclease generates "sticky ends." The 564-, 2027-, 2322-, 4361-, and 6557-bp segments coelute if not denatured prior to injection (peak 24).

(Fig. 8). This nuclease cleaves GGCC sequences, leaving blunt ends. The restriction nuclease cleavage of λ DNA with *Hind*III (A $\downarrow$ AGCTT) results in fragments with "sticky" ends and lengths from 125 to 23,100 bp. The top trace in Fig. 8 shows that the sticky ends will limit resolution of some fragments. In this trace, peaks at 125, 9416, and 23,100 bp are observed; however, five fragments (564, 2027, 2322, 4361, and 6557 bp) are observed to coelute. This can be minimized by exposing the solution to elevated temperatures (e.g., 65° for 10 min) and quick cooling just prior to injection. The heat dissociates the 4-base hydrogen bonding, and quick cooling helps prevent reannealing before analysis. This task can be automated using the heating function of the system autosampler.

The products of PCR reactions vary considerably in base composition. The sensitivity of anion-exchange chromatography to this parameter pre-

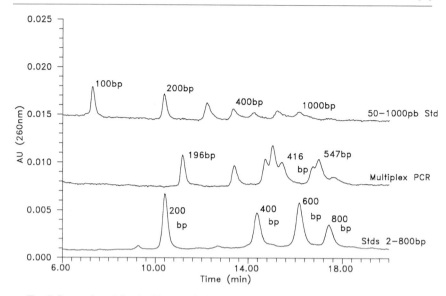

Fig. 9. Separation of the double-stranded products of PCR amplification uses the conditions in the Appendix, method 7. Two different standards (50–1000 and 200–800 bp) are shown above and below the trace from a multiplex PCR reaction employing nine sets of primers for Duchenne muscular dystrophy exons. Partial separation of eight products is demonstrated. The larger peak at ~15 min presumably contains two products.

cludes the direct assessment of product length by elution position. However, PCR products differing by more than ~10% in length are often resolved by chromatography under the conditions described for restriction fragments. Figure 9 shows a simultaneous analysis for nine different exons found in Duchenne muscular dystrophy and demonstrates resolution for eight of the nine amplified sequences known to be present in this sample.

Capillary gel electrophoresis is also a powerful tool for separating double-stranded DNA fragments of limited length. Figure 10 shows high-resolution separations performed on a low-percentage polyacrylamide gel-filled capillary (top trace), and in a sieving buffer (bottom trace). The separation of small DNA fragments up to 1 kbp in length can be resolved in gel-filled capillaries by using low-field strength. Although the separation was achieved for most of the fragments at the baseline level, more than 60 min is required to accomplish this task. The same fragments could be analyzed in half that

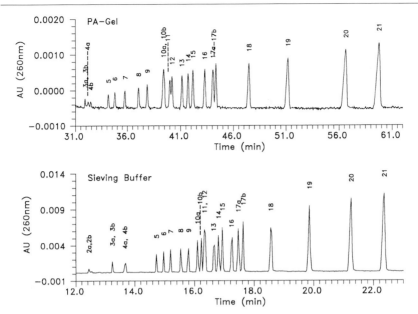

FIG. 10. Separation of low base pair DNA fragments using low-percentage acrylamide gel-filled and sieving buffer-filled capillaries. Small DNA fragments up to 1 kbp long can be separated in gel-filled capillaries at low field strengths, but analysis time is long relative to the same separation on sieving buffer capillary. The same fragments can be readily separated at higher field strengths, resulting in short analysis times using sieving buffers, because they tolerate higher field strengths without significant loss of resolution for small DNA fragments. Separation conditions for gel-filled capillary: capillary length, 45 cm total, 40 cm effective; capillary inner diameter, 75 μm; gel, polyacrylamide gel; injection, 12 sec at 0.7 kV, cathode electroinjection; separation field strength, 125 V/cm, 2 μA, 50 mW; capillary cooling, forced air cooling, ambient; detection, 260-nm absorbance; Separation conditions for sieving buffer-filled capillary: capillary length, 50 cm total, 45 cm effective; capillary inner diameter, 100 μm; sieving buffer, NucleoPhor SB1.5 kB (Dionex); running buffer, same as sieving buffer; injection, 10 sec at 1 kV, cathode electroinjection; separation field strength, 250 V/cm, 21 μA, 260 mW; capillary cooling, forced air cooling, ambient temperature; detection, 260-nm absorbance; sample, *Msp*I digest of pBR322 DNA, 500 μg/ml in TE (gel filled) or 250 μg/ml in TE (sieving buffer filled). Peaks (putative length in base pairs): 2a,b (15 + 15), 3a,b (26 + 26), 4a,b (34 + 34), 5 (67), 6 (76), 7 (90), 8 (110), 9 (123), 10a,b (147 + 147), 11 (160), 12 (169), 13 (180), 14 (190), 15 (205), 16 (217), 17a,b (238 + 238), 18 (307), 19 (404), 20 (527), 21 (622).

time using the Dionex sieving buffer. Thus, sieving buffers appear to be the better choice for dsDNA separations, and both HPLC and capillary electrophoresis are justified for assessing the purity of various single- and double-stranded DNA molecules. Perhaps the major step in determining which technique to apply in a given situation will depend on whether or not one wants to collect the separated fractions. Other than that, these two analytical techniques are more or less equivalent.

Plasmids

Reasons for purifying plasmids are (1) for subsequent use as cloning vectors, (2) as substrates for enzyme activity (e.g., topoisomerases), (3) as templates for transcription studies, (4) to determine plasmid copy number, (5) for DNA biophysical studies (e.g., superhelical organization), (6) for hybridization analysis, and (7) for sequencing. Plasmids are usually obtained from bacterial strains by transformation, followed by growth in media allowing selection of some characteristic conferred by the plasmid (e.g., drug resistance). Plasmids currently employed are small (0.9 to ~5 kbp) to accommodate inserted sequences of a large size or number before becoming too large for efficient transformation (limiting at ~15 kbp). Most also contain readily selectable markers (e.g., drug resistance), and well-defined recognition sites for several or many restriction nucleases. Many have had the sequences controlling (restricting) the copy number inactivated, so numerous copies are present under normal growth conditions. This approach permits recovery of more than a microgram of plasmid DNA in each milliliter of bacterial culture.

Preparation of "Cleared Lysates"

The starting material used for plasmid purification is similar for many different methods. To understand the complexity of the matrix obtained before the final purification is initiated, we briefly describe a widely used method for crude plasmid preparation.

Escherichia coli-containing plasmids are incubated overnight at 37° in 10 ml of Luria–Bertani (LB) growth medium.[18] This culture is added to 990 ml of fresh, sterile LB in a 5-liter Erlenmeyer flask, with the appropriate antibiotic to select for plasmid-containing cells [e.g., chloramphenicol (170 μg/ml)]. The culture is grown to an A_{600} of 0.4 by orbital mixing (300 cycles/min) at 37°. For plasmid amplification, or to limit chromosomal DNA synthesis and hence bacterial growth, without limiting plasmid replication, 5 ml of chloramphenicol (34 mg/ml in ethanol, stored at −20°) is added to the culture, and incubation is extended for 12 to 16 hr. Cells are harvested

---

[18] J. Sambrook, E. F. Fritsch, and T. Maniatis, "*Molecular Cloning: A Laboratory Manual*," 2nd Ed. Cold Spring Harbor Laboratory Press, Cold Spring Harbor, NY, 1989.

by centrifugation at 6300 g for 15 min at 4°. The supernatant is discarded and the pellet resuspended in 10 ml of STE buffer (0.1 M NaCl, 10 mM Tris-HCl, 0.1 mM EDTA, pH 8.0), which helps minimize release of cell wall components during later lysis. The centrifugal step is repeated (for 5 min here), and the pellet resuspended in 8 ml of ice-cold 50 mM Tris-HCl containing 10% sucrose, and transferred to a clean 30-ml centrifuge tube. Several methods for cell lysis can be used; here, we describe a modification of the method of Birnboim and Doly.[19]

Add 1 ml of lysozyme (10 mg/ml solution in 10 mM Tris-HCl, pH 8.0) to the resuspended culture, and incubate for 10 to 15 min. Add 2 ml of 0.5 M EDTA (pH 8.0) to chelate divalent cations and inhibit nucleases. The suspension is mixed by carefully inverting the tube several times, and is placed on ice for 10 min (vigorous mixing at this step will shear chromosomal DNA, rendering its separation from plasmid DNA difficult). Slowly add 2.75 ml of 4% (w/v) sodium dodecyl sulfate (SDS) in 0.8 M NaOH. Mix this solution with a glass rod to disperse the SDS completely. Add 6.9 ml of 3 M potassium acetate in 2 M acetic acid, and mix with the glass rod as before. Incubate at 4° for 5 min. A 5-min centrifugation at 12,000 g at 4° pellets the cell debris with the potassium precipitate of chromosomal DNA and protein. Transfer the supernatant to a fresh 50-ml centrifuge tube, add 20 ml of phenol–chloroform (1 : 1), and mix gently but thoroughly. Pellet the precipitated protein and transfer the supernatant to two fresh 50-ml centrifuge tubes (20 ml each). Add 12 ml of 2-propanol to each tube and let stand for 5 min to precipitate the nucleic acids. Harvest the precipitate by centrifugation at 12,000 g at 4° for 15 min. Carefully remove all residual solution from the sides of the tubes, and rinse with 1 ml of ice-cold 70% (v/v) ethanol. Dry in air for 15 min.

The precipitate now contains supercoiled and nicked plasmid DNA, cellular RNA, and possibly some chromosomal DNA. Major contamination by high molecular weight RNA is not uncommon and can dramatically affect the resulting plasmid purification by HPLC. To eliminate coelution of RNA with plasmid DNA, treat the "cleared" lysate with DNase-free RNase. This is prepared by boiling RNase for 15 min.[15] Treatment of the cleared lysate with this preparation for 1 to 2 hr minimizes the length of residual RNA to sizes that are readily separated from plasmid DNA in the extract (Fig. 11). Resolution of supercoiled from linear and relaxed forms of this plasmid (~7.2 kbp) was confirmed by agarose gel electrophoresis. The elution of supercoiled after the relaxed and linear forms has been

[19] H. C. Birnboim and J. Doly, *Nucleic Acids Res.* **7,** 1513 (1979).

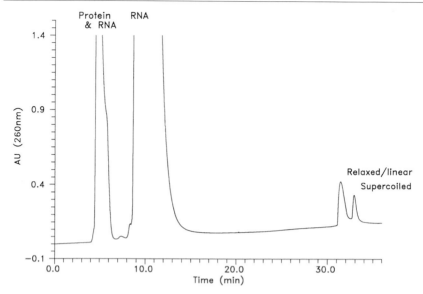

Fig. 11. Purification of a pBR322 derivative harboring an IgG insert. Gradient method 8 of the Appendix was used to separate the protein/RNA, RNA, and relaxed and supercoiled plasmids using a modified flow rate of 0.4 ml/min. This cleared lysate preparation exhibits a substantial percentage of nicked or linear plasmid. Agarose gel electrophoresis of the plasmid peaks reveals that the collected peaks remain as indicated after collection.

demonstrated for plasmids pBR322 (Fig. 12) and pUC19 (data not shown). These plasmids are 4.36 and 2.69 kbp, respectively.

The most widely accepted method for plasmid DNA isolation and purification is centrifugation in a cesium chloride gradient. Sambrook *et al.*[18] reported that it is necessary to centrifuge between 36 and 72 hr to obtain a reasonable separation between the supercoiled and linear DNA. However, major advances have been made in this field, and we briefly review the procedures available today to accomplish this task.

Isolation of Supercoiled Plasmid DNA by Isopycnic Ultracentrifugation. An excellent theoretical treatment of this subject can be found in articles by Minton[20] and Marque.[21] The base compositions, and hence densities, of plasmid and chromosomal DNAs are normally so similar that they cannot be separated by virtue of a density difference. However, it is possible to produce such a difference using the fluorescent dye ethidium bromide, which complexes with DNA by intercalation between base pairs, thus reduc-

---

[20] A. P. Minton, *Biophys. Chem.* **42,** 13 (1992).
[21] J. Marque, *Biophys. Chem.* **42,** 23 (1992).

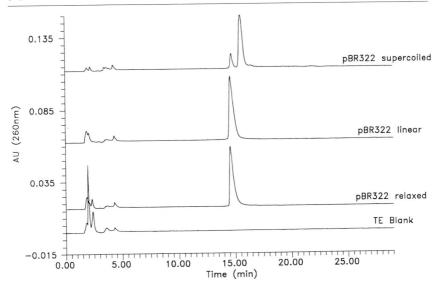

FIG. 12. Separation of supercoiled plasmid DNA (form I) from linear (form III) and nicked or relaxed (form II) DNA. A commercial preparation of supercoiled pBR322 having ~15% form II is chromatographed using the conditions described in the Appendix, gradient method 8 at 1.0 ml/min. Linear and relaxed forms were generated by treating the supercoiled stock with *Bam*HI restriction nuclease (one cleavage site on pBR322) or topoisomerase I. Agarose gels of each peak collected from these chromatograms reveal that the later eluting peak in the commercial preparation remains as form I (supercoiled) after collection, and the earlier eluting component remains as form II (relaxed).

ing its buoyant density. Intercalation of dye can occur only if the DNA double helix unwinds slightly, a process that results in strain in supercoiled plasmid DNA (SC or form I DNA), but which is unhindered in linear (form III) or nicked (form II) molecules. Consequently, in saturation concentrations of ethidium bromide, SC DNA will bind less dye per unit length, and will have a greater buoyant density than linear plasmid or chromosomal DNA as well as form II DNA of the same base composition. RNA molecules have considerably different buoyant densities and will, in principle, be easy to separate from DNA of SC, nicked, or linear DNA.

Adjustment of the CsCl concentration to 1.55 g/ml and ethidium bromide concentration to 0.6 g/ml in a plasmid preparation, followed by centrifugation to equilibrium in an appropriate rotor/ultracentrifuge (Table I), will result in the migration of SC and nicked DNA, chromosomal and linear plasmid DNA, and RNA to distinct positions (densities) in the ultracentrifuge tube. Supercoiled DNA will form a band below chromosomal DNA,

TABLE I
RECOMMENDED MINIMUM PLASMID SEPARATION TIMES[a]

Rotor	Tube size (ml)	Speed (rpm)	Time (hr)
Type 80 Ti	13.5	48,000	24
	6.3	59,000	7
	4.2	65,000	5
Type 75 Ti	13.5	49,000	24
	6.3	61,000	7
	4.2	71,000	5
Type 70.1 TI	13.5	49,000	24
	6.3	61,000	7
	4.2	70,000	5
VTi80	5.1	80,000	3
VTi65	5.1	65,000	4
VTi65.1	13.5	65,000	4.5
	6.3	65,000	4.5
VTi65.2	5.1	65,000	4

[a] Reprinted with permission from S. E. Little and D. K. McRorie, Rapid Separation of Plasmid DNA in Preparative Ultracentrifuge Rotors. Applications Data, DS-734A. Beckman Instruments, Inc., Fullerton, CA, 1989.

and RNA will migrate to near the bottom of the tube. Note that polysaccharides derived from cell wall components may also be present and will not bind ethidium bromide.

Anion-Exchange Chromatography. Supercoiled (form I) plasmid DNA is separated from proteins, RNA, and linear (form III) and nicked (form II) plasmid DNA using gradient method 8 (Appendix). Fractions eluting from the NucleoPac column were collected, concentrated, and desalted by ethanol precipitation, and analyzed by agarose gels in the presence of ethidium bromide. A cleared lysate, containing a pBR322-derived plasmid harboring an immunoglobulin G (IgG) insert, exhibited a large fraction (>99% of A_{260}-absorbing material) of RNA that eluted between 4 and 14 min. Two late-eluting peaks were identified as linear (eluting at ~31 min using a 0.4-ml/min flow rate; Fig. 11) and supercoiled (eluting at ~33 min from the same run). Evaluation of the shearing of pBR322 and pUC19 during chromatography at 0.4 and 1.0 ml/min by agarose gel electrophoresis in differing concentrations of ethidium bromide revealed no evidence of shearing at the higher flows (all tubing from the injector to fraction collector had an inner diameter of 0.020 in.).

Other New Microparticulate Columns

Two other microparticulate media have been described.[1] Oefner *et al.* packed columns with nonporous fused silica [2-μm particle diameter (d_p), polyethyleneimine (PEI) coated] or poly(styrene-divinylbensene (PS-DVB) grafted with poly(vinyl alcohol) (2.3-μm d_p, 0.1% PVA) and used these for anion-exchange and reversed-phase chromatography of nucleic acids (Fig. 13). These columns were compared with capillary electrophoresis, using the ability to resolve oligonucleotides and restriction fragments as benchmarks. For dsDNA, CE in the presence of hydroxyethylcellulose and ethidium bromide were judged to provide better resolution than the HPLC columns.

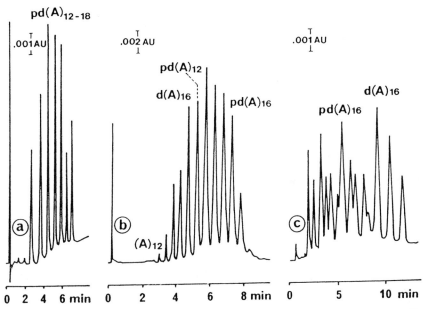

FIG. 13. HPLC separations of oligodeoxyadenylic acids. (a) Column Progel-TSK DEAE-NPR, 2.5 μm, 35 $\times$ 4.6-mm i.d.; buffer, 0.020 M Tris-HCl (pH 7.5); gradient, 0.15–0.5 M (NH$_4$)$_2$SO$_4$ in 15 min; flow rate, 1 ml/min; sample, 0.75 μg of p[d(A)$_{12-18}$]. (b) Column, PEI-silica, 2 μm, 30 $\times$ 4.6-mm i.d.; buffer, 0.05 M phosphate (pH 5.9)–30% (v/v) methanol; gradient, 0–0.5 M (NH4)$_2$SO$_4$ in 10 min; flow rate, 2 ml/min; sample, 1.75 μg of p[d(A)$_{12-18}$]. (c) Column, PS-DVB-PVA, 2.3 μm, 50 $\times$ 4.6-mm i.d.; buffer, 0.1 M TEAA (pH 7.0); gradient, 12.5–20% acetonitrile in 20 min; flow rate, 1 ml/min; sample, 0.5 μg of p[d(A)$_{12-18}$] and 0.083 μg of p[d(A)$_{16}$]. All chromatograms were obtained by means of UV detection at 254 nm at room temperature. [Reprinted by permission from P. J. Oefner *et al.*, J. Chromatogr. **625**, 334 (1992).]

Appendix: Gradient Files for Separation of Nucleic Acids Using Perchlorate Eluents

Program 1 for "Normal" Length (8- to 30-mer) Oligonucleotides at pH 8

Time (min)	Flow (ml/min)	E1 (%)	E2 (%)	E3 (%)	E4 (%)	Curve number	Comments[a]
0.0	1.50	88	0	10	2	5	Load at 7.5 mM ClO$_4^-$
20.0	1.50	52	0	10	38	3	7.5–143 mM/20 min convex-3
21.5	1.50	0	0	0	100	9	143–375 mM concave-9
25.5	1.50	0	0	0	100	5	Hold 4 min (flush column)[b]
25.6	1.50	88	0	10	2	5	Regenerate with 7.5 mM ClO$_4^-$

[a] Regeneration and reequilibration of the column are completed by setting the cycle time of the autosampler to 36 min.

[b] When different oligonucleotides are sequentially chromatographed, and carryover from one injection to another must be scrupulously avoided, several consecutive 2-ml manual injections of 0.2 M HCl will convert residual oligonucleotide to essentially uncharged species that are eluted by the chloride anion. When this is done, a 5-min regeneration of the stationary phase to the ClO$_4$ form is accomplished with 100% E4. When automated cleaning is required, a fifth eluent container may be plumbed directly into the chromatographic system under control of an electrical or pneumatic valve that converts eluent 4 from 0.375 M NaClO$_4$ to 0.2 M HCl. The Dionex system supports both electrical and pneumatic control. This allows washing and regeneration steps to be added to the method. With the automated column cleaning step, the autosampler cycle time is changed to 40 min.

Program 2 for "Normal" Length (8- to 30-mer) Oligonucleotides at pH 12.4

Time (min)	Flow (ml/min)	E1 (%)	E2 (%)	E3 (%)	E4 (%)	Curve number	Comments[a]
0.0	1.50	86	12	0	2	5	Load at 7.5 mM ClO$_4^-$
20.0	1.50	43	12	0	45	4	7.5–169 mM/20 min convex-5
21.5	1.50	0	0	0	100	9	169–375 mM concave-9
25.5	1.50	0	0	0	100	5	Hold 5 min (flush column)
25.6	1.50	86	12	0	2	5	Regenerate with 25 mM OH$^-$

[a] Regeneration and reequilibration of the column are completed by setting the cycle time of the autosampler to 36 min.

Program 3 for "Extended" Length (30- to 70-mer) Oligonucleotides at pH 8

Time (min)	Flow (ml/min)	E1 (%)	E2 (%)	E3 (%)	E4 (%)	Curve number	Comments[a]
0.0	1.50	88	0	10	2	5	Load at 7.5 mM ClO$_4^-$
0.1	1.50	86	0	10	4	5	Step to 15 mM ClO$_4^-$
20.1	1.50	47	0	10	43	3	15–161 mM/20 min convex-3
21.5	1.50	0	0	0	100	9	161–375 mM concave-9
25.5	1.50	0	0	0	100	5	Hold 5 min (flush column)
25.6	1.50	88	0	10	2	5	Regenerate with 25 mM OH$^-$

[a] Regeneration and reequilibration of the column are completed by setting the cycle time of the autosampler to 36 min.

Program 4 for "Extended" Length (30- to 70-mer) Oligonucleotides at pH 12.4

Time (min)	Flow (ml/min)	E1 (%)	E2 (%)	E3 (%)	E4 (%)	Curve number	Comments[a]
0.0	1.50	86	12	0	2	5	Load at 7.5 mM ClO$_4^-$
20.0	1.50	36	12	0	52	3	7.5–195 mM/20 min convex-3
21.5	1.50	0	0	0	100	9	195–375 mM concave-9
25.5	1.50	0	0	0	100	5	Hold 5 min (flush column)
25.6	1.50	86	12	0	2	5	Regenerate with 25 mM OH$^-$

[a] Regeneration and reequilibration of the column are completed by setting the cycle time of the autosampler to 36 min.

Program 5 for Phosphorothioates (Antisense Oligonucleotides) at pH 8

Time (min)	Flow (ml/min)	E1 (%)	E2 (%)	E3 (%)	E4 (%)	Curve number	Comments[a]
0.0	1.50	88	0	10	2	5	Load at 7.5 mM ClO$_4^-$
0.1	1.50	75	0	10	15	5	Step to 56 mM ClO$_4^-$
20.1	1.50	4	0	10	86	4	56–323 mM/20 min convex-4
21.5	1.50	0	0	0	100	7	323–375 mM concave-7
25.5	1.50	0	0	0	100	5	Hold 4 min (flush column)
25.6	1.50	88	0	10	2	5	Regenerate with 7.5 mM ClO$_4^-$

[a] Regeneration and reequilibration of the column are completed by setting the cycle time of the autosampler to 36 min.

Program 6 for Phosphorothioates (Antisense Oligonucleotides) at pH 12.4

Time (min)	Flow (ml/min)	E1 (%)	E2 (%)	E3 (%)	E4 (%)	Curve number	Comments[a]
0.0	1.50	86	12	0	2	5	Load at 7.5 mM ClO$_4^-$
0.1	1.50	73	12	0	15	5	Step to 56 mM ClO$_4^-$
20.1	1.50	0	12	0	88	4	56–330 mM/20 min convex-4
21.5	1.50	0	0	0	100	7	330–375 mM concave-7
25.5	1.50	0	0	0	100	5	Hold 5 min (flush column)
25.6	1.50	86	12	0	2	5	Regenerate with 25 mM OH$^-$

[a] Regeneration and reequilibration of the column are completed by setting the cycle time of the autosampler to 36 min.

Program 7 for Linear Double-Stranded DNA (RFs and PCR Products, pH 8)

Time (min)	Flow (ml/min)	E1 (%)	E2 (%)	E3 (%)	E4 (%)	Curve number	Comments[a]
0.0	1.00	88	0	10	2	5	Load at 7.5 mM ClO$_4^-$
0.1	1.00	47	0	10	43	5	Step to 161 mM ClO$_4^-$
26.1	1.00	32	0	10	58	3	161–218 mM/26 min convex-3
27.0	1.00	0	0	0	100	9	323–375 mM concave-9
32.0	1.00	0	0	0	100	5	Hold 5 min (flush column)
33.0	1.00	88	0	10	2	5	Regenerate with 7.5 mM ClO$_4^-$

[a] Regeneration and reequilibration of the column are completed by setting the cycle time of the autosampler to 45 min.

Program 8 for "Supercoiled" versus Nicked/Linear DNA (pH 8)

Time (min)	Flow (ml/min)	E1 (%)	E2 (%)	E3 (%)	E4 (%)	Curve number	Comments[a]
0.0	1.00	88	0	10	2	5	Load at 7.5 mM ClO$_4^-$
0.1	1.00	47	0	10	43	5	Step to 161 mM ClO$_4^-$
26.1	1.00	30	0	10	60	3	161–225 mM/26 min convex-3
27.0	1.00	0	0	0	100	9	225–375 mM concave-9
32.0	1.00	0	0	0	100	5	Hold 5 min (flush column)
33.0	1.00	88	0	10	2	5	Regenerate with 7.5 mM ClO$_4$

[a] Regeneration and reequilibration of the column are completed by setting the cycle time of the autosampler to 45 min.

Acknowledgment

The authors acknowledge the editorial assistance of Sylvia Morris.

Section II

Electrophoresis

A. Slab Gel Electrophoresis: High Resolution
Articles 8 through 10

B. Capillary Electrophoresis
Articles 11 through 14

[8] Two-Dimensional Polyacrylamide Gel Electrophoresis

By Michael J. Dunn and Joseph M. Corbett

Introduction

Techniques of polyacrylamide gel electrophoresis (PAGE) are capable of high-resolution separations of proteins, but even under optimal conditions, any of the current one-dimensional methods can resolve only about 100 distinct zones. However, it is often necessary to analyze samples of a much higher complexity, such as those from whole cells or tissues. For example, the human genome consists of 3×10^9 base pairs (bp) of DNA forming an estimated 50,000 to 100,000 genes. Perhaps as many as 5000 of these genes may be expressed as proteins in any particular cell type, so that an electrophoretic technique capable of separating this number of proteins is required for global analysis of protein expression at the cell or tissue level.

This has resulted in the search for electrophoretic methods with the potential to separate complex protein mixtures containing several thousand proteins and to resolve proteins sharing similar physicochemical properties. The best approach to this problem is to combine two different one-dimensional electrophoretic procedures into a two-dimensional (2D) technique. The first 2D electrophoretic separation probably can be ascribed to Smithies and Poulik,[1] who in 1956 described a 2D method combining paper and starch gel electrophoresis for the separation of serum proteins. Subsequent developments in electrophoretic technology, such as the use of polyacrylamide as a support medium, the use of discontinuous buffer systems, and the use of polyacrylamide concentration gradients, were rapidly applied to 2D separations. These developments have been reviewed by Dunn.[2] Of particular importance was the application of isoelectric focusing (IEF) techniques to 2D separations, as it then became possible for the first-dimension separation to be based on charge. The coupling of IEF with polyacrylamide gel electrophoresis in the presence of the anionic detergent sodium dodecyl sulfate (SDS–PAGE) in the second dimension resulted in a 2D method that separated proteins according to two independent parameters, i.e., charge and size. These developments culminated in 1975

---

[1] O. Smithies and M. D. Poulik, *Nature* (*London*) **177**, 1033 (1956).
[2] M. J. Dunn, *in* "Advances in Electrophoresis" (A. Chrambach, M. J. Dunn, and B. J. Radola, eds.), Vol. 1, p. 1. VCH, Weinheim, 1987.

with the description by O'Farrell[3] of a method of 2D PAGE optimized for the separation of the total cellular proteins of *Escherichia coli* (*E. coli*). This approach uses a combination of IEF in cylindrical 4% T, 5% C polyacrylamide gels containing 8 *M* urea and 2% (w/v) of the nonionic detergent, Nonidet P-40 (NP-40), with the discontinuous gradient SDS–PAGE system of Laemmli.[4] This method resulted in the separation of about 500 polypeptides of *E. coli*.

The technique of O'Farrell has formed the basis for most developments in 2D PAGE since 1975 and the popularity of this methodology can be judged from the fact that in excess of 1000 papers using this technique have been published. In this chapter the classic method of O'Farrell is described. Limitations of this technique are discussed and procedures designed to overcome these problems described. Readers interested in a more detailed discussion of 2D PAGE and its application to diverse areas of biology and biomedicine are referred to a series of reviews and books dedicated to this technology.[2,5–8]

Sample Preparation

The diversity of samples that are analyzed by 2D PAGE precludes the recommendation of a universal method of sample preparation. The method is considered to be able to resolve proteins differing by as little as 0.1 pH unit in p*I* and by 1 kDa in their molecular mass. It is, therefore, essential to minimize protein modifications during sample preparation as these will inevitably result in the presence of artifactual spots on the final 2D maps. In particular, samples containing urea must not be heated as this can introduce considerable charge heterogeneity due to carbamoylation of the proteins by isocyanate formed from the decomposition of urea. In addition, protease activity within the sample can readily result in the degradation of sample proteins and result in artifactual low molecular weight spots. Samples should be subjected to minimum handling and must be kept cold at all times. It

[3] P. H. O'Farrell, *J. Biol. Chem.* **250,** 5375 (1975).
[4] U. K. Laemmli, *Nature* (*London*) **227,** 680 (1970).
[5] J. E. Celis and R. Bravo (eds.), "Two-Dimensional Gel Electrophoresis of Proteins: Methods and Applications." Academic Press, Orlando, FL, 1984.
[6] B. S. Dunbar, "Two-Dimensional Electrophoresis and Immunological Techniques." Plenum, New York, 1987.
[7] A. T. Endler and S. Hanash (eds.), "Two-Dimensional Electrophoresis: Proceedings of the International Two-Dimensional Electrophoresis Conference, Vienna." VCH, Weinheim, Germany, 1989.
[8] M. J. Dunn (ed.), "2-D PAGE '91: Proceedings of the International Meeting on Two-Dimensional Electrophoresis, London." NHLI, London, 1991.

is possible to add protease inhibitors such as phenylmethylsulfonyl fluoride (PMSF), but such reagents can also result in deleterious protein modification and introduce charge artifacts. Four broad categories of sample types can be considered.

Samples of Soluble Proteins

Soluble samples available in liquid form can often be analyzed by 2D PAGE with no or minimal pretreatment. This type of sample is exemplified by body fluids such as serum, plasma, urine, cerebrospinal fluid, semen, and amniotic fluid. Serum and plasma samples have an inherently high protein concentration and can be analyzed directly by 2D PAGE after dilution with sample solubilization buffer. However, a major problem with serum samples is the high abundance of a small number of proteins such as albumin and the immunoglobulins and these can obscure many of the minor components of the sample. These high-abundance proteins can be depleted from the sample prior to 2D separation using affinity chromatography, but there is a significant risk that this will result in the nonspecific removal of other protein components.

Other types of body fluids have a low protein content but contain a relatively high concentration of salts, which can interfere with the IEF dimension. Such samples must usually be desalted prior to 2D PAGE by techniques such as dialysis or gel chromatography. The samples must then be concentrated by methods such as lyophilization or precipitation with trichloroacetic acid (TCA) or ice-cold acetone. The latter approach has been found to be particularly useful for the concentration of a variety of samples for analysis by 2D PAGE.[9] The development of sensitive detection methods, such as silver staining, has made it possible to analyze some body fluid samples without recourse to extensive desalting and concentration procedures.

Samples of Solid Tissues

Samples of solid tissues should be disrupted in the presence of sample solubilization buffer. Small fragments of tissue should be wrapped in aluminum foil, frozen in liquid nitrogen, and crushed to a fine powder between two cooled metal blocks. Large samples of tissue can be processed by homogenization in solubilization buffer using a Polytron (Beckman, Fullerton, CA) or Ultra-Turrax homogenizer, but heating and foaming must be minimized if the risk of protein modification is to be minimized.

[9] J. Pipkin, J. F. Anson, W. G. Hinson, E. R. Burns, and G. L. Wolff, *Electrophoresis* **6,** 306 (1985).

Samples of Isolated Cells

Circulating cells (e.g., erythrocytes, leukocytes, platelets) and cells grown *in vitro* in suspension culture can be simply harvested by centrifugation, washed in phosphate-buffered saline (PBS), and dissolved in sample solubilization buffer. Cultured cells grown as monolayers on glass or plastic substrates also require little processing. First, the culture medium should be aspirated and the cell layer washed with PBS to minimize contamination of the sample with medium constituents, in particular serum proteins. The cell layer can then be scraped, washed again in PBS, and harvested by centrifugation. The use of proteolytic enzymes to release the cells should be avoided as this will result in degradation of sample proteins. Alternatively, the cells can be lysed while still attached to the culture substrate by the addition of a small volume of solubilization buffer. Samples containing high levels of nucleic acids should be treated by adding one-tenth of a volume of a protease-free DNase/RNase mixture [containing DNase I (1 mg/ml), RNase A (500 μg/ml), 0.5 M Tris (pH 7), 50 mM MgCl$_2$] and incubating in the cold until the sample is no longer viscous.[10]

Samples of Plant Tissues

The majority of plant proteins can be treated like animal tissues, but leaf proteins must first be extracted with acetone to remove phenolic pigments.

Sample Solubilization

The most widely used solubilization procedure used for processing samples to be analyzed by 2D PAGE is that described by O'Farrell[3] using a mixture of 4% (w/v) NP-40, 9.5 M urea, 1% (w/v) dithiothreitol (DTT), and 2% (w/v) synthetic carrier ampholytes. This method gives excellent results for the majority of samples. However, not all protein complexes are fully disrupted using this mixture. This results in incomplete sample protein entry into the first-dimension IEF gels and the presence of artifactual spots in the 2D pattern representing persistent protein complexes.

The anionic detergent, SDS, is able to disrupt most noncovalent protein interactions, but as it is highly charged it cannot be used in IEF gels. However, it is possible to presolubilize samples in 1% (w/v) SDS. In our experience most samples do not require heating in SDS to achieve solubilization. It should be noted that protein degradation can occur when proteins are heated in SDS, so that the time and temperature must be optimized for the particular sample under investigation. After presolubilization, the

---

[10] J. I. Garrels, *J. Biol. Chem.* **254**, 7961 (1979).

sample is diluted with the standard NP-40–urea mixture prior to 2D PAGE. This is intended to displace the SDS from the proteins, replacing it with nonionic detergent, while retaining the proteins in a soluble state. It is essential to control the ratio of SDS to protein (1:3) and SDS to NP-40 (1:8) if effective solubilization is to be achieved while minimizing the deleterious effect of SDS on the IEF separation.[11]

Several other procedures for the solubilization of specific types of sample for subsequent analysis by 2D PAGE have been described (reviewed by Dunn and Burghes[12]), but most of these have not gained general acceptance. The zwitterionic detergent 3-[(3-cholamidopropyl)dimethylammonio]-1-propane sulfonate (CHAPS) appears to be particularly effective for the separation of membrane proteins[13] and is now often used in preference to NP-40 for the solubilization of proteins for analysis by 2D PAGE. Rabilloud and colleagues[14] have described the synthesis of a series of amidosulfobe-taine detergents that have been found to give effective solubilization of samples for 2D PAGE, but these are unlikely to gain general popularity unless they become available commercially.

First Dimension

Isoelectric Focusing Using Synthetic Carrier Ampholytes

The first dimension of 2D PAGE as described by O'Farrell[3] is carried out in cylindrical rod IEF gels (3 to 5% T) cast in glass capillary tubes (1- to 1.5-mm i.d.). The gels contain $8 M$ urea and nonionic or zwitterionic detergent. The detergent is usually included at 2% (w/v), but this can often be reduced to 0.5% without any adverse effect on the separation pattern. The IEF gels also contain 2% (w/v) synthetic carrier ampholytes that gener-ate the pH gradient required for the separation. A variety of commercial synthetic carrier ampholyte preparations are available, and most of these are suitable for use in 2D PAGE. It should be noted that considerable variability can occur between batches of ampholyte from the same manufac-turer, and this can result in problems in the long-term reproducibility of 2D patterns. A broad-range ampholyte (pH 3–10) is often used, but a pH 4–8 ampholyte mixture can also give excellent results as the final pH gradi-ents rarely extend above pH 7.5 using rod IEF gels. A procedure for rod gel IEF is presented that has been found to yield consistently good 2D

[11] G. F. Ames and K. Nikaido, *Biochemistry* **15,** 616 (1976).
[12] M. J. Dunn and A. H. M. Burghes, *Electrophoresis* **4,** 97 (1983).
[13] G. H. Perdew, H. W. Schaup, and D. P. Selivonchick, *Anal. Biochem.* **135,** 453 (1983).
[14] T. Rabilloud, E. Gianazza, N. Cattò, and P. G. Righetti, *Anal. Biochem.* **185,** 94 (1990).

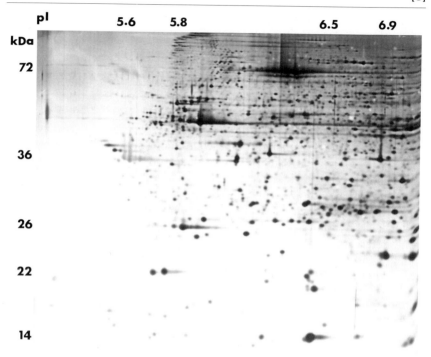

Fɪɢ. 1. Typical separation profile obtained by 2D PAGE using a cylindrical IEF gel containing synthetic carrier ampholytes in the first dimension. The sample contained 15 μg of human heart proteins. The pattern was visualized by silver staining. The protein molecular masses, indicated on the left-hand side, were calculated using standard marker proteins.

separations of a variety of samples in our laboratory. A typical 2D separation of human heart proteins using this technique is shown in Fig. 1.

Casting of Isoelectric Focusing Gels

1. Glass tubes (20 cm in length, 1.5-mm i.d.) for IEF should be cleaned in hydrochloric acid, then rinsed thoroughly in distilled deionized water and acetone before drying in a hot air cabinet.

2. Mark the required gel length (14.5 cm) on the tubes with a felt pen and then place them in a clean graduated measuring cylinder.

3. Prepare 20 ml of IEF gel mixture containing 10 g of urea, 7.4 ml of double-distilled H_2O, and 3.0 ml of acrylamide stock solution (30% T, 2.6% C). Dissolve the urea and deionize by stirring with 0.2 g of Amberlite MB-1 resin for 1 hr. Filter and degas the mixture before the addition of 0.3 g of CHAPS, 1 ml of Resolyte 4-8 (Merck, Rahway, NJ), 15 μl of

N,N,N',N'-tetramethylethylenediamine (TEMED) and 30 μl of 10% (w/v) ammonium persulfate (freshly prepared).

4. Pour the gel mixture into the graduated cylinder containing the tubes until the acrylamide rises to the level marked. The top 0.5 cm of gel solution will not polymerize as it is in contact with the air and serves as an overlay. It is essential that the tubes be filled to the same level in order to maximize reproducibility.

5. Allow the gels to polymerize for 1.5 hr.

6. Remove the tubes from the cylinder and trim off excess acrylamide.

Isoelectric Focusing

1. Apparatus for running gels in glass tubes is available from a number of commercial suppliers and two chambers designed specifically for the IEF dimension of 2D PAGE are available as part of integrated 2D PAGE equipment systems [the IsoDalt system from Hoefer (San Francisco, CA) and the Investigator system from Millipore (Bedford, MA)].

2. Insert the rod gels in the top chamber of the electrophoresis tank.

3. Fill the lower anode chamber with 10 mM phosphoric acid and flush any trapped air bubbles from the ends of the tubes using a buffer-filled syringe.

4. Degas the cathode buffer (20 mM sodium hydroxide) to remove dissolved CO_2 and add it to the top chamber. Remove air remaining in the upper part of the tubes using a Hamilton syringe filled with cathode buffer, ensuring that the unpolymerized urea solution is not disturbed.

5. The gels can be prefocused at 200 V for 15 min, 300 V for 30 min, and finally 400 V for 60 min for a total of 600 V · hr. This step is optional, and we have found in our studies that it can be safely omitted without any deleterious effect on the final 2D separation patterns.

6. Apply the samples (20–50 μl) using a Hamilton syringe. The amount of total sample protein to be loaded depends on the diversity and relative abundance of the proteins constituting the particular sample and the visualization method to be employed. In our studies of the proteins of the human heart, we normally load 10–30 μg of protein for silver staining, while for staining with Coomassie Brilliant Blue R-250 a load of 100–300 μg of protein is recommended.

7. Carry out isoelectric focusing at a constant setting of 800 V for an appropriate number of volt-hours. In our studies of human myocardial proteins the run time is 16,000 V · hr. The run time required is dependent on the length of the IEF gel, the type of sample to be analyzed, and the particular batch of ampholytes used for the separation. It is essential, then, that the IEF separation be optimized and this can be achieved by including a series of commercially available pI markers in the sample mixture. These

markers are produced by the carbamoylation of a purified preparation of a standard protein such as creatine kinase. When applied to a linear pH gradient, the markers appear as a row of equally spaced spots (Fig. 2). A time course can be performed to determine the optimal number of volt-hours required to give the best separation across the entire length of the IEF gel.

8. Extrude the gels onto labeled lengths of Parafilm, using a water-filled syringe attached to the gel tube with a short piece of rubber tubing, and store at −70°. Care must be taken not to damage, stretch, or break the delicate gels. To overcome this problem, in the Investigator 2D PAGE

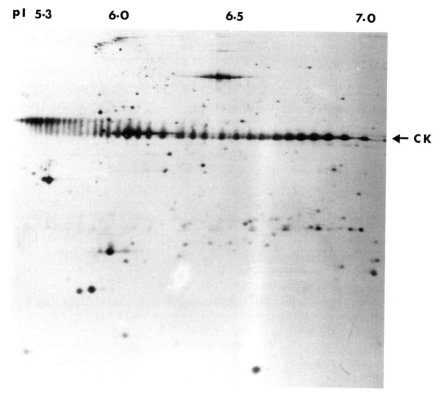

pI 5.3   6.0   6.5   7.0   ← CK

Fig. 2. Two-dimensional separation showing carbamoylated creatine kinase (CK) p*I* markers visualized by silver staining. The sample also contained a small amount of human heart proteins to provide reference points for comparison and matching of other 2D patterns. The relative p*I* values indicated were estimated from the known isoelectric point values of the charge train proteins.

system (Millipore), a 0.08-mm thread is incorporated into the IEF gels at the time of polymerization.[15] In addition a modified Luer adapter is provided to connect the IEF tube to a syringe for gel extrusion.

Problems of Isoelectric Focusing Using Synthetic Carrier Ampholytes

As we have already discussed, considerable variability can be encountered between batches of ampholyte from the same manufacturer and this can result in problems of reproducibility in the first dimension. In addition there is evidence that some proteins can complex with ampholyte molecules and these protein–ampholyte interactions can result in artifactual spots on 2D maps. Perhaps more seriously, pH gradients generated using synthetic carrier ampholytes are subject to severe cathodic drift. This problem is exacerbated in the case of rod IEF gels as a result of the very high electroendosmotic flow caused by charged groups on the glass walls of the capillary tubes. As a consequence, pH gradients are unstable and rarely extend above pH 8.0, leading to loss of basic proteins from 2D maps.

Various approaches to this problem have been described to produce gradients extending to pH 10, including special treatment of the glass IEF tubes[16] and the use of horizontal flat-bed IEF.[16] However, the most commonly used method is again attributable to O'Farrell et al.[17] and is known as nonequilibrium pH gradient electrophoresis (NEPHGE). In this approach the polarity of the IEF apparatus is reversed so that the lower buffer chamber becomes the cathode and the upper reservoir forms the anode. A prefocusing step cannot be used; the sample is applied at the anode and a short run time, on the order of 4000 V·hr, is used. Few proteins will attain their pI positions in the gel under these conditions, so that separation occurs on the basis of protein mobility in a rapidly forming pH gradient. The reproducibility of this method is difficult to control as it is a nonequilibrium method that is sensitive to experimental conditions, synthetic carrier ampholyte composition, run time, gel length, and sample composition.[12] Thus considerable care must be exercised if samples are to be analyzed by 2D PAGE using the NEPHGE technique. Of course, the more acidic proteins are absent from the resulting 2D patterns, so that a combination of 2D patterns using equilibrium IEF and NEPHGE must be used for analysis of each sample.

[15] W. F. Patton, M. G. Pluskal, W. M. Skea, J. L. Buecker, M. F. Lopez, R. Zimmerman, L. M. Belanger, and P. D. Hatch, *BioTechniques* **8**, 518 (1990).
[16] A. H. M. Burghes and M. J. Dunn, *Electrophoresis* **3**, 354 (1982).
[17] P. Z. O'Farrell, M. M. Goodman, and P. H. O'Farrell, *Methods Cell Biol.* **12**, 1133 (1977).

Isoelectric Focusing Using Immobilized pH Gradients

The development of immobilized pH gradients (IPGs)[18] has provided a solution to the problems associated with the use of synthetic carrier ampholytes, as this technique is able to generate reproducible, stable pH gradients of any desired pH range. The Immobilines (Pharmacia, Piscataway, NJ) are a series of seven substituted acrylamide derivatives with different pK values. A thorough review of this topic appears in the book by Righetti.[19,19a] IPG–IEF gels are made by generating a gradient of the appropriate Immobiline solutions, so that during polymerization the buffering groups forming the pH gradient are covalently attached and immobilized via vinyl bonds to the polyacrylamide backbone. This immobilization results in the elimination of pH gradient drift, but not electroendosmosis, making pH gradients reproducible and infinitely stable. These properties of IPG–IEF gels should theoretically make them ideal for use as the first dimension of 2D PAGE.

However, serious problems were encountered when attempts were made to use IPG–IEF for the first dimension of 2D PAGE. These problems were not due to problems with the IPG–IEF dimension itself. Rather, they were caused by difficulties in elution and transfer of proteins from the IPG–IEF gels to the second-dimension SDS–PAGE gels, resulting in severe streaking of spots on the 2D patterns. This problem was found to be due to the presence of fixed charges on the Immobiline matrix, leading to increased electroendosmosis in the region of contact between the IPG–IEF gel and the SDS–PAGE gel and resulting in disturbance of migration of proteins from the first- to the second-dimension gel.

The problems associated with the use of IPGs for 2D PAGE have now been overcome, largely due to the work of Görg and colleagues in Munich, and a standardized protocol of 2D PAGE has been described.[20] In this method, IPG gels (0.5 mm thick) are cast on GelBond PAG support films (Pharmacia) and subsequently dried. The dried gels are cut into strips that are rehydrated in a solution containing urea and nonionic or zwitterionic detergent. These IPG gel strips, which are analogous to cylindrical IEF gels of the classic O'Farrell technique, are then used for the IEF dimension of 2D PAGE using a horizontal flat-bed IEF apparatus.

Immobilized pH gradients were originally optimized for the generation

[18] B. Bjellqvist, K. Ek, P. G. Righetti, E. Gianazza, A. Görg, R. Westermeier, and W. Postel, *J. Biochem. Biophys. Methods* **6**, 317 (1982).

[19] P. G. Righetti, "Immobilized pH Gradients: Theory and Methodology." Elsevier, Amsterdam, 1990.

[19a] P. G. Righetti, *Methods Enzymol.* **270**, Chap. 10, 1996.

[20] A. Görg, W. Postel, and S. Günther, *Electrophoresis* **9**, 531 (1988).

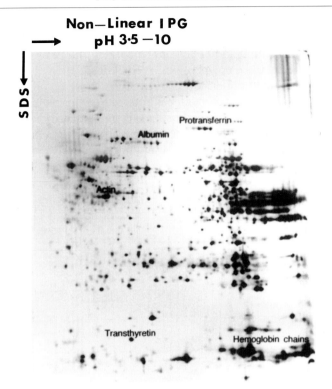

FIG. 3. Separation profile obtained by 2D PAGE using a nonlinear pH 3.5–10 IPG–IEF gel strip in the first dimension. The sample contained human liver proteins. The pattern was visualized by silver staining. [Photograph reproduced by courtesy of D. Hochstrasser (Hôpital Cantonal Universitaire, Geneva, Switzerland).]

of narrow and ultranarrow pH gradients not generally suitable for 2D applications. Recipes are now available for producing wide-pH gradients, so that extended pH gradients spanning the range pH 2.5–11 have been applied to 2D separations.[21] Very wide pH gradients are ideal for maximizing the number of proteins resolved on a single 2D gel. In addition, nonlinear pH gradients spanning a wide pH range can be exploited to increase further the resolution of sample proteins.[22] An example of such a separation of human liver proteins is shown in Fig. 3. However, 2D maps generated using wide-range IPGs are generally complex, making analysis, both visual and computer aided, difficult. For many applications it is better to use pH

[21] P. Sinha, E. Köttgen, R. Westermeier, and P. G. Righetti, *Electrophoresis* **13,** 210 (1992).
[22] G. J. Hughes, S. Frutiger, N. Paquet, F. Ravier, C. Pasquali, J.-C. Sanchez, R. James, J.-D. Tissot, B. Bjellqvist, and D. F. Hochstrasser, *Electrophoresis* **13,** 707 (1992).

gradients spanning three or four pH units. For 2D PAGE studies of the proteins of the human heart we routinely use pH 4–8 IPG–IEF gels for the separation of acidic and neutral polypeptides and pH 6–10 IPG–IEF gels for neutral and basic polypeptides. A brief description of the IPG–IEF procedure for the first dimension of 2D PAGE is presented here, but for further details the interested reader is referred to the review by Görg *et al.*[20] Typical separations obtained using this technique for the acidic and neutral (pH 4–8 IPG) or the neutral and basic (pH 6–10 IPG) proteins of the human heart are shown in Figs. 4 and 5, respectively.

Preparation of Immobilized pH Gradient–Isoelectric Focusing

1. Immobilized pH gradient gels, 0.5 mm thick, 25 cm wide, and with a pH gradient separation distance of 18 cm, are cast on GelBond PAG plastic supports (Pharmacia) using an IPG gel-casting cassette (Pharmacia).

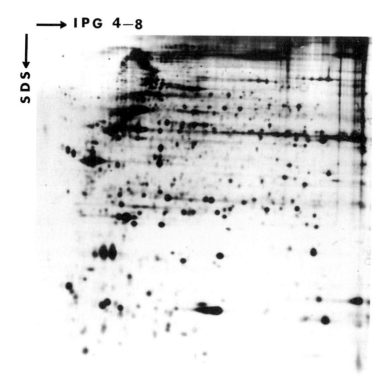

FIG. 4. Separation profile obtained by 2D PAGE using a linear pH 4–8 IPG–IEF gel strip in the first dimension. The sample contained human cardiac proteins. The pattern was visualized by silver staining.

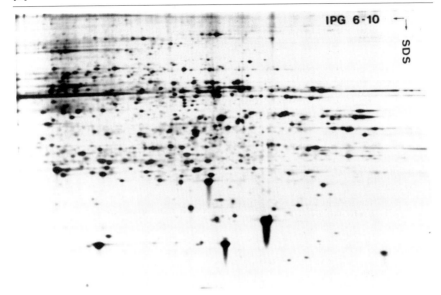

IPG 6-10

SDS

Fig. 5. Separation profile obtained by 2D PAGE using a linear pH 6–10 IPG–IEF gel strip in the first dimension. The sample contained human cardiac proteins. The pattern was visualized by silver staining. [Photograph reproduced by courtesy of A. Görg (Technische Universität, Munich, Germany).]

GelBond PAG film must be washed six times (10 min each) with deionized water prior to use to minimize spot streaking on the 2D patterns. The sheet of GelBond PAG film is applied hydrophilic side uppermost (to which the gel will adhere) to one of the glass plates of the cassette wetted with water. A roller is used to eliminate air bubbles and to seal the film to the plate. The glass plates and the 0.5-mm U-frame are then clamped together to form the cassette.

2. The two starter solutions appropriate for constructing the pH gradient are then prepared. Recipes for pH 4–8 and pH 6–10 IPGs are given in Tables I and II, respectively. Recipes for a diverse range of wide and narrow IPGs can be found in the literature.[19] The IPG is then cast from the top using a suitable gradient mixer, with the dense solution in the mixing chamber and the light solution in the reservoir. The gel is allowed to polymerize for 1 hr at 50°.

3. After polymerization, the gel on its plastic support film is removed from the cassette, washed six times (10 min each) with deionized water, once for 10 min with 20% (w/v) glycerol, and then dried at room temperature. The dried gel is covered with a sheet of plastic film and can be stored at −20° for up to 1 year.

TABLE I
PREPARATION OF pH 4–8 IMMOBILIZED pH GRADIENT–ISOELECTRIC FOCUSING GEL

Step and component	Acidic dense solution, pH 4.0 (10 ml)	Basic light solution, pH 8.0 (10 ml)
1. Immobiline pK_a		
3.6	392 μl	—
4.6	169 μl	369 μl
6.2	157 μl	240 μl
7.0	78 μl	95 μl
8.5	113 μl	223 μl
9.3	—	192 μl
2. Deionized water	5.8 ml	7.6 ml
3. Adjust pH[a]		
4. Add acrylamide and glycerol:		
Acrylamide solution (30% T, 4% C)	1.3 ml	1.3 ml
Glycerol (100%)	2.5 g	—
5. Add just before use:		
TEMED	6 μl	6 μl
Ammonium persulfate (40%)	10 μl	10 μl

[a] For effective polymerization both solutions are adjusted to pH 7.0 before addition of acrylamide and glycerol (acidic solution with 1 N NaOH; basic solution with 1.5 N acetic acid).

4. A range of dried IPG gels is also available commercially as DryPlates (covering the range pH 4–7) or DryStrips (pH 4–7 or pH 3–10.5) from Pharmacia.

Immobilized pH Gradient–Isoelectric Focusing Dimension

1. For the IEF dimension of 2D PAGE, the IPG gel plate is cut into 3-mm wide strips, using a paper cutter. The strips are then inserted into a reswelling cassette that is similar to the casting cassette, except that the thickness of the U-frame must be increased to 0.7 mm to account for the thickness of the GelBond PAG film.

2. The strips are then rehydrated in a solution containing 8 M urea, 0.5% CHAPS (or NP-40), 0.2% (w/v) DTT, and 0.2% (w/v) Pharmalyte 3-10 for a minimum of 6 hr (or overnight).

3. Following rehydration, the IPG strips are lightly blotted and placed side by side, 2 mm apart, on the cooling plate of a horizontal IEF apparatus maintained at 20°.

4. Paper electrode strips are soaked with 10 mM glutamic acid (anode) and 10 mM lysine (cathode) and applied across the ends of the IPG strips. Distilled water can alternatively be used for both electrolytes.

TABLE II

PREPARATION OF pH 6–10 IMMOBILIZED pH GRADIENT–ISOELECTRIC FOCUSING GEL

Step and component	Acidic dense solution, pH 6.0 (10 ml)	Basic light solution, pH 10.0 (10 ml)
1. Immobiline pK_a		
3.6	627 μl	67 μl
6.2	182 μl	222 μl
7.0	162 μl	241 μl
8.5	173 μl	159 μl
9.3	188 μl	217 μl
2. Deionized water	5.4 ml	7.8 ml
3. Adjust pH[a]		
4. Add acrylamide and glycerol:		
Acrylamide solution (30% T, 4% C)	1.3 ml	1.3 ml
Glycerol (100%)	2.5 g	—
5. Add just before use:		
TEMED	6 μl	6 μl
Ammonium persulfate (40%)	10 μl	10 μl

[a] For effective polymerization both solutions are adjusted to pH 7.0 before addition of acrylamide and glycerol (acidic solution with 1 N NaOH; basic solution with 1.5 N acetic acid).

5. Samples, typically around 20 μl, are then applied into silicone rubber frames placed on the surface of the gel strips. Samples can be applied at either the cathodic or anodic side of the gels. The optimal position for sample application should be determined for each particular type of sample to be analyzed. For our samples of human heart proteins, we find it best to apply the samples near the cathode for pH 4–9 IPG–IEF gels and near the anode for pH 6–10 IPG–IEF gels.

6. The IPG–IEF gel strips are run at 0.05 mA/strip, and 5 W limiting. For improved sample entry, the initial voltage should be limited to 150 V for 30 min and to 300 V for a further 60 min. Focusing is then continued at 3500 V until constant protein patterns are obtained. We use 42,000 V · hr for pH 4–8 IPG strips and 35,000 V · hr for pH 6–10 IPG gels.

7. After completion of IEF, the gel strips are removed from the apparatus and frozen at −80°.

Equilibration between Dimensions

An equilibration step is usually employed between the two dimensions of 2D PAGE. The purpose of this step is to allow the proteins to become saturated with SDS so that they migrate correctly during second-dimension SDS. Diffusion during equilibration can result in broadening of protein zones and loss of protein. However, omission of this step often results in streaking, particularly of the high molecular weight proteins.

Cylindrical tube IEF gels are incubated for a few minutes at room temperature in 50 mM Tris buffer, pH 6.8, containing 2% (w/v) SDS, 20 mM DTT, and 0.05% (w/v) bromphenol blue. More vigorous equilibration conditions are required for IPG–IEF gels, which are incubated for 15 min in 50 mM Tris buffer, pH 6.8, containing 2% (w/v) SDS, 30% (w/v) glycerol, 6 M urea, and 20 mM DTT, and then for a further 15 min in the same solution containing 5% (w/v) iodoacetamide in place of DTT. This last step reduces point streaking, which can otherwise occur if silver staining is used to visualize the 2D patterns.[20]

Second Dimension

SDS–PAGE using the discontinuous buffer system of Laemmli[4] is almost universally used for the second dimension of 2-D PAGE. It is possible to use gels of either a single polyacrylamide concentration or containing a linear or nonlinear polyacrylamide concentration gradient to extend the range over which proteins of different molecular weights can be effectively separated. Stacking gels can be used, but in our experience these can be safely omitted as the protein zones within the IEF gels are concentrated and the nonrestrictive IEF gel matrix tends to act as a stacking gel. An important exception here is that for IPG–IEF strips a stacking gel must be used if the second-dimension separation is to be carried out using horizontal, flat-bed SDS–PAGE gels. The preparation of SDS–PAGE slab gels is not described in detail here as this is accomplished using standard procedures in use in most laboratories for one-dimensional SDS–PAGE. It is advantageous to cast gels in batches, rather than individually, as this improves the reproducibility of the 2D patterns. Vertical slab gel systems are usually employed for the SDS–PAGE dimension, but it is possible to use horizontal SDS–PAGE gels, particularly when IPG–IEF gels are used in the first dimension.[20] In the latter case the SDS–PAGE gels are cast on GelBond PAG plastic support films as described by Görg and colleagues.[20]

First-dimension IEF gels are applied, after equilibration, to the origin of the second-dimension SDS–PAGE gels. To obtain good separations, it is essential to ensure that good contact is made between the two gels over their entire length.

Cylindrical IEF gel rods are usually cemented in place on top of the vertical SDS–PAGE slab gel, using 1% (w/v) agarose made up in equilibration buffer. However, many commercial preparations of agarose have been found to be contaminated with impurities that result in artifacts on 2D patterns when these are visualized by silver staining. In practice it is often possible to avoid the use of agarose provided that the thickness of the second-dimension SDS–PAGE slab gels is matched to that of the IEF gels, as the IEF gels become bonded to the surface of the SDS gels during electrophoresis. Care must be taken to ensure that the fragile cylindrical IEF gels are not stretched or otherwise damaged during the transfer step as this will result in distortion of the 2D patterns.

It is much easier to handle first-dimension IPG–IEF gel strips as they are bonded to plastic supports. These strips can be simply applied to the top of the vertical SDS–PAGE slab gels and cemented in place with agarose. Using horizontal flat-bed second-dimension SDS–PAGE gels, the IPG strips are transferred, gel side down, onto the surface of the stacking gel parallel to the cathodic electrode wick and separated from it by 1 mm.

The second-dimension SDS–PAGE gels are then run under suitable constant current conditions until the bromphenol blue tracking dye reaches the bottom of the slab.

Standards

Estimation of pH Gradients

The pH gradient in the IEF dimension using cylindrical rod gels can be determined by slicing the gels into thin transverse segments, eluting them into a small volume of 10 mM KCl, and measuring the pH of the resulting solution. However, this approach is complicated by the effects of urea and temperature on pH. Moreover, direct measurement of pH is not possible using IPG–IEF gels. Another approach is to use a set of marker proteins of known pI, but most commercial kits of pI marker proteins are not suitable as they often contain multimeric proteins that are dissociated under the denaturing conditions of 2D PAGE. The best method for calibrating pH gradients is by using carbamoylated charge standards. In this approach a protein, such as creatine kinase, is heated in the presence of urea under conditions that ensure the generation of carbamoylated derivatives. This results in the generation of multiple carbamoylated forms of the protein producing a continuous series or "train" of protein spots across the 2D pattern (Fig. 2), each spot differing from the previous one by the addition of a single negative charge. The disadvantage of these standards is that the pH gradient cannot be calibrated in terms of absolute pH, but only in terms of relative charge changes from the unmodified protein.

Molecular Weight Standards

The best standards for the SDS–PAGE dimension are usually a mixture of purified proteins of known molecular weight that are electrophoresed down one side of the second-dimension SDS–PAGE gels. Several kits of such markers, either radiolabeled, nonradioactive, colored, or prestained, are available commercially.

Gel Size, Resolution, and Reproducibility

The resolution capacity of 2D PAGE is dependent on the separation length in both dimensions and is, therefore, dependent on the area of the 2D gels. Using a standard 2D gel format, typically measuring 16 to 20 cm in each dimension, between 1000 and 2000 proteins can be resolved, whereas only a few hundred protein spots can be resolved using minigel formats. This dependence of resolution on gel area has prompted some investigators to adopt large formats, for example, 3000–4000 polypeptides can be resolved using 32 × 43 cm 2D gels.[23]

Another important factor in 2D PAGE is the ability of the technique to generate reproducible 2D protein patterns. All steps in the procedure must be carried out in a careful and controlled manner to ensure the required reproducibility. It is also advantageous to prepare and electrophorese large numbers of 2D gels simultaneously. This has resulted in the development of two dedicated commercial systems for 2D PAGE. The first system is based on the work of N. and L. Anderson, initially at the Argonne National Laboratory and later at Large Scale Biology Corporation (Washington, D.C.).[24,25] This equipment, known as IsoDalt, is available through Hoefer Scientific Instruments and is described in detail in Ref. 26. Two gel formats are supported (7 × 7 in. and 10 × 8 in.) and batches of 20 gels can be processed at the same time. The second apparatus, using a 22 × 22 cm format, is available as the Investigator system from Millipore.[15] Either 15 analytical (1-mm i.d.) or 8 preparative (3-mm i.d.) IEF gels can be run in the first dimension and 5 SDS–PAGE gels can be electrophoresed in the second dimension. We have both types of apparatus in our laboratory and have found that if used with care they are capable of producing reproducible high-quality 2D separations.

[23] R. M. Levenson and D. A. Young, *in* "2-D PAGE '91: Proceedings of the International Meeting on Two-Dimensional Electrophoresis, London" (M. J. Dunn, ed.), p. 12. NHLI, London, 1991.

[24] N. G. Anderson and N. L. Anderson, *Anal. Biochem.* **85,** 331 (1978).

[25] N. G. Anderson and N. L. Anderson, *Anal. Biochem.* **85,** 341 (1978).

[26] N. L. Anderson, "Two-Dimensional Electrophoresis." Large Scale Biology, Washington, D.C., 1988.

Although the positional reproducibility of protein spots on 2D gels has been the subject of much debate at international meetings on 2D PAGE, there are few publications dealing with this topic. Using the IsoDalt system, Taylor et al.[27] have reported that positional reproducibility for most spots in gel-to-gel comparisons is better than a spot width (0.5–2.0 mm). However, it should be pointed out these data were generated using the TYCHO computer analysis system to bring the 2D gel images into registration. Fosslien et al.[28] have reported that variation in absolute protein location on 2D maps generally increases with increasing migration distance in both axes, with values of 6–21 and 5–11 mm, respectively, in the IEF and SDS–PAGE dimensions. There has been considerable interest in the use of IPG–IEF for the first dimension of 2D PAGE, as this technique should improve the stability and reproducibility of 2D patterns (see above). Using this approach, Gianazza et al.[29] have measured the positional reproducibility of 12 selected spots and found the variability of spot position to be 2–6 and 3–7 mm, respectively, for the IPG–IEF and SDS–PAGE dimensions. Only a limited number of spots could be analyzed in these studies,[28,29] as computer analysis systems were not available, necessitating the manual measurement of spot positions. Using the PDQuest computer analysis system, we have compared the positional reproducibility of more than 300 protein spots on 2D maps of human heart proteins. The variability of spot position was found to be 0.5–5 mm and 0.5–12 mm, respectively, in the IPG–IEF and SDS–PAGE dimensions, demonstrating the high level of reproducibility that can be obtained using IPG–IEF as the first dimension of 2D PAGE.[29a]

Detection Methods

Fixation

The best general-purpose fixative is 20% (w/v) trichloroacetic acid (TCA). Alcoholic solutions, such as 45% (v/v) methanol, 45% (v/v) water, and 10% (v/v) acetic acid, are popular, but low molecular weight polypeptides, basic proteins, and glycoproteins may not be efficiently fixed by this procedure.

[27] J. Taylor, N. L. Anderson, and N. G. Anderson, Electrophoresis **4**, 338 (1983).
[28] E. Fosslien, R. Prasad, and J. Stastny, Electrophoresis **5**, 102 (1984).
[29] E. Gianazza, S. Astrua-Testori, P. Caccia, P. Giacon, L. Quaglia, and P. G. Righetti, Electrophoresis **7**, 76 (1986).
[29a] J. M. Corbett, M. J. Dunn, A. Posch, and A. Görg, Electrophoresis **15**, 1205 (1994).

General Protein Stains

The most popular general detection method for 2D gels is by staining in 0.1% Coomassie Brilliant Blue R-250 in 45% (v/v) methanol, 45% (v/v) water, 10% (v/v) acetic acid, followed by destaining in the same solution without dye. However, this method lacks sensitivity, being able to visualize spots containing about 0.2–0.5 μg of protein. A more sensitive method has been developed by Neuhoff and colleagues[30] using 0.1% (w/v) Coomassie Brilliant Blue G-250 in 2% (w/v) phosphoric acid, 10% (w/v) ammonium sulfate and 20% (v/v) methanol. This method utilizes Coomassie Brilliant Blue in a colloidal form so that no destaining step is required. This method can detect spots on 2D gels containing about 0.5–1.0 ng of protein.

Silver Staining

The requirement to separate and visualize more and more proteins by 2D PAGE analysis of complex samples such as represented by whole cells and tissues has resulted in increasing use of silver-staining methods. Silver staining of proteins following electrophoresis was introduced by Switzer *et al.* in 1979.[31] More than 100 publications describing variations on this methodology have appeared subsequently. The sensitivity of these methods is claimed to be between 20 and 200 times more sensitive than that of procedures based on the use of CBB, and can visualize spots on 2D gels containing about 0.1 ng of protein. These methods are reviewed in Ref. 32 and the sensitivity of the various procedures is discussed in Ref. 33.

Radioactive Detection Methods

There is no doubt that, if the sample to be analyzed by 2D PAGE is available in a radiolabeled form, a high sensitivity of detection can be obtained. Proteins can be radiolabeled synthetically by the incorporation of radioactive amino acids. This method is largely restricted to use with tissue culture systems, but it is also possible to radiolabel the proteins of small pieces of tissue in this way. Synthetic radiolabeling should not introduce artifacts into 2D patterns. In contrast, postsynthetic labeling of proteins, for example by iodination or methylation, often results in protein charge modifications, leading to artifacts on 2D maps. The most popular method of detecting radiolabeled proteins on 2D gel patterns has, until

[30] V. Neuhoff, N. Arold, D. Taube, and W. Ehrhardt, *Electrophoresis* **9,** 255 (1988).
[31] R. C. Switzer, C. R. Merril, and S. Shifrin, *Anal. Biochem.* **98,** 231 (1979).
[32] T. Rabilloud, *Electrophoresis* **11,** 785 (1990).
[33] T. Rabilloud, *Electrophoresis* **13,** 429 (1992).

recently, been by autoradiographic or fluorographic exposure of dried gels to X-ray film (reviewed by Dunn[2]). However, these methods suffer from the disadvantage that long exposure times are often necessary and problems of nonlinearity of film response, limited dynamic range, autoradiographic spreading, and fogging complicate quantitative analysis.

To overcome these problems, several electronic methods of detecting radioactive proteins directly in 2D gels have been developed (reviewed by Sutherland[34]). An exciting development is the application of photostimulable storage phosphor imaging plates to the detection of radiolabeled proteins following gel electrophoresis[35] and should now be considered to be the method of choice for quantitative imaging of β-emitting radionuclides. These imaging plates are composed of a thin (ca. 500 μm) layer of very small crystals of $BaFBr:Eu^{2+}$ in a plastic binder. Like other insulators, BaFBr crystals have a valence "band" of closely spaced energy states, all of which are occupied by electrons that move freely through the crystal and a conduction band of unoccupied energy states. Passage of a β particle through a crystal results in the transfer of an electron from the Eu^{2+} ion to a localized energy state at a site in the crystal where a Br^- ion is missing, that is, a color or "F" center. This process converts the europium to Eu^{3+}. Electrons are trapped on F centers that are metastable. Photons of wavelengths less than 800 nm have sufficient energy to transfer an electron from the F center into the conduction band, from which it can recombine with a Eu^{3+} ion, which is formed in an excited state. The excited state of the Eu^{3+} returns to its ground state by emitting a photon of a wavelength of 390 nm.

In practice, a phosphor-imaging plate is exposed to the dried 2D gel containing separated radiolabeled proteins. After the plate is exposed, it is transferred to a scanner where light from a HeNe (helium–neon) laser (633 nm) is absorbed by the F centers, resulting in the emission of a blue (390 nm) luminescence proportional to the original amount of radiation incident on the plate. Several commercial instruments exploiting this technology [e.g., Molecular Dynamics 400A PhosphorImager, Fuji BAS 1000, Bio-Rad (Richmond, CA) Molecular Imaging System] are available. These systems require relatively short exposure times, as a phosphor plate can capture the image of a highly labeled gel in about 10% of the time required for conventional autoradiography, and have a high dynamic range.

[34] J. C. Sutherland, *in* "Advances in Electrophoresis" (A. Chrambach, M. J. Dunn, and B. J. Radola, eds.), Vol. 6. p. 1, VCH, Weinheim, 1996.
[35] R. F. Johnston, S. C. Pickett, and D. L. Barker, *Electrophoresis* **11,** 355 (1990).

Photostimulable phosphor-imaging systems can also be used in double-label experiments[36] involving two radionuclides with energies that differ substantially, for example ^{35}S and ^{32}P. Two images are recorded: one with the imaging plate in direct contact with the gel and the other with a thin metal foil interposed between them. A 36-μm copper foil attenuates β emissions from ^{35}S by 750-fold, while reducing β emissions from ^{32}P by only 30%. Detection efficiencies are determined in control experiments and used to deconvolute the two experimental images to determine the distribution of each radionuclide in the gel. This method has been successfully applied to the simultaneous analysis of phosphoproteins and total cellular proteins of PC12 cells.[37]

Computer Analysis and Protein Databases

The methods of high-resolution 2D PAGE described in this chapter are capable of generating highly complex 2D protein maps. The vast amount of information contained in these 2D patterns can only be extracted, analyzed, catalogued, and exploited using an automated computer system. Such a system must be able to (1) extract both qualitative and quantitative information from individual gels, (2) provide pattern matching between gels, and (3) allow the construction of databases for different types of sample.

The essential first step in 2D gel analysis is the imaging and digitization of the 2D pattern. For stained gels, the most popular options are the use of flat-bed scanning densitometers based on the use of a HeNe (633 nm) light source, charge-coupled device (CCD) array scanners, or more recently imaging devices based on document scanners. Laser densitometers are high-resolution devices (typically down to 50 μm) capable of scanning 2D gels at high speed and with a high dynamic range, but they are insensitive to objects whose color approaches that of the laser light. This can be a particular problem in silver staining if orange to red spots are obtained rather than the optimal dark brown color. CCD array scanners are essentially cameras fitted with one- or two-dimensional arrays of photodiodes mounted in the focal plane behind a lens. They allow rapid image acquisition at high dynamic range, but their disadvantages include problems in maintaining normalization of the photodiodes to each other over extended periods of time, lens flare, and correcting for nonuniformity in the background. These

---

[36] R. F. Johnston, S. C. Pickett, and D. L. Barker, *in* "Methods: A Companion to Methods in Enzymology" (M. Harrington, ed.), Vol. 3, p. 128. Academic Press, San Diego, CA, 1991.

[37] M. G. Harrington, L. Hood, and C. Pickett, *in* "Methods: A Companion to Methods in Enzymology" (M. Harrington, ed.), Vol. 3, p. 135. Academic Press, San Diego, CA, 1991.

types of device are also used if 2D gels of radiolabeled protein samples are visualized by autoradiography or fluorography using X-ray film. However, as discussed above, direct detection of radioactive proteins in 2D gels using photostimulable storage phosphor-imaging plate technology should be considered the method of choice for quantitative imaging of β-emitting radionuclides.

Several systems for the analysis of 2D gel patterns are now available. The larger systems (e.g., PDQuest, Kepler, ELSIE, MELANIE, GELLAB, BioImage, and HERMES), which run on powerful microcomputer workstations (e.g., Sun, VAX), provide a full 2D gel analysis capability including scanning, spot extraction, quantitation, pattern matching, database construction, and analysis tools. Systems based on personal microcomputers (e.g., IBM-PC and compatibles) are at present limited in their capabilities, especially in terms of their ability to match large sets of 2D gel patterns to establish protein databases. This topic is discussed in detail by Miller[38] and Dunn.[39] Such large-scale databases[40] can be readily interfaced to other databases such as those containing protein and DNA sequences, and provide an important interface between the techniques of protein biochemistry and those of molecular biology.

Identification and Characterization of Proteins

Western Blotting

Techniques of 2D PAGE have an almost unrivalled capacity to separate and resolve the components of complex protein mixtures. Using this methodology proteins can be characterized in terms of their charge, size, and abundance. However, 2D PAGE analysis does not provide any direct information on the identity or functional properties of the separated proteins. The best approaches to this problem are based on Western blotting, involving the electrophoretic transfer of proteins separated by 2D PAGE onto the surface of an inert membrane. The high sensitivity and affinity of polyclonal and monoclonal antibodies, and a diverse range of other ligands such as lectins, can then be used as highly sensitive and specific reagents for the identification and characterization of proteins separated by 2D

[38] M. J. Miller, *in* "Advances in Electrophoresis" (A. Chrambach, M. J. Dunn, and B. J. Radola, eds.), Vol. 3, p. 181. VCH, Weinheim, 1989.

[39] M. J. Dunn, *in* "Microcomputers in Biochemistry: A Practical Approach" (C. F. A. Bryce, ed.), p. 215. IRL Press, Oxford, 1992.

[40] J. E. Celis, P. Madsen, B. Gesser, S. Kwee, H. V. Nielsen, H. H. Rasmussen, B. Honoré, H. Leffers, G. P. Ratz, and B. Basse, *in* "Advances in Electrophoresis" (A. Chrambach, M. J. Dunn, and B. J. Radola, eds.), Vol. 3, p. 1. VCH, Weinheim, 1989.

PAGE. Techniques of immunoblotting of proteins separated by electrophoresis are reviewed in detail by Baldo and Tovey.[41]

The most important step in immunoblotting procedures is the reaction with the specific antibody (or other ligand) and the sensitive visualization of the proteins in the sample with which it interacts. The required sensitivity is usually achieved by an indirect approach utilizing a secondary (or tertiary) antibody conjugated with a suitable reporter group, the most popular being an enzyme such as horseradish peroxidase or alkaline phosphatase. Subsequent detection is usually achieved using a substrate that generates an insoluble, stable, colored reaction product at the site on the blot where the secondary antibody is bound. Extra sensitivity, down to the 1-pg level or less, can be gained at this stage using methods based on chemiluminescence. Methods are available for use with (1) peroxidase-conjugated antibodies based on the oxidation of luminol in the presence of hydrogen peroxide,[42] and (2) alkaline phosphatase-conjugated antibodies based on the use of a dioxetane substrate, disodium (3-(4-methoxyspiro[1,2-dioxetane-3-(2'-tricyclo[3.3.1.1^{3,7}]decan-4-yl)]phenyl phosphate) (AMPPD).[43] An additional advantage of these techniques is that primary and secondary antibodies can be completely removed ("stripped") after immunodetection and the blots reprobed several times with different primary antibodies.

Protein Sequence Analysis

The development of highly sensitive micromethods of amino acid analysis and protein sequencing has now made it possible to obtain direct chemical characterization of proteins separated by 2D PAGE. Indeed, gel electrophoretic procedures can now be regarded as the method of choice for the purification of proteins for subsequent chemical characterization. The current generation of gas–liquid phase and solid-phase sequenators are capable of determining limited sequence information from as little as 10–20 pmol of a highly purified protein, corresponding to 0.5–1 μg of pure sequencable protein of 50-kDa molecule mass. The success of the method is, therefore, dependent on sufficient sample protein being applied to the 2D gels. A load suitable for visualization with a general protein stain such as CBB R-250 is required, typically several hundred micrograms for a complex protein mixture, and as a general rule of thumb is that a spot that

[41] B. A. Baldo and E. R. Tovey (eds.), "Protein Blotting: Methodology, Research and Diagnostic Applications." Karger, Basel, 1989.

[42] M. W. Cunningham, I. Durrant, S. J. Fowler, J. A. Guilford, M. Moore, and R. M. MacDonald, *Int. Lab.* **22**, 36 (1992).

[43] I. Bronstein, J. C. Votya, O. J. Murphy, L. Bresnick, and L. J. Kricka, *BioTechniques* **12**, 748 (1992).

is clearly visible with such a stain is potentially amenable to microsequence analysis. Only a brief account of this methodology is given here and the interested reader is recommended to read the detailed review of this topic by Aebersold.[44]

In this approach, Western blotting is again used to transfer the separated proteins from the 2D gel onto the surface of an inert membrane. The choice of the membrane is critical as it must combine the property of a good blotting membrane (that is, it must have a good protein binding capacity) with that of a good sequencing support (that is, it should perform well during automated sequence analysis). Nitrocellulose cannot be used as it is incompatible with the reagents and organic solvents used for sequencing. A range of other membranes has been used,[44] but the best supports currently available for protein sequencing are based on polyvinylidenedifluoride (PVDF) membranes.[45] Transfer buffers containing glycine (or other amino acids) should be avoided. The two most popular buffer systems are (1) 50 mM Tris, 50 mM borate (pH 8.5), 20% (w/v) methanol, or (2) 10 mM CAPS (pH 11.0), 10% (w/v) methanol.

After the Western blotting step is complete, the proteins to be sequenced are visualized by a general protein stain; for example, CBB R-250 is compatible with PVDF membranes. The protein spot of interest is then excised and placed directly into the reaction cartridge of the protein sequenator to obtain the corresponding partial N-terminal sequence for that protein. The protein can then be identified, or homologies suggested, by comparative sequence analysis using the international protein and DNA sequence databases. If the particular protein of interest has not previously been characterized, the availability of partial amino acid sequence information facilitates its further characterization by techniques of protein chemistry or molecular biology.

Although proteins prepared by 2D PAGE and Western blotting can often be sequenced at high efficiencies, many proteins subjected to analysis are found to yield no N-terminal sequence information. This is because they lack a free α-amino group due to blockage of the N terminus occurring either as a result of posttranslational modification *in vivo* (e.g., acetylation, acylation, or cyclization) or as an artifact during sample workup (i.e., during sample preparation, electrophoresis, or blotting). The best approach to this problem is to use chemical (e.g., cyanogen bromide or 2-[2'-nitrophenylsulfenyl]-3-methyl-3-bromoindolenine [BNPS-skatole]) or enzymatic (e.g., trypsin, subtilisin, or papain) methods to cleave the protein of interest, while still on its inert membrane support (e.g., nitrocellulose or PVDF).

[44] R. Aebersold, *in* "Advances in Electrophoresis" (A. Chrambach, M. J. Dunn, and B. J. Radola, eds.), Vol. 4, p. 81. VCH, Weinheim, 1991.
[45] C. S. Baker, M. J. Dunn, and M. H. Yacoub, *Electrophoresis* **12**, 342 (1991).

Cleavage fragments are then released using trifluoroacetic acid (TFA) and the peptides separated by a suitable procedure. Small peptides are best separated by narrow- or microbore reversed-phase HPLC, in which technique the peptide peaks of interest can be collected directly on glass fiber filter disks for direct application to the protein sequenator. Alternatively, larger peptide fragments can be separated by one-dimensional SDS–PAGE followed by electroblotting to a PVDF membrane for subsequent sequence analysis.

Mass Spectrometric Analysis

An important development in protein biochemistry has been the application of methods of mass spectrometry. In particular, the development of matrix-assisted laser desorption ionization (MALDI) has overcome the problem of desorption and ionization of large and labile biomolecules such as proteins. This technology is also compatible with Western blotting, so that it is now possible to measure accurately the mass of proteins separated by 2D PAGE simply by soaking pieces of PVDF or polyamide membranes containing the spots of interest in matrix solution (e.g., succinic or malic acid) and placing them directly in a time-of-flight (TOF) mass spectrometer.[46]

Another exciting recent innovation has been the development of a peptide mass fingerprinting technique for the rapid identification of proteins.[47,48] In this approach, protein spots electroblotted from a 2D gel onto a PVDF membrane are subjected to in situ enzymatic or chemical cleavage. The masses of the cleavage products are determined by MALDI/TOF mass spectrometry. A computer program is then used to identify the protein by matching the molecular masses of the peptide fragments obtained from the cleavage with all fragments in the protein sequence database. Surprisingly, sample proteins can be uniquely identified using as few as three or four experimentally determined peptide masses when screened against the fragment database.[47] This technique of peptide mass fingerprinting can prove as discriminating as N-terminal or internal protein sequence analysis, but can be obtained in a fraction of the time, using less protein. This technique will allow the rapid identification of proteins separated by 2D PAGE, and will facilitate the development of comprehensive 2D gel protein databases,

[46] C. Eckerskorn, K. Strupat, M. Karas, F. Hillenkamp, and F. Lottspeich, *Electrophoresis* **13,** 664 (1992).

[47] D. J. C. Pappin, P. Hojrup, and A. J. Bleasby, *Curr. Biol.* **3,** 327 (1993).

[48] W. J. Henzel, T. M. Billeci, J. T. Stults, S. C. Wong, C. Grimley, and C. Watanabe, *Proc. Natl. Acad. Sci. U.S.A.* **90,** 5011 (1993).

as only previously unidentified proteins will need to be subjected to the more laborious techniques of protein sequence analysis.

Concluding Remarks

The techniques of 2D PAGE described here now make it possible to separate reproducibly complex mixtures of proteins containing several thousand components. The availability of computer systems capable of rigorous qualitative and quantitative analysis of the resulting 2D protein profiles to establish large-scale 2D protein databases,[49] now makes it possible to apply this technology to the study of patterns of protein expression in various areas of biology and biomedicine. Moreover, the development of techniques for the chemical characterization (i.e., amino acid analysis, protein sequencing, mass spectrometry) of proteins separated by gel electrophoresis can be seen as providing an interface between studies of gene expression at the protein level by techniques such as 2D PAGE and computer databases with the rapidly expanding body of information concerning the complexity of gene expression at the DNA level.[43]

Acknowledgments

The work in the authors' laboratory is supported by the British Heart Foundation. We thank Dr. Angelika Görg and colleagues for help and encouragement in the establishment of the IPG–IEF strip technique for 2D PAGE in our laboratory.

[49] J. E. Celis, H. Leffers, H. H. Rasmussen, P. Madsen, B. Honoré, B. Gesser, K. Dejgaard, E. Olsen, G. P. Ratz, J. B. Lauridsen, B. Basse, A. H. Andersen, E. Walbaum, B. Brandstrup, A. Celis, M. Puype, J. Van Damme, and J. Vandekerckhove, *Electrophoresis* **12,** 765 (1991).

[9] Affinophoresis: Selective Electrophoretic Separation of Proteins Using Specific Carriers

By Kiyohito Shimura and Ken-ichi Kasai

Introduction

The combination of biospecific affinity with electrophoretic separation yields useful tools for separation purposes, making it possible to modify the mobility of a target substance present within a complex mixture. Such procedures are categorized as affinity electrophoresis (AE). There are variations based on the manner in which AE is used and the source of the affinity ligands. The methods can be divided primarily into two categories, according to whether the ligand is immobilized or free. Electrophoresis using immobilized ligands is analogous to affinity chromatography and results in the stoppage or retardation of a target molecule. A report on electrophoresis of lectins in starch gel demonstrating such an effect appeared in the late 1960s.[1] Development of techniques for immobilization of biological molecules to polymer matrices promoted use of artificial supporting gels for electrophoresis bearing a ligand for the target substance. A number of reports have appeared that describe applications to various proteins such as lectins and antibodies.[2–4] The term *affinity electrophoresis* refers almost exclusively to this particular mode.

However, the change in electrophoretic mobility of a target substance due to formation of a soluble complex with its specific ligand has also been successfully utilized, especially for analytical purposes; however, procedures based on this principle have rarely been understood as variations of affinity electrophoresis. They were called by different names such as cross-electrophoresis.[5] This approach was effective in detecting interactions between a protein and other biomolecules. The gel-shift assay[6] is also based on the mobility change of DNA fragments on complex formation with DNA-binding proteins. This procedure is useful for various purposes such as the detection of DNA-binding proteins, analysis of interaction, and assignment of binding region on DNA fragments. In these cases, the counterparts have generally been biological substances.

Affinophoresis is one of the procedures belonging to the soluble complex category, but it has a distinct feature in that it uses an artificial mobility modifier called an *affinophore* (Fig. 1).[7,8] An affinophore is a specific carrier and is prepared by attaching affinity ligands for a target substance to a soluble macromolecule with a large number of positive or negative charges. When the affinophore molecule migrates toward the electrode of opposite polarity, the target substance, which binds to the ligand, is also pulled in the same direction, regardless of its own net charge. Creation of such a new mode has made it possible for the first time to introduce a rationale-based selectivity to electrophoresis. This mode seems to have much wider

[1] G. Entlicher, M. Tichá, J. V. Koštíř, and J. Kocourek, *Experientia* **25**, 17 (1969).
[2] K. Takeo and S. Nakamura, *Arch. Biochem. Biophys.* **153**, 1 (1972).
[3] T. C. Bøg-Hansen, *Anal. Biochem.* **56**, 480 (1973).
[4] V. Hořejší and J. Kocourek, *Biochim. Biophys. Acta* **336**, 338 (1974).
[5] S. Nakamura, "Cross Electrophoresis." Igaku Shoin, Elsevier, Amsterdam, 1966.
[6] M. M. Garner and A. Revzin, *Trends Biochem. Sci.* **11**, 395 (1986).
[7] K. Shimura and K. Kasai, *J. Biochem.* **92**, 1615 (1982).
[8] K. Shimura, *J. Chromatogr.* **510**, 251 (1990).

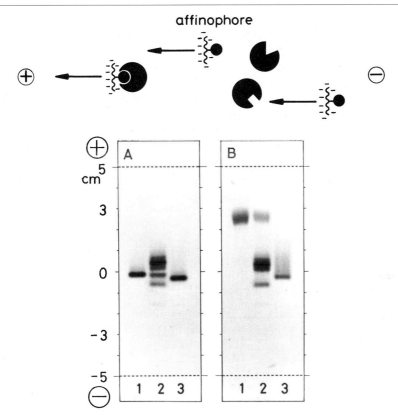

FIG. 1. *Top:* Principle of affinophoresis. The affinophore is a specific carrier that is prepared by attaching affinity ligands to a soluble macromolecule with a large number of positive or negative charges. When the affinophore migrates toward the electrode of opposite polarity in an electric field, a target biomolecule is also pulled in the same direction, regardless of its own net charges. *Bottom:* An example of specific separation of *Streptomyces griseus* trypsin by using anionic affinophore. Examples show agarose gel electrophoresis of native trypsin (lane 1), pronase (lane 2), and inactivated trypsin with TLCK (lane 3) in the absence (A) or in the presence (B) of the affinophore. The affinophore has sulfonic acid as ionic groups and benzamidine moieties as the specific ligand for trypsin. In the presence of the affinophore, significant migration of both native trypsin and that contained in pronase toward the anode was observed (B, lanes 1 and 2), although it scarcely migrated in the absence of the affinophore at the pH of the experiment (A, lanes 1 and 2). Inactivated trypsin did not migrate even in the presence of the affinophore (B, lane 3). (Reprinted from Refs. 8 and 9 with permission.)

application in comparison to immobilized affinity ligand electrophoresis, because it can make full use of the advantages of electrophoretic procedures.

Matrices for affinophores include several kinds of polyelectrolytes: e.g., DEAE-dextran for cationic affinophores,[7] and polyacryloyl-β-alanyl-β-alanine[9] and succinyl-poly-L-lysine for anionic affinophores.[10] The utility of affinophoresis has been demonstrated by application to various proteins such as proteases,[7,9,10] lectins,[11] and antibodies,[12] and even to erythrocytes.[13] Although affinophoresis was initially carried out in one dimension, two-dimensional affinophoresis has also been found useful.[11,12,14] The two-dimensional application enables one to detect and to separate a particular substance contained in a complex mixture with great sensitivity and efficiency (see [8] in this volume[15]). A unique feature of affinophoresis is that it is applicable to a target molecule that does not move in an electric field, such as a neutral substance or a protein at its isoelectric point. It is also applicable to suspended particles such as cells, because this procedure does not require an insoluble support.[13]

This chapter focuses on this highly selective electrophoretic procedure, providing the reader with both the fundamentals and a practical guide to carrying out affinophoresis.

Basic Principles of Affinophoresis

To perform affinophoresis, an affinophore for a target substance must be prepared. Target substances are assumed to be proteins; however, the method is not restricted only to these biopolymers.

Mobility Change in Affinophoresis

The change in target electrophoretic mobility under affinophoresis is based on the difference between the original mobility of the free protein (μ_o) and that of the affinophore–protein complex (μ_c). In the presence of affinophore molecules, the ratio of the affinophore–protein complex to the free protein is determined by the dissociation constant. The mobility of a protein subjected to affinophoresis is microscopically discontinuous. Only when it forms a complex does it migrate at a velocity of μ_c. Therefore, the observed mobility of the protein (μ) is on average as follows:

$$\mu = \mu_o(P_f/P_t) + \mu_c(P_c/P_t) \tag{1}$$

[9] K. Shimura and K. Kasai, *Biochim. Biophys. Acta* **802**, 135 (1984).
[10] K. Shimura and K. Kasai, *J. Chromatogr.* **376**, 323 (1986).
[11] K. Shimura and K. Kasai, *J. Chromatogr.* **400**, 353 (1987).
[12] K. Shimura and K. Kasai, *Electrophoresis* **8**, 135 (1987).
[13] K. Shimura, N. Ogasawara, and K. Kasai, *Electrophoresis* **10**, 864 (1989).
[14] K. Shimura and K. Kasai, *Anal. Biochem.* **161**, 200 (1987).
[15] M. J. Dunn and J. M. Corbett, *Methods Enzymol.* **271**, Chap. 8, 1996 (this volume).

where P_f and P_c are the concentration of the free protein and that of the affinophore–protein complex, respectively, and P_t is the total concentration of the protein, i.e., $P_t = P_f + P_c$.

In the following discussion, the concentration of the affinophore is assumed to be much larger than that of the protein. The apparent dissociation constant (K_d) of the affinophore–protein complex is given by

$$K_d = [A](P_f/P_t) \tag{2}$$

where [A] is the concentration of the free affinophore. From Eqs. (1) and (2), the following equation describing the change in mobility of a protein subjected to affinophoresis can be obtained:

$$\mu - \mu_o = (\mu_c - \mu_o)[A]/(K_d + [A]) \tag{3}$$

Equation (3) has the same form as the Michaelis–Menten equation of enzyme kinetics and Langmuir's adsorption isotherm. The observed mobility of the protein (μ) asymptotically approaches that of the complex (μ_c) at infinite concentration of the affinophore. Two important features can be seen from Eq. (3). First, the difference between the mobility of the complex and the original mobility of the protein determines the maximum change in protein mobility on complexation, while the mobility of the affinophore itself does not directly influence protein mobility, only indirectly via its effect on the mobility of the complex. Second, it is desirable to use an affinophore concentration comparable to or greater than the dissociation constant in order to result in sufficient mobility change for the target protein.

Equation (3) also indicates that the dissociation constant (K_d) between a target biomolecule and its affinophore can be determined from the dependence of mobility change of the biomolecule on the concentration of the affinophore. The concentration of the affinophore that gives a mobility change of half of the maximal change corresponds to the K_d. For this purpose, it is convenient to use the double-reciprocal form of Eq. (3), which is analogous to the Lineweaver–Burk plot of enzyme kinetics, because the K_d value can be calculated from the intercept of a straight line on the abscissa. K_d values of an anionic affinophore bearing tryptophan for chymotrypsin and anhydrochymotrypsin were successfully determined by this procedure.[8,10]

Requirements for Ligands

A ligand having adequate affinity for a target protein, e.g., a dissociation constant less than 10^{-4} M, will be advantageous for affinophoresis. The ligand molecule should have either an amino group or a carboxyl group available for coupling to the polymer matrix; of course, such groups are not required for specific binding.

Number of Ligands on Affinophore Molecule

As a rough estimate, the dissociation constant for an affinophore bearing n ligands $[K_d(n)]$ should decrease from that for the affinophore bearing a single ligand $[K_d(1)]$ by a factor of $1/n$, because there are n ways of complex formation. Increasing the number of ligand groups on an affinophore would be effective in raising its affinity for a target protein. If a target protein has multiple binding sites and binds to more than two ligands simultaneously, the interaction should become very strong.

Number of Electric Charges on Affinophore Molecule

The electrophoretic mobility in a free solution of a linear polyionic polymer is known to be independent of its degree of polymerization. However, a portion of its charges are neutralized when it forms a complex with a target molecule having the opposite net charge. Therefore, it should have a sufficient number of charges not to be significantly affected by complex formation. More than 100 charged groups on an affinophore molecule seem to be sufficient for most cases.

Heterogeneity of Affinophore

Polymers used as the matrices in affinophoresis are heterogeneous in size, which would then make the electrophoretic mobility of the affinophore–protein complex heterogeneous. When the affinophore and the target protein are in equilibrium with rapid association–dissociation kinetics, each protein molecule undergoes many association–dissociation reactions with many different affinophore molecules in the course of affinophoresis. As a result, the mobility of the affinophore–protein complex should approach an average value. If the value of k_{-1} is assumed to be 10^2 min^{-1}, a value considerably lower than the k_{-1} usually proposed for enzyme–substrate complexes, an affinophoresis time of 30 min will allow a protein to change ligands about 3000 times, and this number seems sufficiently large for close approach to the average mobility.

Effects of Nonspecific Ionic Interaction

The nonspecific interaction of a protein with its affinophore should be as small as possible. However, there will be ionic interactions if the net charge of the target protein is the opposite of the charge on the affinophore. Nevertheless, such interactions were found to be suppressed if electrophoresis was carried out in a buffer of relatively high ionic strength, e.g., 0.1 M phosphate.[16]

[16] K. Shimura and K. Kasai, *Electrophoresis* **10**, 238 (1989).

Supporting Matrix

In principle, affinophoresis can be carried out in a free solution, but the use of an insoluble gel support greatly facilitates its application in practice. The gel support should be sufficiently porous to allow migration of the affinophore–protein complex. An agarose gel of about 1% (w/v) is suitable for free migration of affinophore–protein complexes and can be easily impregnated with affinophore molecules by adding the affinophore to a warm solution of agarose before casting a gel.

Preparation of Affinophore

Although DEAE-dextran and polyacrylyl-β-alanyl-β-alanine were initially used,[7,9] polylysine has been found to be the most convenient starting material for affinophores.[10] It is commercially available as fractionated polymers differing in the degree of polymerization. Both cationic and anionic affinophores can be prepared, and the ε-amino groups of the polylysine can be used in a coupling reaction with ligand molecules. No significant differences in performance were found based on the size of the affinophore when polylysines with an average degree of polymerization of 120 and 190 were used.[10,11]

An affinophore can be either cationic or anionic; however, the use of cationic affinophores posed some practical problems because they were found to be adsorbed on agarose gel,[8] due to an ionic interaction with sulfate or carboxyl groups on agarose. The adsorbed affinophore stained with dyes such as Coomassie blue, resulting in an interference with the detection of proteins in the gel. If the ligand was a cationic or aromatic molecule, even anionic affinophores with carboxyl groups were stained with Coomassie blue. Conversion of the carboxyl groups to sulfate made the affinophores no longer stainable with the dye.[9]

Ligands are generally coupled to polylysine by the formation of amide bonds by means of a water-soluble carbodiimide, 1-ethyl-3-(3-dimethylaminopropyl)carbodiimide (EDC). Ligands having an amino group are reacted with succinylpolylysine at pH 4.5–5. For aliphatic amines with high pK_a values, use of a buffer of higher pH is recommended. For ligands having a carboxyl group, activation in advance of the carboxyl group as the N-hydroxysuccinimide ester and coupling to ε-amino groups of the polylysine are recommended. Direct reaction, e.g., by means of EDC, did not result in efficient coupling yield. The product is succinylated, and, if necessary, further coupled with aminomethanesulfonic acid by using EDC at pH 4.5–5. The reaction is generally completed rapidly with a high coupling yield. Electrophoretic mobility at pH 8 of affinophores prepared from succinyl-polylysine was found to be 1.0–1.3 relative to bromphenol blue.

Experimental Procedure

Preparation of Anionic Affinophore Having Amino Acid Coupled with α-Amino Group

An affinophore for anhydrochymotrypsin is prepared by coupling L-tryptophan to polylysine.[10] Solid succinic anhydride (100 mg) is added all at once to a solution of poly(L-lysine hydrobromide) (100 mg) in 5 ml of 0.1 M NaCl. The pH of the solution is maintained between pH 8 and 10 with 6 M NaOH over 10 min. Unreacted amino groups are determined by reaction with fluorescamine, and amount to less than 1% of the original. The solution is dialyzed against 0.1 M NaCl (three times against 1000 ml). To the dialysate (6.1 ml) containing 340 μmol of lysine residue (determined by amino acid analysis), L-tryptophan methyl ester hydrochloride (17.2 mg, 68 μmol) is added and the pH is adjusted to 4.75 with 1 M HCl. EDC-HCl (65 mg, 340 μmol) is then added, and the mixture is left for 15 min without any adjustment of pH. Almost all of the L-tryptophan methyl ester is bound to the polymer. Aminomethanesulfonic acid (57 mg, 510 μmol) and EDC-HCl (130 mg) are added to the solution at pH 5, and the pH is maintained between pH 4.5 and 5.0 for 1 hr by using 1 M HCl or 1 M NaOH. The solution is then dialyzed against 0.1 M NaCl (three times against 1000 ml). Then 0.13 ml of 6 M NaOH is added to the dialysate (7.4 ml), and the mixture is left for 30 min at 24° to hydrolyze the ester. After neutralization with 6 M HCl, the solution is dialyzed against water (three times against 1000 ml) and freeze-dried (112 mg). This affinophore is effective in separating anhydrochymotrypsin.[10]

Preparation of Anionic Affinophore for Lectins Having D-Mannoside Groups

To a solution of succinylpoly(L-lysine) (5 ml) containing 334 μmol of lysine residue, p-aminophenyl-α-D-mannoside · 1.5H$_2$O (10 mg, 34 μmol) is added, and the pH is adjusted to 4.75 with 1 M HCl.[11] EDC-HCl (32 mg, 170 μmol) is added at once, and the mixture is stirred for 10 min. After a further 10 min, aminomethanesulfonic acid (7.4 mg, 67 μmol) and EDC (32 mg, 170 μmol) are added, and the pH is maintained for 20 min between pH 4.5 and 5.0 with 1 M NaOH. The pH is then adjusted to 7 with 1 M NaOH, and the solution is dialyzed against 0.1 M NaCl (once against 1000 ml) and water (twice against 1000 ml). The dialysate (9 ml) is used as a stock solution of the affinophore. Analysis of the amino acid and sugar content of this solution showed the presence of 31 mM lysine residue and 2.9 mM mannoside.

Preparation of Anionic Affinophore Having Peptide Coupled with Carboxyl Group

An antihypertensive peptide, N-(dibenzyloxyphosphinoyl)-L-alanyl-L-prolyl-L-proline (PAPP) is used as a hapten, and rabbit antisera are prepared.[12] An anionic affinophore bearing this hapten is prepared with the goal of specific separation of the antibody by affinophoresis. PAPP (26 mg, 48 μmol), N-hydroxysuccinimide (5.5 mg, 48 μmol), and dicyclohexylcarbodiimide (9.9 mg, 48 μmol) are dissolved in 0.1 ml of dimethylformamide and left at room temperature overnight. The supernatant of this mixture is added to 50 mg of poly(L-lysine hydrobromide) (degree of polymerization, 190) in 2.7 ml of 0.1 M sodium phosphate buffer (pH 7.5) and stirred for 1 hr. Succinic anhydride (50 mg, 0.5 μmol) is added at once, and the pH is maintained at pH 7–8 for 10 min with 6 M NaOH. The reaction mixture is then dialyzed three times against 500 ml of 0.1 M NaCl. The carboxyl groups are converted to sulfonic acid groups as described. Amino acid analysis of the stock solution of the final product shows the presence of 196 μmol of lysine (27 mM), 30 μmol of alanine (4.2 mM), and 64 μmol of proline (8.9 mM) residues. Therefore, introduced ligand groups are present at about one per 6.5 lysine residues.

Procedure for Affinophoresis

Procedure for One-Dimensional Affinophoresis

An agarose gel plate (1%), 1 mm thick, is made on a GelBond film (8 $\times$ 12.5 cm; FMC, Marine Colloids Division, Rockland, ME).[9] On the film, warm agarose solution (1%) in an appropriate buffer containing the affinophore (0.01–0.1%) is poured and spread. The gel plate is left in a moisture box for at least 1 hr. The gel together with the film is placed on the cooling plate (through which ice-cold water is circulated) of a flat-bed electrophoretic apparatus (Atto Co., Tokyo, Japan). A sample mixture (2 μl) containing an affinophore and a protein is applied on the middle of the plate. Electrophoresis is generally carried out at constant current (e.g., 100 mA/plate for 30 min) in a cold room (4°). The gel plate together with the film is then soaked in a solution of 1.2% (w/v) picric acid and 20% (v/v) acetic acid for 10 min to fix the proteins. After the solution contained in the gel is almost adsorbed with sheets of blotting paper, the gel is dried with hot air and stained for 5 min with 0.5% Coomassie Brilliant Blue R-250 in methanol–acetic acid–water (4:1:5, v/v).

Procedure for Two-Dimensional Affinophoresis

Because two-dimensional electrophoresis simplifies the identification of a specific component contained in a complex mixture (see [8] in this volume[15]), its application to affinophoresis has been attempted.[11,12,14] The first electrophoresis is carried out without affinophore, and the second is carried out at right angles to the first under identical conditions except for the presence of an affinophore. Proteins devoid of affinity for the affinophore should lie on a diagonal line while the target protein should be found away from the line, as its mobility is exclusively changed in the second electrophoretic step. Even if the change in mobility is not large, the deviation can be detected with great sensitivity.

Before agarose solution is poured on a sheet of GelBond film, a strip of adhesive plastic tape 10 cm long is placed on the film so as to leave a square open area with sides of 8 cm (Fig. 2, the square ABCD). In this area, 6.5 ml of 1% agarose solution in an appropriate buffer is poured and spread. The tape is removed after the gel has formed. The gel is next placed on the cooling plate of the electrophoretic apparatus. Electrodes are set parallel to the longer edges of the gel plate (Fig. 2, sides AH and BG). A

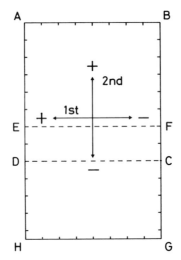

FIG. 2. Construction of agarose gels for two-dimensional affinophoresis. The first-dimensional electrophoresis was carried out in a square gel without an affinophore (the square ABCD, 80 × 80 mm). A sample solution was applied in a hole 25 mm from edge CD. Prior to the second-dimensional electrophoresis, part of the gel (the rectangle CDEF) was cut away and the gel containing an affinophore was formed (the rectangle EFGH). The gels were 1 mm thick. The edges of the diagram are notched at 10-mm intervals. (Reprinted from Ref. 14 with permission.)

hole for sample application is made in the agarose gel at a point 2.5 cm from edge CD, which has been defined by the adhesive tape. Sample solution (2 μl) is applied in the hole. Bridging sponges are placed on sides AD and BC of the gel with 5-mm overlap.

After the electrophoresis in the first dimension, the part of the gel (2-cm width) that faces the open area of the support film is cut away (Fig. 2, the area CDEF). On the enlarged open area (EFGH) of the film, 5 ml of 1% agarose solution containing the affinophore, which has been added after melting the agarose, is poured and spread. Immediately after gel formation, the gel plate is placed in the direction perpendicular to that of the first electrophoresis. In the case of an anionic affinophore, the side containing the affinophore (Fig. 2, side HG) faces the cathode. Electrophoresis in the second dimension is carried out under the same conditions as before.

Results of Affinophoresis

One-Dimensional Affinophoresis of Microbial Trypsin

An affinophore for trypsin is prepared by coupling p-aminobenzamidine to polymerized acrylyl-β-alanyl-β-alanine.[9] Because this affinophore is slightly stained with Coomassie blue, the carboxyl groups are converted to sulfonic acid groups as described. The amount of ligand represents about one-fifth of the anionic groups of the polymer.

The mobilities of *Streptomyces griseus* trypsin and bovine trypsin are significantly changed in the presence of the anionic affinophore (Fig. 3B). Because both enzymes are basic proteins, they migrate toward the cathode in ordinary electrophoresis (Fig. 3A). However, in the presence of the affinophore, they migrate toward the anode. The effect of the affinophore is not observed if trypsins have been inactivated by tosyl(L-lysine) chloromethyl ketone (TLCK). This clearly indicates that the active sites of the trypsins are responsible for the interaction with the affinophore. Migration of *Streptomyces erythreus* trypsin, an anionic species, is not influenced by the affinophore, owing to ionic repulsion. Affinophoresis of trypsins by means of a cationic affinophore (DEAE-dextran matrix) is also successful.[7]

Two-Dimensional Affinophoresis of Proteases

Pronase, a mixture of several extracellular proteases produced by *S. griseus,* is subjected to two-dimensional affinophoresis[14] by using the same anionic affinophore described above for the trypsins (Fig. 4). First- and second-dimensional electrophoresis are carried out in the absence and in the presence of the affinophore, respectively. Such a two-dimensional affi-

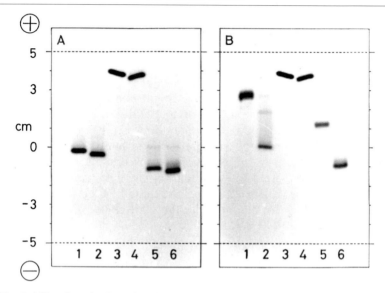

FIG. 3. Affinophoresis of trypsins with the anionic affinophore. Electrophoresis of trypsins and TLCK-treated trypsins (4 μg each) was carried out in the absence (A) or presence (B) of the affinophore (0.1%). Lane 1, *S. griseus* trypsin; lane 2, TLCK–*S. griseus* trypsin; lane 3, *S. erythreus* trypsin; lane 4, TLCK–*S. erythreus* trypsin; lane 5, bovine trypsin; lane 6, TLCK–bovine trypsin. (Reprinted from Ref. 9 with permission.)

nophoresis could be completed within 1.5–2 hr. Coomassie blue staining reveals that a spot of protein significantly deviates from the diagonal line formed by other protein spots (Fig. 4A). Without the affinophore, as expected, all the protein spots are found on the diagonal line (Fig. 4B). It is evident that the separate spot in Fig. 4A is carried by the affinophore. Prior to protein staining, a sheet of filter paper impregnated with a fluorogenic substrate for trypsin is placed on the gel for a short period in order to absorb a part of the solution in the gel. Incubation of the paper developed a fluorescent spot at the position corresponding to the deviated protein spot revealed by protein staining (data not shown). When the matrix polyionic polymer without ligand is used instead of the affinophore, the deviation of trypsin from the diagonal line is not observed.

Two-Dimensional Affinophoresis of Lectins

An extract of legume seeds is subjected to two-dimensional affinophoresis with an anionic affinophore bearing α-D-mannoside as an affinity ligand in order to separate mannose-binding lectins.[11] Pea seed contains a lectin specific for D-mannose (molecular weight of about 50,000, with two

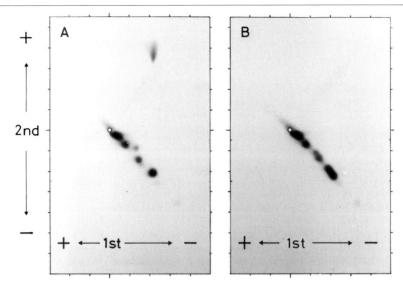

Fig. 4. Separation of *S. griseus* trypsin from pronase by two-dimensional affinophoresis. Pronase (40 μg in 2 μl of 0.1 *M* Tris–acetic acid buffer, pH 7.9) was applied to the sample application hole (left of center in each plate). Electrophoresis was carried out for 30 min at 60 mA/plate (about 25 V/cm) for each dimension. (A) Affinophoresis with the affinophore (0.04%); (B) electrophoresis without affinophore. (Reprinted from Ref. 14 with permission.)

sugar-binding sites). Two-dimensional affinophoresis of the extract of pea seed separates a spot from the diagonal line (Fig. 5A). Immunostaining of the protein blotted onto a nitrocellulose membrane after the affinophoresis with anti-pea lectin antibody shows that the spot is lectin (Fig. 5B). Blotting of separated proteins from the agarose gel is easily performed and 10 ng of the lectin can be detected. In the presence of a free ligand, methyl-α-D-mannoside, the spot does not migrate away from the diagonal line (Fig. 5C and D). Lectins are also separated from the extracts of fava bean and jack bean in the same way.

Two-Dimensional Affinophoresis of Antibodies

An antibody for a hapten peptide, PAPP, is separated by two-dimensional affinophoresis directly from rabbit antiserum.[12] The hapten is a tripeptide with a blocked amino terminus. Coomassie blue staining of the gel plate shows a diffuse spot that deviates from the diagonal line (Fig. 6A). This spot does not appear when the affinophore is omitted in the second-dimension electrophoresis. Therefore, this spot should be a group of hapten-specific antibodies. This is also supported by the result of immunostaining

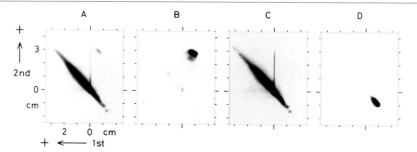

FIG. 5. Two-dimensional affinophoresis of an extract of pea seed with mannose affinophore. The sample was applied at position 0 and electrophoresis was carried out for 30 min at 25 V/cm (40–50 mA/plate) in each direction by using 0.1 M Tris–acetic acid buffer (pH 7.9) in the absence (A and B) or the presence (C and D) of 0.1 M methyl-α-D-mannoside. The concentration of the affinophore was 5.2 μM (58 μM for the ligand). (A and C) Two microliters of the extract was applied and stained with Coomassie Brilliant Blue R-250; (B and D) 2 μl of the 10-fold-diluted extract was applied and immunostaining was carried out after blotting onto a nitrocellulose membrane. Only the central part of gels containing proteins is shown. (Reprinted from Ref. 11 with permission.)

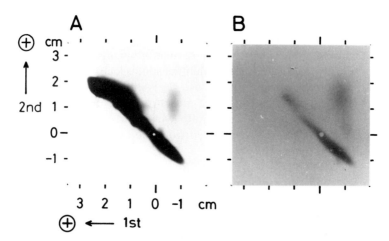

FIG. 6. Two-dimensional affinophoresis of anti-PAPP serum. (A) Serum (2 μl) was applied to an agarose gel plate at position 0. Electrophoresis was carried out for 30 min at 20 V/cm in 0.1 M Tris–acetate buffer, pH 7.9, in each direction. Second-dimensional electrophoresis was carried out in the presence of PAPP-affinophore. (A) Protein staining; (B) IgG on the nitrocellulose blot visualized by immunostaining. (Reprinted from Ref. 12 with permission.)

by goat anti-rabbit immunoglobulin G (IgG) antiserum. IgG is found at the position corresponding to the diffused spot visualized by protein staining (Fig. 6B).

Conclusion

Affinophoresis is an electrophoretic procedure that uses a specially designed carrier molecule aimed at modifying the mobility of a target substance. The power of the method as an analytical procedure is well demonstrated by the linkage of two-dimensional affinophoresis with immunoblotting or enzyme activity measurements. Other modes of two-dimensional application, e.g., combination with isoelectric focusing, should also be effective. The same affinophoretic principle can be applied by using polyionic biomolecules such as polynucleotides and mucopolysaccharides as natural equivalents for affinophores. Sodium dodecyl sulfate-polyacrylamide gel electrophoresis of proteins and electrokinetic chromatography can be considered analogous to affinophoresis, although they do not depend on biospecific affinity.

While affinophores bearing only ligands of low molecular weight were described in this chapter, various binding proteins such as antibodies and lectins can be used as ligands as well. Possible ligands are also not limited to biomolecules. A variety of interacting systems that have been utilized in affinity binding, e.g., protein–dye, sugar–boronate, chelating group–

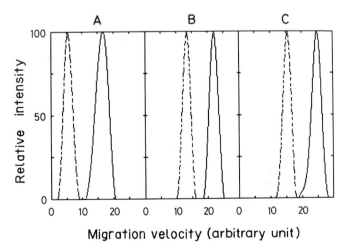

FIG. 7. An anionic affinophore was attached to the red blood cells from a rabbit (A), human (B), and a rat (C) by the aid of specific antibodies for each species of cells. The control populations are represented by dotted lines.

metalloprotein, and hydrophobic interaction, are candidates. It would be useful if neutral complex carbohydrates could be separated according to their structure, for example, by using affinophores bearing specially oriented boronate groups. Although preparation of a special affinophore for each target may appear troublesome, a more general affinophore can be prepared by linking with the avidin–biotin system.[13]

From the technical viewpoint, a combination of the principle of affinophoresis with advanced electrophoretic techniques such as capillary electrophoresis (see [13] in this volume[17]) should result in the creation of high-performance affinophoresis. Several reports on capillary electrophoresis for the purpose of analyzing specific interaction have appeared.[18–21b] More recently, a combination called "affinity probe capillary electrophoresis" was developed. This technique is based on the mobility change of a fluorescent-labeled monoclonal antibody caused by formation of an immune complex. It enables specific detection and quantitation of small amounts of human growth hormone (detection limit, $5 \times 10^{-12}\ M$) within 20 min.[22] Therefore, it is a possible candidate to be used instead of enzyme immunoassay. Application of the principle of affinophoresis to cell separation is worth exploring. Some preliminary results shown in Fig. 7 seem promising. Electrophoretic mobility of cells can be modified specifically by labeling with anionic affinophores. Although the apparatus used in these experiments could give only analytical results, when a simple and effective electrophoretic apparatus for preparative cell separation is developed, affinophoresis will become a method of choice for separating specific cells. Combination with free-flow electrophoresis is also of interest.

[17] S. Hjertén, *Methods Enzymol.* **271,** Chap. 13, 1996 (this volume).
[18] R. G. Nielsen, E. C. Rickard, P. F. Santa, D. A. Sharknas, and G. S. Sittampalam, *J. Chromatogr.* **539,** 177 (1991).
[19] S. Honda, A. Taga, K. Suzuki, and K. Kakehi, *J. Chromatogr.* **697,** 377 (1992).
[20] J. C. Kraak, S. Buschand, and H. Poppe, *J. Chromatogr.* **608,** 257 (1992).
[21] N. H. H. Heegaad and F. A. Robey, *Anal. Chem.* **64,** 2479 (1992).
[21a] J. L. Carpenter, P. Camilleri, D. Dhanak, and D. M. Goodall, *J. Chem. Soc. Chem. Commun.* 804 (1992).
[21b] Y.-H. Chu, L. Z. Avila, H. A. Biebuyck, and G. M. Whitesides, *J. Med. Chem.* **35,** 2915 (1992).
[22] K. Shimura and B. L. Karger, *Anal. Chem.* **66,** 9 (1994).

[10] High-Speed Automated DNA Sequencing in Ultrathin Slab Gels

By Lloyd M. Smith, Robert L. Brumley, Jr., Eric C. Buxton, Michael Giddings, Michael Marchbanks, and Xinchun Tong

Introduction

The demands of the Human Genome Initiative for improved DNA sequencing technology are considerable.[1] At this time virtually all DNA sequencing is based on the separation of DNA fragments in high-resolution polyacrylamide gels. One reasonable approach to improved sequencing is thus to increase the performance of such gel-based sequencing methods. It has been shown using the technique of capillary gel electrophoresis (CGE)[2–7] that the time required for gel electrophoretic separations of DNA fragments can be greatly reduced. In CGE, electrophoresis is performed in very thin (typically 50-μm i.d.) capillaries filled with denaturing polyacrylamide gel. The efficiency of heat transfer in these capillaries permits the use of much larger electric fields without deleterious thermal effects. The high fields result in separation speeds increased as much as 26-fold over conventional electrophoresis.

However, capillary gel electrophoresis in a single capillary is necessarily a serial technique. Although separations are much faster, the overall throughput of DNA sequencing by this method is at best comparable to that of existing instruments for automated DNA sequencing, which are capable of processing as many as 24 samples in parallel.[8] To increase the overall throughput of automated sequencing instruments by a comparable factor, it is necessary to obtain a similar parallelism in conjunction with

[1] L. E. Hood, M. W. Hunkapiller, and L. M. Smith, *Genomics* **1,** 201 (1987).

[2] H. Drossman, J. A. Luckey, A. J. Kostichka, J. D'Cunha, and L. M. Smith, *Anal. Chem.* **62,** 900 (1990).

[3] J. A. Luckey, H. Drossman, A. J. Kostichka, D. A. Mead, J. D'Cunha, T. B. Norris, and L. M. Smith, *Nucleic Acids Res.* **18,** 4417 (1990).

[4] A. Guttman, A. S. Cohen, D. N. Heiger, and B. L. Karger, *Anal. Chem.* **62,** 137 (1990).

[5] H. Swerdlow and R. Gesteland, *Nucleic Acids Res.* **18,** 1415 (1990).

[6] L. M. Smith, *Nature (London)* **349,** 812 (1991).

[7] D. Chen, H. R. Harke, and N. J. Dovichi, *Nucleic Acids Res.* **20,** 4873 (1992).

[8] C. Connell, S. Fung, C. Heiner, J. Bridgham, V. Chakerian, E. Heron, B. Jones, S. Menchen, W. Mordan, M. Raff, M. Recknor, L. Smith, J. Springer, S. Woo, and M. Hunkapiller, *BioTechniques* **5,** 342 (1987).

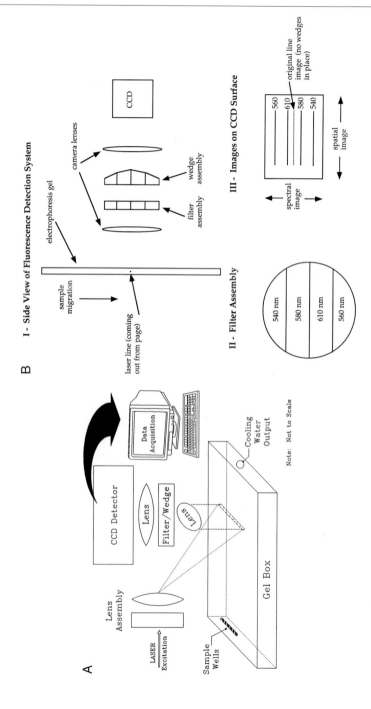

these extremely rapid separations. There are two obvious approaches to this problem: (1) employ many gel-filled capillaries in parallel, or (2) perform the separations in a slab gel format. The former approach has been described by Mathies and co-workers.[9-11] We describe here the instrumentation and methods we have developed for high-speed automated DNA sequencing by ultrathin slab gel electrophoresis.[12,13]

System Overview

A sketch of the system, discussed in detail in the following section, is shown in Fig. 1. It consists of a horizontal gel electrophoresis cell, a laser and optical components to provide a line of fluorescence excitation across the gel near its bottom, collection optics, a charge-coupled device (CCD) detector system to measure the emitted fluorescence, and a computer system to control the CCD camera as well as to store and analyze the resultant data. The fluorescence system is designed for sequencing by the four-fluorophore approach developed previously by one of the authors.[14]

[9] X. C. Huang, M. A. Quesada, and R. A. Mathies, *Anal. Chem.* **64,** 967 (1992).

[10] X. C. Huang, M. A. Quesada, and R. A. Mathies, *Anal. Chem.* **64,** 2149 (1992).

[11] R. A. Mathies and X. C. Huang, *Nature (London)* **359,** 167 (1992).

[12] R. L. Brumley and L. M. Smith, *Nucleic Acids Res.* **19,** 4121 (1991).

[13] A. J. Kostichka, M. L. Marchbanks, J. Robert, L. Brumley, H. Drossman, and L. M. Smith, *Bio/Technology* **10,** 78 (1992).

[14] L. M. Smith, J. Z. Sanders, R. J. Kaiser, P. Hughes, C. Dodd, C. R. Connell, C. Heiner, S. B. H. Kent, and L. E. Hood, *Nature (London)* **321,** 674 (1986).

FIG. 1. (A) Overview diagram of the high-speed automated DNA-sequencing instrument. A Series 200 CCD camera system from Photometrics, Ltd. (Tucson, AZ) using a Thomson 7863 CCD configured for frame transfer was employed. This system includes control electronics, a thermoelectric cooler, and a Macintosh interface. Software was written in Think-C (Symantec, Cupertino, CA) and the experiment was controlled and data anlayzed on a Mac IIci computer. (B) More detailed diagram of the fluorescence detection system employed for imaging of the line of emitted fluorescence onto the CCD detector at four wavelengths. (I) Side view of optical system; (II) on axis view of the filter assembly, a custom-fabricated piece consisting of four components. Each filter element is a 10-nm bandpass filter centered at the indicated wavelengths; (III) diagram indicating the nature of the images formed on the CCD detector surface. The dotted line shows where the image of the excitation line in the gel would have formed on the detector in the absence of the wedge assembly; each of the four solid lines is an equivalent image formed at a discrete wavelength due to the presence of the filter and wedge assemblies in the optical path. (Reprinted by permission from Kostichka *et al., Bio/Technology* **10,** 78–81, copyright © 1992 by Nature Publishing Co.)

TABLE I
FLUORESCENT PRIMERS

Dye–primer	Base	Excitation (Ex_{max}) (nm)	Emission (Em_{max}) (nm)
REG	A	528	555
RHO	C	502	530
TAMRA	G	552	580
ROX	T	580	610

Sequencing Chemistry

This fluorescence detection system is designed for use with the four-fluorophore sequencing strategy.[14] As detection sensitivity in fluorescence experiments is always fundamentally limited by the photostability of the dyes,[15] the use of more photostable dyes can provide a significant advantage. Most of our experiments to date have employed a set of four rhodamine dyes coupled to primer oligonucleotides. These dye–primers were generously provided to us by L. Lee of Applied Biosystems (Foster City, CA). The four fluorophores employed are rhodamine 110 (RHO), rhodamine 6G (REG), tetramethylrhodamine (TAMRA), and X-rhodamine (ROX), which are coupled to the −21 M13 sequencing primer at the 5' end (see Table I). The RHO and REG dyes are substantially more photostable than the fluorescein and JOE (5- (and -6) -carboxy-4',5'-dichloro-2',7'-dimethoxyfluorescein) fluorophores available commercially from Applied Biosystems.

Table II provides a protocol for the performance of DNA-sequencing reactions. DNA sequencing reactions are conducted by mixing 0.8 pmol (2 μg) of single-stranded template DNA, and 1.5 μl of 10× SB (SB, or sequence buffer, is 10 mM MgCl$_2$ and 10 mM Tris-HCl, pH 8.5). Water is added to a total volume of 8.5 μl, which is then divided into four equal volumes, each of which is combined with 0.5 μl of a 0.4-pmol/μl solution of one of the four fluorescent primers. The four mixtures are placed at 65° (heating block) for 2 min, at 37° (heating block) for 5 min, and at room temperature (air) for 10 min to anneal the primer to the template. *Bst* DNA polymerase (0.5 μl) (0.25 unit/μl; Bio-Rad, Richmond, CA) and 2.0 μl of a deoxy (d)- or dideoxy (dd)NTP mixture (see below) are added to each of the four tubes. The sequencing reactions are placed at 65° for 5 min. The four sets of A, C, G, T reactions are combined and 12.0 μl of 10 mM ammonium acetate and 80 μl of 100% ethanol are added. The tube is spun in a microcentrifuge for 15 min (15,800 g at 4°) and the supernatant

[15] R. A. Mathies, K. Peck, and L. Stryer, *Anal. Chem.* **62,** 1786 (1990).

TABLE II
PROTOCOL FOR DNA-SEQUENCING REACTIONS

1. Label tubes A, C, G, and T and aliquot 0.5 μl of the appropriate fluorescent primers to each tube. Keep on ice

   d/ddATP = REG–primer (0.4 pmol/μl)
   d/ddCTP = RHO–primer (0.4 pmol/μl)
   d/ddGTP = TAMRA–primer (0.4 pmol/μl)
   d/ddTTP = ROX–primer (0.4 pmol/μl)

2. Mix the following cocktail (10 μl/set of four reactions):

   1.0 μl of ssM13mp19 template DNA (2.0 mg/μl = 0.8 pmol/μl)
   1.5 μl of 10× SB (100 mM MgCl$_2$, 100 mM Tris, pH 8.5)
   6.0 μl of H$_2$O

3. Add 2.0 μl of the above cocktail to the fluorescent primer tubes. Anneal at 65° (heating block) for 2 min, 37° (heating block) for 5 min, room temperature (air) for 10 min

4. Add 0.5 μl of Bst DNA polymerase (0.25 unit/μl) and 2.0 μl of d/ddNTP mix to each annealed dye–primer/template tube, respectively, and immediately place at 65° for 5 min

5. Combine A, C, G, and T reactions and add 12.0 μl of 10 M ammonium acetate and 80 μl of ethanol. Spin for 15 min and then withdraw the supernatant with a Pipetman

6. Add 80 μl of ethanol to the reaction sample and spin it for another 5 min. Withdraw the supernatant and dry the reaction for 3–4 min in a vacuum desiccator

7. Resuspend the reaction in 12.0 μl of 10 mM EDTA–95% (v/v) formamide

8. Denature the reaction at 90° for 2 min prior to loading

discarded with a Pipetman. The sample is rinsed with 80 μl of 100% ethanol and then centrifuged for 5 min. After removing the supernatant, the sample is dried in a vacuum desiccator for 3 to 4 min. The purified DNA pellets are resuspended in 12 μl of 10 mM EDTA–95% (v/v) formamide and heated at 90° for 2 min to denature. The sample is loaded immediately onto a preelectrophoresed acrylamide gel. The following nucleotide mixtures were used: a 250 μM concentration of each of three dNTPS, one of 240 μM ddGTP, 480 mM ddATP, 500 μM ddTTP, or 200 μM ddCTP, and a 25 μM concentration of the corresponding dNTP.

Electrophoresis Cell

Schematic diagrams of the electrophoresis apparatus are shown in Fig. 2A–C. The apparatus base (3.5 × 18.0 × 34.0 cm) is composed of polycarbonate machined to provide space for a series of clamps, guide blocks, water jacket, and inlet and outlet manifolds.

Temperature regulation is provided by circulating coolant from a water bath (GeneSprinter thermal regulator; Fotodyne, New Berlin, WI) through

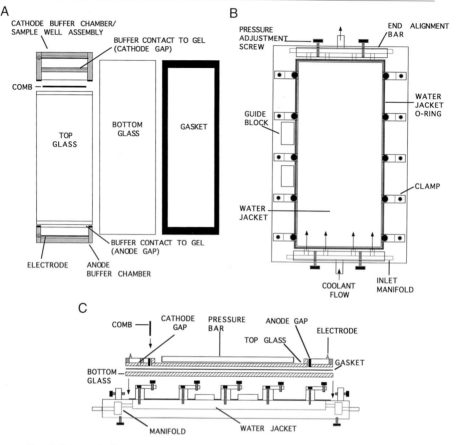

Fig. 2. Schematic diagram of horizontal electrophoresis apparatus. (A) Top view of glass and plastic components. (B) Top view of horizontal apparatus base. (C) Side view of horizontal apparatus base and glass components.

the water jacket under the glass plates. Fluid flow is dispersed across the width of the glass by the inlet and outlet manifolds, which mix the fluid in the water jacket and thus aid in obtaining a uniform temperature distribution across the glass plates.

Correct alignment of the glass components on the apparatus is provided by the end alignment bar and two guide blocks. These aid in the correct positioning of the glass components on the water jacket O ring. In addition, pressure-adjustment screws on the end alignment bar apply pressure at the end of the buffer chamber/sample well assembly. This ensures that the comb fits properly between the glass components and is essential for the

consistent formation of sample wells. Electrodes are constructed by string-
ing platinum wire across the back of the buffer chambers. The wire is
secured with silicone glue in small holes drilled in the sides of the buffer
chambers. The wires are connected to "banana"-type plugs (Figs. 2A, 2C).

The buffer chamber and buffer chamber/sample well assembly (Fig.
2A) were designed to hold up to 5 ml of liquid buffer. Figure 2A shows
the slot formed by the juxtaposition of the top glass with the rubber spacers
on the buffer chamber (anode gap) as well as the slot in the buffer chamber/
sample well assembly (cathode gap), each of which provides electrical
contact between the gel and the buffer. The comb fits in the gap created
by the rubber spacers between the buffer chamber/sample well assembly
and the top glass plate.

The comb is made of 0.75-mm polymethylmethacrylate (PMMA) (Acry
Fab, Inc., Sun Prairie, WI), and is 3.8 × 7.2 cm. The bottom glass plate is
10.0 × 30.5 × 0.5 cm, whereas the top glass plate is 10.0 × 25.0 × 0.5 cm.
To ensure uniformity of glass surfaces, optical-quality (fused silica or BK-7)
glass polished to four-wave (~2 μm) flatness over any 2-in. surface, and
cut with opposite sides parallel (obtained from American Precision Glass,
Duryea, PA), was employed. The glass face of the buffer chamber/sample
well assembly and end of the top glass that form the gap for the comb were
polished flat to within 5 μm. The gasket, which determines the thickness
of the gel, is cut from a polyester sheet (Acry Fab, Inc.) of the desired
thickness. The final gel area is 205 cm^2 (8.2 × 25 cm).

Gel Preparation

The most critical aspects in ultrathin slab gel preparation are the elimination of dust particles, mechanical stresses, and avoidance of air bubble
formation during pouring. It is crucial to work in an area kept as clean as
possible. The following procedures have the aim of limiting problems due
to these or other factors.

Cleaning

The glass plates are first cleaned using Alconox detergent (Fisher Scientific, Pittsburgh, PA). Stubborn grease residues are taken off with RBS 35
(Pierce, Rockford, IL) prepared according to the instructions on the label
and mixed with methanol (2 : 1). The plates are washed with distilled water
and dried with paper towels. They are then wiped three times with ethanol,
again with paper towels. It is essential that the paper towels be lint free to

avoid lint or dust from the towels accumulating on the plates. Laboratory tissues such as Kimwipes leave particles that subsequently interfere with the electrophoresis. Common single-fold paper towels (e.g., Second Nature single-fold towels; Wisconsin Tissue, Menasha, WI), however, work fairly well and do not leave small fragments behind. These steps are repeated for all of the other glass pieces being used. If reusing a plastic spacer (see below), it should be cleaned with Alconox, rinsed with distilled water, and dried with the same-type towels used in drying the glass plates. The gel apparatus should be wiped clean of any urea, grease, or other materials making certain that the silicone gasket is clean and in its designated slot. It has been found that using a small brush to wipe off dust particles on the upper and lower plates before final assembly minimizes the number of particles trapped between the plates.

Preparing Gel Plates and Assembly

The end of the top plate and the end piece where the comb is inserted (and the top of the bottom plate if using radioactive labels) are treated with γ-methacryloxypropyltrimethoxysilane[16] (Sigma Chemical, St. Louis, MO) to bond the polyacrylamide to the glass surface. In addition, the top plate may be siliconized to facilitate the flow of the gel solution during pouring. To assemble the gel box, the bottom plate is positioned on the apparatus using guide blocks. The gasket is then put onto the glass plate using the guide blocks for correct positioning. The spacers on the end pieces (buffer chamber and well assembly) are coated with a small amount of petroleum jelly or grease to prevent leakage in the seal with the top plate. The end pieces and the top piece are then positioned on the bottom plate and clamped down. At this point the top piece and the buffer chamber are clamped in place, care being taken not to put too much pressure on the clamps. The well end is only lightly clamped because it needs to be adjusted later when the comb is secured.

Gel Pouring and Final Assembly

A variety of different gel and buffer compositions may be employed. The 6% (w/v) gel in 1× Tris–borate–EDTA (TBE) has proven robust

---

[16] H. Garoff and W. Ansorge, *Anal. Biochem.* **115**, 450 (1981).

TABLE III
GEL COMPOSITION

Component	Composition[a]
6% Acrylamide	1.5 ml of acrylamide stock
1× TBE	1 ml of 10× TBE stock
7.5 M Urea	7.5 ml of urea stock

[a] Preparation of stock solutions: 40% acrylamide stock solution [38% acrylamide, 2% N,N'-methylenebisacrylamide to volume in 8.3 M urea; 19 g of acrylamide and 1 g of bisacrylamide in 8.3 M urea to volume (50 ml)]; 10× TBE stock [890 mM Tris–borate, 20 mM EDTA; 10.8 g of Tris base, 5.5 g of boric acid, and 0.4 ml of EDTA in water to volume (100 ml)]; and Urea stock [8.3 M; 320 g of urea to volume (650 ml) in water].

in radioactivity-based work and is the gel formulation used in this laboratory.

Make up 10 ml of polyacrylamide gel solution as described in Table III. Add 50 μl of freshly prepared 10% ammonium persulfate and 5 μl of N,N,N',N'-tetramethylethylenediamine (TEMED). Pour approximately 5 ml of the gel solution into the buffer chamber and elevate that end of the apparatus by 15–30°. The gel should flow, in a fairly even front, toward the lower chamber; however, dust or imperfections on the glass can retard the flow, resulting in potential formation of air bubbles. This can be alleviated by lightly tapping on the top glass plate at points that are moving slower than the solution front. After the solution front reaches the lower chamber, approximately 1 ml of the gel solution is added to it and the apparatus is leveled. Finally, the comb is inserted into the slot between the top glass and the end piece, with the "teeth" sitting on the bottom plate. The comb is secured by turning the pressure adjustment screws just enough to eliminate gel solution between the comb and the top glass. The gel is then allowed to polymerize for at least 90 min prior to use.

Gel Loading and Electrophoresis

After the gel has polymerized, excess unpolymerized polyacrylamide is removed from the buffer chambers with a syringe (10 ml) and absorbent laboratory tissue (i.e., Kimwipes). Approximately 5 ml of the electrophoresis buffer (TBE at the same concentration as is in the gel) is then added to the buffer chambers. The comb is then removed by gently pulling it up and out of the gel. Care should be taken to pull it straight up such that the well walls are not damaged. The wells are then rinsed by pipetting approximately 100 μl of the fluid into each well. This step is repeated three or four times. To avoid spills, the buffer is removed from the buffer chambers before attaching the apparatus to the detection system.

To set up a run, the apparatus is connected to a preheated water circulator, set to the desired temperature, and water is circulated through the water jacket for approximately 5 min to allow for uniform temperature in the glass plates. The gel box is then positioned in the detector assembly and the power supply is connected to the banana clips on the apparatus. The buffer chambers are then refilled with buffer and the gel is prerun for about 10–20 min by turning on the electric field (generally at the same voltage/wattage at which the samples are run). Just prior to sample loading, the power is turned off and the wells are flushed out again, using the same method discussed above. The samples are then loaded into the wells as described below.

Sample Injection Technique

The top plate is cut into four pieces. Between each piece there is a small (~1.5 mm) gap. This means that there are three liquid–gel interfaces. The interface at each end is where the electric field is coupled to the gel, whereas the internal gap is where the sample is loaded (see Fig. 2). In this configuration the electric field lines flow fairly linearly through the sample wells. To load the sample, the buffer in the sample wells and compartment is replaced with pure water. The wells are rinsed with pure water and 200 nl of sample is loaded per millimeter of well width, using a 1.0-μl Hamilton syringe; 2500 V is applied for about 30 seconds and then (after turning off the power supply) the wells are flushed out with water and the water replaced with TBE at the same concentration as was used in preparing the gel. The field is reapplied and data collection is begun.

A field of 100 V/cm (2500 V for the 25-cm length of the gel utilized) is typically employed. Higher fields may be applied, but a loss of resolution is encountered for reasons that have been explored in detail elsewhere.[17,18] After the run is complete, the power supply and water bath are turned off and the water jacket is emptied back into the bath. The apparatus is then disassembled and cleaned before another gel is poured.

[17] J. A. Luckey and L. M. Smith, *Electrophoresis* **14**, 492 (1993).
[18] J. A. Luckey and L. M. Smith, *J. Phys. Chem.* **97**, 3067 (1993).

Optical System

Input Optics

The function of the input optics is to provide a line of excitation light at a desired wavelength across the gel. The excitation source employed in these studies is the 514-nm line of an argon ion laser, which excites the fluorophores used in the system reasonably well. Some of the design issues that need to be considered include (1) the width and uniformity of the line, (2) the power density of the excitation light, and (3) the background fluorescence and scattered light generated both in the gel and in the rest of the electrophoresis apparatus (i.e., the glass or quartz gel plates, cooling liquid, and Plexiglas assembly). There are three basic approaches to introduction of the excitation light: fanning the beam, scanning the beam, and introducing the beam from the side.

Figure 1A shows fanning of the excitation beam. A cylindrical lens permits a line to be made from the circular profile of a Gaussian laser beam. This line will retain a Gaussian intensity profile along its linear dimension, peaked in the center and falling off to the sides. Selection of the appropriate lens permits the desired line dimensions to be obtained. Introduction of the light into the upper gel plate at Brewster's angle minimizes the light reflected from the surface and maximizes the light transmitted into the gel. The advantage of this approach is its simplicity. Disadvantages include the fact that the light passing through the gel plates and cooling fluid excites nonspecific background fluorescence, increasing system noise and thereby decreasing sensitivity. This can be partially alleviated by use of synthetic fused silica gel plates with extremely low levels of fluorescence contaminants. Another disadvantage is that the excitation power density in the gel is lower than that which can be obtained using from-the-side excitation (described below). Therefore, when the beam is fanned it is necessary to use a much higher power laser source.

Another approach to introduction of the excitation light is beam scanning. This can be done in a number of ways, including use of an acoustooptic modulator, scanning mirror, or a mechanical scanning assembly. The latter approach is employed in the commercial automated sequencing instruments available from Applied Biosystems. In beam scanning, the laser beam, focused down to a desired diameter with an appropriate lens, is scanned across the sample. However, the method has the same two major drawbacks as fanning the beam; i.e., the excitation of background fluorescence and scattering in the surrounding media, as well as the lower time-averaged power density obtainable with an excitation source of a given power.

The third approach is to bring the excitation beam in from the side.

TABLE IV
DEPTH OF FOCUS CALCULATIONS

Beam size at focus (μm)	F (calculated)[a] (mm)	Depth of focus[b] (mm)	Size at edge of DoF (μm)
1	2.0	0.0031	1.4
10	20	0.310	14
25	50	1.9	35
50	99	7.6	71
75	150	17	110
100	200	31	140
150	300	69	210
200	400	120	280

[a] F is the focal length of the lens, which is dependent on the spot size required.
[b] Depth of focus (DoF) is defined as the distance at which the beam is the square root of 2 times larger than the beam waist, as in the diagram below.

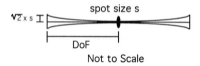

Not to Scale

This approach, described by Ansorge and co-workers[19] and utilized in commercial sequencing instruments from both Pharmacia (Piscataway, NJ) and Hitachi, has the significant advantages of eliminating background light from the gel plates and surrounding media, as well as providing a much higher excitation power density from a source of a given power than either of the previous two approaches. It has the disadvantage, however, of raising both fundamental and technical difficulties in this implementation in ultra-thin gels. The fundamental difficulties derive from the physics of Gaussian laser beams themselves, in particular the limitations of Gaussian beam optics on the narrowness of the Gaussian beam radius that can be obtained.[20] The well-known effect in standard camera systems of a tradeoff between the size of the lens aperture and the depth of field (large aperture, small depth of field) also applies to Gaussian beams. The more tightly one focuses a Gaussian beam, the shorter the distance over which one can maintain that focus. For example, by solving the equations presented in Ref. 20 one can obtain the results presented in Table IV, giving the depth of focus of a 514-nm Gaussian laser beam focused to a specified size. It

[19] W. Ansorge, B. Sproat, J. Stegemann, C. Schwager, and M. Zenke, *Nucleic Acids Res.* **15,** 4593 (1987).
[20] S. A. Self, *Appl. Optics* **22,** 658 (1983).

may be seen from Table IV that a beam focused to a waist of 50 μm will stay below a diameter of 71 μm only over a region 1.5 cm wide. As the slab gels of the present system are 10 cm wide, of which about 7.5 cm can be used, this is clearly unacceptable. It is thus necessary to trade off the narrowness of the beam versus the width of the region employed for the electrophoresis.

The technical issue that arises concerns the introduction of the beam into the gel. It is necessary to bring the beam in between the very closely spaced gel plates through a good-quality optical interface, so that the beam is not distorted or scattered. A gap must be introduced into the spacer material between the gel plates. The disruption in the gel uniformity, if not accomplished carefully, can lead to problems with gel stability. We have obtained reasonably good results simply by cutting a gap in the spacer material, and putting a flat glass window on the side of the gel at the position of the gap. When the gel is poured the polymerizing gel "glues" the window in place, holding it there, and it provides a good-quality optical interface for bringing the beam into the gel. However, this method still encounters occasional problems, and until improved is probably not suitable for routine use in sequencing.

Output Optics

The fluorescence detection system was designed to address speed and sensitivity constraints of these thin gel systems. Whereas in existing automated sequencers the bands of DNA take about 1 min to pass through the fluorescence detector monitoring the gel, with the high fields employed in the ultrathin gel system peaks may take only seconds to pass through the detection region. Thus it is necessary to take data points far more frequently than in an apparatus employing conventional electrophoresis. Because much less material can be loaded on the thin gels (scaling in proportion to the volume of the gel through which the sample passes), the absolute detection sensitivity required in this system is much greater than in existing automated sequencers.

To meet these constraints, the fluorescence detection system diagrammed in Fig. 1 was designed and constructed. This system uses a cooled CCD array detector operated in frame transfer mode.[21] A camera lens (Nikkor 105 mm, $f1.8$) is positioned with its focal point coincident with the line of excitation light impinging on the gel, thereby collecting and collimating the emitted fluorescence. The collimated fluorescence is passed through a four-element interference filter (Fig. 1B, custom fabricated by

[21] J. J. Linderman, L. J. Harris, L. L. Slakey, and D. J. Gross, *Cell Calcium* **11,** 131 (1990).

Omega Optical, Inc., Brattleboro, VT) in which each element is a 10-nm bandpass filter. This filter rejects scattered excitation light and also selects four discrete wavelength regions from the emitted fluorescence. The light then passes through a four-component wedge prism assembly (custom fabricated by Broomer, Islip, NY), in which each wedge is aligned with one of the four filter elements. Thus each of the four selected wavelength regions is diverted angularly in a different direction. A second camera lens (Nikkor 50 mm, $f1.2$) collects this light and produces four images of the original line on the surface of the CCD, each corresponding to a different wavelength region. These two lenses yield a system demagnification of 2.1; the imaged area is 1.85 cm in width, giving a line image on the CCD of 0.88 cm. To image a larger area other lenses may be chosen. It is important, however, that the lenses be of adequate quality to image the larger area without significant distortion.

Each of these four lines is a spatial image of the original excitation line. The four-point fluorescence spectrum corresponding to any given physical point along the excitation line in the gel is encoded in the intensity of the measured signal at the corresponding position in each of the four images. Figure 3[21a] shows data present in the 560-nm line obtained in the parallel analysis of 18 samples on this system. The time axis shows these plots obtained in successive 700-msec integrations. The peaks evident in these plots correspond to the different fragments of fluorescently tagged DNA migrating through the excitation beam during the measurements.

Data Collection and Analysis

Data from the imaging optics are collected in frames (snapshots) at intervals of 0.2 to 1 sec (typically 0.5 sec) in length on a Series 200 CCD camera system from Photometrics, Ltd. (Tucson, AZ) using a Thomson 7863 CCD configured for frame transfer. This system includes control electronics, a thermoelectric cooler, and a Macintosh interface. The CCD chip is a rectangular array of 384×576 pixels, each measuring 23×23 μm. Each frame of data collected on the CCD system requires several steps, including exposure, frame transfer, binning, and readout. The CCD chip is divided into two equal regions for this purpose—one for image collection, and a masked portion for binning and readout. This scheme is necessary owing to the large amount of image data that must be processed rapidly.

In frame transfer mode, an exposure is made on the unmasked portion of the CCD, and then rapidly transferred to the masked region where the

[21a] D. A. Mead, J. A. McClary, J. A. Luckey, A. J. Kostichka, F. R. Witney, and L. M. Smith, *BioTechniques* **11**, 76 (1991).

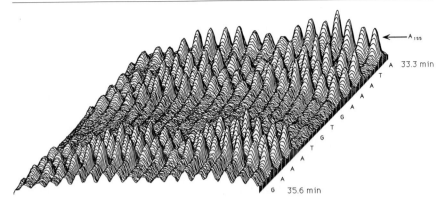

FIG. 3. Data obtained at one of the four wavelengths in the parallel analysis of 18 samples on the CCD-based automated sequencing system. Each horizontal tracing corresponds to the integrated fluorescence intensity obtained in the 560-nm line image over a 700-msec period. Successive tracings represent such data obtained between 33.3 and 35.6 min of electrophoresis time, as indicated. The migration time and DNA sequence corresponding to these data are indicated on the right-hand side; the sequence covers the region A155 to G167. The conditions employed in this separation were as follows: 75-μm thick 6% polyacrylamide–5% bisacrylamide gel in 2× TBE (where TBE is 89 mM Tris–borate, 2 mM disodium EDTA, pH 8.3) and 4 M urea; 40-W electrophoresis power, constant power mode; 40° water bath temperature. The higher buffer concentration and lower urea were found experimentally to give better gel stability and fragment resolution than the standard 1× TBE, 8 M urea. Samples (600 nl) containing 200 ng of template DNA were loaded using a Hamilton syringe in a region 1.85 cm wide, using a comb with eighteen 0.6-mm slots and 0.4-mm divisions between slots. Adequate signal intensity is obtained from 100-ng samples, but a twofold greater amount was employed here to give better visualization of the peaks. Sequencing reactions were performed on M13mp19 DNA as described[21a] using *Bst* polymerase and dye-conjugated primers, except that the samples were resuspended after ethanol precipitation at a twofold higher concentration than normal. (Reprinted by permission from Kostichka *et al., Bio/Technology* **10**, 78–81, copyright © 1992 by Nature Publishing Co.)

slower readout process is performed. During readout of the masked portion, a new frame can be collected on the unmasked portion. Specifically, each frame is integrated for (typically) 500 msec on the 384 × 288 pixel unmasked region, and then transferred to the masked portion of the CCD in 0.58 msec, leaving the remaining integration time for binning and readout. In addition to allowing time for readout, the integration period allows collection of enough light from the sample that the effects of background and detector noise become much less significant.

Binning,[22] which is simply a hardware-based "summation" over defined pixel regions of the CCD, allows much faster readout at the expense of

[22] P. M. Epperson and M. B. Denton, *Anal. Chem.* **61**, 1513 (1989).

spatial resolution. This binning is performed over four regions of the CCD, corresponding to the four spectral line projections from the filter-wedge assembly described above. Little useful information is lost in this process because the binning occurs only across the "spectral" axis of the CCD chip. Full resolution is retained on the "spatial" axis for each of the four binned regions, allowing 384 pixels of spatial information to resolve the individual lanes in the gel. This means that at each 0.5-sec sample interval, 384×4 data points are collected and stored, resulting in a final "image" file consisting of ~11 million data points.

To determine the nucleotide sequences present in each of the samples loaded on the gel, analysis of the resulting data is broken into several steps, including lane finding, multicomponent analysis, noise reduction, mobility corrections, and base calling.

Lane finding is presently performed through manual interaction with the computer, whereby the user is presented with one of the 384 point "spatial" cross-sections collected early in the run and allowed to demarcate the peaks that indicate the presence of lanes. To compensate for any lane drift, the user is then presented with a similar cross-section from a later point in the run to demarcate lanes in a similar fashion, on which the computer performs a linear interpolation between the points chosen from each cross-section. Then, for each lane marked, a four-channel signal file is produced that represents the intensity variations along the time axis of the run at a particular lane location at each of the four spectral wavelengths. For each sample loaded, this type of four-channel "raw" data file is produced representing the spectral variations as the various components (nucleotides) are separated and travel through the detection region.

Each raw data file must then be processed by multicomponent analysis to produce a "trace" file, in which each channel represents the relative quantity of each of the fluorophores present at that moment in the detector region. Multicomponent analysis is performed by determining a 4×4 matrix in which the four entries in each column represent the relative spectral response of one particular dye at each of the four wavelengths.[23] Thus, four columns are produced corresponding to each of the four dyes. This matrix can be inverted, and then multiplied by each four-channel spectral point in time (represented as a vector for this purpose) to produce a four-channel trace point at the corresponding moment in time. This process is fast and relatively accurate, although it can be somewhat sensitive to noise or spectral variations in the data being collected.

The trace file is then bandpass filtered both to reduce high-frequency

[23] L. M. Smith, R. J. Kaiser, and L. E. Hood, *Methods Enzymol.* **155**, 260 (1987).

noise components and to eliminate baseline drift by reducing low-frequency information. We are using a fast Fourier transform-based bandpass filter. The data are first transformed to a frequency domain signal in 1024-point windows on a Motorola digital signal processor, and then multiplied by a filter that is the product of two Gaussian functions. The filter is designed to cause a rapid cutoff at the low-frequency end, and a more gradual cutoff at the high-frequency end. The resulting (complex) data set is then conjugated and transformed back to the time domain in 1024-point windows on the digital signal processor.

Finally, compensation for differing mobilities in the gel matrix of each of the four dyes used to label the DNA fragments must be performed.[23,24] The differing mobilities appear as a slight and usually constant offset of one or several of the channels from their expected position relative to one another. This results in overlap and uneven spacing of the peaks corresponding to different bases. Adding (or subtracting) a constant offset to the necessary channels for a good alignment compensates for the "mobility shift." For a particular instrument and set of run conditions, the mobility offsets are constant, allowing one calibration to be done that can then be used repeatedly.

Base calling can then be performed on the trace file resulting from these various signal-processing steps. This process can be quite accurate when the underlying data are high in quality. The data quality depends on factors ranging from the template preparation method to the care with which the mobility shift and multicomponent analysis are calibrated. Variations in the quality of data can account for more than a 20% variation in the accuracy of the base-called fragments. With our current algorithm applied to "quality" data, base-calling accuracy is approximately 96–98% out to 400 bases.[25] Under certain run conditions, this accuracy has been extended to more than 500 bases. An example of sequence data, along with the called bases, is shown in Fig. 4.

The base-calling routine is based on a device-independent, object-oriented peak-filtering algorithm. The algorithm is started with a large list of all potential "bases" in the trace, and then assigns a confidence level to each, removing those that fall below some confidence level. This is an iterative process whereby on successive iterations, information from high-confidence peaks can be used to hone the base calling in areas where data quality is poorer. Thus as iteration ensues and low-confidence peaks are

[24] L. M. Smith, S. Fung, M. Hunkapiller, T. Hunkapiller, and L. E. Hood, *Nucleic Acids Res.* **13,** 2399 (1985).
[25] M. Giddings, R. L. Brumley, Jr., M. Haker, and L. M. Smith, *Nucleic Acids Res.* **21,** 4530 (1993).

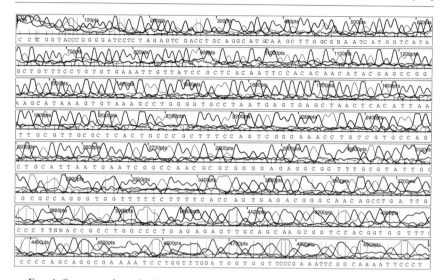

FIG. 4. Sequence data obtained on the CCD-based automated sequencing system. The steps of data extraction, multicomponent analysis, digital filtering, and base calling described in text have been performed. These data were obtaned on a 50-μm thick 4% acrylamide gel in 2× TBE and 4.2 M urea, with a coolant temperature of 50°. The run was performed at a constant voltage of 2500 V. pts, Data points, which were taken at 0.5-sec intervals.

removed, a final base list is arrived at where each base is labeled with some confidence level.

For this process, the initial list of potential bases is created by locating all maxima in each of the four traces, and creating a list sorted by location. Confidence levels are assigned based on peak spacing and height. However, there are no predetermined spacing or height values; instead, on the basis of the initial peak list, standard curves for the particular data set are automatically determined for both height and spacing using a polynomial curve-fitting technique. Starting with these "rough" calibration curves, peaks can be assigned a confidence level, and many unlikely peaks removed. With the resulting base list, new and more accurate spacing and height calibration curves can be obtained by weighting these according to the confidence levels of the peaks used for calibration. With these curves, more accurate confidences can be assigned to the base list. This process is iterated until the average confidence measure of the peaks stabilizes, or until a predefined number of iterations is reached. The process generally stabilizes within several iterations.

The algorithm is object oriented in the sense that each of the "feature detectors" is wrapped up in an object that can then be "plugged" in. Thus, there is one object that specializes in determining calibrations for spacing and assigning confidences based on spacing, and likewise for peak height. In this way, new features of the data can be identified (such as peak width, shape, etc.) and encapsulated in an object that can then simply be added to the program. This promotes a high degree of modularity and adaptability of the code to different types of data.

Finally, the fragments resulting from all these steps are aligned to produce the entire sequence of a clone. Currently such alignments are done statistically, based only on the redundancy of overlap in fragments. This process has failings because it utilizes no information about the quality of the underlying fragments being aligned, and thus in locations where there is disagreement it is generally a "majority rules" determination of the final sequence. If there are only two fragments overlapping, or if there is not a clear majority, it becomes particularly difficult to assign a sequence accurately. We hope to improve this process by utilizing the confidences assigned to the bases by the base-calling algorithm during the alignment process, particularly in regions of disagreement between the fragments. The algorithm to be used will be a simple adaptation of one of the commonly used fragment alignment routines, which are based on the algorithm of Needleman and Wunsch.[26] Through this process the final aligned sequences can also be assigned confidences at each base. This information can be used to indicate the overall quality of sequences stored in databases for future reference. Such indicators of quality would be valuable to other researchers utilizing the sequence information in their work.

[26] S. B. Needleman and C. D. Wunsch, *J. Mol. Biol.* **48,** 443 (1970).

[11] Applications of Capillary Zone Electrophoresis to Peptide Mapping

By Eugene C. Rickard and John K. Towns

Introduction

A variety of methods is needed to assess the identity and purity of macromolecules, owing to their size and structural complexity. Careful selection and use of several methods based on unrelated principles allow evaluation of various aspects of the macromolecular structure. For example, orthogonal methods that demonstrate identity and purity give added assurance and credibility to the testing program that defines the quality of a protein. Peptide mapping is one of the fundamental methods that can be used to characterize the important chemical parameters of macromolecular structure, especially the primary amino acid sequence.

The general concept of peptide mapping (or peptide fingerprinting) is relatively simple. When proteins are subjected to enzymatic digestion or cleavage by chemical reagents, they will be broken down into discrete fragments. These peptide fragments are then differentiated via a suitable separation technique to give a distinct pattern. Matching of the pattern most often includes comparison of the relative retention times in a chromatographic system, but may also include a comparison of peak areas or other parameters. If the pattern of the fragments from the protein being tested matches the pattern of a digest prepared in the same manner from a reference protein, then the two proteins are most likely the same. In characterization of pharmaceutical products, this procedure is used to confirm the identity of a protein or to pinpoint the differences in primary structure between the drug substance and a related species.

The initial step in peptide mapping is to select an appropriate digestion step. The digestion should (1) produce enough fragments to provide a positive identity but not so many that separation and comparison are unnecessarily difficult; (2) be robust, reproducible, and easy to perform; and (3) be as "clean" as possible—that is, cleave nearly all of the protein at each of the selected cleavage sites with only minimum cleavage at other sites. These requirements frequently lead to the choice of an enzyme, such as trypsin, for the digestion. However, chemical cleavage with reagents such as cyanogen bromide is also possible. For glycoproteins, it is helpful to cleave the carbohydrate moieties prior to peptide mapping to reduce the complexity of the map. Carbohydrate heterogeneity is then characterized as a separate step.

After digestion conditions have been established and optimized, the peptides must be separated for analysis. Various separation modes have been used to differentiate between the peptides. These have included sodium dodecyl sulfate-polyacrylamide gel electrophoresis (SDS–PAGE),[1,2] capillary zone electrophoresis (CZE),[3,4] two-dimensional (2D) gel electro-

[1] W. C. Plaxton and G. B. C. Moorhead, *Anal. Biochem.* **178,** 391 (1989).
[2] R. C. Judd, *Methods Enzymol.* **182,** 613 (1990).
[3] K. A. Cobb and M. Novotny, *Anal. Chem.* **61,** 2226 (1989).
[4] R. G. Nielsen, R. M. Riggin, and E. C. Rickard, *J. Chromatogr.* **480,** 393 (1989).

phoresis,[5] and either reversed-phase[6,7] or ion-exchange[8] high-performance liquid chromatography (HPLC). High-resolution reversed-phase HPLC is the most popular choice owing to its ability to provide the resolution needed to identify and quantitate individual peptides, as well as for its robustness and reproducibility (see also [2] in this volume[8a]). In addition, HPLC detectors do not have to rely on staining or labeling of the protein—tedious and problematic steps—prior to quantification as is required in slab gel separations (see also [10] in this volume[8b]).

However, chromatographic methodology also has limitations. These include long analysis times and the possibility that not all of the peptide fragments will be separated by any one technique, no matter how powerful.[9] In addition to these concerns, an important point to consider in the development of the chromatographic method is the "wedding" of the particular peptide map to a specific chromatographic column used in the method development phase. Often, much effort is expended on optimizing the experimental conditions using a specific column, but chromatographic columns vary over time owing to manufacturing changes in the coating chemistry or stationary support. These changes can drastically alter the peptide map and may result in the need to rework the analytical method completely. For the above reasons, it is desirable to have an additional separation technique that (1) provides unique specificity (also known as selectivity) that complements conventional chromatographic methods, (2) provides fast, efficient separations of the peptide fragments, and (3) eliminates the need to use a specific column to achieve the desired separation for a given peptide map. Capillary zone electrophoresis may be the answer to these concerns.

This chapter describes several separation strategies that exploit differences in mobility for the separation of digest fragments in peptide mapping by CZE. Optimization strategies and appropriate "recipes" are discussed to provide optimum specificity and efficiency for maximum resolving power of the peptide map. These recipes are largely based on buffer composition,

[5] M. Zivy and F. Granier, *Electrophoresis* **9,** 339 (1988).

[6] S. Renlund, I.-M. Klintrot, M. Nunn, J. L. Schrimsher, C. Wernstedt, and U. Hellman, *J. Chromatogr.* **512,** 325 (1990).

[7] G. F. Lee and D. C. Anderson, *Bioconjugate Chem.* **2,** 367 (1991).

[8] D. L. Crimmins, R. S. Thoma, D. W. McCourt, and B. D. Schwartz, *Anal. Biochem.* **176,** 255 (1989).

[8a] E. R. Hoff and R. C. Chloupek, *Methods Enzymol.* **271,** Chap. 2, 1996 (this volume).

[8b] L. M. Smith, R. L. Brumley, Jr., E. C. Buxton, M. Giddings, M. Marchbanks, and X. Tong, *Methods Enzymol.* **271,** Chap. 10, 1996 (this volume).

[9] G. W. Becker, P. M. Tackitt, W. W. Bromer, D. S. LeFeber, and R. M. Riggin, *Biotech. Appl. Biochem.* **10,** 326 (1988).

but also include the use of coating chemistries and auxiliary methods to achieve separations. These separation strategies as well as alternative detection modes are illustrated with selected examples to show the power of CZE in the separation of digestion fragments for peptide mapping. These peptide maps demonstrate the unique differences in separation specificity that make CZE a viable alternative that complements existing chromatographic techniques.

Separation Strategy

Mobility

Electrophoretic Mobility. Electrophoretic mobility, the first contribution to the total mobility, describes the movement produced by an electric field acting on a charged species in solution. Differences in electrophoretic mobilities produce separations as species migrate through the capillary. For each species, the electric field produces a force that is directly proportional to its net charge. That force is balanced by the frictional drag experienced by the species as it moves through a viscous medium. Thus, charge and size are the principal factors that determine mobility. Because drag is dependent on viscosity, electrophoretic mobility increases by about 2% per degree Celsius increase in temperature for aqueous buffers. Therefore, it is important to control temperature and to correct for buffer viscosity when absolute, rather than relative, mobility comparisons are made.

Electrophoretic mobility can be manipulated by changes in pH that produce a change in the net charge. Differences in charge are maximized by manipulating the pH to a value near the midpoint of the pI (isoelectric point) range of the peptide mixture. Thus, selection of the separation buffer pH is one of the most important variables to consider when optimizing separation conditions.

Because the velocity for each species depends on its mobility, a late-eluting (slow moving) species spends more time within the detector window than a corresponding early-eluting zone. Thus, the response of concentration-sensitive (flow dependent) detectors, such as ultraviolet (UV) detectors, must be corrected for the migration velocity to give true relative abundancies.[10]

Electroosmotic Mobility. Electroosmotic mobility, the second contribution to the total mobility, describes the movement of the bulk solution that occurs when a voltage is applied to a capillary that has a surface charge. Electroosmotic flow in an uncoated silica capillary ranges from very low

[10] R. G. Nielsen, G. S. Sittampalam, and E. C. Rickard, *Anal. Biochem.* **177,** 20 (1989).

values when the surface has a small charge (separation buffers with a pH less than about pH 3) to much higher values in basic solutions. In addition to pH, the electroosmotic flow depends on ionic strength (increasing the ionic strength decreases the flow rate) and temperature (increasing the temperature decreases the viscosity).[11] The electroosmotic flow may also be modified by adding substances such as ethylene glycol, cellulose derivatives, or organic solvents that decrease the effective surface charge and, therefore, the flow rate. Coating of the capillaries can change or eliminate the surface charge. Finally, the electroosmotic flow may be changed slightly (especially in dilute, acidic buffers) by applying an external electric field that changes the surface charge.

The electroosmotic flow has two effects on the total mobility. First, it tends to sweep all species past the detector because its magnitude is greater than most electrophoretic velocities, especially at high pH values in silica capillaries. The electroosmotic flow on uncoated silica capillaries sweeps the solution toward the negative electrode. However, when the capillary wall has been modified to produce a net positive surface, then the electroosmotic flow will sweep all species toward the positive electrode. For this reason, instruments generally are operated so that the electrode polarity at the detector end is the same as the polarity of the capillary wall. When the electroosmotic flow is very small or absent, neutral species or species whose electrophoretic mobility is toward the injector end of the capillary may not migrate past the detector. For example, negative or neutral species in low-pH buffers with uncoated silica capillaries may not be swept past the detector.

The second effect of electroosmotic flow is observed because the resolution is inversely related to how long the species remain within the electric field. That is, much higher resolution can be obtained when the total mobility is very small as long as band-broadening effects are minimal. Low total mobilities are obtained when the electroosmotic flow nearly counterbalances the electrophoretic flow; however, the analysis time will be much longer.

Peptide Charge. Ionization of the acidic and basic side chains as well as of the carboxy- and amino-terminal groups of determines the net charge on the peptide. The calculation of the net charge of an ionic species using the Henderson–Hasselbach equation is simple when the buffer pH and the ionization constants are known. For this calculation, it is usually assumed that the charge on each ionizable group is independent of that contributed by other groups in the molecule. However, estimation of the appropriate ionization constants is difficult for peptides and proteins. The ionization

---

[11] K. Salomon, D. S. Burgi, and J. C. Helmer, *J. Chromatogr.* **559**, 69 (1991).

TABLE I

ADJUSTED VALUES OF IONIZATION CONSTANTS USED FOR CALCULATION
OF CHARGE[a]

Amino acid residue	C-Terminal	N-Terminal	Side chain
Ala (A)	3.20	8.20	
Arg (R)	3.20	8.20	12.50
Asn (N)	2.75	7.30	
Asp (D)	2.75	8.60	3.50
Cys (C)	2.75	7.30	10.30
Gln (Q)	3.20	7.70	
Glu (E)	3.20	8.20	4.50
Gly (G)	3.20	8.20	
His (H)	3.20	8.20	6.20
Ile (I)	3.20	8.20	
Leu (L)	3.20	8.20	
Lys (K)	3.20	7.70	10.30
Met (M)	3.20	9.20	
Phe (F)	3.20	7.70	
Pro (P)	3.20	9.00	
Ser (S)	3.20	7.30	
Thr (T)	3.20	8.20	
Trp (W)	3.20	8.20	
Tyr (Y)	3.20	7.70	10.30
Val (V)	3.20	8.20	

[a] Calculated from values given in Shields.[12a]

constants are well known for isolated amino acids. However, the effects of neighboring groups, specific microenvironments (especially changes in the dielectric constant related to the hydrophobicity of the environment), temperature, and ionic strength all cause shifts of the ionization constants within peptides.[12] These complex relationships make it nearly impossible to determine the correct ionization constant for each group in the protein. We have developed the set of ionization constants given in Table I[12a] to calculate the average net charge on the peptide as a function of pH.

In addition to the considerations described, the charges on the peptide will be partially shielded by the counterions that surround the charged group, especially in a high ionic strength solution. In practice, this effect is less than about 10% at ionic strengths less than about 20 mM in 1:1

[12] E. C. Rickard, M. M. Strohl, and R. G. Nielsen, *Anal. Biochem.* **197,** 197 (1991).
[12a] J. Shields, Eli Lilly and Company, personal communication, 1988.

electrolytes.[13,14] Note that shielding will affect all peptides in a similar manner, so that it will have little effect on selectivity in a specific separation but will affect correlation of charge and size to mobility when measurements are made in different environments.

Peptide Size and Shape. If the shapes of peptides are assumed to be spherical, then their size will be proportional to their molecular weight. This assumption leads to a model in which the radius will be directly proportional to the cube root of the molecular weight. Unquestionably, peptides do not exist as balls; any secondary structure present in larger peptides will tend to distort the shape assumption, and the hydration sphere surrounding an ion will increase its effective size.[13] However, using molecular weight as a property that is proportional to size is justified by the difficulty in obtaining actual sizes and by the apparently successful correlations obtained using this approximation.

Mobility Relationships. As noted above, the electrophoretic mobility is directly proportional to the net charge of the species and inversely proportional to the drag. Although it is not necessary to know the mobility–charge–size relationship to find adequate separation conditions, it would be helpful to predict the conditions that would achieve the desired selectivity for complex mixtures such as peptide maps. For instance, the calculated charges combined with the known weights could be used to construct a predicted mobility-versus-pH map. Then, experimental efforts could be allocated to buffers in the most promising pH region(s).

Several models have been proposed for the quantitative relationship of size and shape to drag.[12,13,15,16] These models generally predict a mobility inversely proportional to the molecular weight to a power between 1/3 and 2/3. Within limited ranges, several of these models have been successful. For example, it has been shown that the electrophoretic mobilities of a wide variety of peptides produced in peptide-mapping experiments are highly correlated to the charge (calculated using the adjusted pK_a values as described above) divided by the molecular weight to the 2/3 power.[12]

Efficiency and Resolution

The efficiency of the separation is described by the plate number analogous to its use in chromatography. Many phenomena can cause band broadening (dispersion) with loss of efficiency. These include diffusion, sample

[13] P. D. Grossman, *in* "Capillary Electrophoresis, Theory and Practice" (P. D. Grossman and J. C. Colburn, eds.), p. 111. Academic Press, San Diego, CA, 1992.

[14] P. D. Grossman, *in* "Capillary Electrophoresis, Theory and Practice" (P. D. Grosman and J. C. Colburn, eds.), p. 3. Academic Press, San Diego, CA, 1992.

[15] B. J. Compton, *J. Chromatogr.* **559,** 357 (1991).

[16] B. J. Compton and E. A. O'Grady, *Anal. Chem.* **63,** 2597 (1991).

interaction with the capillary surface, poor heat dissipation that causes changes in the viscosity and a nonuniform velocity profile, large injector or detector volumes, laminar solution flow due to residual pressure differences, etc. When band broadening is due primarily to longitudinal diffusion, efficiency is proportional to the applied voltage.

The resolution of two peaks is directly proportional to the square root of the applied voltage and inversely proportional to the square root of the difference in electrophoretic mobilities. Resolution also depends on the difference between the electrophoretic and electroosmotic velocities. A minimum value for this difference—achieved when the electrophoretic flow and electroosmotic flow are equal in magnitude but opposite in direction— will give the greatest resolution at the sacrifice of increased analysis time. Therefore, the keys to achieving an optimum separation are eliminating conditions that cause a loss in efficiency, using a high separation voltage, and controlling the electroosmotic velocity as well as selecting conditions that maximize the mobility differences between peptides.

Both electrophoretic and electroosmotic mobilities are inversely related to viscosity, and viscosity is highly dependent on temperature. Efficient heat dissipation gives uniform temperatures within all regions (radial and longitudinal) of the capillary and, therefore, reduces variance in viscosity. Spatial gradients in viscosity would cause band broadening and poorer resolution; changes in viscosity with time would give drifts in migration times.

The injection volume must be kept small to prevent excessive peak broadening.[17,18] Theoretical values for essentially no increased dispersion are about 0.1% of the capillary length (this corresponds to about 1 nl for 50-μm i.d. or to about 5 nl for 75-μm i.d. in a 100-cm length column). In practice, injection volumes 5 to 10 times that value produce only minor increases in peak width.

Peak efficiency in CZE is relatively sensitive to the sample matrix, especially its pH and ionic strength. Preparation of the sample in the separation buffer minimizes changes in migration times, peak shapes, and efficiencies. Efficiency can be improved for dilute samples by using focusing (sample stacking) to concentrate the analyte in a narrow zone. Focusing occurs when the ionic strength of the sample is lower than that of the buffer or when the analyte has the same charge as the sign of the applied voltage at the injector end of the capillary. The opposite effect, defocusing, causes a

[17] X. Huang, W. F. Coleman, and R. N. Zare, *J. Chromatogr.* **480,** 95 (1989).
[18] E. V. Dose and G. A. Guiochon, *Anal. Chem.* **63,** 1063 (1991).

large loss in peak efficiency that is particularly troublesome when the sample solution has a higher ionic strength than the separation buffer.

The buffer strength needs to be adequate to maintain the same pH in the sample zone as in the bulk electrolyte. Buffer concentrations at least 100 times the sample concentration are necessary to meet this requirement. This translates into concentrations in the 100 mM range for samples that are 1 mg/ml and have a molecular mass of 1000 Da. However, high concentrations of ionic buffers produce high conductivity, and therefore high currents and high heat loads. Use of zwitterionic buffers allows the analyst to meet these conflicting criteria. Fortunately, zwitterionic buffers are available for a wide pH range (about pH 5.5 to 11). More traditional buffers such as phosphate, citrate, and acetate can be used in moderately low-pH regions. In some cases, concentrations will need to be kept small to minimize heat problems. Buffers with very high pH (above about pH 12) or very low pH (below about pH 2) are not as useful because of their high conductivities. Typical buffer concentrations are 10 to 200 mM to balance buffer capacity with heat generation. Common buffers utilized for CZE separations are listed in Table II.

Summary

An optimal separation is achieved when all the factors that affect selectivity, efficiency, and resolution have been balanced. The first step is to choose a pH that is likely to achieve the desired selectivity. Factors that affect efficiency and resolution can then be optimized. These include instrumental parameters such as separation voltage and capillary dimensions, separation conditions such as pH, buffer selection, and buffer concentration, and sample solution composition. The interrelationships are complex, but the best selectivity, efficiency, and resolution are achieved with the following:

   Maximum charge differences between species
   High field strengths
   Small injection volumes
   Efficient heat dissipation
   Minimum total velocity (achieved by balanced, but opposite, electrophoretic and electroosmotic flow velocities or by increases in viscosity)
   Buffer capacity much greater than sample concentration
   Minimal dispersion from sources such as capillary wall interactions, laminar flow, injection technique, and column overload.

The selection of the instrumental parameters to achieve the desired effects involves choices in two major parameters—capillary dimensions

TABLE II
BUFFERS USEFUL FOR CAPILLARY ELECTROPHORESIS

Buffer components[a]	pK$_a$ value	Useful pH range
Phosphate	2.2	
	7.2	1.2–3.2
	12.4	
Citrate	3.1	
	4.7	1.5–4.5
	6.4	
Glycine[b]	2.4	1.4–3.4
	9.8	
Formate	3.8	2.8–4.8
Acetate	4.8	3.8–5.8
MES: 2-(N-Morpholino)ethanesulfonic acid[b]	6.1	5.5–6.7
PIPES: Piperazine-N,N-bis(2-ethanesulfonic acid)[b]	6.8	6.1–7.5
Phosphate	2.2	
	7.2	6.2–8.2
	12.4	
MOPS: 3-(N-Morpholino)propanesulfonic acid[b]	7.2	6.5–7.9
HEPES: N-(2-Hydroxyethyl)piperazine-N'-(2-ethanesulfonic acid)[b]	7.5	6.8–8.2
Tricine: N-Tris(hydroxymethyl)methylglycine[b]	8.1	7.4–8.8
Tris: Tris(hydroxymethyl)aminomethane[b]	8.1	7.4–8.8
Borate	9.2	8.2–10.2
CHES: 2-(N-Cyclohexylamino)ethanesulfonic acid[b]	9.3	8.6–10.0
Glycine	2.4	8.8–10.8
	9.8	
CAPS: 3-(Cyclohexylamino)-1-propanesulfonic acid[b]	10.4	9.7–11.1

[a] Mixed buffers such as Tris–phosphate, citrate–phosphate, or borate–phosphate can be used to broaden the pH range available for a given set of buffer components.
[b] Denotes zwitterionic buffers.

and applied voltage. A summary of the effects obtained from changing their values is given below.

1. Increasing the capillary length or diameter produces the following results when there is no electroosmotic flow; similar trends occur in the presence of electroosmotic flow:

Increased length increases migration time (proportional to length to detector)

Increased length increases the number of theoretical plates (proportional to length to detector) and resolution (proportional to square root of length to detector)

Increased length decreases the heat load (inversely proportional at constant voltage)

Increased diameter produces a greater heat load (proportional to diameter squared)

Increased diameter gives more sensitive on-column detection (higher signal-to-noise ratio because signal increases with diameter whereas the noise remains nearly constant)

Increased diameter gives a higher mass loading (proportional to diameter squared).

2. Increasing the applied voltage or field strength produces the following results when there is no electroosmotic flow; similar trends occur in the presence of electroosmotic flow:

Increased plate number and resolution (proportional to field strength).

Increased heat production (proportional to voltage squared).

Shorter analysis times (inversely proportional to field strength).

Optimization Strategy

The utility of capillary zone electrophoresis for peptide mapping will depend on the rapid development of optimized separations and the reproducible performance of those methods. The crucial separation variables include pH, buffer concentration, ionic strength, buffer species, temperature, and capillary construction. An example related to optimization of the buffer composition—pH, buffer concentration, and ionic strength—is given below. The selection of the buffer species is determined by the desired pH range, any preference for zwitterionic species to moderate currents at high buffer concentrations, and any known interactions of buffer ions with the analyte or capillary. Decreases in temperature usually produce slightly lower efficiencies, but this is a minor effect. Changes in temperature also can affect conformation, but this is rarely observed in peptide-mapping experiments. An effective wall coating may be important because the adsorption of species onto the capillary wall can adversely affect separation performance. These interactions tend to occur most frequently with highly charged peptides, especially when the peptide has the opposite charge as the capillary wall, or with hydrophobic peptides. Adsorption of these fragments can reduce recovery, alter peak shape, reduce efficiency, diminish resolution, and alter electroosmotic flow. Column technologies borrowed from liquid chromatography and capillary gas chromatography have provided surface modification techniques that greatly reduce solute adsorption. Buffers such as phosphate and Tris bind to silica to reduce interactions of analytes with the wall. Dynamic buffer modifiers and static modifications of capillaries designed to reduce interaction of the analyte with the wall are discussed in the section on coated capillaries.

Buffer Composition

The buffer composition was optimized to achieve the best specificity and resolution in a peptide-mapping experiment.[19] The effects of pH, buffer strength, ionic strength, and morpholine concentration (a buffer additive intended to minimize wall interactions) were explored in an uncoated silica capillary. The large number of human growth hormone (hGH) tryptic cleavage fragments and their corresponding diverse properties were ideal for this optimization. Furthermore, a reversed-phase HPLC (RP-HPLC) mapping method was available for comparison and as an aid in peak identification.

As described above, the pH should be chosen to maximize the differences in charge between species. For a peptide-mapping experiment, this pH will tend to be near the midrange of the pI values for the peptides. Calculation of charges and/or predicted mobilities would further refine the most promising pH regions. The trypsin digestion of hGH produces acidic, neutral, and basic peptides. On the basis of prior experiences and the pI range for the fragments, four pH values were chosen for study: pH 2.4, 6.1, 8.1, and 10.4. A calculation of the peptide charges revealed that maximum differences at low pH (pH 2–4) or at about pH 7–11 would be expected although the average pI is about 6.1. Experimentally, the efficiency and resolution at pH 2.4 and 8.1 were better than those at pH 6.1 and 10.4, as predicted. The fact that the optimum pH value is not at the average pI is simply a reflection of the diverse nature of this group of peptides. The specific buffer composition at pH 8.1 was investigated further using a design that systematically varied the buffer composition (Fig. 1). The buffer components were Tricine (buffer), sodium chloride (ionic strength adjuster), and morpholine (buffer additive).

The buffer concentration must be sufficiently high to maintain constant conductivity and pH within the analyte zone to prevent band broadening. High buffer strength also increases the ionic strength that, in turn, reduces the electroosmotic flow velocity and minimizes ion-exchange interactions with the silica capillary surface. Both of these effects tend to increase further the efficiency of the separation. On the other hand, high buffer strength leads to an increase in current and greater heating problems. Zwitterionic buffers such as Tricine contribute good buffer capacity with a minimum increase in ionic strength. Tricine concentrations of at least 50 mM, or 300 times the analyte concentration, gave the best resolution in this study.

The ionic strength was adjusted by the addition of sodium chloride. As

[19] R. G. Nielsen and E. C. Richard, *J. Chromatogr.* **516**, 99 (1990).

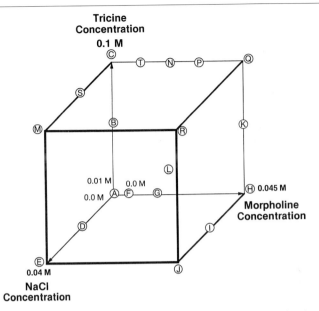

FIG. 1. Buffer composition diagram for tryptic digest method optimization by CZE.

stated above, increasing the ionic strength will tend to increase efficiency by reducing the electroosmotic flow and decreasing ion-exchange interactions with the silica. However, the addition of sodium chloride in concentrations up to 40 mM proved to be deleterious to the separation efficiency.

The effect of morpholine was investigated because it is known that some proteins interact with the silica capillary. Morpholine was added up to 45 mM. Morpholine contributes ionic strength and, at pH 8.1, it also contributes to buffer strength. The addition of about 20 mM morpholine increased the resolution of some peaks in the peptide-mapping experiment.

The optimum separation conditions from this series were found to be 100 mM Tricine and 20 mM morpholine. Overall, the selectivity of the CZE experiment is comparable to, but different from, the selectivity of the RP-HPLC experiment.[4] This difference is illustrated in Fig. 2 for the tryptic digest of human growth hormone. The orthogonality of these maps arises because the selectivity of the CZE experiment is based on the charge and size of the peptides whereas that of the RP-HPLC is based on their hydrophobicity. This difference is dramatically illustrated in Fig. 3. A separation that resulted in a comparable, but different, selectivity was achieved in 0.1 M glycine buffer at pH 2.4.

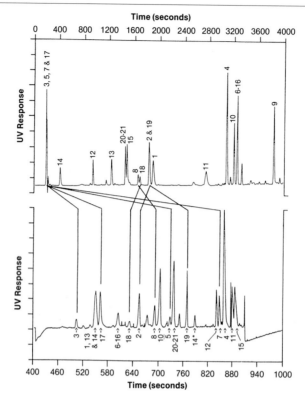

Fig. 2. Comparison of RP-HPLC and CZE separations. Differences in selectivity for these two techniques are illustrated for an hGH digest. *Top:* Tryptic map by RP-HPLC. *Bottom:* Tryptic map by CZE. Conditions given in Ref. 4.

Resolution Mapping

Resolution maps have been used to help identify optimum separations for complex mixtures. For example, Yeo *et al.*[20] developed a resolution map by systematically varying the concentrations of three cyclodextrins (α, β, γ) for nine acidic plant growth regulators, then determined the optimum concentrations to give the best resolution using Venn diagrams. In another example, Kenndler and Friedl[21] developed an approach that allowed them to produce contour plots for the resolution as a function of pH and separation voltage for seven test compounds (benzoic acid and phenol derivatives). These approaches allow selection of conditions that give the maximum

[20] S. K. Yeo, C. P. Ong, and S. F. Y. Li, *Anal. Chem.* **63,** 2222 (1991).
[21] E. Kenndler and W. Friedl, *J. Chromatogr.* **608,** 161 (1992).

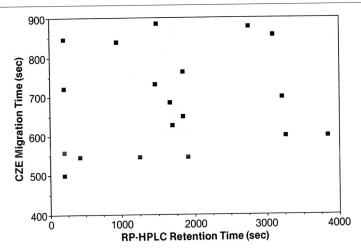

FIG. 3. Correlation of RP-HPLC retention times and CZE migration times for tryptic digest of human growth hormone. Separation conditions same as in Fig. 2.

resolution for a group of components. The latter approach also allows the analyst to minimize the separation time for a minimal acceptable resolution. Resolution mapping is one way to visualize the effect of experimental parameters on the separation. When combined with theoretical predictions, it may simplify finding conditions for acceptable separations.

Experimental Considerations

Mobility Measurement. Positive identification of peaks requires either spiking the unknown (or test component) with the known component or comparing the electropherogram of the unknown with that of the known. Experimentally, it is usually easier to do the latter. Electrophoretic mobility is more reproducible than migration time because any variation in electroosmotic flow velocity is removed. Also, the ratio of electrophoretic mobilities or migration times tends to be significantly more reproducible than absolute mobilities.

Electrophoretic mobilities usually are measured versus a neutral marker in high-pH buffers with uncoated fused silica capillaries. Typical neutral electroosmotic markers include mesityl oxide, acetone, dimethyl sulfoxide, benzyl alcohol, and acetonitrile. The electrophoretic velocity (v_{ep}, cm/sec) is equal to the difference between the total velocity (v_{tot}, cm/sec) and the electroosmotic velocity (v_{eo}, cm/sec). It can be calculated from the migration time for a species (t_m, sec), the required time to elute a neutral marker (t_{nm}, sec), and the length of the capillary from the injection end to the detector (L_{det}, cm). The electrophoretic (μ_{ep}, cm^2/V $\cdot$ sec) and electroosmotic (μ_{ep}, cm^2/V $\cdot$ sec) mobilities are calculated from the corresponding velocities and the electric field strength (E, V/cm). The electric field strength is equal to the applied voltage (V, volts) divided by the total capillary length (L_{tot}, cm). These relationships are given by

$$v_{ep} = v_{tot} - v_{eo} = L_{det}/t_m - L_{det}/t_{nm} \tag{1}$$
$$\mu_{ep} = v_{ep}/E = v_{ep}(L_{tot}/V) = (L_{det}L_{tot}/V)(1/t_m - 1/t_{nm}) \tag{2}$$
$$\mu_{eo} = v_{eo}/E = v_{eo}(L_{tot}/V) \tag{3}$$

When the analyst uses silica capillaries with a low-pH buffer or buffers or columns designed to minimize electroosmotic flow, electrophoretic mobilities are frequently calculated versus a species of known mobility because it is not practical to use a neutral marker. When using a mobility marker, its electrophoretic mobility (including the appropriate sign) must be added to find the true electrophoretic mobility.

Peak Identification. Peak identification is one of the major tasks in developing peptide maps; it is required to verify the structure of the protein. If the laboratory has the capability to do CZE–mass spectrometry (MS), it is possible to obtain molecular weight information directly. However, it is difficult to obtain structural information from MS/MS experiments because of the small sample injections. Because of this limitation and the fact that most laboratories do not have on-line MS capability, it is frequently necessary, at present, to verify structures in an off-line mode.

Samples for off-line analysis can be collected by any suitable technique. It is difficult, but not impossible, to collect sufficient quantities of eluted samples from CZE separations. RP-HPLC is a common isolation technique that offers the advantage of relatively high mass throughput. After any isolation, the integrity of the material following isolation should be verified by a suitable separation procedure.

Structure elucidation or confirmation can be obtained by a combination of mass spectrometry (molecular weight), amino acid analysis, and protein sequencing. At least two of these techniques should be applied to confirm a structure unequivocally. However, a coupled MS/MS experiment might be sufficient to establish identity and structure in some cases.

When material of known structure is available, the corresponding peak in the CZE map is best identified by spiking the known material into the digest. Direct comparison of migration times from electropherograms of isolated materials and the digest are difficult because the complex digest matrix perturbs the electrophoretic environment.

Capillary Conditioning and Washing. Most instrument manufacturers suggest that new capillaries be washed with dilute sodium hydroxide (usually 0.1 M), then with water and buffer. Similar procedures can be used to remove adsorbed proteins or peptides. If desired, a wash with dilute hydrochloric acid (usually 0.1 M) can be incorporated to remove acid soluble residues or to equilibrate the silanol surface when subsequent experiments will be in acidic solutions. The frequency of washing is dependent on the specific separation conditions and analytes. Frequently, a short wash with the separation buffer between samples is sufficient. Drifting in migration times or change in peak shapes with multiple injections indicate that a column is poorly equilibrated. Monitoring mobilities (which are independent of changes in the electroosmotic flow) will generally be more precise and less sensitive to changes in the silica surface than monitoring the corresponding migration times.

Examples

Capillary zone electrophoresis has been used by a variety of groups for peptide mapping. Many of these are listed in a review.[22] Some of these provided comparisons between RP-HPLC maps and CZE maps; those studies usually identified the peak identities. Representative examples are discussed below. Many studies simply present a map without any attempt to identify the peaks. In some cases, separations of a set of similar peptides (either from isolation of naturally occurring variants or from synthetic preparation) were reported; these are not discussed in this chapter, but pursuit of this topic will give a greater understanding of those factors that influence separation and how separation procedures are developed. The use of two-dimensional techniques is discussed below.

Figure 3 shows a specific comparison of the CZE and RP-HPLC tryptic maps for human growth hormone; it has a molecular mass of about 25 kDa and gives 19 major digest fragments. Figure 4 gives a similar comparison for the tryptic map of the β-globin subunit from human hemoglobin.[23] It has a molecular mass of about 16 kDa and gives about 17 digest fragments. In both cases, the CZE and HPLC separations have similar resolution although some fragments give overlapping peaks by either technique.

Peptide mapping by CZE can discriminate species variants, individual variants, or variation in posttranslational modifications of proteins. Ferranti et al.[23] also demonstrated the ability to detect a variation in human hemoglobin between individuals (Fig. 5). Wheat et al.[24] examined the effect of buffer

[22] W. G. Kuhr and C. A. Monnig, Anal Chem. **64**, 389R (1992).
[23] P. Ferranti, A. Malorni, P. Pucci, S. Fanali, A. Nardi, and L. Ossicini, Anal. Biochem. **194**, 1 (1991).
[24] T. E. Wheat, P. M. Young, and N. E. Astephen, J. Liq. Chromatogr. **14**, 987 Marcel Dekker, Inc., N.Y. (1991).

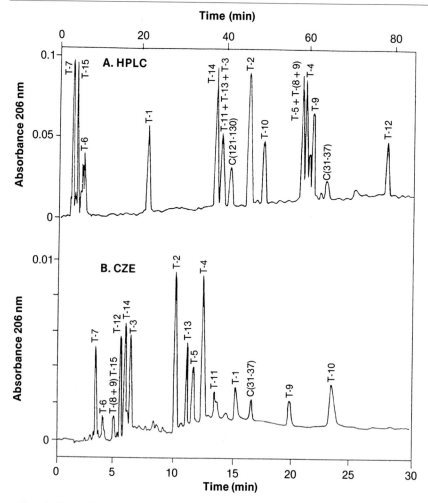

FIG. 4. Comparison of (A) RP-HPLC and (B) CZE tryptic maps of human β-globin. Peptide peaks are identified by a number indicating the position in the protein sequence. (Data from Ref. 23, with permission.)

pH and then presented the tryptic maps of beef (Fig. 6), chicken, rabbit, and horse heart cytochrome c by CZE.

Cobb and Novotny[3] demonstated the detection of a single amino acid modification in a protein by analysis of the difference between tryptic maps of phosphorylated and dephosphorylated β-casein (Fig. 7). This system utilized trypsin immobilized on agarose gel and placed in a small reactor column. Trypsin also had been immobilized onto the surface of an aminoal-

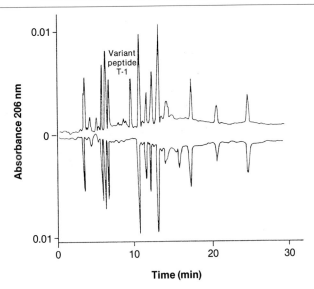

FIG. 5. Tryptic maps by CZE of a variant San Jose β-globin (top) compared with the normal β-globin. (Data from Ref. 23, with permission.)

kylsilane-treated fused silica capillary via biotin–avidin–biotin technology.[25] The enzyme-modified capillary was used to digest β-casein with the peptide fragments being analyzed by CZE, either by collecting the effluent from the capillary[25] or on-line.[26] Cobb and Novotny[27] demonstrated the use of several different protein digestion reagents—trypsin, chymotrypsin, and cyanogen bromide—in the analysis of human serum albumin. Each reagent gave quite different peptide maps because they clip at different aminio acid residues.

Capillary zone electrophoresis has also been used to analyze tryptic maps of proteins that are posttranslationally modified by glycosylations at either or both N or O sites. Affinity CZE in the analysis of recombinant human erythropoietin[28] has enabled the simultaneous analysis of both the translationally expressed protein and the posttranslationally modified protein. Figure 8 shows the total tryptic map segregated into the nonglycosylated and glycosylated peptide regions. Eighteen tryptic peptides were identified within the first 30 min from a theoretically possible 21 peptides, assuming complete digestion. Considerable microheterogeneity was associ-

[25] L. N. Amankwa and W. G. Kuhr, *Anal. Chem.* **64,** 1610 (1992).
[26] L. N. Amankwa and W. G. Kuhr, *Anal. Chem.* **65,** 2693 (1993).
[27] K. A. Cobb and M. Novotny, *Anal. Chem.* **64,** 879 (1992).
[28] R. S. Rush, P. L. Derby, T. W. Strickland, and M. F. Rohde, *Anal. Chem.* **65,** 1834 (1993).

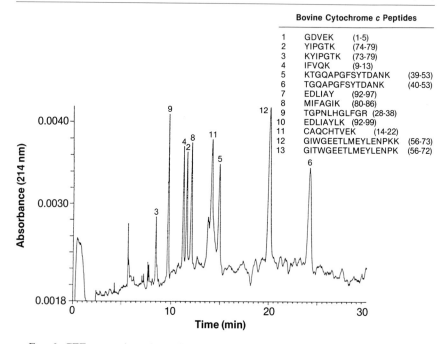

	Bovine Cytochrome c Peptides		
1	GDVEK	(1-5)	
2	YIPGTK	(74-79)	
3	KYIPGTK	(73-79)	
4	IFVQK	(9-13)	
5	KTGQAPGFSYTDANK	(39-53)	
6	TGQAPGFSYTDANK	(40-53)	
7	EDLIAY	(92-97)	
8	MIFAGIK	(80-86)	
9	TGPNLHGLFGR	(28-38)	
10	EDLIAYLK	(92-99)	
11	CAQCHTVEK	(14-22)	
12	GIWGEETLMEYLENPKK	(56-73)	
13	GITWGEETLMEYLENPK	(56-72)	

Fig. 6. CZE separation of tryptic peptides of bovine heart cytochrome c. Separation conditions were 25 mM sodium citrate, pH 4.0, 75 μm × 60 cm capillary, and 28 kV. (Reprinted from Ref. 24, pp. 987–996, by courtesy of Marcel Dekker, Inc.)

ated with the carbohydrate structure(s), as indicated by the number of peaks observed (at least 12 glycopeptide forms partially or totally separated) in the glycosylated region of the electropherogram. An ion-pairing agent was used to increase peptide resolution, decrease analyte wall interactions, and evaluate glycopeptide microheterogeneity.

Instrumentation

Detection

Detection of narrow zones of low analyte concentrations in peptide mapping within capillaries of 10- to 75-μm i.d. places a heavy burden on the sensitivity of the CZE detection system. Numerous detection modes are available due to extensive investigation of detectors for column liquid chromatography. The most popular detector for peptide mapping is UV–Vis, which is available in all of the commercial instruments. The additional

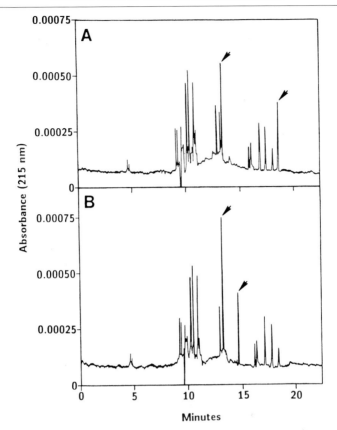

FIG. 7. Comparison of tryptic digests from (A) phosphorylated and (B) dephosphorylated forms of β-casein. Arrows point to two peaks that exhibit different migration times. (Reprinted with permission from Ref. 3. Copyright 1989 American Chemical Society.)

chemical specificity offered by multiwavelength UV–Vis absorbance detection has also been exploited. For example, Schlabach and Sence simultaneously monitored molecular absorption at two wavelengths (200 and 280 nm) to analyze a β-lactoglobulin A digest.[29]

Most reported applications of CZE for peptide separations utilize UV detection at wavelengths close to 200 nm. The use of UV detection, however, has severe limitations for high-sensitivity peptide mapping owing to the short path length afforded by the capillary. Extending the optical path length enhances UV absorbance detection. This approach offers a

[29] T. Schlabach and R. Sence, *Spectra* **149**, 47 (1990).

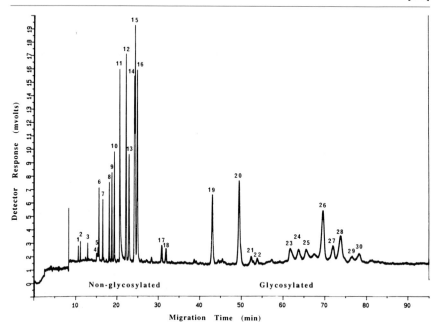

FIG. 8. CZE profile of trypsin-digested recombinant human erythropoietin, showing the nonglycosylated and glycosylated sections of the map. Peaks are numbered according to migration time. (Reprinted with permission from Ref. 28. Copyright 1993 American Chemical Society.)

means of enhancing sensitivity without significant hardware modifications or chemical derivatization. For example, rectangular borosilicate glass capillaries have been used to increase the detector path length and boost the detection sensitivity.[30] Detection across the long cross-sectional axis of the capillary provided a substantial increase in the sensitivity. Two capillary manufacturers have introduced modifications—Z-shaped[31] and bubble[32] cells—to increase detection path length. These approaches improved sensitivity with little loss in electrophoretic resolution and separation efficiency.

Fluorescence detection is one of the most sensitive methods for monitoring analytes, especially in the laser-induced fluorescence (LIF) mode. Native

[30] T. Tsuda, J. V. Sweedler, and R. N. Zare, *Anal. Chem.* **62,** 2149 (1990).

[31] S. E. Moring, C. Pairaud, M. Albin, S. Locke, P. Thibault, and G. W. Tindall, *Am. Lab.* **25**(11), 32 (1993).

[32] D. M. Heiger, *in* "High Performance Capillary Electrophoresis—An Introduction," p. 100. Hewlett-Packard Company, Avondale, PA, 1992.

fluorescence has been applied to monitor the digest of conalbumin and recombinant factor III using trypsin[33] and in the on-column digestion of β-lactoglobulin using pepsin.[34] However, relatively few molecules fluoresce naturally.[35] Fluorescent derivatization can be used to tag an analyte. For example, Jorgenson and Lukacs[36] first reported the use of fluorescamine in CZE separations of tryptic peptides. However, it is not always possible to find acceptable derivatization reagents and separation conditions. Also, the derivatization reaction alters the structure of the analyte permanently. Indirect fluorescence detection partially transfers the sensitivity advantage of fluorescence detection to nonfluorescent analytes without the need for sample derivatization. Detection is based on charge displacement and not on any adsorption or emission property of the analyte. The utility of indirect fluorescence, however, may be limited by noise arising from instability of common laser sources. In an example of the use of indirect fluorescence detection, 13 of the 16 expected peptides were detected in an analysis of the tryptic digest of β-casein.[37]

The drive to obtain additional chemical information about molecules separated by capillary zone electrophoresis has led to the coupling of CZE with mass spectrometry. Mass spectrometry is potentially an ideal detector for CZE. Typical flow rates associated with CZE are in the range of 500 nl/min and lower, making this technique compatible with MS. An advantage for MS detection is that molecular weight information may be obtained in addition to the migration time for each component peak in a mixture. The most common CZE–MS interfacing methods are based on either electrospray ionization (ESI) or fast atom bombardment (FAB).

Electrospray ionization is a favorite method for interfacing CZE instruments to mass analyzers. New interfacing methods for ESI instruments have greatly extended the utility of CZE–MS by allowing operation over a range of flow rates and buffer compositions. Thus, CZE separations are not degraded by a need to match the separation conditions to conditions, especially buffer composition, suitable for sample introduction into the mass spectrometer. These developments also allowed the on-line combination of CITP–MS (capillary isotachophoresis–mass spectrometry) so that larger sample sizes could be used. The CITP-MS separation and detection of

[33] T. T. Lee, S. J. Lillard, and E. S. Yeung, *Electrophoresis* **14,** 429 (1993).
[34] H.-T. Chang and E. S. Yeung, *Anal. Chem.* **65,** 2947 (1993).
[35] J. W. Jorgenson and K. D. Lukacs, *Anal. Chem.* **53,** 1298 (1981).
[36] J. W. Jorgenson and K. D. Lukacs, *J. High Resolut. Chromatogr. Chromatogr. Commun.* **4,** 230 (1981).
[37] B. L. Hogan and E. S. Yeung, *J. Chromatogr. Sci.* **28,** 15 (1990).

enzymatic digests of glucagon, a 3483-Da polypeptide, have been demonstrated.[38]

The interfacing of CZE to ESI–MS has been greatly enhanced by the advancement of the sheath flow design.[39] Using this interface, a small coaxial flow (1–5 μl/min) of liquid is employed to facilitate the electrospraying of buffers that could not be directly electrosprayed with previous interface designs. In addition to the sheath flow, the use of an etched conical tip has served to increase the stability of the electrospray process and minimize the effective mixing volume between the sheath liquid and the CZE effluent. Another approach for the CZE–MS interface is the use of a liquid junction. In this design, the junction is used to establish electrical contact with the separation capillary and to provide makeup flow of buffer.[40] The disadvantage of both the sheath flow and liquid junction designs is that the additional flow of liquid introduces charge-carrying species into the process that can lead to a decrease in detection sensitivity.

Matrix-assisted laser desorption/ionization (MALDI) mass spectrometry also has been used for the characterization of enzymatic digests of proteins. MALDI offers high sensitivity and allows for the direct analysis of unfractionated mixtures owing to a high tolerance for buffer components. This technique, however, often suffers from the presence of a large matrix background ion intensity that often requires more than a single matrix preparation. By using various comatrices to reduce the background, Billeci and Stults[41] were able to produce 96 and 88% of the expected tryptic map fragments for recombinant human growth hormone and recombinant human tissue plasminogen activator, respectively.

Coupling electrophoretic data with analysis of enzymatic digests by mass spectrometry using fast atom bombardment (FAB–MS) has allowed simple and rapid analytical methods for the structural identification of abnormal human hemoglobins.[23] FAB–MS was able to confirm the site of modification. A further MS application was reported for the peptide mapping of hemoglobins using on-line CZE coupled to atmospheric pressure ionization mass spectrometry for a tryptic digest of human hemoglobin.[42] In addition to the molecular weight determination, amino acid sequence information for peptides can be obtained by utilizing on-line tandem MS. After the tryptic digest sample components enter the atmospheric pressure ionization

[38] R. D. Smith, S. M. Fields, J. A. Loo, C. J. Barinaga, H. R. Udseth, and C. G. Edmonds, *Electrophoresis* **11**, 709 (1990).
[39] R. D. Smith, C. J. Barinaga, and H. R. Udseth, *Anal. Chem.* **60**, 1948 (1988).
[40] E. D. Lee, W. Muck, J. D. Henion, and T. R. Covey, *J. Chromatogr.* **458**, 313 (1988).
[41] T. M. Billeci and J. T. Stults, *Anal. Chem.* **65**, 1709 (1993).
[42] I. M. Johansson, E. C. Huang, J. D. Henion, and J. J. Zweigenbaum, *J. Chromatogr.* **554**, 311 (1991).

(API–MS) system, the molecular ion species of individual peptides may be focused and transmitted into the collision cell of the tandem triple-quadrupole mass spectrometer. Collision-induced dissociation of protonated peptide molecules yielded structural information for their characterization following injection of 10 pmol of a tryptic digest from human hemoglobin.

Thus, CZE in combination with MS is making increasing contributions to the study of biopolymers. The concurrent developments in MS and CZE provide complementary tools for manipulation and separation of very small sample volumes. Developments with tandem MS suggest that effective fingerprinting, and perhaps even partial sequence information, may be obtainable for proteins at the attomole level.

Coated Capillaries

For HPLC separations, the larger peptides in a map often exhibit poor peak shapes and low recoveries as a result of adsorption to the column surface. The high surface-to-volume ratio of small-diameter silica capillaries used in CZE creates a strong potential for sample–wall interactions and impairs performance in peptide mapping. To combat this limitation, often it is sufficient to add a species such as morpholine[19] or lysine[43] to compete for sites on the capillary wall. Additives, however, may not always be effective and it may be necessary to reduce wall interaction by deactivating the capillary wall. Coating the interior surface of the capillary with an uncharged hydrophilic material reduces or eliminates electroosmotic flow and minimizes larger peptide fragment adsorption. The disadvantage is that the capillary coating must be extremely stable under typical separation conditions, because a loss of coating leads to degraded efficiency and poor reproducibility.

Castagnola et al. found that initial peptide separations of myoglobin by CZE on unmodified capillaries gave nonreproducible results.[44] The measured migration time variability was greater than 8%, and resolution parameters (theoretical plate number and resolution) were poor. Separation buffer additives designed to compete for charged sites on the capillary wall provided better resolution, but the observed reproducibility was still unsatisfactory. Modification of the capillary by a monolayer of acrylamide provided high resolution and reproducible conditions for the separation of approximately 8 pmol of tryptic peptides from horse myoglobin.

In the CZE tryptic mapping and submapping of human α_1-glycoprotein, all separations were performed on a capillary with hydrophilic coating on

[43] M. M. Bushey and J. W. Jorgenson, J. Chromatogr. **480**, 301 (1989).
[44] M. Castagnola, L. Cassiano, R. Rabino, and D. V. Rossetti, J. Chromatogr. **572**, 51 (1991).

the inner walls. The hydrophilic coating minimized solute–wall interactions and also permitted the electrophoresis of basic proteins at acidic pH with high separation efficiencies. Furthermore, the electroosmotic flow decreased by a factor of 3.5 with the coated capillaries so that higher resolution and better reproducibility were obtained.[45] Submapping of glycosylated and nonglycosylated tryptic fragments of the glycoprotein by CZE was facilitated by selective isolation of the glycopeptides via solid-phase extraction. In addition, the electrophoretic map and submaps of the whole tryptic digest and its concanavalin A (ConA) fractions allowed the elucidation of the microheterogeneity of the glycoprotein. Thibault *et al.* analyzed tryptic peptides of glucagon using noncovalently coated capillaries that minimized sample adsorption onto the walls of the capillary.[46]

Auxiliary Methodology

Separation of complex peptide mixtures was one of the early demonstrations of the analytical power of capillary zone electrophoresis. There are, however, methods that may be utilized to further enhance specificity of the tryptic map. While migration rates of various peptides can be optimized through an appropriate pH adjustment, using micellar electrokinetic capillary chromatography (MEKC) can be beneficial in separating substances with similar net charges. Through the addition of micelle-forming surfactants or inclusion-forming compounds (such as cyclodextrins) to the separation buffer, a dynamic partitioning mechanism of solute separation is established. Cyclodextrins gave beneficial results such as narrowing peaks and enhancing fluorescence intensities for separations of fluorescamine-labeled peptides.[47] This reference demonstrated highly efficient separation of peptides obtained from the tryptic digestion of cytochrome *c* utilizing cyclodextrins.

A system that further enhanced the specificity of the tryptic mapping of proteins is a comprehensive two-dimensional (2D) separation system.[48,49] This system utilized reversed-phase chromatography as the first-dimension separation. The 2D system has much greater resolving power and peak capacity than either of the two systems used independently.

The system is used for the analysis of fluorescent-labeled peptide products from a tryptic digest of ovalbumin[48] and cytochrome *c*.[49] Figure 9 shows the three-dimensional data representation of the 2D tryptic map for

[45] W. Nashabeh and Z. El-Rassi, *J. Chromatogr.* **536,** 31 (1991).
[46] P. Thibault, C. Paris, and S. Pleasance, *Rapid Commun. Mass Spectrom.* **5,** 484 (1991).
[47] J. Liu, K. A. Cobb, and M. Novotny, *J. Chromatogr.* **519,** 189 (1990).
[48] M. M. Bushey and J. W. Jorgenson, *Anal. Chem.* **62,** 978 (1990).
[49] M. M. Bushey and J. W. Jorgenson, *J. Microcolumn Sep.* **2,** 293 (1990).

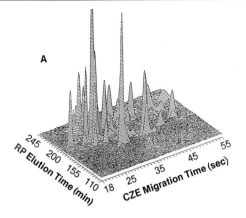

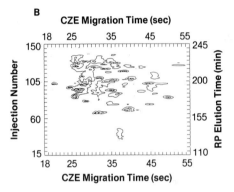

FIG. 9. Chromatoelectropherogram of fluorescamine-labeled tryptic digest of ovalbumin. (A) Three-dimensional plot and (B) contour plot of same data set. (Reprinted with permission from Ref. 48. Copyright 1990 American Chemical Society.)

ovalbumin. These data provide a means of viewing peak profiles in either separation dimension, and contour mapping of the 2D data provides a fingerprint of the protein digest.

Conclusion

Capillary zone electrophoresis has been used for peptide mapping of a wide variety of proteins. It has demonstrated unique features that make it valuable for that purpose. These include the following:

1. Fast and efficient separations are possible.
2. Separations orthogonal to RP-HPLC are obtained; that is, separa-

tions are achieved by differences in electrophoretic mobility (dependent on charge and size) rather than differences in hydrophobicity.
3. Separation conditions, especially pH, can be easily varied to achieve the desired separation.
4. Separation principles are well enough established to employ strategies to optimize separations.
5. Coupling to other separation techniques, such as RP-HPLC, can be utilized for even greater separation power.
6. Approaches are available to identify the structure of the fragment peaks observed in the peptide-mapping experiments.

Most of the remaining challenges for using CZE in peptide mapping are being met as the technique continues to develop. These include the following:

1. High-sensitivity techniques suitable for detection of low quantities of material; substantial progress has been made in the use of on-line CZE–MS, derivatization reagents for fluorescence detection, and indirect fluorescence detection.
2. Changes in column surfaces are being developed to minimize adsorption of peptides and changes in the electroosmotic flow rate.

Overall, capillary zone electrophoresis has been demonstrated to be a versatile tool that has been widely utilized in peptide mapping. Its specificity complements conventional chromatographic mapping techniques. Capillary zone electrophoresis provides fast, efficient separations that can be optimized to the desired specificity. Speed of separation can be traded for resolution for even greater enhancement of its rapid separations.

[12] Capillary Electrophoresis Analysis of Recombinant Proteins

By Glen Teshima and Shiaw-Lin Wu

Introduction

Naturally occurring proteins present at low levels in humans can be cloned and expressed in large quantities in bacteria, yeast, or mammalian vectors, as the result of recombinant DNA technology. The desired protein can then be purified to near homogeneity using a variety of column chromatographic steps (i.e., affinity, ion exchange, hydrophobic interaction, and size exclusion). While the purification process reduces quantities of DNA and host proteins to parts per million (ppm) levels, significant amounts of closely related protein variants may still be present. These variants are often the result of single amino acid modifications (i.e., deamidation, oxidation, and proteolytic cleavage). Quantitative methods are needed to monitor the levels of these protein variants in order to assure lot-to-lot consistency and product quality with regard to issues of safety and efficacy. Capillary electrophoresis (CE) would seem to be an ideal technique for separating these protein variants differing by a single residue, due to the high-efficiency separations that can be achieved.

There are several different modes of capillary electrophoresis that are applicable to protein analysis (see Table I). As is the case in high-performance liquid chromatography (HPLC) the most complete picture with regard to the structural integrity of a protein is obtained by applying as many complementary techniques as are available. In this chapter, we discuss the application of these CE techniques to the analysis of protein variants.

Capillary Zone Electrophoresis

Introduction

In the capillary zone electrophoresis (CZE) technique, the protein migrates in free solution (typically a low to moderate ionic strength buffer, i.e., 10 to 100 mM sodium phosphate) through a narrow-diameter (20–100 μm), coated or uncoated fused silica capillary under the presence of an electric field (i.e., 10 to 30 kV). Proteins are separated on the basis of their charge-to-mass ratio. Applied field strengths vary from 200 to more than 1000 V/cm or approximately one to two orders of magnitude higher than in slab gel electrophoresis. It is these high applied field strengths that are responsible for the potentially high separation efficiencies of CE. The use

TABLE I
MODES OF CAPILLARY ELECTROPHORESIS SUITABLE FOR
PROTEIN ANALYSIS

Technique	Principle of separation
Capillary zone electrophoresis (free solution)	Charge to mass
Isoelectric focusing	Isoelectric point
Capillary gel electrophoresis (entangled polymer network)	Molecular size

of narrow-diameter capillaries, which dissipate heat efficiently owing to their high surface-to-volume ratios, allows the application of high electric fields.[1] Therefore, better and faster analytical separations can generally be achieved using very narrow-diameter capillaries (10–50 μm) while preparative separations require capillaries of at least 75 μm.

Instrumentation

The CE system became commercially available in the late 1980s. Prior to this, capillary electrophoresis was performed using homemade systems consisting of a high-voltage power supply with the electrodes and capillary immersed in a semiconductive buffer. Now most CE instruments are computer controlled with respect to sample injection, analysis, and data processing.[2] In addition, most are capable of capillary thermostatting by either forced air or water cooling. This serves to minimize Joule heating, which can result in the formation of radial temperature gradients manifested by non-Gaussian peaks. As a result of these features, capillary electrophoresis analysis is becoming more reliable and reproducible. Autosamplers, available on most higher end systems, allows for unattended analysis. This is an extremely important feature in the routine analysis of quality control samples as well as in method development (i.e., in evaluating the effect of running buffer pH).

Sample Preparation

Ideally, proteins should be at a high starting concentration (greater than 1 mg/ml) because only a small volume can be injected (typically 1–10 nl) without perturbation of the electric field. If the protein solution is very dilute (less than 0.1 mg/ml), the sample should be concentrated by dialysis or filtration. Alternatively, coupled column isotachophoresis (ITP)[3] or transient on-column ITP[4,4a] can be used to concentrate dilute samples. Samples should be dissolved in water (if the protein is soluble) or contain only approximately one-tenth of the running buffer concentration in order to take advantage of the enriching signal intensity through sample stacking

[1] J. W. Jorgenson and K. D. Lukacs, *Anal. Chem.* **53,** 1298 (1981).
[2] R. Kuhn and S. Hoffstetter-Kuhn, *in* "Capillary Electrophoresis: Principles and Practice." Springer-Verlag, Berlin, 1993.
[3] D. Kaniansky and J. Marak, *J. Chromatogr.* **498,** 191 (1990).
[4] F. Foret, E. Szoko, and B. L. Karger, *J. Chromatogr.* **608,** 3 (1992).
[4a] L. Křivánková, P. Gebauer, and P. Boček, *Methods Enzymol.* **270,** Chap. 17 (1996).

phenomena.[5] Nevertheless, some protein samples that must be dissolved in high ionic strength buffer can also be analyzed through a pressure injection mode with a relatively smaller volume size.[6]

Analysis

Samples can be injected either electrophoretically or by pressure. Pressure loading is usually preferred in cases in which the quantification of components in a mixture is desired because loading of the various charged species in the sample is more uniform. A disadvantage of pressure loading is that counterions present in the sample matrix are also loaded, resulting in a somewhat noisier background. Electrophoretic loading introduces sample bias, as components with greater charge are preferentially loaded. The positive sample bias can be useful in the detection of proteins present at a low concentration relative to the background electrolytes. Samples loads of up to 1 to 5% of the capillary volume can be introduced by increasing the injection time and/or voltage, assuming the resolution of the components of interest is not adversely affected. Separations can be performed on either uncoated fused silica capillaries or coated capillaries. Recovery of proteins, in particular basic proteins, is generally less problematic on a neutral, hydrophilic coated capillary. Narrow-diameter capillaries (less than 50 μm) yield better separation efficiencies owing to their more efficient dissipation of heat but are more likely to plug, especially when using high-salt or viscous buffers.

A wide variety of buffers can be used in the CZE analysis of proteins. A list of various buffers and their properties is shown in Table II. The two main requirements of a CZE buffer are that it have good buffering capacity at the relevant pH and that it have a low ultraviolet (UV) absorbance at 200 to 220 nm. Typically, the signal is monitored at 200 nm owing to the aforementioned low sample load and short path length for detection. A linear polymer, i.e., methylcellulose or hydroxypropylcellulose, is sometimes added to the buffer to further reduce electroosmotic flow in the capillary and to minimize protein–wall interactions.[7,8]

Polarity can be set either in the normal mode, with the cathode at the detector end of the instrument, or in the reversed mode, anode at detector end, depending on the isoelectric point of the protein and the pH of the running buffer. For example, a basic protein would typically be analyzed in the normal polarity mode with the direction of electrophoresis from the

[5] R. Chien, and D. S. Burgi, *Anal. Chem.* **64,** 489A (1992).
[6] M. Herold and S.-L. Wu, *LC-GC* **12,** 531 (1994)
[7] S. Hjertén, *Arkiv. Keni* **13,** 151 (1958).
[8] F. A. Chen, *J. Chromatogr.* **559,** 445 (1991).

TABLE II
USEFUL BUFFERS FOR CAPILLARY ZONE ELECTROPHORESIS

Buffer/counterion	pK_a	Minimum useful wavelength (nm)
Phosphate	2.2, 7.2, 12.3	195
Formate	3.75	
Citrate	3.1, 4.7, 5.4	260
Acetate	4.76	250
Bis–Tris	6.50	
PIPES	6.8	215
Ammonium acetate		
Tris	8.2	230
Tricine	8.1	230
Bicine	8.35	230
Borate	9.2	190
CHES	9.5	190
CAPS	10.4	210

positive to negative pole because it is cationic except at extremely high pH. A strongly acidic protein would be analyzed in the reversed polarity mode because it is anionic except at extremely low pH. Run times vary depending on many factors including solute charge, the presence of modifiers such as methylcellulose in the buffer, the applied field strength, and capillary length. The optimized analysis conditions often represent a compromise between resolution and speed of analysis.

Capillary Regeneration

Following analysis, an uncoated capillary can be regenerated by purging with 0.1 M sodium hydroxide and then 0.1 M sodium phosphate at pH 2. This is effective at removing adsorbed proteinaceous material. This wash procedure is not possible with coated capillaries because the coated material can be easily deteriorated by a strong base. In this case, purging with 5 capillary volumes of a low-pH phosphate buffer (i.e., pH 2) is recommended.

Applications in Uncoated Capillaries

One of the major problems in the analysis of proteins by CZE is their adsorption to the walls of the fused silica (uncoated) capillary. Basic residues on the protein interact electrostatically with ionized silanols on the wall of the capillary. One approach to minimize protein adsorption is to perform separations at extremely high pH.[9] Above pH 11, the basic residues are

[9] H. H. Lauer and D. McManigill, *Anal. Chem.* **58,** 166 (1986).

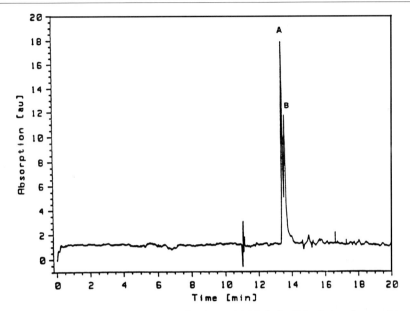

FIG. 1. Capillary zone electrophoresis of a misfolded disulfide form of recombinant IGF-I (A) and intact IGF-I (B). Conditions: Capillary, fused silica uncoated (75 μm × 120 cm, 105 cm to detector); running buffer, 10 m*M* CAPS, 5 m*M* sodium tetraborate, 1 m*M* EDTA, pH 11.11; field strength, 250 V/cm; detection, 215 nm. (From Ref. 10.)

uncharged. In addition, the protein has a net negative charge and should be repelled from the wall. A disadvantage of this approach is that native protein conformation will be compromised.

As an example of high-pH buffer operation, two forms of insulin-like growth factor I (IGF-I) differing in their disulfide arrangements were separated by CZE using a buffer consisting of 10 m*M* 3-(cyclohexylamino)-1-propanesulfonic acid (CAPS), 5 m*M* sodium tetraborate, 1 m*M* EDTA, pH 11.1, on an uncoated fused silica capillary (Fig. 1).[10] IGF-I is a polypeptide of 70 amino acids with 3 intrachain disulfide bridges. There are two adjacent cysteine residues at positions 47 and 48. In the native molecule, disulfides form between residues 6 and 48 and between residues 47 and 52, whereas in the variant form, disulfides form between residues 6 and 47 and between residues 48 and 52. Thus, the native conformations of the two IGF-I forms would be expected to be similar. In this case, the denaturing conditions caused by exposure to the high-pH buffer might have been advantageous in exploiting any minor differences between the molecules. Other ap-

[10] H. Ludi, E. Gassman, H. Grossenbacher, and W. Marki, *Anal. Chim. Acta* **213,** 215 (1988).

TABLE III
COATED CAPILLARIES FOR CAPILLARY ZONE ELECTROPHORESIS

Manufacturer (location)	Coating chemistry	Usable pH range	Available diameter (μm)
Supelco (Bellefonte, PA)	C_1	2–10	50, 75
Supelco	C_8	2–10	50, 75
Supelco	C_{18}	2–10	50, 75
Isco (Lincoln, NE)	C_{18}/Brij 35	4–7	75
Isco	Glycerol	2–9	75
Bio-Rad (Richmond, CA)	Polyacrylamide	2–9	25, 50
Beckman (Fullerton, CA)	Polyacrylamide	2–9	50, 75
Beckman	Amine, positively charged	2–7	50, 75
MicroSolv	Polyethylene glycol	2–9	50, 75
Hewlett-Packard	Poly(vinyl alcohol)	2–9.5	50, 75

proaches to minimize protein adsorption on fused silica capillaries include the use of high ionic strength buffers,[9,11] buffer additives such as zwitterions, detergents, or polyamines,[12,13] and coated capillaries as discussed below.

Applications in Coated Capillaries

Coated capillaries have been successfully used in the analysis of proteins (see reviews by Kohr and Engelhardt[14] and El Rassi and Nashabeh.[15] The number and availability of various types of hydrophilic coated capillaries suitable for the analysis of proteins has expanded (see Table III). One of the main advantages of using coated capillaries is that analysis can be performed between pH 2 and 9 without the use of high ionic strength buffers, which typically limit the usable electric field strength due to Joule heating, or buffer modifiers, which can affect run-to-run reproducibility due to the dynamic nature of their interaction with the silanols on the walls of the capillary. The other main advantage of coated capillaries is that proteins in general and basic proteins in particular, can be better recovered. A disadvantage of coated capillaries is that buffers above pH 9 cannot be

[11] J. S. Green and J. W. Jorgenson, *J. Chromatogr.* **478,** 63 (1989).
[12] M. M. Bushey and J. W. Jorgenson, *J. Chromatogr.* **480,** 301 (1990).
[13] J. E. Wiktorowicz and J. C. Colburn, *Electrophoresis* **11,** 769 (1990).
[14] J. Kohr and H. Engelhardt, *in* "Capillary Electrophoresis Technology" (Norberto A. Guzman, ed.), Chromatography Science Series, Vol. 64. Marcel Dekker, New York, 1993.
[15] Z. El Rassi and W. Nashabeh, *in* "Capillary Electrophoresis Technology" (Norberto A, Guzman, ed.), Chromatography Science Series, Vol. 64. Marcel Dekker, New York, 1993.

used, because the coated material can deteriorate easily. There, a coated capillary that is stable at higher pH buffer is needed. Another disadvantage is that both anions and cations cannot be mobilized simultaneously and therefore both species types cannot be analyzed in a single run owing to the lack of significant electroosmotic flow. This is not a problem with respect to the analysis of closely related recombinant protein variants differing by a single charge. Four examples of this case are demonstrated below.

Soluble CD4. The CD4 receptor is located on a subset of peripheral T cells of the immune system. Human immunodeficiency virus type 1 (HIV-1), the retrovirus implicated in acquired immune deficiency syndrome (AIDS), binds to CD4, thereby gaining entry into the host cell.[16,17] Two different forms of CD4 have been produced by recombinant DNA technology for testing as potential AIDS therapeutics.[18,19] Soluble CD4 contains the extracellular portion of this receptor and has a molecular mass of approximately 45 kDa. It contains two sites of glycosylation, both of which are mostly sialylated,[20] and the isoelectric point ranges from 8.8 to 9.3, depending on the number of sialic acid groups present. The two major sources of charge heterogeneity in soluble CD4 are the result of deamidation at Asn-52[21] and sialylation at the two glycosylation sites.

The most obvious way to manipulate solute mobility in CZE is to change the buffer pH. The effect of buffer pH on the separation of recombinant soluble CD4 is shown in Fig. 2.[22] Separations were performed on a capillary coated with linear polyacrylamide (Bio-Rad, Richmond, CA), thermostatted to 20°. It should be noted that soluble CD4 could not be recovered from an uncoated capillary due to its high p*I*. The pH optimum for the separation of soluble CD4 was pH 5.5, using a 100 m*M* sodium phosphate buffer containing 0.04% methylcellulose. At values greater than pH 6.0, the protein was not mobilized owing to the lack of electroosmotic flow by virtue of the neutral coated capillary, and insufficient solute charge as the pH approached the p*I* of CD4 (pH 8.8–9.3). In general, CZE separations on coated capillaries should not be performed within 2 units of the p*I* owing

[16] A. G. Dalgleish, P. C. L. Beverly, P. R. Clapham, D. H. Crawford, M. F. Greaves, and R. A. Weiss, *Nature (London)* **312**, 763 (1984).
[17] D. Klatzman, E. Champagne, S. Chameret, J. Gruest, D. Guetard, T. Hercend, J. C. Gluckman, and L. Montagnier, *Nature (London)* **312**, 767 (1984).
[18] J. S. McDougal, M. X. Kennedy, J. M. Sligh, S. P. Cort, A. Mawle, and J. K. A. Nicholson, *Science* **231**, 382 (1986).
[19] D. H. Smith, R. A. Byrn, S. A. Marsters, T. Gregory, J. E. Groopman, and D. J. Capon, *Science* **238**, 1704 (1987).
[20] M. W. Spellman, C. K. Leonard, and L. J. Basa, *Biochemistry* **30**, 2395 (1991).
[21] G. Teshima, J. Porter, K. Yim, V. Ling, and A. Guzzetta, *Biochemistry* **30**, 3916 (1991).
[22] S.-L. Wu, G. Teshima, J. Cacia, and W. S. Hancock, *J. Chromatogr.* **516**, 115 (1990).

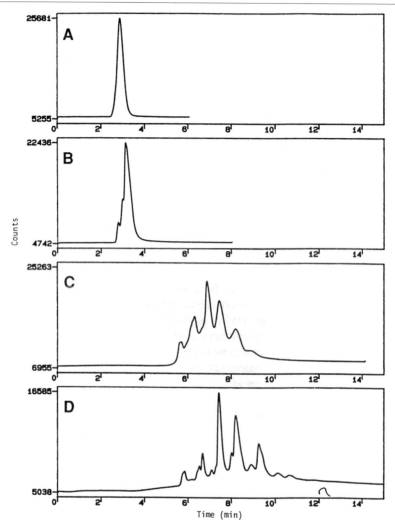

FIG. 2. Effect of buffer pH on the separation of rCD4 variants by capillary zone electrophoresis. Conditions: Capillary, fused silica coated with linear polyacrylamide (25 μm $\times$ 24 cm, Bio-Rad); running buffer, 100 mM sodium phosphate with 0.04% methylcellulose; field strength, 521 V/cm; detection, 200 nm. The pH conditions were pH 2.5 (A), 3.5 (B), 4.5 (C), and 5.5 (D). Recombinant CD4 concentration was 0.5 μg/μl. (From Ref. 22.)

to charge (mobility) considerations. In addition, proteins have reduced solubility at pH values close to their pI. The resolution of the various sialylated forms of a glycoprotein improves dramatically as the buffer exceeds pH 3.5. The variable number of sialic acids at the two N-linked glycosylation sites results in a ladder of charged species.

To minimize the complexity of the electropherogram in Fig. 2, the CD4 sample was treated with neuraminidase to remove the sialic acid groups. Two acidic minor peaks were observed (Fig. 3). Peak B (Fig. 3) has been identified as the deamidated variant by spiking in a standard sample collected from ion-exchange HPLC. Peak (Fig. 3) has not yet been identified but may result from incomplete removal of sialic acid.

Identification of CE-separated peaks is under rapid development. Capillary electrophoresis–mass spectrometry (CE–MS) shows promise with respect to peptides and small, nonglycosylated proteins where the mass accuracy is sufficient to distinguish [for example, a deamidated protein from

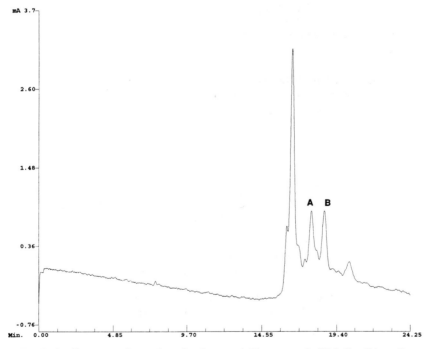

FIG. 3. Capillary zone electrophoresis of neuraminidase-treated rCD4. Conditions: Same as in Fig. 2; the pH of the running buffer was 5.5. Peaks A and B are acidic variants of rCD4. Peak B was identified as the deamidated (Asn-52 to aspartate) variant by spiking in a standard collected from ion-exchange HPLC.

the unmodified form (a 1-Da difference)]. The fraction collection technique with off-line identification (matrix-assisted laser desorption ionization with mass spectrometry and Edman sequencing) also shows promise in identifying peptides and proteins. However, at this stage of development, identification of variants of larger proteins (>25kDa) still relies on cross-correlation with standards isolated from ion-exchange HPLC and fully characterized by the usual array of protein chemistry techniques.[23]

CD4–IgG. Although soluble CD4 blocked HIV infection *in vitro*, it was not efficacious in human clinical trials. One factor contributing to the poor efficacy of this protein was its short plasma half-life (15 min).[19] Its molecular mass is 55 kDa [determined by sodium dodecyl sulfate-polyacrylamide gel electrophoresis (SDS–PAGE)], making it just small enough to be cleared by glomerular filtration in the kidney. In an attempt to develop a more effective therapeutic, two of the four immunoglobulin (Ig)-like domains of soluble CD4 were fused to the constant domain (Fc) of human IgG. The Fc region has a long plasma half-life; therefore it was expected that the hybrid molecule would also be long-lived. The molecular mass of this molecule is 113 kDa, well above the 70-kDa limit for glomerular filtration, suggesting that it would have a significantly longer half-life compared to soluble CD4. Pharmacokinetic studies have shown that the plasma half-life of CD4–IgG in rabbits is 48 hr compared to only 15 min for soluble CD4.[19]

CD4–IgG is a disulfide-linked dimer with a molecular mass of 100 kDa and a pI of 8.9. It contains two sites of glycosylation in Fc, neither of which is sialylated. The major source of charge heterogeneity is due to deamidation at Asn-52 (our unpublished data, 1995); interestingly, this is also the primary site of deamidation in soluble CD4.

CD4–IgG was analyzed on a Supelco (Bellefonte, PA) H50 coated capillary (see Table III) with a running buffer consisting of 25 mM ammonium acetate, pH 5.3 (Fig. 4). Phosphate buffer at low pH (e.g., pH 2) with preconditioning (washes performed for several minutes between runs) was found to be essential in using low ionic strength buffers in conjunction with the Supelco coated capillaries. While coated capillaries differ in the extent of surface coverage, all have some significant number of free silanols that can interact with protein. Phosphate is known to bind strongly to silica.[24] Therefore, using phosphate to precondition the capillary may improve separation performance and reproducibility. The preconditioning step worked well for several different proteins on the Supelco H50 capillary and could possibly be effective for other types of coated capillaries bonded with carbon chain linkers.

[23] S.-L. Wu, *LC-GC* **10,** 430 (1992).
[24] R. M. McCormick, *Anal. Chem.* **60,** 2322 (1988).

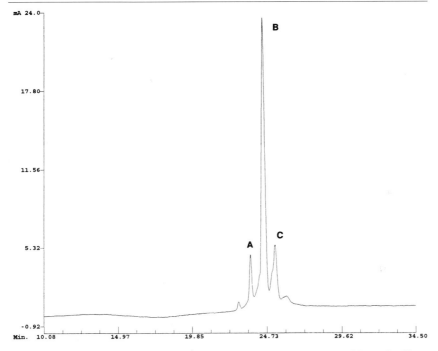

FIG. 4. Analysis of CD4–IgG by capillary zone electrophoresis. Conditions: Capillary, fused silica coated with C_1 (Supelco H50, 50 μm $\times$ 36 cm); running buffer, 25 mM ammonium acetate, pH 5.3; field strength, 347 V/cm; detection, 200 nm. Peak A was an unidentified basic species; peak B was intact CD4–IgG; peak C was deamidated (Asn-52 to aspartate) CD4–IgG.

The lack of sialic acid heterogeneity in CD4–IgG simplified the analysis with regard to detecting deamidation. The slower migrating component (peak C, Fig. 4) was identified as the deamidated variant. The faster migrating species has not yet been determined. One advantage of the low ionic strength buffer used in this application was that the analysis time could be reduced by applying higher field strength.

Tissue-Type Plasminogen Activator. Recombinant tissue-type plasminogen activator (tPA) is a glycoprotein of 60 kDa with five distinct functional domains. The major source of charge heterogeneity is the result of a variable number of sialic acid groups at either one or two of the N-linked glycosylation sites. Tissue-type plasminogen activator is an example of a protein in which solubility is a major consideration. Capillary zone electrophoresis is often performed under conditions (low ionic strength buffers) that promote protein–protein interactions and that may result in aggregation or precipitation of tPA. Detergents, chaotropes, and glycols have been used successfully to maintain protein solubility.

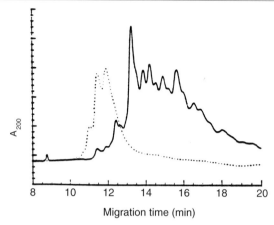

Fig. 5. Capillary zone electrophoresis of recombinant tPA (rtPA) (—) and desialylated rtPA (·····). Conditions: Capillary, fused silica coated with linear polyacrylamide (25 μm × 24 cm, Bio-Rad); running buffer, 100 mM ammonium phosphate, pH 4.6, with 0.01% Triton X-100 and 200 mM ε-aminocaproic acid; field strength, 250 V/cm; detection, 200 nm. (From Ref. 25.)

A buffer consisting of 100 mM ammonium phosphate, pH 4.6, containing the nonionic detergent Triton X-100 and 200 mM ε-aminocaproic acid was used to analyze tPA (Fig. 5).[25] As many as 20 peaks can be observed, consistent with the known carbohydrate charge heterogeneity of tPA. Treatment with neuraminidase greatly reduces the complexity of this pattern, as expected (Fig. 5). A list of UV-transparent additives for maintaining protein solubility is shown in Table IV.

Insulin-Like Growth Factor Type I. Some protein variants may not be amenable to separation strictly on the basis of charge. In the case of IGF-I, an improperly folded form and an oxidized form have the same net charge as the native form and are not resolved under typical CZE conditions. Nashabeh *et al.*[26] used a buffer system consisting of 20 mM β-alanine–critic acid, 5 mM zwitterion detergent, pH 3.8, in 50% acetonitrile on a neutral coated capillary to separate these IGF-I variants (Fig. 6). The addition of the acetonitrile modulated the binding of the zwitterion detergent to the protein variants, leading to optimized separation based on a combination of hydrophobic and charge-based differences. The capability for adding a variety of modifiers that alter selectivity makes capillary zone electrophoresis very flexible for the analysis of proteins.

[25] K. W. Yim, *J. Chromatogr.* **559**, 401 (1991).
[26] W. Nashabeh, K. F. Greve, D. Kirby, F. Foret, and B. L. Karger, *Anal. Chem.* **66**, 2148 (1994).

TABLE IV
CAPILLARY ZONE ELECTROPHORESIS BUFFER ADDITIVES FOR ENHANCING
PROTEIN SOLUBILITY

Additive	Manufacturer (location)
$C_{12}E_8$ (octaethylene monododecyl ether)	CalBiochem (LaJolla, CA)
Triton X-100, reduced	Aldrich (Milwaukee, WI)
Glycerol	J. T. Baker (Phillipsburg, CA)
Ethylene glycol	J. T. Baker
Propylene glycol	J. T. Baker
Urea	Aldrich

Capillary Isoelectric Focusing

Introduction

In the isoelectric focusing mode of capillary electrophoresis (cIEF), separation of proteins is based on differences in isoelectric points. Capillary IEF has two major advantages: first, resolving power is potentially very high in cIEF owing to the zone-sharpening effect occurring during the focusing step; second, quantification can be accomplished by peak area

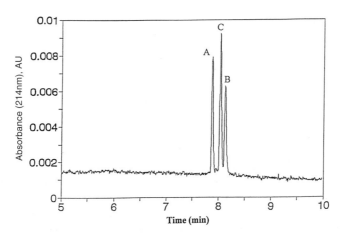

FIG. 6. Capillary zone electrophoresis of IGF-I variants using a modified buffer system. Conditions: Capillary, fused silica coated with siloxanediol–linear polyacrylamide (75 μm $\times$ 47 cm; Beckman); running buffer, 20 mM β-alanine–citric acid, pH 3.8, with 5 mM zwitterionic detergent, N-dodecyl-N,N-dimethyl-3-amino-1-propane sulfonate (DAPS), and 50% acetonitrile in the buffer; field strength, 750 V/cm; detection, 214 nm. Elution order: (A) misfolded IGF-1, (C) correctly folded IGF-I, and (B) Met-sulfoxide IGF-I. (From Ref. 26.)

integration. It is interesting to note that minor bands on a slab gel are particularly difficult to detect and quantify by densitometric methods.

Sample Preparation

The sample is mixed with wide-range ampholytes (pH 3–10) or a mixture of narrow- and wide-range ampholytes, if better resolution is desired, to a final concentration of 0.1 to 0.5 mg/ml. The protein concentration must be sufficiently high (i.e., 0.1 to 0.5 mg/ml) for detection at 280 nm. Ultraviolet detection is usually performed at 280 nm because the ampholytes have significant absorbance at lower wavelengths. Precipitation is a particular concern in cIEF because proteins are focused to their pI, at which solubility is at a minimum, and because the proteins become very concentrated (300-fold) during the focusing step. Nonionic detergents such as chemically reduced Triton X-100 (1 to 2%), glycols such as ethylene glycol (10 to 20%), or chaotropes such as urea (2 to 6 M) can be added to the sample–ampholyte mixture to reduce the risk of precipitation. The final salt concentration in the sample should be less than 50 mM. The presence of excess salt can perturb the electric field and interfere with the IEF process. Therefore, samples of high ionic strength should be dialyzed against 1% ampholytes prior to analysis. The ampholytes typically contain a basic spacer, N,N,N',N'-tetramethylethylenediamine (TEMED), to prevent basic proteins from focusing past the on-column point of detection. The concentration of TEMED can be adjusted depending on the pI of the protein being analyzed. The ampholyte reagent may also contain methylcellulose to suppress electroosmotic flow and thereby allow the formation of a stable pH gradient. The presence of methylcellulose also minimizes protein–wall interactions by forming a hydrophilic layer on the capillary surface. For a similar purpose, the use of neutral hydrophilic coatings [e.g., polyacrylamide or poly(vinyl alcohol)] is also recommended.

Analysis

A typical cIEF protocol consists of three steps. In the first step, the capillary is filled with the sample and ampholyte mixture by pressure injection. Because the entire capillary is filled with sample, cIEF has potential as a micropreparative technique.

In the second step, a pH gradient is formed by placing an acidic solution, e.g., 10 mM phosphoric acid, at the anodic end of the capillary and a basic solution, e.g., 20 mM sodium hydroxide, at the cathodic end. A coated capillary is preferred to establish a stable pH gradient because the presence of electroosmotic flow will prevent focusing of ampholyte and sample components into narrow bands. Focusing is then accomplished by applying high voltage. The current decreases as sample components migrate to the point in the pH gradient where they are electrically neutral. Focusing is determined to be complete when the current stabilizes at a low value (e.g., less than 1 μA). It is the zone-sharpening effect occurring during the focusing step that is responsible for the high resolving power of cIEF. If a protein molecule diffuses out of the focused zone it acquires a net charge and therefore migrates back to the focused zone. This self-correcting effect ultimately results in very narrow bands at the conclusion of the focusing step.

In the third step, bands are mobilized past the detector. This can be accomplished by several means. In chemical mobilization,[27,28] the composition of one of the solutions is changed, resulting in a pH shift. Sample components acquire a net charge and are thus mobilized past the point of detection. For example, addition of sodium chloride at the cathodic end results in a competition between chloride ions and hydroxide ions entering the capillary. Consequently, the hydroxide ion concentration and pH decrease. The pH gradient thus formed in conjunction with the application of high voltage causes focused bands to be mobilized past the detector in order of decreasing isoelectric points. However, these methods did not produce linear pI-vs-migration plots and the range of the focused pH gradient was limited. By incorporating various additives (e.g., TEMED) in the ampholyte and mobilization buffer (e.g., zwitterions), the range of the pH gradient extended beyond pH 8.6.[29] Resolution of proteins with pI differences of as little as 0.05 pH unit has been reported.[29]

One alternative strategy utilizes electroosmotic flow (EOF),[30–32] present at a reduced but still significant level for mobilization. The magnitude of EOF is controlled by adjusting the concentration of linear polymer (methylcellulose or hydroxypropylmethylcellulose) in the sample–ampholyte mixture. The EOF is slow enough to allow bands to focus but sufficiently fast to mobilize focused zones past the detector. The analysis uses a neutral coated capillary (e.g., polyacrylamide) in the reversed polarity mode (negative to positive). The pH gradient is formed by placing 20 mM sodium hydroxide at the inlet end and 10 mM phosphoric acid at the outlet end. The capillary is filled with the sample–ampholyte mixture, which

[27] S. Hjertén, J. Liao, and K. Yao, *J. Chromatogr.* **387,** 127 (1987).
[28] F. Kilar and S. Hjertén, *Electrophoresis* **9,** 589 (1989).
[29] M. Zhu, R. Rodriguez, and T. Wehr, *J. Chromatogr.* **559,** 479 (1991).
[30] J. R. Mazzeo and I. S. Krull, *Anal. Chem.* **63,** 2852 (1991).
[31] J. Mazzeo and I. S. Krull, *in* "Capillary Electrophoresis Technology" (Norberto A. Guzman, ed.). Marcel Dekker, New York, 1993.
[32] T. Pritchett, Beckman Application Information Bulletin A-1769. Beckman, Fullerton, CA.

includes TEMED and hydroxypropylmethylcellulose (HPMC). A relatively high concentration of TEMED is used such that the sample and ampholytes focus in the distal end of the capillary (between the detector window and capillary outlet). This distance varies considerably among CE manufacturers and is 7 cm in the Beckman (Fullerton, CA) cartridge. Ideally, this distance should be relatively long (>7 cm) for high-resolution separations. Application of high voltage causes bands to focus and mobilize in the direction of the cathode (inlet end) and past the detector. An example of this is shown in Fig. 7.[32] Three major and three minor glycoforms of a murine monoclonal antibody, anticarcinoembryonic (anti-CEA) MAb, were resolved in less than 6 min. Analysis times are fast because sample components are focused within a 7-cm distance directly adjacent to the detector window. Furthermore, the method was shown to be precise (migration time RSD = 0.2%). Lack of precision, a common problem with typical cIEF methods, is due in part to the uncertainty of knowing when to stop focusing and start mobilization. However, the EOF (for mobilization) of

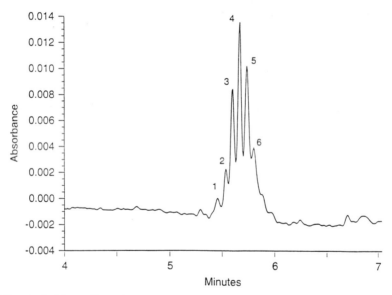

FIG. 7. Capillary IEF analysis of anti-CEA monoclonal antibody. Sample preparation: The anti-CEA MAb was mixed with a 3/10-range ampholyte solution containing TEMED (*N,N,N',N'*-tetramethylethylenediamine) and HPMC (hydroxypropylmethylcellulose). Conditions: Capillary, Beckman neutral coated; anolyte, 10 mM phosphoric acid; catholyte, 20 mM sodium hydroxide; polarity, reversed (negative to positive pole); detection, 280 nm. Peaks 3, 4, and 5 were three major glycoforms; peak 1, 2, and 6 were three minor glycoforms. (From Ref. 32.)

the capillaries from lot to lot is usually difficult to maintain consistently and that might generate some differences in migration and resolution from capillary to capillary.

Another mobilization scheme involves the use of pressure or vacuum to flow the contents of the capillary past the detector.[33] In those procedures, samples must be focused to a somewhat arbitrary end point (i.e., a particular length of time after which the current reaches a steady low value). Because electroosmotic flow has been virtually eliminated, pressure or vacuum instead of EOF is used to mobilize the focused protein zones past the detector, while maintaining the high voltage. Pressure (or vacuum) is applied gently, so as not to disrupt the integrity of the bands, while the application of high voltage maintains the pH gradient and zone sharpness. The advantages of this technique are that not only can extremely acidic proteins be mobilized efficiently, but a linear pI-vs-migration plot can also be produced.[33–35] This approach does not need TEMED to focus proteins in the distal end of the capillary. Proteins will be focused with ampholytes as in a typical IEF procedure (e.g., positive to negative) and then mobilized past the detector by applying voltage (e.g., 30 kV) and pressure (e.g., 50 mbar) simultaneously.[34,35] A list of typical analysis conditions for this method is shown in Table V. In this approach, high resolution and high precision as well as the linear pI-vs-migration plots can be achieved.

In all of the IEF procedures previously described, the capillary is exposed to high pH from the cathode, which can result in hydrolysis of the coating. Separation performance may deteriorate after a few runs, forcing the analyst to replace the capillary. Therefore, a stable coating is necessary for these analyses. At a certain point, the cost of purchasing coated capillaries may become prohibitive. An alternative approach is to perform cIEF in uncoated fused silica capillaries. In this technique, a linear polymer such as methylcellulose is used to control EOF.[30,34]

Capillary Regeneration

Following analysis, the coated capillary should be washed with a phosphoric acid solution to minimize hydrolysis of the coating resulting from exposure to high pH (20 mM sodium hydroxide, pH 12) at the cathodic end. Continual stripping of the coating will result in a buildup of electroos-

[33] M. Chen and J. E. Wiktorowicz, *Anal. Biochem.* **206,** 84 (1992).

[34] R. Grimm, "Capillary Isoelectric Focusing (cIEF)." Application Note, Hewlett-Packard, 1995.

[35] T. H. Huang, P. Shieh, and N. Cooke, A Quantitative Approach of Capillary Isoelectric Focusing for Protein Separation. Presented at the Seventh International Symposium in High Performance Capillary Electrophoresis, Wurzburg, Germany, January 29–February 2, 1995.

TABLE V

CAPILLARY ISOELECTRIC FOCUSING ANALYSIS CONDITIONS FOR PRESSURE MOBILIZATION

Parameter	Comments
Capillary	Coated or fused silica, 20–60 cm × 25–50 μm (depends on instruments)
Sample buffer preparation	Sample + ampholyte (e.g., 1.25% Servalyt pH 3–10) ± additives (e.g., 4 M urea or 0.2% Triton)
Polarity	Normal (+ to −)
Anolyte	20 mM phosphoric acid
Catholyte	20 mM sodium hydroxide
Focus end point	Less than approximately 1 μA
Focusing voltage	10 to 30 kV (depends on coating)
Mobilizer	Vacuum or pressure (e.g., 50 matm) with voltage
Mobilization voltage	Same as focusing voltage
Wavelength	280 nm (depends on proteins)
Preconditioning	Flush 2 min water, 2 min 1 M phosphoric acid, 2 min water
Loading condition	Flush 5 min ampholyte, then 3.5 min sample buffer

motic flow. The presence of electroosmotic flow interferes with the focusing process, as previously discussed. In addition to the linear polyacrylamide-coated capillaries (BioRad, Beckman), a capillary with a GC-immobilized coating (dimethylsiloxane, DB-1 wax; J & W), a polyethylene glycol (PEG)-coated capillary (MicroSolv), and a poly(vinyl alcohol) (PVC)-coated capillary (Hewlett-Packard, Palo Alto, CA) have been used for cIEF.[29,32–36] A list of coated capillaries used in the analysis of proteins by cIEF is shown in Table VI. The main requirements are that the capillaries be stable over the range of the pH gradient (usually pH 2 to 12) and virtually without electroosmotic flow.

Applications

The most attractive feature of cIEF in the analysis of recombinant proteins is the high resolving power, which is useful in the separation of recombinant variants differing by a single charge. In the following sections, separations of charge variants of two recombinant proteins, soluble CD4 and CD4–IgG are discussed.

Soluble CD4. In addition to deamidation, a second major source of charge heterogeneity in soluble CD4 is the variable number of sialic acid groups at the two N-linked glycosylation sites. The extent of sialylation is known to affect both the activity and clearance of glycoproteins. Therefore,

[36] T. Kasper and M. Melera, 27th Eastern Analytical Syposium, New York, October 2–7, 1988.

TABLE VI
COATED CAPILLARIES USED IN ANALYSIS OF PROTEINS BY CAPILLARY
ISOELECTRIC FOCUSING

Supplier	Capillary	Type of coating
Bio-Rad (Richmond, CA)	Coated	Neutral, hydrophilic polyacrylamide
Beckman (Fullerton, CA)	Neutral; charged	Neutral, hydrophilic polyacrylamide; polyamine
J & W	DB-1	GC-immobilized dimethyl siloxane
Hewlett-Packard	Neutral	Poly(vinyl alcohol)

it was important to develop a quantitative technique for monitoring the levels of the various sialylated species in different production lots of soluble CD4. Capillary IEF was investigated because a demonstrated strength of the technique is in the resolution of sialylated species.[25,36]

The cIEF analysis of soluble CD4, using a mixture of ampholytes (a 3:7 ratio of 8/10 range to 3/10 range), is shown in Fig. 8. The separation was performed on a 17 cm × 25 μm coated capillary (Bio-Rad). Sample bands were mobilized with a proprietary p*I* 3.2 zwitterion (Bio-Rad). The separation pattern appears similar to the corresponding IEF gel (data not shown). However, a minor basic peak can be detected that is not present in the gel. In addition, a few shoulder peaks can be observed, indicating the higher resolving power of the capillary technique. The resolution of the various sialylated forms is sufficient to permit quantification by peak area integration. Sample preparation is simple and commercial instrumentation allows for automated sample and data analysis. Thus, the cIEF technique looks promising as an alternative to slab gel electrophoresis.

CD4–IgG. The cIEF electropherogram of CD4–IgG is shown in Fig. 9. The analysis conditions were the same as for soluble CD4 (Fig. 8). Two acidic and two basic minor bands were resolved. The more acidic band (peak d) is most likely the deamido-52 variant. The pattern of peaks is similar to the IEF slab gel although the basic minor peak (a) is not detectable by IEF gel analysis (data not shown). The high efficiencies of CE manifested by sharp peaks facilitates the detection of minor protein impurities. The two basic species evidently appear as a single unresolved, diffuse band on the gel.

Summary

Capillary IEF is an inherently high-resolution separation technique for the analysis of closely related protein variants. Quantification can be accomplished in cIEF by peak area integration while slab gels are only semiquanti-

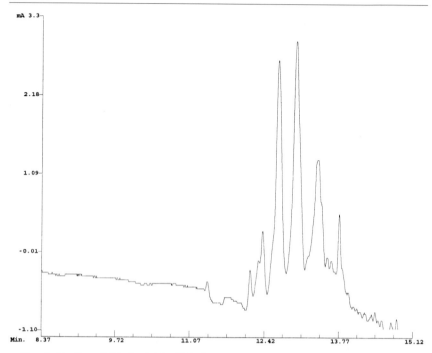

Fig. 8. Capillary IEF of soluble CD4. Sample preparation: rCD4 (1 mg/ml) was mixed with an equal volume of 2% ampholytes (Biolyte; Bio-Rad). The ampholyte mixture used in the separation consisted of a 3:7 ratio of 8/10- to 3/10-range ampholytes. Triton X-100 (1%) was added to minimize the risk of precipitation. Conditions: Capillary, fused silica coated with linear polyacrylamide (17 cm × 25 μm; Bio-Rad); anolyte, 10 mM phosphoric acid; catholyte, 20 mM sodium hydroxide; mobilizer, p*I* 3.22 zwitterion (Bio-Rad); focusing limit, 240 sec; focusing voltage, 10 kV; mobilization voltage, 10 kV; polarity, + to −; detection, 280 nm.

tative owing to the diffuseness of the bands. Still, some improvements (in particular, more stable coated capillaries and better mobilization techniques) are needed before capillary isoelectric focusing can overtake IEF slab gels as the method of choice.

Capillary Gel Electrophoresis

Introduction

High-resolution separations of polynucleotides and proteins have been achieved using cross-linked polyacrylamide gel-filled capillaries.[37] The ma-

---

[37] A. S. Cohen, D. R. Najarian, A. Paulus, A. Guttman, J. A. Smith, and B. L. Karger, *Proc. Natl. Acad. Sci. U.S.A.* **85,** 9660 (1988).

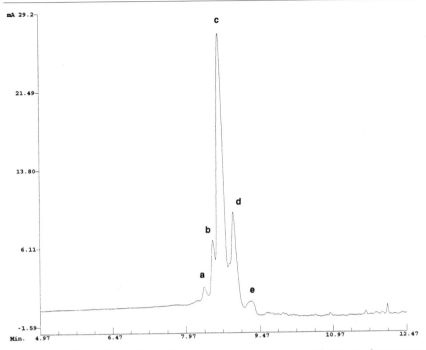

Fig. 9. Capillary IEF of CD4–IgG. Sample preparation and conditions were the same as for rCD4 (Fig. 8). Peaks a and b were unidentified basic variants; peak c was intact CD4–IgG; peak d was deamidated (Asn-52 to aspartate) CD4–IgG; peak e was an unidentified acidic variant.

jor disadvantages of this technique are the poor stability of the gel matrix and limited sensitivity due to the UV absorption of polyacrylamide. A sieving technique, in which SDS–protein complexes migrate through a capillary filled with a replaceable linear polymer, has been introduced.[38,39] When the polymer reaches a critical concentration, the polymer chains overlap, forming a mesh. The size of the mesh is determined by the length of the linear polymer and its concentration. Mobility is determined solely by molecular size because complexing of SDS to protein results in an equivalent charge per unit mass. This mode of capillary electrophoresis has alternatively been referred to as polymer network, entangled polymer, and non-gel-sieving electrophoresis by some manufacturers to distinguish it from methods using rigid cross-linked gels. For simplicity, all these analyses are referred as CGE here. Because the capillary is flushed and refilled prior

---

[38] M. Zhu, D. L. Hansen, S. Burd, and F. Gannon, *J. Chromatogr.* **480,** 311 (1989).
[39] A. M. Chin and J. C. Colburn, *Am. Biotech. Lab/News Ed.* **7,** 10A (1990).

to an analysis many more runs can be performed before deterioration of the separation and with greater reproducibility. Use of UV-transparent polymers such as dextran or polyethylene glycol permit UV detection at 214 nm, resulting in improved sensitivity. Runs can be rapid (<10 min) depending on the length of capillary used, the field strength, and the relative molecular mass of the protein of interest. Although 10–15 samples can be analyzed on a single slab gel, the total time required to pour a gel and run it can be several hours. Commercial capillary electrophoresis (CE) instrumentation with autosamplers allows unattended analysis, making CE sieving analysis (CGE) competitive with SDS–PAGE.

Sample Preparation

The protein is diluted to a final concentration of 0.1 to 0.5 mg/ml with the sample preparation buffer. If the salt concentration exceeds 50 mM, dialysis is required. The sample preparation buffer is an alkaline buffer (i.e., Tris-HCl, pH 9) containing 0.5% SDS. Dithiothreitol or 2-mercaptoethanol can be added if necessary to reduce disulfide bonds. The sample is usually heated at 95° for 2 to 10 min to ensure complete complexation of SDS to protein. In the case of antibody analysis, it may be helpful to run a time course study in order to determine the minimal time required. Longer reaction times can result in the generation of artifacts (i.e., free light chain or dimer). Prior to analysis, samples should be stored at 5°.

Analysis

The sample is injected electrophoretically or by pressure for a relatively long time owing to the viscosity of the running buffer. This time may need to be optimized depending on the purging capabilities of the particular commercial instrument. The running buffer consists of a linear polymer, typically dextran or polyethylene glycol, the chains of which overlap at a critical density, forming a mesh. Polarity is set in the reversed mode, negative to positive. The SDS–protein complexes have a net negative charge and therefore migrate electrophoretically toward the positive pole. On an uncoated capillary there is some electroosmotic flow counter to the direction of electrophoresis; however, it is minimized by the effect of the linear polymer in solution. Analysis can also be performed on a coated capillary if protein adsorption is suspected. Migration times on a coated capillary will typically increase. This occurs because of the hydrolysis of the coating by the alkaline SDS-containing running buffer and the concomitant increase in electroosmotic flow. If shifting of migration times is an issue, i.e., in the quality control of materials for lot release, an internal marker could be used.

Applied electric field strengths can vary from 500 to over 1000 V/cm.

TABLE VII

ANALYSIS CONDITIONS FOR CAPILLARY ELECTROPHORESIS THROUGH POLYMER NETWORK

Parameter	Comments
Capillary	36 cm × 50 μm uncoated
Sample buffer preparation	Samples diluted or dissolved in water containing 1% SDS and 1% 2-mercaptoethanol and heated at 90° for 15 min
Temperature	20 to 30°
Polarity	Reverse ($-$ to $+$)
Running buffer	Alkaline buffer containing linear polymer (proprietary)
Preconditioning	Flush with 0.1 M sodium hydroxide (no more than 10 min) Flush with water (2 min) Fill with sieving buffer (2 min)
Loading condition	-10 kV for 20 sec (electrophoretic) or 5 psi (or 330 matm) for 12 sec (pressure)
Detection	214 nm
Running voltage	15–30 kV, negative to positive polarity
Run time	Approximately 10–30 min

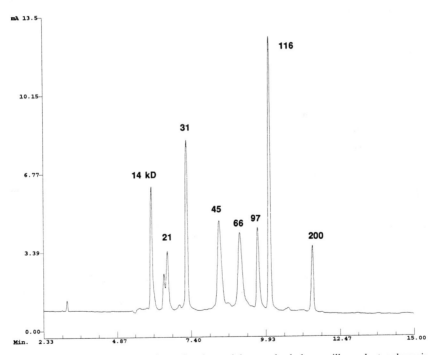

Fig. 10. Separation of protein molecular weight standards by capillary electrophoresis through a polymer network. Conditions: See Table VII; field strength, 625 V/cm.

Capillary temperature is 20° or ambient if no thermostatting is available. Capillary length can vary from 20 to 75 cm and often represents a compromise between resolution and analysis time. Following each analysis, the capillary (uncoated only) can be regenerated by purging with 0.1 M sodium hydroxide, water, and then sieving buffer. For analysis conditions, refer to Table VII.

Applications

The purification of novel proteins from biological fluids is a multistep process that often results in the isolation of very small quantities. SDS–PAGE is an important analytical technique for verifying protein identity. Molecular weight estimates are made by comparison to molecular weight standards, the bands of which are often broad and diffuse. Molecular weight can be determined with greater ease and accuracy by CE through a polymer network owing to the sharpness and symmetry of the peaks. A 12-min separation of protein molecular weight standards ranging from 14,000 to 200,000 is shown in Fig. 10. A relatively short length of capillary (24 cm) with a high electric field (625 V/cm) was used in this analysis. A plot of mobility versus log molecular weight yields a linear standard curve that can be used to calculate molecular weight (Fig. 11). Shown in Fig. 12 is a

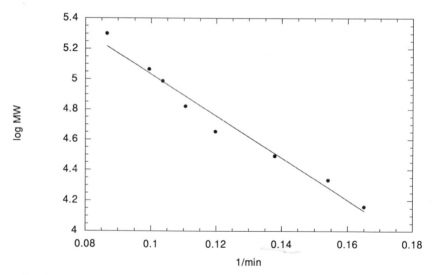

Fig. 11. Linear standard curve derived from migration times of protein molecular weight standards. A plot of mobility versus log molecular weight was derived from the data shown in Fig. 10. The linear standard curve generated was used to calculate the apparent molecular weight of protein samples.

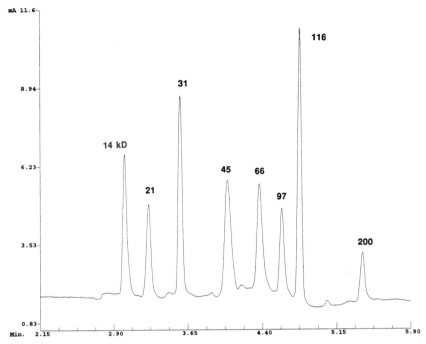

FIG. 12. Separation of protein molecular weight standards by capillary electrophoresis through a polymer network at high field strength (1050 V/cm). Conditions: Same as in Fig. 10, except that the applied field strength was increased from 625 to 1050 V/cm.

separation of the protein standards using the same capillary but at an even higher field strength (1050 V/cm). The total analysis time is about 5 min with no discernible loss in performance. Analysis time is a critical issue if CE is to gain acceptance as a replacement technique for SDS–PAGE. Although 10 to 15 samples can be loaded onto a single gel, 50 to 100 samples could be run overnight by CE in unattended mode. Furthermore, sample preparation time is much shorter with the CE technique. A disadvantage is that the molecular weight range of the gel-sieving method is limited to between 10,000 and 200,000. Biomolecules outside this molecular weight range will require a sieving buffer optimized in terms of the nature, length, and density of the linear polymer. Also, dried salts from the sieving buffer can cause plugs at the ends of the capillaries; therefore the tips should always be immersed in water between analyses. However, uncoated fused silica capillaries are inexpensive if replacement is necessary.

Pegylated Recombinant Human Superoxide Dismutase. Molecular weight determinations by molecular sieving techniques are not possible

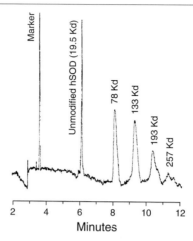

Fig. 13. Separation of pegylated species of human superoxide dismutase (hSOD) by capillary electrophoresis through a polymer network. Sample preparation: hSOD was denatured in an alkaline buffer containing 1% SDS and 1% 2-mercaptoethanol, and injected electrokinetically. Conditions: Capillary, 42 cm × 55 μm (ABI); running buffer, ProSort SDS-protein analysis reagent; field strength, 285 V/cm; detection, 215 nm. Underivatized hSOD and pegylated species with one, two, three, or four polyethylenes (50-kDa molecular mass for each PEG) attached were resolved. (From Ref. 40.)

for many classes of proteins, i.e., glycoproteins, antibodies, and pegylated proteins, that do not behave as undeformable spheres according to the Ogston model. In the case of pegylated proteins, determination of relative molecular mass provides some insight into its pharmacokinetic clearance properties. Modification of a protein with polyethylene glycol (PEG) results in an increase in the hydrodynamic radius depending on the number of PEG units attached and a concomitant decrease in kidney clearance as molecular size exceeds the limit for glomerular filtration (approximately 70 kDa for a globular protein). Determination of relative molecular mass provides a means of predicting the circulating half-life of samples that are pegylated to varying degrees. A separation of recombinant human superoxide dismutase (rhSOD) modified with PEG (molecular weight 50,000) is shown in Fig. 13.[40] The peak at 19.5 kDa represents unmodified hSOD. The peaks at 78, 133, 193, and 257 kDa represent hSOD with one, two, three, and four PEG units attached, respectively. Human SOD with one PEG attached is close to the 70-kDa limit for renal filtration whereas hSOD fractions with two or more PEGs attached are well above the 70-

[40] "Analysis of Protein Pegylation by Capillary Electrophoresis." Application Note 127444-1. Applied Biosystems, Inc., Foster City, CA, 1993.

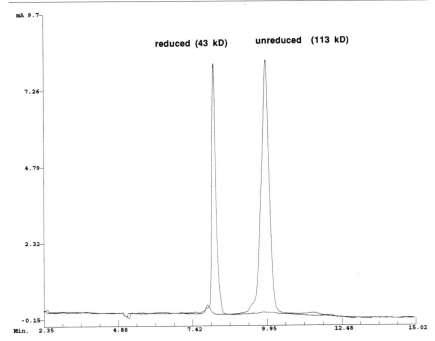

FIG. 14. Analysis of subunit interaction in CD4–IgG by capillary electrophoresis through a polymer network. Samples were prepared as described in the section Capillary Gel Electrophoresis. 2-Mercaptoethanol (2.5%) was added to the reduced sample. Conditions: Same as in Fig. 10.

kDa limit and would be expected to have long plasma half-lives compared to unmodified hSOD.

Soluble CD4 and CD4–IgG. Soluble CD4, an extracellular portion of the CD4 receptor, has a relative molecular mass of 45 kDa as determined by SDS–PAGE and CGE. The relative molecular mass of the hybrid molecule, CD4–IgG, is 113 kDa by polymer network CE under nonreducing conditions and 43 kDa under reducing conditions (Fig. 14). This verified that the hybrid was a disulfide-linked dimer.

Recombinant Human Growth Hormone. Recombinant human growth hormone (rhGH) is expressed in *Escherichia coli* as a polypeptide of 191 amino acids with a molecular mass of 22,125 Da. Cleavage between Thr-142 and Tyr-143 has been associated as a side reaction occurring during secretion into the periplasmic space.[41] The two chains generated by proteo-

[41] E. Canova-Davis, G. Teshima, J. Kessler, P. J. Lee, A. Guzzeta, and W. S. Hancock, *in* "Analytical Biotechnology" (C. Horvath and J. G. Nikelly, eds.). ACS Symposium Series No. 434, p. 90. American Chemical Society, Washington, DC, 1990.

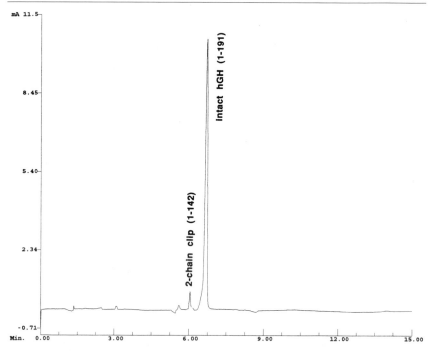

FIG. 15. Separation of a two-chain variant of recombinant human growth hormone by capillary electrophoresis through a polymer network. The sample was reduced by treatment with 2.5% 2-mercaptoethanol. Conditions: Same as in Fig. 10.

lysis remain connected by a disulfide bridge between Cys-53 and Cys-165. A sample of intact hGH was spiked with two-chain hGH and analyzed by CE through a polymer network under reducing conditions (Fig. 15). The major peak at 22 kDa represents intact hGH. The minor peak (5%) at 14 kDa represents the longer of the two chains (residues 1–142) generated by cleavage at Thr-142–Tyr-143 and reduction of the disulfide bonds. Thus, CE shows promise in the detection of proteolytically cleaved forms of recombinant proteins.

Summary

SDS–PAGE is one of the most widely used techniques for the analysis of proteins, owing to the importance of determining molecular weight. The capillary electrophoresis sieving technique described here can also provide molecular weight data and with greater ease, speed, and accuracy. In addition, CE through a polymer network shows promise as a high-resolution separation tool for determining protein purity. Impurities present at extremely low levels can be quantitated with relative ease compared to SDS–PAGE. The technology is currently available in the form of kits sold by Applied Biosystems, Beckman, and Bio-Rad and should provide an attractive alternative to SDS–PAGE, especially in applications where high resolution and quantitation are required.

Conclusion

The three CE techniques described can provide complementary information with regard to the structural integrity of proteins. Use of narrow-diameter capillaries (for efficient heat dissipation) with stable coatings (to prevent adsorption of protein) has furthered the development of CE as a high-resolution, high-speed technique for the analysis of proteins. Future directions include the characterization of separated components by off-line (fraction collection followed by Edman sequencing/mass spectrometry) and on-line approaches (CE–MS).

[13] Capillary Electrophoresis with Polymer Matrices: DNA and Protein Separation and Analysis

By Barry L. Karger, Frantisek Foret, and Jan Berka

Introduction

Many separations performed by slab gel electrophoresis are based on size differences between the molecules. For example, in free solution, double-stranded DNA (dsDNA) and single-stranded DNA (ssDNA) molecules have the same mobility, independent of the length of the molecule.[1] To achieve separation of DNA molecules, it is necessary to create a length-based discrimination of the molecules via a meshlike network of a gel, typically cross-linked polyacrylamide for short molecules and agarose for longer molecules.[2] As another example, the most widely used method of separation of proteins on slab gels is sodium dodecyl sulfate-polyacrylamide gel electrophoresis (SDS–PAGE). In free solution, reduced and denatured protein–SDS complexes cannot be separated; however, in

---

[1] B. M. Olivera, P. Baine, and N. Davidson, *Biopolymers* **2**, 245 (1964).
[2] B. D. Hames and D. Rickwood (eds.), "Gel Electrophoresis of Proteins—A Practical Approach." IRL Press, Oxford, 1986.

a gel network, separation based on size is achieved, and a first-order estimate of molecular weight obtained.[2,3] It is clear that transferring biopolymer applications from slab gel to capillary electrophoresis requires modes of size-based separation in the column.

Previous chapters in this volume and in Volume 270 have dealt with the resolving power of capillary electrophoresis. The use of narrow-bore capillaries (50–100 μm) with high column surface area-to-volume ratios permits the application of high electric fields at low currents, at least one order of magnitude greater than for slab gels, resulting in rapid separation. Furthermore, when band broadening is controlled by diffusion, fast separations result in less time available for diffusion, and thus sharper peaks and higher resolution. Besides resolution and speed, capillary electrophoresis offers an automated instrumental approach to separation and analysis. In this format, quantitative analysis is readily achievable, as well as isolation of individual bands.

This chapter explores various approaches to size separation of DNA and SDS–protein complexes by capillary electrophoresis. We first briefly discuss cross-linked gel-filled capillary columns, followed by non-cross-linked polymer network-based columns, the latter being in predominant use today, especially in a replaceable format. In this latter case, the polymer solution is removed from the capillary tube by pressure, and fresh polymer solution is reloaded into the column after each analysis. A survey of typical matrices that have been successfully employed is provided, as well as some practical details on their implementation. Fortunately, many commercial vendors have kits available to conduct such separations, and these kits can often be directly used in commercial equipment. This discussion then turns to providing specific examples and practical details of the application of such systems to DNA and protein analysis.

Capillary Gel Electrophoresis

Given the wide use of gel electrophoresis, it was natural that the initial efforts at size-based separations be focused on transferring slab gel materials to the capillary format. Cross-linked gel matrices of polyacrylamide for both SDS–PAGE[4] and DNA analysis[5] were first demonstrated. In terms of SDS–protein complexes, while successful separations were achieved,[4,6]

[3] A. Chrambach, "The Practice of Quantitative Gel Electrophoresis." VCH, Deerfield Beach, FL, 1985.

[4] A. S. Cohen and B. L. Karger, *J. Chromatogr.* **397**, 409 (1987).

[5] A. S. Cohen, D. R. Najarian, A. Paulus, A. Guttman, J. A. Smith, and B. L. Karger, *Proc. Natl. Acad. Sci. U.S.A.* **85**, 9660 (1988).

[6] K. Tsuji, *J. Chromatogr.* **550**, 823 (1991).

TABLE I

COMMERCIAL KITS FOR PROTEIN AND DNA ANALYSIS

Company	Kit	Comment
Applied Biosystems, Inc. (Foster City, CA)	ProSort SDS–Protein Analysis kit	Replaceable matrix
	GeneScan Polymer	Replaceable matrix–dsDNA
	Sequencing Polymer	Replaceable matrix–DNA sequencing
	MicroGel Oligonucleotide Analysis kit	Replaceable matrix
Beckman Instruments, Inc. (Fullerton, CA)	eCAP dsDNA 1000 kit	Replaceable matrix
	eCAP ssDNA 100 kit	Nonreplaceable linear polymer
	LIFluor dsDNA 1000	Replaceable matrix, fluorescent DNA intercalator for laser induced fluorescence detection
	eCAP SDS 14-200 kit	SDS–protein analysis
Bio-Rad (Hercules, CA)	PCR* Products Analysis kit	Replaceable matrix
Dionex (Sunnyvale, CA)	NucleoPhor SB1.5kB sieving buffer	For double-stranded DNA fragments
J&W Scientific (Fisons) (Folsom, CA)	μPAGE 3% T, 3% C– 5% T, 5% C	Oligonucleotides
	μSIL, DB-17, DB-1, DB-WAX	Coated capillaries for use with DNA sieving matrices containing cellulose derivatives
Scientific Resources, Inc. (Eatontown, NJ)	MicroSolv CE 3% T, 3% C PAA	Oligonucleotides

the lifetime of the gel column was short owing to the high salt concentration imposed on the system, along with the high electric fields. However, cross-linked gel separations of DNA were further developed, and indeed at least one commercial vendor for such columns is available (see Table I).

Remarkable separations of single- and double-strand DNA molecules have been achieved using cross-linked gels. For example, the largest efficiency reported to date, 30 million plates/m, has been published using a 3% T, 5% C matrix.[7,7a] This high resolving power is a direct consequence

---

[7] A. Guttman, A. S. Cohen, D. N. Heiger, and B. L. Karger, *Anal. Chem.* **62,** 137 (1990).

[7a] T = [wt(acrylamide + bis)]/100 ml (wt per volume %); C = wt bis/[wt(acrylamide + bis)] (wt %).

of the reduction of diffusion in the gel matrix (relative to free solution diffusion), because molecules are constrained to move in specific directions rather than freely in three dimensions.

The problem with cross-linked polyacrylamide (or agarose) gels in a capillary column is the limited stability they possess. First, small changes in osmotic pressure within the porous structure resulting from increases in salt concentration or temperature can cause collapse of the gel with the formation of bubbles and loss of electrical connection.[8] Indeed, cross-linked polyacrylamide gel columns are difficult to operate at 30° and above. Second, owing to mobility differences of ions in solution and in the gel matrix, zones of low conductivity can form at the head of the column.[9] The resulting high-voltage drop over a short distance can lead to bubble formation. Third, the high electric fields desired in capillary electrophoresis can stress the gels. Finally, the columns cannot be easily regenerated and must be replaced each time after a failure occurs.

For these and other reasons, most workers have converted from gel-filled capillaries to columns with non-cross-linked polymer networks or solutions. Polymer networks for size-based separations in slab gel electrophoresis were introduced in the mid-1970s[10] and followed in the 1980s.[11] The concentration of the polymer solution had to be sufficiently high (10%, w/v) or contain an anticonvective agent (agar) in order to maintain the matrix on the plate. Importantly, polymer solutions can be quite dilute and still maintain stability in capillary electrophoresis, as a consequence of the anticonvective properties of the walls of the capillary.

The dilute nature of the polymer network results in a sufficiently low viscosity that the matrix can be replaced by pressure on commercial capillary electrophoresis (CE) equipment after each run or after a few consecutive runs. This characteristic is important in that a fresh column is available at the start of each run. It is interesting to note that the original CE instruments were designed for sample injection and buffer wash purposes, not for replacing polymer matrices. As a consequence, some present-day instruments have only low pressure available, roughly 20 psi. More recent instrument designers have appreciated the value of higher pressure systems (100 psi or greater) to allow a broader range of polymer concentrations and thus solution viscosities to be used.

---

[8] T. Tanaka, *Sci. Am.* **244,** 124 (1981).

[9] H. Swerdlow, K. E. Dew-Jager, K. Brady, N. J. Dovichi, and R. Gesteland, *Electrophoresis* **13,** 475 (1992).

[10] H. J. Bode, "Electrophoresis '79," p. 39. Walter de Gruyter & Co., New York, 1980.

[11] D. Tietz, M. H. Gottlieb, J. S. Fawcett, and A. Chrambach, *Electrophoresis* **7,** 217 (1986).

Capillary Electrophoresis with Polymer Solutions

High performance separations, in many cases comparable to those on gel-based systems, are obtained with non-cross-linked polymer matrices. Figures 1A and B illustrate the resolution possible in the separation of a restriction digest of pBR 322 with 6% (w/v) linear polyacrylamide (LPA) and SDS complexes of standard proteins using a dextran solution, respectively. In Fig. 1A, single bp resolution with plate counts of several million per meter are observed.[12] In Fig. 1B, the migration velocity is found to be linear with the log of the molecular weight of the protein, as expected for SDS–PAGE.[13] Dextran is used in Fig. 1B rather than linear polyacrylamide in order to detect the proteins at 214 nm (acrylamide absorbs at this wavelength). Other low-ultraviolet (UV) wavelength transparent polymers such as polyethylene oxide (PEO) can be used as well.

Table I provides a list of commercial sieving kits, some with proprietary polymers, that have been employed in capillary electrophoresis. Among the most popular polymers today are linear polyacrylamide and various cellulose derivatives.

In general, successful column operation with polymer solutions has involved use of a coating, covalently attached or adsorbed, on the walls of the capillary to reduce electroosmotic flow to a minimum. Electroosmotic flow can lead to stress on the polymer matrix structure with a consequent increase in bandwidths and loss in resolution or, in extreme cases, flow of the polymer out of the column. To date, the most widely used covalently attached coating is linear polyacrylamide, bonded to the walls through a bifunctional agent.[14] Table II presents a procedure to synthesize this coating. Linear polyacrylamide works well as a coating except that it slowly hydrolyzes in basic solution, especially above pH 9, to produce negatively charged acrylic acid.[2] Other acrylamide-based polymers have been suggested in its place.[15] Another problem that has emerged is the photochemical degradation of the coating with the use of high-intensity UV sources employed with diode array detectors. If a diode array detector is to be used, a filter blocking the light below 250 nm can be employed or a non-UV-absorbing coating selected. Nevertheless, for normal UV detectors and laser-based fluorescence systems, the polyacrylamide coating has been widely used.

The application of somewhat more hydrophobic polymers as a sieving

[12] Y. F. Pariat, J. Berka, D. N. Heiger, T. Schmitt, M. Vilenchik, A. S. Cohen, F. Foret, and B. L. Karger, *J. Chromatogr.* **A652,** 57 (1993).
[13] K. Ganzler, K. S. Greve, A. S. Cohen, and B. L. Karger, *Anal. Chem.* **64,** 2655 (1992).
[14] S. Hjertén, *J. Chromatogr.* **347,** 191 (1985).
[15] M. Chiari, C. Micheletti, M. Nasi, M. Fagio, and P. G. Righetti, *Electrophoresis* **15,** 177 (1994).

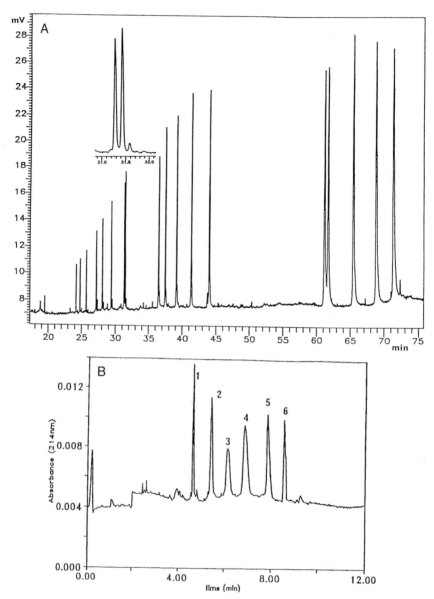

FIG. 1. (A) Electropherogram of the pBR 322/*Hae*III sample separation from peak 18 bp to 587 bp using 6% T LPA with ethidium bromide (1 μg/ml). Conditions: 1× TBE running buffer, $E = 150$ V/cm, l (effective length) = 30 cm, column temperature 30°, electrokinetic sample injection at 10 kV/2.5 sec. Inset: Blow-up of separation of 123 bp from 124 bp. (From Ref. 12.) (B) Separation of standard SDS–protein complexes in a dextran polymer network. Buffer: 0.06 *M* AMPD-cacodylic acid (pH 8.8), 0.1% SDS, 10% (w/v) dextran (MW 2×10^6). Capillary, $l = 18$ cm; 400 V/cm. Peaks: 1, myoglobin; 2, carbonic anhydrase; 3, ovalbumin; 4, bovine serum albumin; 5, β-galactosidase; 6, myosin. (Reprinted with permission from Ref. 13. Copyright 1992 American Chemical Society.)

TABLE II
PROTOCOL FOR PREPARATION OF LINEAR POLYACRYLAMIDE-COATED CAPILLARIES

Step I.   Activation of fused silica surface:
           Fill capillaries with 20% HCl and heat at 110° for 2 hr. Flush acid out with deion-
           ized water and dry the capillaries with an inert gas
Step II.   Attachment of bifunctional reagent:
           Methacryloxypropyltrimethoxy silane (2 ml) is hydrolyzed with 0.4 ml of 0.1 M
           HCl. The reaction is allowed to proceed for 60 min for condensation of oligosila-
           nols and cooling of the hot solution to room temperature. The solution is slowly
           flushed through the capillary and left in the column overnight for covalent attach-
           ment to the surface. The capillaries are then rinsed with methanol and dried with
           an inert gas
Step III. Polymerization and acrylamide attachment to the bifunctional agent:
           Make up 2 ml of 7% monomer and place in a 4-ml vial. Degas the solution with
           helium for 30 min and cool to 0° with magnetic stirring. Add 20 μl of a 10% solu-
           tion of ammonium peroxosulfate and 20 μl of a 10% solution of tetramethyleth-
           ylenediamine. After 5–10 sec of mixing, push the solution rapidly into the capil-
           lary. The polymerization is allowed to occur overnight. The resultant product is
           linear polyacrylamide covalently attached to the fused silica wall

matrix such as hydroxypropyl cellulose or polyethylene oxide results in some adsorption on the walls of the capillary, contributing to dynamic coating of the surface. One popular approach is the use of a gas chromatographic coating, e.g., DB-1, a wax, with hydroxycellulose matrices.[16] The cellulose adsorbs on the moderately nonpolar surface, reducing electroosmotic flow to a small value. Several commercial kits operate without coating, relying simply on the adsorption of the polymer or a trace amount of an additive, e.g., an aminoacrylamide polymer, to the bare surface. Our laboratory has the most experience with covalent coatings and we have found them to be highly reproducible, yielding high performance.

Columns containing polymer solutions should in general not be coiled to a small radius of curvature.[17] With coiling, the distance traveled by a molecule on the outside track will be greater than that on the inside. Because radial diffusion is limited owing to the intervening polymer, band broadening can result from the longer time necessary for molecules to travel in the outer track. A second factor contributing to broadening is the distortion of the polymer matrix structure caused by the shear forces resulting during the change in column shape.[17] Thus, loading a polymer network in a straight capillary followed by coiling leads to poor column

[16] H. E. Schwartz and K. Ulfelder, *Anal. Chem.* **64,** 1737 (1992).
[17] S. Wicar, M. Vilenchik, A. Belenkii, A. S. Cohen, and B. L. Karger, *J. Microcol. Sep.* **4,** 339 (1992).

performance, or vice versa. The conclusion of these studies is that straight capillaries or columns with a low radius of curvature should be used with polymer networks. In addition, there should be little or no pressure drop on the columns, otherwise the polymer matrix will again be distorted. Either horizontal columns or columns with their ends at equal heights should be used.

As mentioned previously, workers either employ laboratory-made polymer solutions or commercial kits. The kits can be directly applied and are backed up by the company in terms of reliability; however, laboratory-made solutions provide flexibility in terms of a wider range of concentrations, molecular weights, buffer systems, etc. The polymer is often purchased from a manufacturer, such as PolySciences (Warrington, PA), Sigma (St. Louis, MO), Aldrich (Milwaukee, WI), etc. The polymer concentration and molecular weight have a significant effect on solution viscosity.[18]

Prior to use, either the kit or the homemade polymer solution should be degassed by vacuum. Care should be taken in the degassing step that evaporation does not occur, or the concentration of polymer solution will change. Dilute polymer solutions or the buffer alone are typically used in buffer reservoirs. As with all methods utilizing capillary electrophoresis, it is important to change the buffer frequently, after roughly 10 runs or so, because electrolysis products in the reservoirs can affect the reproducibility of migration and perhaps peak shape.

The polymer solution is delivered by pressure into the capillary. It is necessary to flow at least 2–3 column volumes of solution through the column in order to ensure that a representative polymer network is present. The electric field is then applied and, as often happens, the current slowly decreases to a lower fixed value. This change in current results from the removal of charged impurities from the polymer solution. While the polymer solution can be reused, it is a good strategy to replace the matrix after each run or a few runs in order to present the sample with a fresh column. We have found that migration reproducibility can be higher when replacing the matrix after each run.[13]

Injection

There are two basic procedures for sample introduction in capillary columns containing polymer solutions: pressure or electrokinetic injection. In pressure injection, the injection side of the separation capillary is inserted into the sample vial and a pressure difference applied between the capillary ends, causing the sample to enter the capillary. For a constant pressure

[18] D. Wu and F. E. Regnier, *J. Chromatogr.* **608**, 349 (1992).

difference, the sample volume introduced into the capillary can be controlled by the time of injection. Typical pressure differences of a few pounds per square inch are used to inject 1–100 nl of the sample over a few seconds of injection time. This injection mode, suitable for low-viscosity replaceable polymer matrices, provides a true aliquot of the original sample. The reproducibility is typically better than 1% RSD (relative standard deviation). An advantage of pressure injection is that sample clean-up is often not required, i.e., desalting may not be necessary.

Electrokinetic injection uses electromigration of the sample ions into the separation capillary. The capillary is inserted into the sample vial containing an electrode for connection to the high-voltage power supply. When the voltage is applied, sample ions electromigrate into the separation capillary. The amount injected is dependent on the electric current, time of injection, and mobility of individual ions present in the sample, with highly mobile ions being preferentially transferred into the capillary. Thus, a bias of the amount of individual sample components injected can result.[19] For DNA and SDS–protein complexes, this bias should in principle not occur because the free solution mobilities are independent of molecular size; however, electroosmotic flow in the column can result in sample bias.[19a]

Electrokinetic injection can easily be automated with cross-linked gels or viscous polymer matrices. The best results are obtained with a well-desalted sample, because the electric current will then be transported by the sample ions. In addition, sample preconcentration of several orders of magnitude can be achieved. Such a focusing step provides for sharp injection plugs and very narrow peaks for high resolution. DNA-sequencing samples and polymerase chain reaction (PCR) reactions can be effectively desalted by Sephadex G25 spin columns (Centrisep; Princeton Separations, Adelphia, NJ) with size exclusion of 30-base lengths. Of course, desalting is not possible with SDS–protein complexes. Alternatively, standard ethanol precipitation procedures can be used for desalting as well.

A few practical considerations are important for good separation performance with electrokinetic injection. When multiple analysis of sample volumes ($<10 \ \mu l$) is performed, a decrease in detection signal can frequently be observed.[20] The cause of this problem is sample contamination by the background electrolyte ions migrating from the separation capillary into the sample solution. A simple procedure that provides reproducible multiple electrokinetic injections uses pressure injection of a short plug (~5 mm)

[19] X. Huang, M. J. Gordon, and R. N. Zare, *Anal. Chem.* **60**, 375 (1988).
[19a] K. Kleparnik, M. Garner, and P. Bocek, *J. Chromatogr.* **A698**, 375 (1995).
[20] H. E. Schwartz, K. Ulfelder, F. J. Sunzeri, M. P. Busch, and R. G. Brownlee, *J. Chromatogr.* **559**, 267 (1991).

of distilled water prior to the sample injection.[21] (Obviously, the matrix must withstand pressure in this case.) In addition, because the injection electrode is in direct contact with the sample solution, electrochemically labile sample components can be degraded, and the resultant electrolysis products may contaminate the sample.

Polymer Molecular Weight and Concentration

There have been a number of studies, particularly in the area of DNA separations, dealing with the role of the molecular weight (MW) of the polymer (M_n and M_w, number average MW and weight average MW, respectively) and concentration on size separation. Because polymers can be commercially obtained with different MWs or appropriately polymerized in house, it is important to understand the role of these parameters in order to optimize separations.

It is first to be noted that both of the above parameters can influence the pressure drop necessary for replacing the matrix in the column. As we noted, commercial equipment permits pressures up to 100 psi for polymer replacement (one instrument has pressures in excess of 1000 psi to replace a more viscous medium). With respect to instrumentation permitting pressures up to 100 psi, the viscosity of the polymer matrix should generally be less than 100–200 cP. With higher replacement pressures, one can elevate the viscosity to 500–1000 cP or more.

It was noted in the earliest days of polymer solution separations in capillary electrophoresis that low-MW polymers seem well suited to low-MW solutes while high-MW polymers seem best for separation of long biopolymers.[22] In terms of DNA separations, it has been shown for both single- and double-stranded molecules that for a given concentration, the peaks are sharpest for low base number molecules with low-MW polymers and for high base number molecules with high-MW polymers.[23,24] While the exact reason for this is not fully understood, it is likely that this behavior is related to the resistance the polymer presents to individual DNA molecules during their migration in the electric field (e.g., see Ref. 25). It is known that all linear polymers will, above a specific concentration, achieve entanglement of their chains, i.e., the entanglement threshold. Typically, the column matrices are operated well above this concentration. We can envision that high-MW polymers will be more fully entangled with other

---

[21] A. Guttman and H. E. Schwartz, *Anal. Chem.* **67**, 375 (1995).

[22] M. Zhu, D. L. Hansen, S. Burd, and F. Gannon, *J. Chromatogr.* **480**, 311 (1989).

[23] Marie C. Ruiz-Martinez, Ph.D. thesis. Northeastern University, Boston, MA, 1995.

[24] H. T. Chang and E. S. Yeung, *J. Chromatogr.* **B669**, 113 (1995).

[25] B. Kozulic, *Anal. Biochem.* **231**, 1 (1995).

polymer chains than low-MW molecules,[25a] thus presenting a greater degree of resistance than the polymer of the lower MW.

With respect to polymer concentration, the trends follow closely that observed in gel electrophoresis, i.e., higher concentrations lead to better band spacing (selectivity) of individual fragments, albeit at the price of time. One can envision this result to be a consequence of a higher density of polymer in the cross-section of the capillary as the concentration is raised, leading to a greater extent of interaction between the biopolymer molecules migrating through the medium and the polymer network. Alternatively, workers have suggested that higher concentration matrices lead to smaller pores or mesh sizes in the polymer network, and thus a more selective influence on the migration of biopolymers through the medium. Of course, in this discussion one assumes that the same MW is utilized for the various concentrations. In summary, for the most part, the concentration of the polymer will affect the band spacing or selectivity, whereas the MW will affect the peak dispersion or band sharpness.

As a consequence of the above trends, low base number solutes, such as single-strand primers or antisense DNA, will be best separated with low-MW polymers. For example, the use of approximately 50,000 MW linear polyacrylamide provides an excellent means of separating single-strand DNA molecules, as illustrated in Fig. 2. The concentration of the polymer is roughly 15% and yet, owing to the low MW, the viscosity is sufficiently low that the matrix may be replaced after every run on commercial equipment, if desired.

With respect to higher base number solutes, such as may result from Sanger sequencing reaction products more than 500 bases or more, the use of high-MW polymers is recommended. However, because the viscosity is substantially higher in this case, for a given polymer concentration, it is necessary to work with much lower solution concentrations in order to permit facile replacement of the polymer. Thus, for example, linear polyacrylamide polymerized to greater than 5 million MW has been found effective for separating sequencing reaction products in excess of 600–800 bases in length at concentrations of 2–3% (w/v).[26] Polyethylene oxide of 8 million MW and 2.5% (w/v) concentration has also been used for separation of fragments of greater than 400 bases.[24] In addition, in order to enhance separation, particularly for small sequencing fragments, workers have at times added 2–3% (w/v) of low-MW polymer mixed with the high-MW polymer.[24] In this way, both selectivity and efficiency can be manipulated.

[25a] The entanglement threshold is at a lower concentration, the greater the molecular weight of the polymer.

[26] K. Kleparnik, F. Foret, J. Berka, and B. L. Karger, *Electrophoresis* 1996 (submitted).

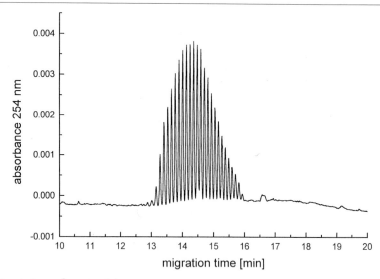

Fig. 2. Separation of pd(A)$_{40-60}$ by capillary electrophoresis with UV detection at 254 nm. Capillary: 20/27 cm × 100-μm i.d. Separation matrix: 15% linear polyacrylamide (MW 60,000) in TBE buffer with 7 M urea. E = 500 V/cm.

It is also important to note that separation of double-stranded DNA fragments using ultradilute polymer solutions is possible.[27] Here, employing high-MW polymers [but well below the entanglement threshold, e.g., 0.025% (w/v) or less], separation of double-stranded fragments has been demonstrated. At such low polymer concentrations, pores are not available, and therefore the separation must result from the interaction of DNA molecules with the polymer, leading to retardation on the basis of size. Indeed, the authors postulate that DNA transiently entangles with the polymer, and this DNA–polymer complex then migrates slower than free DNA alone.[27] The mechanism of separation would then involve a series of entanglements between the DNA and the polymer, as solute molecules migrate through the column. There is even evidence for transient entanglements from fluorescence microscopic imaging of DNA in the ultradilute polymer solutions.[28]

In summary, there is rapidly developing experience on the use of polymer matrices for separation of DNA molecules, as well as SDS–protein complexes. We can anticipate further advances in the coming period. However, already the matrices represent powerful approaches to separation and analysis. It is to be emphasized that the onerous task of gel preparation is

[27] A. E. Barron, W. M. Sunada, and H. W. Blanch, *Electrophoresis* **16,** 64 (1995).
[28] X. L. Shi, R. W. Hammond, and M. D. Morris, *Anal. Chem.* **67,** 1132 (1995).

minimized utilizing the polymer matrices, and automatic replacement of the matrix is readily possible. High efficiencies are generally achieved; for example, 3–5 million theoretical plates per meter for DNA molecules, in part because of the order of magnitude reduction in the solute diffusion coefficient relative to free solution. In the next sections, we present applications for the polymer matrix separations in both the DNA and SDS–protein complex fields.

Nucleic Acid Separations by Capillary Electrophoresis

Advances in molecular biology and biotechnology, e.g., polymerase chain reaction (PCR), solid-phase DNA or RNA synthesis, and fluorescence-based DNA-sequencing chemistries, have resulted in a growing need for fast, quantitative, and automated analytical techniques for the separation of both single- and double-stranded nucleic acid molecules. In analogy to slab gel electrophoresis, both denaturing and nondenaturing buffer/matrix systems have been developed for CE, enabling separations of nucleic acids based on size, conformation, or base composition. (For a review, see Ref. 29.) The wide variety of biological applications requires that specific CE conditions (e.g., polymer matrix, background electrolyte, electrophoresis settings, and sample preparation) be carefully optimized for each situation in order to surpass traditional techniques [high-performance liquid chromatography (HPLC), PAGE]. However, a capillary electrophoresis instrument, equipped with multicolor laser-induced fluorescence detection, can serve these diverse applications.

Oligonucleotide Analysis

Automated DNA synthesis has become a routine and reliable method for production of oligonucleotides used in a vast array of applications. As a result, large numbers of oligonucleotides produced on a daily basis by a DNA synthesis facility must be rigorously examined for purity. Although some methods, such as PCR amplification and sequencing, often perform well without extensive purity control of oligonucleotide primers, there are specific applications, namely antisense DNA or RNA drug synthesis, hybridization probes, and oligonucleotide library production, in which quality assessment is critical. Capillary electrophoresis, with the characteristics of high speed, resolution, and automation, is well suited to analysis of oligonucleotides, DNA analogs, oligoribonucleotides, and their labeled conjugates.

[29] A. E. Barron and H. W. Blanch, *Sep. Purif. Methods* **24**, 1 (1995).

As mentioned previously, denaturing cross-linked polyacrylamide gel columns [3–7% T, 3–5% C, 1× Tris–borate–EDTA (TBE), 7 M urea] were initially employed together with UV detection, and superb separations of model polyadenylic acid mixtures, as well as crude oligonucleotides, in the size range of 20–160 nucleotides were achieved.[5,7] From this work, it became clear that the electrophoretic mobility of oligonucleotides in a gel-filled capillary was not only determined by the mass and charge of the solutes, but also by molecular conformation, charge distribution, and hydrophobic and hydrogen bonding with the gel matrix.[30] Several studies have shown the possibility of prediction of migration behavior by calculating empirical base-specific migration coefficients or by using computer-aided simulations.[31–33]

As replaceable linear polymer matrices gained popularity over traditional cross-linked gels owing to stability issues with the gels, it was found that comparable separations were achieved. For example, a replaceable 10% linear polyacrylamide solution (MW unknown) in denaturing (7 M urea) buffer provided baseline resolution of oligodeoxycytidylic acid standards 10–15 and 24–36 nucleotides long.[34] Low molecular weight linear polyacrylamide (e.g., 50,000 Da), low in viscosity even at high concentration (~15%), was applied for the separation of a poly(dA) mixture (from 40 to 60 nucleotides; shown in Fig. 2; see also Ref. 35). Polymer-filled columns for oligonucleotide analysis are commercially available (see Table I).

Single-stranded, synthetic ribo- or deoxyribonucleotides from 4 to 150 bases can be readily analyzed with polymer solution columns. Applications of such analyses include detection and quantitation of short failure sequences and high molecular weight or less charged species (e.g., incomplete deprotection, branching), as well as assessment of purity of the derivatized oligonucleotides, e.g., 5′-aminolinkers, biotin or fluorescent dye labeled. New developments in the "antisense" DNA drug field have also created new analytical challenges. Antisense oligonucleotides are synthesized using base or internucleotide phosphate analogs (e.g., phosphorothioates), and they are often significantly more hydrophobic than their phosphodiester analogs. Furthermore, the sulfurizing reaction yields a large number of diastereomers, both factors causing peak broadening.[30] Phosphorothioate

[30] A. Andrus, "Methods: A Companion to Methods in Enzymology," Vol. 4, pp. 213–226. Academic Press, San Diego, CA, 1992.
[31] A. Guttman, R. J. Nelson, and N. Cooke, *J. Chromatogr.* **593,** 297 (1992).
[32] Y. Baba, N. Ishimaru, K. Samata, and M. Tsuhako, *J. Chromatogr.* **A618,** 41 (1993).
[33] Y. Cordier, O. Roch, P. Cordier, and R. Bischoff, *J. Chromatogr.* **A680,** 479 (1994).
[34] J. Sudor, F. Foret, and P. Bocek, *Electrophoresis* **12,** 1056 (1991).
[35] C. Heller, *J. Chromatogr.* **A698,** 19 (1995).

analogs are prone to fragmentation or reoxygenation, both during synthesis and metabolic cycles *in vivo,* thus creating challenges for analytical separations.[36]

DNA Sequencing

Over the past decade, slab gel-based, automated fluorescence DNA sequencers have become a standard technology for large-scale sequencing efforts. Capillary electrophoresis, utilizing laser-induced fluorescence, advanced detectors [such as intensified diode arrays or charge-coupled devices (CCDs)] and four-color fluorescence labeling identical to that used on the standard slab gel instruments, has been shown in numerous reports to surpass conventional slab gel methods in single-lane speed and resolution (for review, see Refs. 37a and b). Despite these advantages, the use of CE for real sequencing projects has been hampered by the lack of suitable commercial instrumentation. A step toward better acceptance of CE for DNA sequencing has been made by the introduction of a single-capillary commercial instrument for this purpose (Perkin-Elmer/ABI). Needed as well is a multiple-capillary array to match the multiple lanes on a slab gel; however, much progress toward the commercialization of such instrumentation has occurred, due in large part to funded research supported by the Human Genome Project.

In a single-column approach, in addition to the standard four-color detection scheme, a simple, inexpensive instrument uses two dye-primers and two peak-height base coding,[38] taking advantage of equal signal heights produced by T7 DNA polymerase in the presence of manganese cations.[39] This approach requires only a single laser excitation source (a low-cost helium–neon laser can be selected[40]) and two photomultipliers (PMTs) as detectors. Employing this approach, sequencing of malaria genome clones using cross-linked nonreplaceable gels (4% T, 5% C) has been reported. Read lengths of 575 bases with accuracy of 99% in less than 2 hr of electrophoresis time were achieved.[41] It is noted that the sequencing chemistry

[36] A. Belenkii, D. Smisek, and A. S. Cohen, *J. Chromatogr.* **A700,** 137 (1995).
[37a] B. L. Karger, Y. H. Chu, and F. Foret, *Annu. Rev. Biophys. Biomol. Struct.* **24,** 579 (1995).
[37b] C. Heller, *J. Chromatogr.* **A698,** 19 (1995).
[38] D. Y. Chen, H. R. Harke, and N. J. Dovichi, *Nucleic Acids Res.* **20,** 4873 (1992).
[39] S. Tabor and C. S. Richardson, *J. Biol. Chem.* **265,** 8322 (1990).
[40] H. Lu, E. Arriaga, D. Y. Chen, and N. J. Dovichi, *J. Chromatogr.* **A680,** 497 (1994).
[41] S. Bay, H. Starke, J. Z. Zhang, J. F. Elliot, L. D. Coulson, and N. J. Dovichi, *J. Cap. Electrophoresis* **1,** 121 (1994).

for this system is limited to the use of T7 DNA polymerase, labeled primers, or internal labeling with fluorescent dNTPs.[42]

Our laboratory has focused on the development of the replaceable linear polyacrylamide (LPA) matrices, which, together with other polymer matrices, have become widely accepted as a solution to the reliability problems associated with the gel-filled columns. Using the above-described system of two-color–two peak-height base coding, we introduced replaceable matrices for DNA sequencing, leading to a reproducible and reliable system.[43] This methodology was able to yield routinely 350 bases of sequence data in about 30 min, at an electric field of 250 V/cm, as shown in Fig. 3 (see color insert). The separation matrix was 6% T LPA in 1× TBE–3.5 M urea–30% formamide buffer. These conditions permitted fast separations and efficient resolution of compressed sequences because of the stringent denaturation.

To extend sequencing read lengths toward the 450- to 500-base range, new LPA matrix compositions were sought. It has been found that decreasing the acrylamide concentration to 4% T and adjusting the polymerization conditions (deoxygenation under helium, polymerization at 0° for 48 hr in 1× TBE–3.5 M urea–30% formamide buffer) yields long polymer fibers (MW > 1,000,000). This LPA matrix has provided significantly improved resolution for sequencing fragments up to 500 bases in roughly 60 min, yet maintaining relatively low viscosity for easy replacement into the capillary.[23] Once polymerized, these LPA matrices should be handled carefully, avoiding excessive stirring or other shearing, which could cause breakage of long polymer fibers. Also, storing polymer batches at −20° helps to reduce buffer denaturant (urea, formamide) decomposition. A commercial sequencing kit (see Table I) is also available for sequencing ~450 bases in 1.5–2 hr.

In our work, we have found the use of an intensified diode array detector ideal for four-color sequencing with a single column, providing full spectral data and excellent sensitivity.[44] In combination with a two-laser–two-window system (488-nm argon ion laser plus a 543-nm helium–neon laser), optimal excitation for the four PE/ABI [Perkin-Elmer (Norwalk, CT)/Applied Biosystems, Inc. (Foster City, CA)] dye-labeled primers (FAM, JOE, TAMRA, and ROX) was attained. This sequencing chemistry is

[42] H. R. Starke, J. Y. Yan, J. Z. Zhang, K. Muhlegger, K. Effgen, and N. J. Dovichi, *Nucleic Acids Res.* **22**, 3997 (1994).

[43] M. C. Ruiz-Martinez, J. Berka, A. Belenkii, F. Foret, A. W. Miller, and B. L. Karger, *Anal. Chem.* **65**, 2851 (1993).

[44] S. Carson, A. Belenkii, A. S. Cohen, and B. L. Karger, *Anal. Chem.* **65**, 3219 (1993).

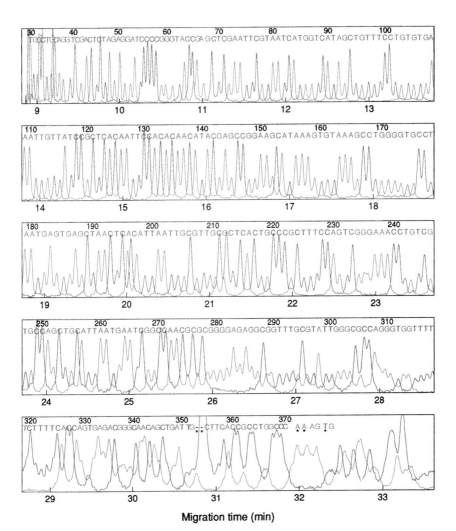

Migration time (min)

Fig. 3. Two-color, peak height ratio sequencing of M13mp18. Red, JOE C > T; blue, FAM A > G. Electrophoretic conditions: 6% T LPA–1× TBE–30% formamide–3.5 M urea running buffer, E = 250 V/cm, column temperature 32°. Asterisks represent software base miscalls. (Reprinted with permission from Ref. 43. Copyright 1993 American Chemical Society.)

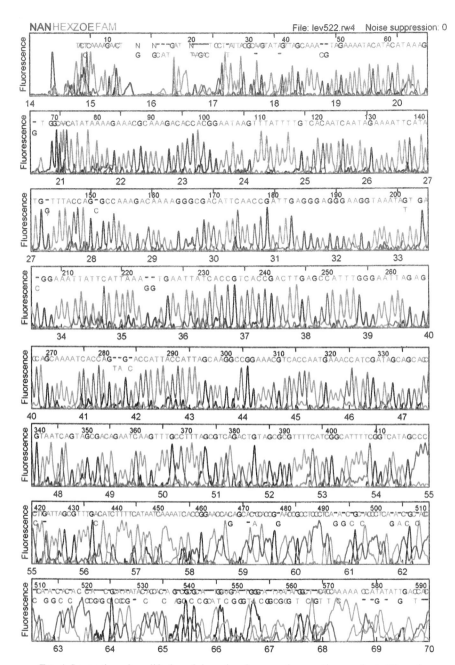

Fig. 4. Separation of modified modular primed sequencing reaction products. The primers used for this reaction were two degenerate, phosphorylated heptamers and one pentamer at position 3′ 2892 on M13mp18. The electrophoretic conditions were as follows: capillary, $l = 30$ cm; column temperature 30°, separation matrix 4% T LPA–30% (w/v) formamide–3.5 M urea; $E = 200$ V/cm. (Reprinted with permission from Ref. 46. Copyright 1993 American Chemical Society.)

widely used in "shotgun" sequencing strategies, in which large numbers of single-stranded or double-stranded templates share the same priming sites, and only two sets (forward and reverse) of dye primers are utilized throughout the project. A more universal labeling scheme is offered by the use of fluorescent dideoxynucleotides (dye terminators), allowing one to select any nonlabeled sequencing primers, or even a combination of short oligonucleotides (e.g., hexamers) from a library, for primer walking applications.[45] The sets of dye terminators are commercially available (PE/ABI) both for T7 DNA polymerase (FAM-T, ZOE-C, HEX-A, NAN-G) and *Taq* polymerase (R110-G, R6G-A, TAMRA-T, ROX-C) and can thus be used with Sequenase or cycle sequencing employing thermostable DNA polymerases for a variety of DNA templates (e.g., single-stranded phagemids, double-stranded plasmids or cosmids, double-stranded PCR products). We have performed sequencing of a variety of model templates with dye-terminators using the above-mentioned diode array detector in combination with an argon ion laser.[46] By way of example, a typical sequencing separation is shown in Fig. 4 (see color insert), using an oligonucleotide library for primers.

Another important consideration for successful separation of sequencing fragments by CE is sample clean-up. Because, as noted previously, DNA samples are generally introduced into a capillary by electrokinetic injection, any ionic species other than DNA fragments would interfere with this process, leading to diminished signals. Sequencing reactions are complex mixtures, containing large quantities (up to 100 mM) of buffer ions, dNTPs, ddNTPs, oligonucleotide primers, and proteins. Efficient desalting of sequencing samples is thus important, followed by vacuum drying and redissolving of DNA in a small volume (20–40 μl) of organic denaturant, typically formamide or dimethyl sulfoxide. Moreover, it is helpful to remove template molecules prior to injection, because their presence in the sample may result in column fouling.[47] (In the case of cycle sequencing, it may not be necessary to remove the template from the sample, because the products are amplified relative to the template.) We have devised a simple, two-step protocol utilizing a size-exclusion spin column and a centrifugal ultrafilter, for efficient sample clean-up, enabling larger quantities of sequencing products to be injected without compromising resolu-

[45] J. Kieleczawa, J. J. Dunn, and F. W. Studier, *Science* **258,** 1787 (1992).
[46] M. C. Ruiz-Martinez, E. Carrilho, J. Berka, J. Kieleczawa, A. W. Miller, F. Foret, S. Carson, and B. L. Karger, *Bio/Techniques* 1996 (in press).
[47] H. Swerdlow, K. E. Dew-Jager, K. Brady, R. Grey, N. J. Dovichi, and R. Gesteland, *Electrophoresis* **13,** 475 (1992).

tion.[46] One may also consider standard ethanol precipitation procedures for removal of salt; however, this procedure would be difficult to automate in a microtiter well.

As noted, future developments in CE for DNA sequencing are directed toward parallel sample processing, i.e., expanding to multicapillary format. Indeed, capillary array prototype instruments have been described[48–50] and represent an active field of instrument development. In addition to multi-capillary instruments, ultrathin horizontal slab gels[51] and gel electrophoresis in a high-density array of micromachined channels[52] are emerging technologies promising high-throughput DNA sequencing. A multiple capillary array instrument with replaceable matrices may well be the next-generation large-scale DNA sequencer.

Double-Stranded DNA Sizing

Size-dependent separations of double-stranded DNA from several tens up to millions of base pairs under nondenaturing conditions are traditionally performed on cross-linked polyacrylamide or agarose slab gels. In CE, however, the anticonvective property of the gels is no longer needed and, as in the case of oligonucleotides and sequencing product separations, gels have been substituted by polymer networks operated at high electric fields. A large variety of polymers has been successfully applied for separation of model dsDNA mixtures (see Table I). For dsDNA, on-column UV detection at 260 nm is often employed, because of its simplicity. However, if ultrasensitive detection, competitive with autoradiography, is required, laser-induced fluorescence (LIF) should be the preferred method. DNA labeling involves either the use of fluorescent intercalation dyes, added to the CE buffer and/or sample, or covalent attachment of a fluorophore, mainly to the 5' end of primers or probes used in PCR-based applications.[53] Novel energy transfer (ET) fluorescent dye-labeled oligonucleotide primers for DNA sequencing and other types of multicolor genetic analysis have also been developed.[54] These ET primers exhibit high fluorescence intensity,

[48] X. C. Huang, M. A. Quesada, and R. A. Mathies, *Anal. Chem.* **64,** 2149 (1992).

[49] K. Ueno and E. S. Yeung, *Anal. Chem.* **60,** 1424 (1994).

[50] N. J. Dovichi, Paper presented at the 17th Int. Symp. on Capillary Chromatogr. and Electrophoresis, Wintergreen, VA, May 1995.

[51] A. J. Kostichka, M. L. Marchbanks, R. L. Brumley, H. Drossman, and L. M. Smith, *Bio/Technology* **10,** 78 (1992).

[52] A. T. Woolley and R. A. Mathies, *Anal. Chem.* **67,** 3676 (1995).

[53] H. E. Schwartz, K. J. Ulfelder, F. T. Chen, and S. L. Pentoney, *J. Cap. Electrophoresis* **1,** 36 (1995).

[54] J. Y. Ju, I. Kheterpal, J. R. Scherer, C. C. Ruan, C. W. Fuller, A. N. Glazer, and R. A. Mathies, *Anal. Biochem.* **231,** 131 (1995).

minimal mobility shifts, and are efficiently excited with a single laser wavelength.

Restriction Fragment Analysis. Capillary electrophoresis with polymer matrices can offer 1- to 2-bp (or more generally 3- to 4-bp) resolution up to 400 bp, and sufficient resolution up to 5 kbp.[55] This represents a suitable range for restriction analysis applications, such as cosmid, plasmid, or bacteriophage mapping, and practical examples have been shown.[56,57] Size-dependent separations and accurate fragment size determinations can be obtained; however, sequence-induced anomalous migration effects must be carefully examined, especially in light of the high electric fields used.[58,59] Secondary DNA structures can cause substantial differences between actual and calculated sizes for specific dsDNA fragments. These effects are mainly dependent on concentration of sieving polymer, separation temperature, field strength, and background electrolyte additives. Experimental conditions for reliable size determination of unknown fragments include the use of lower concentration polymers (e.g., 3% T LPA), electric field strength under 200 V/cm, a column temperature of 50°, and the addition of monomeric intercalation dyes.[58] The use of fluorescent intercalating dyes not only greatly reduces these undesirable effects, but also provides high-sensitivity LIF detection.[60]

To extend the upper limit of resolvable DNA fragments, inverted electric field pulses [in analogy to field inversion gel electrophoresis (FIGE)] can be applied. Pulsed-field CE has demonstrated very high resolution up to 100 kb[61] and the potential for megabase DNA separations using ultradilute polymer matrices.[62]

DNA Typing. DNA-typing methods are used for disease diagnosis, sample validation, parentage testing, and forensic identification. Genetic markers employed in these methods include restriction fragment length polymorphisms (RFLPs), minisatellite markers with a variable number of tandem repeats (VNTRs) ranging in length from 17–35 to hundreds of base pairs, and microsatellite short tandem repeats (STRs) whose

[55] M. S. Liu, J. Zang, R. A. Evangelista, S. Rampal, and F. T. Chen, *Biotechniques* **18**, 316 (1995).

[56] Y. Baba, R. Tomisaki, C. Sumita, I. Morimoto, S. Sugita, M. Tsuhako, T. Miki, and T. Ogihara, *Electrophoresis* **16**, 1437 (1995).

[57] K. Kleparnik, Z. Mala, J. Doskar, S. Rosypal, and P. Bocek, *Electrophoresis* **16**, 366 (1995).

[58] J. Berka, Y. F. Pariat, O. Muller, K. Hebenbrock, D. N. Heiger, F. Foret, and B. L. Karger, *Electrophoresis* **16**, 377 (1995).

[59] H. M. Wenz, *Nucleic Acids Res.* **22**, 4002 (1994).

[60] H. P. Zhu, S. M. Clark, S. C. Benson, H. S. Rye, A. N. Glazer, and R. A. Mathies, *Anal. Chem.* **66**, 1941 (1994).

[61] J. Sudor and M. Novotny, *Nucleic Acids Res.* **23**, 2538 (1995).

[62] Y. S. Kim and M. D. Morris, *Anal. Chem.* **67**, 784 (1995).

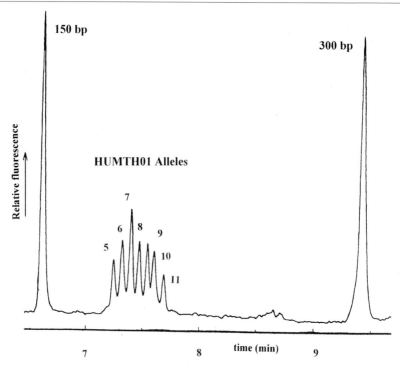

FIG. 5. Rapid separation of HUMTH01 allelic ladder. The 150- and 300-bp markers allow the size determination of each allele. Conditions: l = 37 cm (DB-17 column); 1% HEC–100 mM Tris-borate–2 mM EDTA (pH 8.1); 50 ng/ml YO-PRO-1 (Molecular Probes, Eugene, OR); pressure injection at 0.5 psi for 5 sec water, 45 sec sample; voltage gradient: 0–5.2 min at −15 kV (14.5 μA), 5.2–10 min at −5 kV (4.8 μA); 25°; LIF detection, 520 nm. (Reprinted from Ref. 65, with permission.)

variable repeat length is two to seven nucleotides.[63,64] These polymorphic markers are PCR amplifiable and their sizing by capillary electrophoresis is being studied by several groups. One example is the human HUMTH01 VNTR system, currently being examined for use in human identification. HUMTH01 alleles, ranging in size from 179 to 203 bp, with a 4-bp repeat, can be PCR amplified and separated under 10 min, as shown in Fig. 5.[65] Precise and accurate sizing requires either the use of two internal standards flanking the allele, as shown in Fig. 5 (see also Ref. 66), or the use of

[63] S. Wood and S. Langolis, *J. Chromatogr.* **569,** 421 (1991).
[64] H. A. Hammond and C. T. Caskey, *Methods Mol. Cell. Biol.* **5,** 78 (1994).
[65] D. M. Northrop, B. R. McCord, and J. M. Butler, *J. Cap. Electrophoresis* **2,** 158 (1994).
[66] J. M. Butler, B. R. McCord, J. M. Jung, J. A. Lee, B. Budowle, and R. O. Allen, *Electrophoresis* **16,** 974 (1995).

a two-color assay in which one color codes for the sample and the other for an internal standard of a ladder of the same allele.[67]

Quantitative Analysis of Polymerase Chain Reaction Products. In many biological applications, quantitation of target sequences amplified by PCR is frequently required. Specific applications of quantitative PCR are the detection of viral DNA in hosts infected with HIV or other viruses, or determination of the genome copy number of genes that are amplified in malignant tumors. An automated CE instrument equipped with LIF offers a direct and sensitive post-PCR detection system, in addition to the specificity provided by size determination of PCR products. These advantages utilized to analyze, for example, HIV-1 cDNA or RNA by multitarget PCR and reverse transcription-based PCR (RT-PCR).[68] This work demonstrated the possibility of simultaneous quantitative detection of multiple target sequences together with a molecular weight standard, with linearity between template amounts and PCR products yields. In quantitative PCR, high accuracy can be achieved by using an internal standard in the competitive assay. In this method, the internal control must be well designed and experimentally proven to amplify with the same efficiency as the unknown target. Using this approach, CE/LIF has been demonstrated to be well suited for the analysis and quantitation of a human immunodeficiency virus type 1 (HIV-1) target.[69] It has been further shown that the speed and automation of CE permit rapid checking that the rate of amplification of the internal standard closely matches that of the target.

Mutation Detection Methods. Scanning techniques that would reliably detect any and all possible changes in DNA nucleotide sequences without the need for actual full-length sequencing are in great demand in the field of mutagenesis, human population genetics, and cancer genetics. A battery of slab gel-based techniques is currently used for these purposes, e.g., denaturing gradient gel electrophoresis (DGGE), single-stranded conformation polymorphism (SSCP), heteroduplex analysis (HPA), chemical or enzymatic cleavage of base mismatches (CCM), or RNase A cleavage.[70] These methods differ in terms of detectable mutation fraction and sensitivity to various base substitutions. While in principle all slab gel methods can be done by CE, the first three cited above have been especially modified for CE analysis with polymer solutions.

A particularly powerful approach, related to DGGE, is to maintain a

[67] Y. Wang, J. Ju, B. A. Carpenter, J. M. Atherton, G. F. Sensabaugh, and R. A. Mathies, *Anal. Chem.* **67,** 1197 (1995).

[68] W. Lu, D. S. Han, J. Yuan, and J. M. Andrieu, *Nature (London)* **368,** 269 (1994).

[69] S. Williams, C. Schwer, A. S. M. Krishnarao, C. Heid, and B. L. Karger, *Anal. Biochem.* 1996 (in press).

[70] K. R. Mitchelsonn and J. Cheng, *J. Cap. Electrophoresis* **2,** 137 (1995).

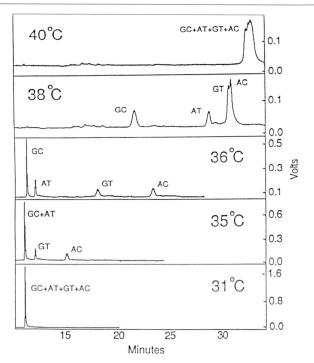

FIG. 6. CDCE separation as a function of column temperature. The sample was prepared using fluorescein-labeled DNA fragments and run on a capillary at the temperatures indicated. An equal amount of sample was injected per run. Column: 6% T linear polyacrylamide–3.3 M urea–20% (v/v) formamide in 1× TBE. $E = 250$ V/cm. (Reprinted from *Nucleic Acids Res.* **22**, 364 (1994), by permission of Oxford University Press.)

constant elevated temperature across the column and add denaturant to the buffer. This approach, called constant denaturant capillary electrophoresis (CDCE), is used to separate partially melted homo- and heteroduplexes.[71] CDCE has been developed for the detection of C → T substitution in a 206-bp human mitochondrial DNA fragment, containing a natural low-melting and high-melting domain (see Fig. 6). At 36°, partial melting of the low-melting domain leads to separation of the four homo- and heteroduplexes. This method has been shown to be very sensitive for single base mismatches (down to less than 1 in 10^6, mutant to wild type[72]) and should be generally applicable for fragments with attached GC clamp. A

[71] K. Khrapko, J. S. Hanekamp, W. G. Thilly, A. Belenkii, F. Foret, and B. L. Karger, *Nucleic Acids Res.* **22**, 364 (1994).
[72] W. Thilly, private communication (1995).

temperature change can also be incorporated into the CE column via Joule heating (i.e., increased electric field) to mimic DGGE.[73]

The ability of nondenaturing linear polyacrylamide to resolve differences in conformations of single-stranded DNA (SSCP) has been demonstrated for point mutations in 372-bp PCR products from the p53 tumor suppressor gene.[74] Improvement in resolving power of the CE/SSCP could be further obtained on addition of glycerol to sieving matrix, as used with slab gels,[75] and two-color LIF can be employed to assist in the identification of the two single strands of the product.[76] Heteroduplex DNA polymorphism is another widely used method, where wild-type and mutant double-stranded fragments are denatured and reannealed, resulting in formation of novel heteroduplex three-dimensional structures in which single base alterations cause conformational changes. Capillary electrophoresis with linear polymer matrices has been employed in several of these types of studies, and single base-mismatched heteroduplexes were separated from homoduplexes in a 125-bp long fragment from the rRNA gene of *H. annosum* strains.[77] More recently, two examples of heteroduplex analysis were shown using 6% (w/v) linear polyacrylamide for detection of the 3-bp variant allele ΔF508 mutation[78] and a tetranucleotide repeat (GATT) polymorphism, both associated with cystic fibrosis.[79] The above examples illustrate the power of capillary electrophoresis for conformation-based DNA separations in mutational analysis. In addition to these methods, mutation detection based on DNA fragment length differences (PCR-RFLP, ARMS, MVR, LCR) have also been readily applied in capillary electrophoresis.[70]

Fraction Collection in Capillary Electrophoresis. Besides being a powerful analytical tool, CE with polymer matrices or open tube also possesses potential as a micropreparative technique. Similar to chromatographic techniques, fractions containing individual DNA fragments, proteins, or other molecules can be collected from the capillary outlet. This collection allows for further analysis, e.g., sequencing, or manipulation (e.g., cloning) of sample fractions of interest.

[73] C. Gelfi, P. G. Righetti, L. Cremonesi, and M. Ferrari, *Electrophoresis* **15,** 1506 (1994).

[74] A. W. H. M. Kuypers, P. M. W. Willems, M. J. Vanderschans, P. C. M. Linssen, H. M. C. Wessels, C. H. M. M. Debruijn, F. M. Everaerts, and E. J. B. Mensink, *J. Chromatogr.* **B621,** 149 (1993).

[75] J. Cheng, T. Kasuga, N. D. Watson, and K. R. Mitchelson, *J. Cap. Electrophoresis* **1,** 24 (1995).

[76] K. Hebenbrock, P. M. Williams, and B. L. Karger, *Electrophoresis* **16,** 1429 (1995).

[77] J. Cheng, T. Kasuga, K. R. Mitchelson, E. R. T. Lightly, N. D. Watson, W. J. Martin, and D. Atkinson, *J. Chromatogr.* **A677,** 169 (1994).

[78] C. Gelfi, P. G. Righetti, V. Brancolini, L. Cremonesi, and M. Ferrari, *Clin. Chem.* **40,** 1603 (1994).

[79] M. Nesi, P. G. Righetti, M. C. Patrosso, A. Ferlini, and M. Chiari, *Electrophoresis* **15,** 644 (1994).

In the simplest case, the fractions can be collected into a microvial on electromigration from the separation capillary. The separation is interrupted at the time the zone is calculated to exit the capillary. The capillary end is then moved from the buffer reservoir into the collection vial, and electric field or pressure is applied to elute the zone from the capillary. This procedure is repeated for other fractions to be collected. Although the collected fractions can be significantly diluted, the collected amount (often in the nanomole to picomole range) is sufficient for further analysis such as microsequencing or PCR amplification.[80] When a larger amount of collected material is required, the procedure can be automated and repeated several times. The advantage of this approach is that commercial instruments can be used without modification. A disadvantage is the uncertainty in prediction of elution times because the migrating zones are typically detected more than 70 mm prior to the capillary exit and their elution time is calculated from the migration speed and capillary length. Also, only several peaks can be collected from an individual run.

A different approach, introduced in the eary 1970s for capillary isotachophoresis (Tachofrac; LKB, Bromma, Sweden) and redeveloped later for CE, utilizes a special collection membrane.[81] In this case, the collection end of the CE capillary is in contact with a wetted surface of a moving collection membrane [e.g., cellulose acetate, polyvinylidene difluoride (PVDF)]. After separation, the collected "spots" can be detected by standard staining procedures or eluted from the membrane for further analysis or use. The advantage of this procedure is that the separation is a single, uninterrupted process.

Another approach for continuous fraction collection uses a sheath of the collection buffer for continuous elution of separated zones. In this arrangement the CE capillary is inserted into a sheath tube supplied with a continuous flow of the collection buffer.[82] The signal of the detector from an optical fiber 10 mm from the exit was used for computer control of a stepper motor-operated fraction collector. The application of the device was demonstrated for precise automated collection of 11 fragments of a ϕX174 plasmid DNA digest separated by capillary electrophoresis using a replaceable sieving matrix. This was followed by PCR analysis of collected fractions.[82] The use of low-volume (20 μl) glass capillaries as collection vials provided for easy handling of submicroliter fraction volumes, resulting in relatively low sample dilution. The approach alleviates the difficulty of

[80] R. Grimm, *J. Cap. Electrophoresis* **3**, 111 (1995).
[81] W. J. Warren, Y. F. Cheng, and M. Fuchs, *LC-GC* **12**, 22 (1994).
[82] O. Muller, F. Foret, and B. L. Karger, *Anal. Chem.* **67**, 2974 (1995).

the manual process of removing sections of the gel from a slab and extruding sample from the gel.

Capillary Electrophoresis of Sodium Dodecyl Sulfate–Protein Complexes

As noted earlier, slab gel electrophoresis is one of the most frequently employed techniques for protein separation and analysis based on the migration of protein–sodium dodecyl sulfate (SDS) complexes through a polyacrylamide gel, i.e., SDS–PAGE. The technique relies on the fact that in the denatured and reduced state, most proteins will bind a constant amount of SDS per unit weight (1.4 g of SDS per gram of protein).[2,3] The binding is based mainly on hydrophobic interactions and results in formation of detergent–protein complexes of approximately constant charge per unit mass. When electrophoresed on a polyacrylamide gel, the SDS–protein complexes, possessing a similar shape, migrate in order of increasing molecular weight. With appropriate calibration, molecular weights can be determined with an accuracy of approximately 10%. Although the technique has been widely used for many years, the procedure is quite time and labor intensive, and only semiquantitative results can be obtained by densitometry of the stained protein spots.[3]

It was logical to develop CE procedures for separation of SDS–protein complexes to overcome the above problems with the slab gel approach. Not only could the analysis be fast, but precise quantitative analysis, including minor impurities, could be attained. Thus, in the initial phases of CE development, the transfer of slab gel and buffer conditions was explored. In early work, the resolution in polyacrylamide gel-filled capillaries proved to be identical to the standard PAGE technique, and typical linear plots of electrophoretic mobility vs log of the molecular weight of the respective proteins were obtained.[4] This work also revealed two disadvantages of cross-linked polyacrylamide gels in capillaries: the previously mentioned difficulty in maintaining stable columns and the limited sensitivity due to the high absorbance of the gel at 210 nm. The use of solutions of linear polyacrylamide as a replaceable sieving matrix was briefly examined to improve stability[13]; however, it was soon substituted by new low-viscosity UV-transparent replaceable matrices.[13] The new matrices allowed for on-column detection at 214 nm, making analysis of trace impurities below the 0.1% level feasible. Several kits for capillary electrophoresis of SDS–protein complexes are now commercially available from commercial sources (see Table I).

Sample preparation procedures for use with these kits are identical to those with standard SDS–PAGE. In general, pressure injection is employed for SDS–protein separations, and capillaries with the inner wall either dynamically or covalently coated with a hydrophilic polymer are used. Separation buffers typically contain 1–10% (w/v) of a soluble dissolved replaceable sieving polymer such as dextran, polyethylene oxide, or polyethylene glycol, or a proprietary matrix. The analysis typically is complete in 10 min or less with an applied electric field of 300–800 V/cm (see Fig. 1B). The reproducibility of migration times of standard substances can be better than 0.5% when the separation matrix is replaced after each analysis. For some samples, the use of an internal standard can improve precision of migration times and peak areas. In cases in which "dirty" samples are injected and impurities adsorb on the capillary wall, a short (1 min) wash with 1 M HCl between runs is recommended.

The accuracy of molecular weight determinations is typically comparable with the accuracy obtained with slab gel procedures, i.e., around 10%. Notable exceptions to this result include highly basic proteins, e.g., histones, with a high net positive charge counterbalancing the charge on the SDS molecules or highly hydrophobic membrane proteins capable of binding larger amounts of SDS.

Another large class of proteins that produces an inaccurate estimation of MW from standard calibration plots consists of the glycoproteins. In this case, a substantial portion of the protein mass is contributed by the carbohydrate residues, which do not bind to SDS. However, an accurate MW of glycoproteins can be obtained using a Ferguson plot.[3] This method is based on plotting the logarithm of electrophoretic mobility of a specific protein vs gel or polymer concentration. The slope of the resulting linear plot yields the retardation coefficient (R), which is proportional to (log MW)^2 of the analyzed protein. From a Ferguson plot of a set of standard proteins, the molecular weight of the unknown glycoprotein can be determined. Interestingly, the use of standard slab gel procedures for this type of measurement is not popular because it requires casting multiple gels with differing concentration. However, capillary electrophoresis with replaceable sieving matrices allows fully automated procedures for analysis of glycoproteins using Ferguson plots. In this case, a set of solutions of a sieving matrix with increasing concentrations can be loaded into the buffer tray of the automated CE analyzer, and the measurement of the electrophoretic mobility of the protein is then automatically made under computer control. An example of an application of SDS capillary electrophoresis for Ferguson analysis of a complex glycoprotein sample can be found in Ref. 83.

[83] W. E. Werner, D. M. Demorest, and J. E. Wiktorowicz, *Electrophoresis* **14,** 759 (1993).

Today, separation and analysis of SDS–protein complexes is a standard that can be easily implemented on a CE instrument. Of course, molecular weight determination by mass spectrometry is more precise (~0.1%) and requires even less material than CE. Nevertheless, SDS–protein analysis by CE provides different separation principles (molecular weight) than other CE modes (charge, pI) and is therefore useful for purity assessment. In addition, the solubilizing power of SDS means that protein recoveries are high. In quality control operations for recombinant proteins, the CE approach can replace laborious SDS–PAGE procedures.

Conclusion

The breadth of applications of CE with polymer matrices has been demonstrated in this chapter. The use of replaceable matrices is critical to long-term column useage in an automated fashion. The type, MW, and concentration of the polymer play an important role in the separation achieved. Capillary electrophoresis is predicted to become an important tool for DNA sequencing, sizing, and mutation analysis in the forthcoming years.

Acknowledgment

The authors thank the NIH under Project GM15847 and the DOE under Human Genome Project Grant DE-FG02-90ER 60985 for support of this work. This is Contribution No. 669 from the Barnett Institute.

[14] Glycoconjugate Analysis by Capillary Electrophoresis

By MILOS V. NOVOTNY

Introduction

Throughout the history of biochemical research, the major steps in creating new knowledge of biochemical processes and structures were almost invariably preceded by the development of new analytical methodologies. This has been particularly evident in the long-term involvement of biochemical researchers and analysts with chromatography and electrophoresis. Since the 1970s, numerous scientific advances in biochemistry have been further accelerated by the availability of high-performance liquid chromatography (HPLC), a major instrumental and methodological improvement in terms of separation efficiency, speed of analysis, and precise measurement capabilities. Since the early 1980s, high-performance capillary electrophoresis (HPCE) has also evolved into a powerful bioanalytical methodology that is now the electrophoresis counterpart of HPLC.

The early pioneering studies in HPCE provided the basis for extensive development of commercial instrumentation, column studies, new detection capabilities, and numerous applications across an amazing range of molecular sizes, from small ions to large biopolymers. During the past few years, HPCE has also been applied to carbohydrates. Although HPCE of glycoconjugates has not yet reached its methodological maturity, the initial studies in the area permit a highly optimistic view of the likelihood that this powerful method may play a central role in the future of the field of glycobiology. Owing to its unprecedented separating power, HPCE can deal effectively with complex mixtures of glycoconjugates that were previously difficult to separate chromatographically. When empowered with ultrasensitive means of detection, such as laser-induced fluorescence or electrospray mass spectrometry, HPCE becomes capable of addressing issues such as the role of glycoproteins in biomolecular recognition, receptor biochemistry, blood coagulation, and repair mechanisms, among others. Owing to its quantitative capabilities, speed of analysis, and easy automation, the method has the potential to be attractive in clinical research and the diagnosis of carbohydrate-related metabolic disorders.

While the methodological aspects of the HPCE of glycoconjugates are comprehensively addressed elsewhere,[1] the orientation of this chapter is mainly toward a practicing biochemical researcher. The chapter emphasizes techniques and methodological developments from the laboratory of the author. First, certain aspects and goals of analytical and structural glycobiology are briefly outlined, and a brief review of the contemporary HPCE instrumentation suitable for the purpose is provided. The issues of glycoform separation raised by the studies of native glycoproteins are addressed, but the major emphasis of this chapter is on glycan components, as the applications and general problems of HPCE in separating proteins have been dealt with elsewhere.[2] Because fluorescence-tagging procedures are becoming central to the effective use of HPCE in glycoconjugate research, the current approaches in this area is reviewed and their applications demonstrated. Finally, an important extension of the current techniques to polysaccharides is discussed.

[1] M. V. Novotny and J. Sudor, *Electrophoresis* **14**, 373 (1993).
[2] M. V. Novotny, K. A. Cobb, and J. Liu, *Electrophoresis* **11**, 735 (1990).

Analytical Issues of Glycobiology

The ubiquity and numerous ways in which nature uses saccharides in biological processes are paralleled in the structural intricacies of glycoconjugate molecules. From the analytical–chemical viewpoint, the major problems of the field are compounded by the complexity of carbohydrate mixtures, either encountered directly or generated by a controlled degradation of the studied biopolymers; and by the limited capacity of an unmodified saccharide molecule to generate a strong, easily recognizable spectroscopic signal for detection. The first problem can be overcome by the use of a separation technique with great resolving power: capillary electrophoresis with its efficiency in an excess of 10^6 theoretical plates is thus attractive for the purpose. It is essential, however, that the analyzed saccharides possess an electric charge for migration. Borate complexation can easily convert most hydroxylated structures to the electrophoretically suitable solutes,[1,3] and in addition, some glycoconjugates (such as glycosaminoglycans) possess a negative charge in their molecules already. To detect carbohydrates at the levels compatible with HPCE,[4] it is essential to convert them to derivatives exhibiting adequate ultraviolet (UV) absorbance or fluorescence properties.

The demands for component resolution and measurement sensitivity may vary with the nature of the glycoprotein under investigation. Certain determinations will undoubtedly require the highest sensitivity that an analytical technique can offer. Processes as important as antigen–antibody interaction, cell-to-cell recognition, protein targeting, and functions of the receptor proteins have been poorly understood because of the limited availability of the material collected for structural analysis. Because protein glycosylation is largely a posttranslational event, biological amplification is not feasible. In contrast, somewhat less stringent sensitivity requirements may exist in certain cases of biotechnologically produced glycoprotein pharmaceuticals. Here, the resolving power of HPCE is likely to be the major advantage of this method in separating various glycoforms of a native glycoprotein, or in mapping the oligosaccharides.

During efforts to elucidate the relationship between bioactivity and the structure of a glycoprotein, or of its glycoforms, it becomes essential to determine the sites of attachment of its oligosaccharide chains to the polypeptide backbone and to characterize the oligosaccharide class (N-linked vs O-linked; high mannose; complex; hybrid; biantennary, triantennary;

[3] S. Honda, S. Iwase, A. Makino, and S. Fujiwara, *Anal. Biochem.* **176**, 72 (1989).
[4] J. W. Jorgenson, *in* "New Directions in Electrophoretic Methods" (J. W. Jorgenson and M. Phillips, eds.), pp. 182–198. American Chemical Society, Washington, D.C., 1987.

etc.) at a specific site. To begin with, glycoproteins must first be isolated from biological materials by means of dialysis, preparative chromatography, gel electrophoresis, isoelectric focusing, lectin binding, etc., or, in the most typical case, through an effective combination of these tools. Owing to the great variety of glycoprotein structures and analytically different situations, it would be illusory to provide a general scheme or a general isolation strategy. It should, however, be noted that the isolation step, if improperly chosen, could become a bottleneck to the overall high-sensitivity carbohydrate analysis. Although this chapter primarily emphasizes the use of HPCE as a powerful end method in analytical and structural glycobiology, one also must stress the importance of appropriate isolation techniques for minimizing the losses that occur through sample degradation and adsorption. Whenever feasible, we advocate[5] sample treatment procedures that involve a minimum number of transfers, the use of microcolumn separation techniques, or miniaturized systems, in general.

Once a glycoprotein has been isolated, it becomes feasible to use capillary isoelectric focusing or zone electrophoresis[6-8] in assessing the complexity of its various glycoforms. For a further structural determination, specific or general degradation steps may be applied: (1) a site-specific proteolysis (tryptic or chymotryptic digest, cleavage with cyanogen bromide, etc.), yielding a mixture of glycopeptides for further characterization, (2) removal of oligosaccharides from the polypeptide through a glycanase, or hydrazine, treatment followed by development of the oligosaccharide map, and (3) chemical hydrolysis yielding a monosaccharide mixture for further separation (compositional analysis). Alternatively, various of these approaches can be combined in micropreparative separations for subsequent structural analysis, a site-of-glycosylation determination,[9] selective sample derivatization, etc. Alternatively, sugar-specific or bond-specific enzymes could be applied in conjunction with highly sensitive analytical tools, such as HPCE/laser-induced fluorescence, to yield the structural details of a given glycan. Figure 1 shows a possible scheme[10] that may be representative, but by no means exhaustive, of the methods of glycoprotein analysis. Specific examples are given as a followup, after the discussion of instrumental aspects of HPCE.

High-performance capillary electrophoresis, with its increasing range of detection and ancillary techniques, is likely to become both complemen-

[5] M. Novotny, *J. Microcol. Sep.* **2**, 7 (1990).

[6] F. Kilár and S. Hjertén, *Electrophoresis* **10**, 23 (1989).

[7] K. W. Yim, *J. Chromatogr.* **559**, 401 (1991).

[8] J. P. Landers, B. J. Madden, R. P. Oda, and T. C. Spelsberg, *Anal. Biochem.* **205**, 115 (1992).

[9] O. Shirota, D. Rice, and M. Novotny, *Anal. Biochem.* **205**, 189 (1992).

[10] O. Shirota, Ph.D. thesis. Indiana University, Bloomington, Indiana, August 1992.

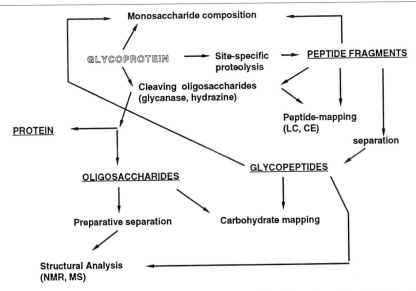

F<small>IG</small>. 1. A representative scheme for glycoprotein analysis. (Reproduced from Ref. 10 with permission of the author.)

tary to and competitive with the more established or previously reported methods of glycoconjugate analysis: gas chromatography/mass spectrometry,[11] HPLC using aminopyridyl precolumn labeling[12,13] or pulsed amperometric detection,[14,15] fast atom bombardment[16,17] and electrospray mass spectrometry, supercritical fluid chromatography/mass spectrometry,[18] and high-field nuclear magnetic resonance.[19] At this stage of development, it is hardly surprising that the most elegant structural studies documented in the literature come from glycoproteins bearing little relevance to "real-world" problems, as they were chosen as model systems primarily because of their availability in fairly large quantities.

[11] J. F. Rocca and J. Rouchousse, *J. Chromatogr.* **117,** 216 (1976).

[12] S. Hase and T. Ikenaka, *Anal. Biochem.* **184,** 135 (1990).

[13] S. Hase, H. Oku, and T. Ikenaka, *Anal. Biochem.* **167,** 321 (1987).

[14] D. C. Johnson, *Nature (London)* **321,** 451 (1986).

[15] M. R. Hardy and R. R. Townsend, *Proc. Natl. Acad. Sci. U.S.A.* **85,** 3289 (1988).

[16] A. S. B. Edge, A. Van Langenhove, V. Reinhold, and P. Weber, *Biochemistry* **25,** 8017 (1986).

[17] F. Maley, R. B. Tremble, A. L. Tarentino, and T. H. Plummer, *Anal. Biochem.* **180,** 195 (1989).

[18] J. Kuei, G. P. Her, and V. N. Reinhold, *Anal. Biochem.* **172,** 228 (1988).

[19] D. A. Cumming, C. G. Hellerqvist, H. Harris-Brandts, S. W. Michnick, J. P. Carver, and B. Bendiak, *Biochemistry* **28,** 6500 (1989).

Instrumental Aspects of High-Performance Capillary Electrophoresis

In principle, capillary electrophoresis (CE) is a relatively simple technique. Its essential element is the separation capillary (up to 1.0 m in length, but most often shorter). The separation capillaries used in carbohydrate analysis are either untreated fused silica tubes, or the columns whose inner surface has been chemically modified. In the great majority of glycoconjugate separations, constant voltage is employed. However, the use of pulsed fields, controlled through a suitable pulse generator, may become of use in the separations of certain polysaccharides.[20] Most CE home-built setups, now in use in numerous research laboratories, have been modeled according to the system described by Jorgenson and Lukacs[21] in 1981. Commercial versions of CE are now also widely available. For the benefits of carbohydrate analysis, different detection techniques may be applied.

As demonstrated first by Jorgenson,[4] capillary zone electrophoresis can attain its separation efficiency only if the input sample mass and volume are kept at low levels. In the common analytical practice of HPCE, this roughly translates into measuring nanogram-to-picogram quantities per component as the maximum amounts in 50- to 80-μm i.d. separation capillaries. Miniaturized UV absorbance detectors, being the most commonly used concentration-sensitive devices of adequate mass sensitivity in microcolumn liquid chromatography,[22] work sufficiently well in HPCE. Thus, detection of glycoproteins and their peptide fragments through UV absorbance at 205–220 nm is just as feasible as with other proteins. Ultraviolet detection of underivatized carbohydrates at 195 nm in HPCE is likely to be confined to the simplest applications, although, as shown by Hoffstetter-Kuhn et al.,[23] borate complexation enhances the absorbance signal somewhat. Sugar derivatization for increased UV absorption is also feasible, as demonstrated with the use of 2-aminopyridine,[3] 6-aminoquinoline,[24] and 3-methyl-1-phenyl-2-pyrazolin-5-one.[25] Considerable attention is still being paid to the development of new detection techniques for HPCE, and it is likely that future improvements in the area will aid glycoconjugate analysis.

Among the detection techniques for HPCE of carbohydrates, three measurement principles appear particularly attractive in terms of high sensitivity: (1) laser-induced fluorescence, (2) electrochemical (amperometric)

[20] J. Sudor and M. V. Novotny, Proc. Natl. Acad. Sci. U.S.A. **90,** 9451 (1993).

[21] J. W. Jorgenson and K. D. Lukacs, Anal. Chem. **53,** 1298 (1981).

[22] M. Novotny, in "Microcolumn High-Performance Liquid Chromatography" (P. Kucera, ed.), pp. 194–259. Elsevier Science Publishers, Amsterdam, 1984.

[23] S. Hoffstetter-Kuhn, A. Paulus, E. Gassman, and H. M. Widmer, Anal. Chem. **63,** 1541 (1991).

[24] W. Nashabeh and Z. El Rassi, J. Chromatogr. **514,** 57 (1990).

[25] S. Honda, S. Suzuki, A. Nose, K. Yamamoto, and K. Kakehi, Carbohydr. Res. **215,** 193 (1991).

detection, and (3) mass spectrometry. Pulsed amperometric detection of sugars[14,15] is a common and effective method used in HPLC and, according to a communication from Colon and co-workers,[26] other electrochemical detection technologies are also applicable to HPCE. Naturally, the use of mass spectrometry extends well beyond detection to the actual structural information on glycoconjugates.

Laser-based detectors have become of much interest to the users of miniaturized separation systems, including HPCE, as the highly collimated laser radiation can be easily focused into a small area. When there is a good match between the radiation of a laser and the spectral properties of the measured solutes, laser-induced fluorescence (LIF) provides a significant improvement in detection sensitivity compared to conventional fluorescence. A typical instrumental setup for HPCE with laser-induced fluorescence detection used in carbohydrate analysis in our laboratory has evolved[27] as a combination of instruments reported by Diebold and Zare[28] at Stanford University (Stanford, CA) and the Yeung group[29] at Iowa State University (Ames, IA). The helium/cadmium laser has been used here as a light source either in its UV mode (325 nm) or the blue mode (442 nm). Through the appropriate optical components, the laser beam is focused onto the end section of a fused silica capillary. The incident beam is passed through a mechanical light chopper and aligned to its optical position on the flow cell by a positional holder. Fluorescence emission is typically collected through a 600-μm fiber optic situated at a right angle to the incident beam. Instrumental aspects of CE/LIF are described in more detail in Chapter 19, volume 270.[29a]

Using chemical species that feature excitation maxima at, or near, the outputs of the most readily available lasers confers a strong advantage on the researcher. From a practical point of view, the modern versions of the helium/cadmium and the air-cooled argon-ion laser are the most desirable match for work with HPCE or the miniaturized forms of chromatography at excitation wavelengths below 500 nm. The intensity of the fluorescence signal produced is directly proportional to that of the laser light illuminating the cell. However, it is not advisable to use powers of more than a few milliwatts in laser-induced fluorescence measurements; most fluorophores tend to photobleach at high intensities and, furthermore, the scattered light may cause some noise enhancement.

[26] L. A. Colon, R. Dadoo, and R. N. Zare, *Anal. Chem.* **65,** 476 (1993).
[27] J. Gluckman, D. Shelly, and M. Novotny, *J. Chromatogr.* **517,** 443 (1984).
[28] G. Diebold and R. N. Zare, *Science* **196,** 1439 (1977).
[29] M. Sepaniak and E. Yeung, *J. Chromatogr.* **190,** 377 (1980).
[29a] T. L. Lee and E. S. Yeung, *Methods Enzymol.* **270,** Chap. 19, 1996.

With the most elaborate laser-detector designs, it has become feasible to achieve sensitivities at the level of single molecules[30] confined to very small sample volumes. An ordinary detection system represents a compromise, emphasizing simplicity and low cost, rather than the maximum achievable sensitivity. However, depending on the fluorophore utilized, detection limits in the subfemtomole-to-subattomole range are quite easy to achieve.[31]

A key component of the HPCE system is the separation column itself. For most analytical work with glycoconjugates, the inner column diameters of fused silica capillaries are around 50–80 μm, although departures from this range may be more often encountered in future studies. Very small column diameters are of interest in connection with manipulating extremely small samples, as for the analysis of single biological cells[32]; naturally, ultrasensitive detection is mandatory. Capillaries with greater inner diameters (100–150 μm) are desirable for efforts at micropreparative isolation and for further investigation of the trapped fragments by physical or chemical means. It should be emphasized that departures from the usual column diameters are not without certain penalties. For both open tubes and gel-filled capillaries in HPCE, the column diameter has an appreciable effect on column performance.[33] For larger columns, slow dissipation of the Joule heat generated during electrophoresis can have adverse effects on both the component resolution and integrity of biological samples.

In the current practice of HPCE, three types of column media are of importance to biomolecular separations: (1) free buffer media, (2) gel-filled capillaries, and (3) entangled polymer solutions. Each type can be crucial to success for a particular application involving glycoconjugates. Simple sugar mixtures, such as those encountered in a typical mono- or disaccharide analysis, can easily be separated in free buffer media (the open tubular format) or by means of electrokinetic (micellar) capillary chromatography.[34] Because differences in hydrophobicity of the individual sugars appear to be insufficient alone for effective resolution, additional buffer additives may be required[35] when the electrokinetic capillary mode of separation is used. Strategies for separation are typically contingent on the type of sugar derivative selected for analysis.[1] Because most sugars are uncharged, neutral molecules with pK values around 11.0, borate complexation is usually utilized to impart a negative charge to the solutes of interest. Serendipitously,

[30] Y. F. Cheng and N. J. Dovichi, *Science* **242,** 562 (1988).
[31] J. Liu, O. Shirota, D. Wiesler, and M. Novotny, *Proc. Natl. Acad. Sci. U.S.A.* **88,** 2302 (1991).
[32] J. B. Chien, R. A. Wallingford, and A. G. Ewing, *J. Neurochem.* **54,** 633 (1990).
[33] J. P. Liu, V. Dolnik, Y.-Z. Hsieh, and M. Novotny, *Anal. Chem.* **64,** 1328 (1992).
[34] S. Terabe, *Trends Anal. Chem.* **8,** 129 (1989).
[35] J. Liu, O. Shirota, and M. Novotny, *Anal. Chem.* **63,** 413 (1991).

the negatively charged saccharides, small or large, are generally repelled by the negatively charged surfaces of the system, very much like the naturally negative oligonucleotides. Consequently, the surface adsorption problems in HPCE are much less severe for carbohydrate analysis than, for example, protein and peptide separations. Unfortunately, glycoproteins may still suffer from the general problems of protein surface adsorption, as discussed elsewhere.[2]

In separation capillaries with unmodified surfaces, strong electroosmosis may sometimes complicate the resolution of even moderately sized oligosaccharides. Borate complexation may also result in unfavorable mass-to-charge ratios. The use of a surface-modified capillary improves the resolution of lower oligosaccharides separated as pyridylamine derivatives.[36] The problem of unfavorable mass-to-charge ratios for higher oligosaccharides can be solved either through the use of gel-filled capillaries,[37,38] or suitable buffer media.[39]

Gel-filled capillaries were introduced to HPCE by Hjertén et al.[40] and Cohen et al.[41] The matrix most widely used is a cross-linked polyacrylamide, and concentration of the monomer and cross-linker (percent T and percent C, respectively, according to the terminology of Hjertén et al.[40]) must be carefully controlled. Using specialized polymerization techniques, numerous laboratories have been able to overcome the problems of polymer shrinkage during column preparation, and gel-filled capillaries have also become available commercially. The most important aspect of these media is their capacity to resolve complex mixtures according to molecular weight. The most widely used monomer and cross-linker concentrations are 3–5% T, and 1–3% C; however, a considerably larger percentage of monomer (high-T capillaries) is essential in oligosaccharide separations.

Electrophoretic separation of a mixture of oligosaccharides in a polyacrylamide column is shown in Fig. 2, together with the relation between the migration times and degree of polymerization.[37] The migration experienced in gels is highly predictable, allowing a fairly accurate estimation of the oligomeric ranges. Figure 3,[38] obtained with an 18% T (3% C) gel and a hydrolytically cleaved sample of a polygalacturonic acid, is indicative of

[36] S. Honda, A. Makino, S. Suzuki, and K. Kakehi, *Anal. Biochem.* **191**, 228 (1990).
[37] J. Liu, O. Shirota, and M. Novotny, *J. Chromatogr.* **559**, 223 (1991).
[38] J. Liu, O. Shirota, and M. Novotny, *Anal. Chem.* **64**, 973 (1992).
[39] M. Stefansson and M. Novotny, *Anal. Chem.* **66**, 1134 (1994).
[40] S. Hjertén, K. Elenbring, F. Kilár, J.-L. Liao, A. J. C. Chen, C. J. Siebert, and M. D. Zhu, *J. Chromatogr.* **403**, 47 (1987).
[41] A. S. Cohen, D. R. Najarian, A. Paulus, A. Guttman, J. A. Smith, and B. L. Karger, *Proc. Natl. Acad. Sci. U.S.A.* **85**, 9660 (1988).

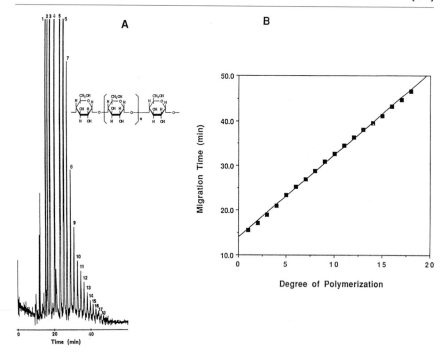

FIG. 2. (A) Electrophoretic separation of partially hydrolyzed Dextrin 15 in a polyacryl-amide gel-filled capillary. (B) Correlation between molecular masses of the maltodextrin oligomers and their migration times. (Reprinted from *J. Chromatogr.*, **559**, J. Lui, O. Shirota, and M. Novotny, p 223, Copyright 1991 with kind permission of Elsevier Science–NL, Sara Burgerhartstraat 25, 1055 KV Amsterdam, The Netherlands.)

the general range and separation efficiency of high-T, gel-filled capillaries. It should be emphasized that in the preparation of such concentrated gels, the isotachophoretic polymerization technique[42] has been essential.

Very large glycoconjugate molecules, such as various polysaccharides and glycosaminoglycans, cannot be passed through the tightly arranged and cross-linked polyacrylamide matrices, yet their separation according to molecular weight is highly desirable. The use of entangled polymer solutions has now become increasingly popular in separating large oligonu-cleotides[43] and proteins.[44,45] In viscous solutions of various polymers, a "dynamic sieving effect" is encountered,[43] which allows relatively unhin-

[42] V. Dolnik, K. A. Cobb, and M. V. Novotny, *J. Microcol. Sep.* **3**(2), 155 (1991).
[43] P. D. Grossman and D. S. Soane, *J. Chromatogr.* **559**, 257 (1991).
[44] K. Ganzler, K. S. Greve, A. S. Cohen, B. L. Karger, A. Guttman, and N. C. Cooke, *Anal. Chem.* **64**, 2665 (1992).
[45] D. Wu and F. E. Regnier, *J. Chromatogr.* **608**, 349 (1992).

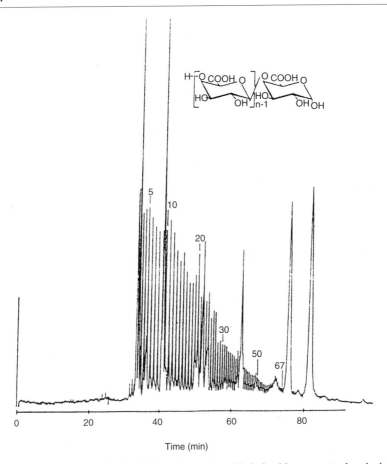

Fɪɢ. 3. Separation of derivatized oligogalacturonic acids derived from an autoclave hydrolysis. The numbers indicate the oligomers differing by a sugar unit. (Reproduced from J. Lui, O. Shirota, and M. Novotny, *Anal. Chem.*, **64,** 973. Copyright 1992 American Chemical Society.)

dered migration of the charged solutes through uncharged, hydrophilic and flexible polymer networks. Such media have shown some promise[20] in separating polysaccharides.

Some attention has been given to achieving satisfactory CE separations of oligosaccharides without the use of a gel medium. Under the optimized conditions of buffer pH, ionic strength, and the nature of tagging molecules, efficiencies in the excess of 1 million theoretical plates per meter can be achieved,[39] enabling some separations of the branched oligomeric species.[46]

[46] M. Stefansson and M. Novotny, *Carbohydr. Res.* **258,** 1 (1994).

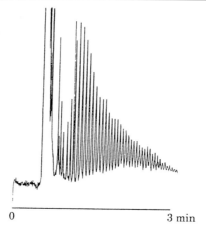

0   3 min

FIG. 4. High-speed separation of corn amylose. Conditions: −750 V/cm (15 μA) using 0.1 M borate–Tris at pH 8.65 as the electrolyte. The effective length of the column was 15 cm. (Reproduced from M. Stefansson and M. Novotny, *Anal. Chem.*, **66**, 1134. Copyright 1994 American Chemical Society.)

The resolving power and speed of such systems are illustrated in Fig. 4[39] with the example of amylose oligomers. Moreover, suitable buffer additives were found effective[47] in modifying the electrophoretic mobility of neutral and charged oligosaccharides: with water-soluble cellulose derivatives, hydrophobic adsorption of detergents on the analyte molecules governs migration, while cationic ion-pairing reagents are effective in moderating highly charged polysaccharides (e.g., heparins).

Mass spectrometry (MS), with its various ionization methods, has traditionally been among the key techniques used for the structural elucidation of proteins and carbohydrates. Various types of mass spectrometers, including quadrupole mass analyzers, ion traps, time-of-flight mass analyzers, and sector instruments, have now been shown technically feasible for HPCE/MS coupling. Besides the accurate molecular weight information supplied by MS, the tandem (MS/MS) instruments can often solve the primary structures of both peptides and oligosaccharides. Thus far, applications to glycoconjugates are much less frequent than those in protein chemistry.

The electrospray ionization principle has truly revolutionized modern MS of biological molecules through its inherent sensitivity and ability to record large molecular entities within a relatively small mass scale. Correspondingly, researchers in the area have reoriented much of their attention to the coupling of HPCE to electrospray MS. The technical aspects of

[47] M. Stefansson and M. Novotny, *Anal. Chem.* **66**, 3466 (1994).

HPCE/electrospray MS are dealt with in Chapter 21, volume 270.[47a] The potential of electrospray MS in glycobiology is exemplified by Duffin *et al.*[48] with both the intact glycoprotein (ovalbumin) and its *N*-glycanase digests being investigated. The oligosaccharides containing sialic acids were primarily studied in the negative-ion mode of detection, while the remaining glycans were cationized by adding sodium acetate or ammonium acetate to the sample for the positive-ion detection. Cleavages of the individual glycosidic bonds were observable in the tandem (MS/MS) instrumental arrangement, yielding much information on the sequence of sugar units. At this stage, the methodology does not differentiate the isomeric sugars and linkage positions. Other ionization and mass separation techniques must be further developed for the CE/MS of carbohydrate derivatives to defuse frustration with perfectly separated, but structurally ill-defined mixtures.

Fluorescence-Tagging Strategies

Fluorescent labeling techniques are common in various studies of biomolecules, because their intrinsic fluorescence is usually weak. Derivatization of carbohydrates with a fluorophoric moiety is particularly appealing, because the spectroscopic properties of carbohydrates are generally nondistinctive, and so for analytical purposes some sort of label is required anyway. Fluorescent labeling finds its optimum utilization when the excitation maximum coincides with the wavelength of a laser source.

Different glycoconjugates possess a variety of hydroxy groups, which, at first glance, would seem to present natural sites for derivatization. In practice, however, this does not appear to work owing to the different reactivities within such groups, the incidence of multiple (and incomplete) tagging, and potential complications arising from the advanced structure of complex carbohydrates. Amino sugars, however, present easily taggable moieties for a variety of reagents that are applicable to high-sensitivity measurements of primary amines. Carbohydrates with a reducing end can be converted to amino derivatives through reductive amination at the analytical scale.[31,37] The reaction that follows with an amine-selective reagent then leads to a suitable derivative. The end groups of certain saccharide structures can also be derivatized directly: the carbonyl groups can be attached to a chromophore through the Schiff-base mechanism, while the open structural form of sialic acids reacts with an aromatic diamine to form a fluorescent aromatic system.[49]

[47a] C. H. Whitehouse and F. Banks, Jr., *Methods Enzymol.* **270,** Chap. 21, 1996.
[48] K. L. Duffin, J. K. Welply, E. Huang, and J. D. Henion, *Anal. Chem.* **64,** 1440 (1992).
[49] O. Shirota, D. Wiesler, and M. Novotny, 1996 (in press).

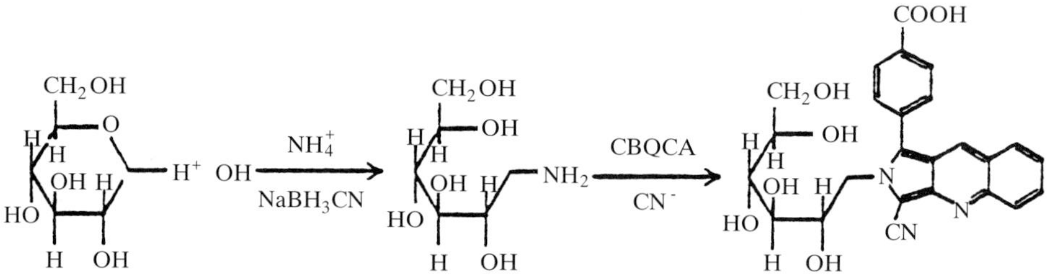

SCHEME I

3-(4-Carboxybenzoyl)-2-quinolinecarboxaldehyde (CBQCA) was developed in our laboratory as a reagent for the ultrasensitive detection of primary amines by laser fluorescence. During its reaction with primary amines, a highly fluorescent isoindole is formed, whose excitation maximum is near the 442-nm blue line of the helium/cadmium laser and the 456-nm secondary line of the argon laser. There is a conveniently large Stokes shift associated with the derivative, as the emission maximum occurs at approximately 550 nm. The reagent has also been effective in work with amino sugars[35] and reductively aminated reducing, neutral sugars.[31] In the derivatization Scheme I used for the latter compound type, sodium cyanoborohydride in the presence of an ammonium salt first produces the amino group, which is, in turn, tagged with CBQCA, a reactive ketoaldehyde.

The high sensitivity of this fluorogenic labeling method for sugars is demonstrated in Fig. 5 with a run of three simple monosaccharides[31] at the low attomole level (several nanoliters of a 3×10^{-9} M solution being introduced into the capillary). With various types of separation capillaries, it has been feasible to use the CBQCA tagging methodology to determine components of the complex mixtures of monosaccharides and the oligosaccharides originating from a controlled degradation of glycoproteins[31] and polysaccharides.[37,38]

Sialic acids, which are among the most important constituents of various glycoproteins, are normally difficult to detect spectroscopically. To address this problem, we have developed[49] a fluorogenic reagent, 4,5-dinitrocatechol-O,O-diacetic acid (DNCDA). The method utilizes the previously explored ability of α-keto acids to form a quinoxaline ring structure via a coupling reaction with o-diamines.[50,51] As shown in Scheme II, an *in situ* reduction of the dinitro compound to a diamine (the active reagent) is accomplished first because of the limited stability of the aromatic diamine

[50] S. Hara, M. Yamaguchi, and M. Nakamura, *J. Chromatogr.* **377,** 111 (1986).
[51] P. A. David and M. Novotny, *J. Chromatogr.* **452,** 623 (1986).

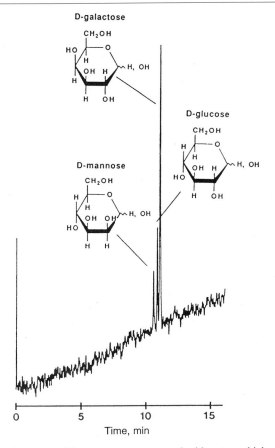

FIG. 5. Electropherogram of three common monosaccharides at very high sensitivity (subattomole detection limits). (Reproduced from Ref. 31.)

during storage; the reaction with *N*-acetylneuraminic acid (NANA) yields a highly fluorescent derivative, with detection limits down to the attomole range[49] when a helium/cadmium laser at the 325-nm output is used as a part of the fluorescence detection system. This derivatization technique is also applicable to reducing carbohydrates, which must first be oxidized by means of Benedict's reagent.

The derivatization method for sialic acids offers the following advantages: (1) the excitation maximum of the fluorescent quinoxaline derivative is very near the 325-nm output of the helium/cadmium laser, (2) charged moieties of the reagent molecule ensure effective "phase transfer" during derivatization, and (3) the derivative is suitable for HPCE. Electrophero-

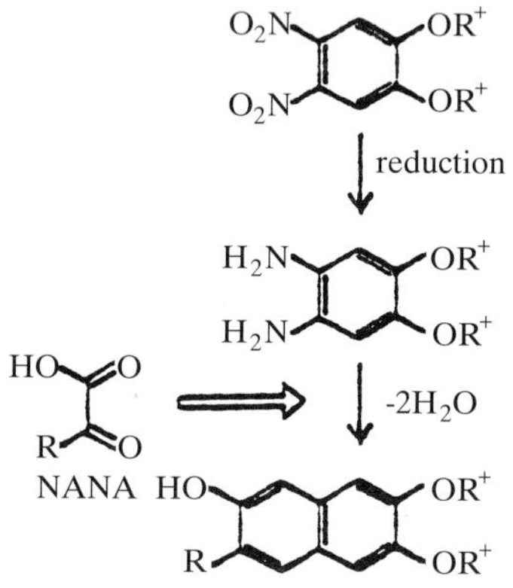

SCHEME II

grams of the derivatized NANA and α-ketoglutaric acid (an internal standard) are shown in Fig. 6.[10]

2-Aminopyridine, a reagent used previously in HPLC of carbohydrates,[13] also produces derivatives that are suitable for HPCE with fluorescence detection (Scheme III). Honda and co-workers[36] used this approach in the analysis of ovalbumin oligosaccharides; conventional fluorescence was employed in their work. The aminopyridylated neutral sugars also appear suitable for HPCE. Because of the close proximity of their excitation maximum to the 325-nm UV line of the helium/cadmium laser, relatively high sensitivity can be achieved with these derivatives.[52]

The amino-substituted naphthalenesulfonic acids are yet another group of useful reagents for HPCE fluorometric analysis of carbohydrates. This has been demonstrated by Lee et al.[53] in measuring glycosyltransferase activity. Various reagents of this class, such as 7-amino-1,3-naphthalenedisulfonic acid or 8-aminonaphthalene-1,3,6-trisulfonic acid (ANTS), form a Schiff base with the neutral sugars, which can be converted to a fluorescent product. 5-Aminonaphthalene sulfonate derivatives have been employed by Stefansson and Novotny[52] to resolve monosaccharide enantiomers by HPCE (see Fig. 7) in borate–oligosaccharide complexation media; laser-induced fluorescence at 325 nm was applied as the detection technique. Another application of this reagent, this time to a fairly complex oligosaccharide mixture, is shown in Fig. 8.[46]

The interest in fluorescent tagging for HPCE analysis of glycoconjugates

[52] M. Stefansson and M. Novotny, J. Am. Chem. Soc. 115, 11573 (1993).
[53] K. B. Lee, U. R. Desai, M. M. Palcic, O. Hindsgaul, and R. J. Linhardt, Anal. Biochem. 205, 108 (1992).

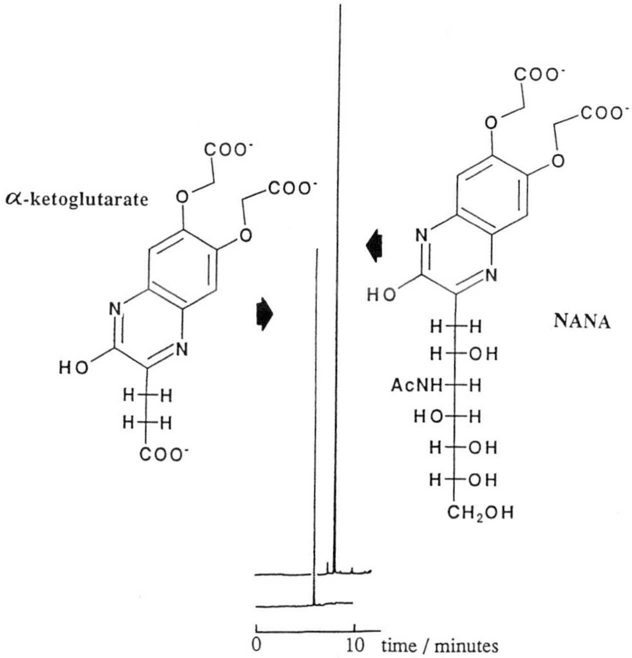

FIG. 6. Electropherograms of α-keto acid/DNCDA derivatives obtained with a coated capillary and detected by laser-induced fluorescence. (Reproduced from Ref. 10 with permission of the author.)

has been substantial. A search for the ideal reagent shall undoubtedly continue. The Schiff base reagents (like ANTS) have the distinct advantage of procedural simplicity over CBQCA, which, in turn, is a far better fluorophore with desirable spectral properties. Additionally, a reagent with charged functional group is highly beneficial from the point of fast electrophoretic migration. Finally, the effectiveness of derivatization with small quantities of glycans cleaved from various glycoproteins must also be assured.

Selected Applications

Intact Glycoproteins

As the glycoprotein sites and the type of glycan at a single site may vary considerably, the microheterogeneity of such molecules represents a considerable analytical challenge. Attempts at separating glycoforms by

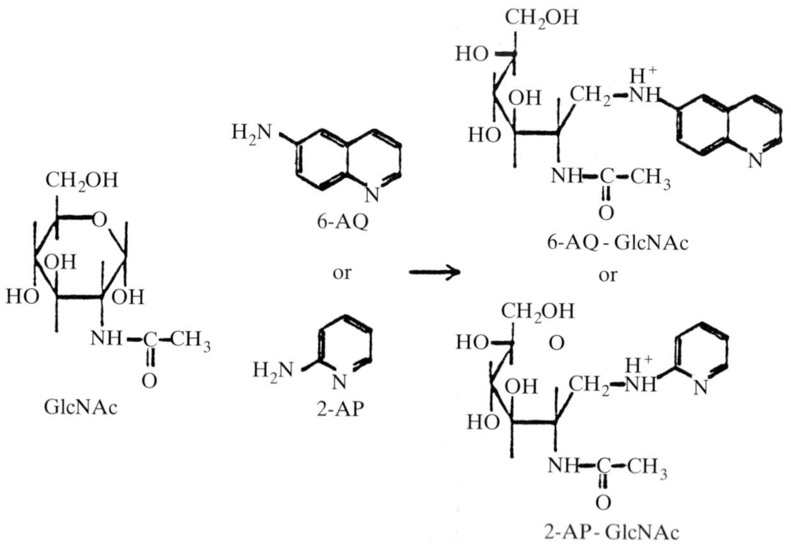

SCHEME III

HPCE have been described in several publications. First, Kilár and Hjertén[6] investigated the relative merits of capillary zone electrophoresis and isoelectric focusing (IEF) for separating the glycoforms of human transferrin. To provide suitably charged molecules for these experiments, the sialic acid residues, with their negative charges, were removed by treatment with neuraminidase. A publication comparing CZE and IEF for a similar purpose was provided by Yim,[7] who studied the human recombinant plasminogen activator with its many possible glycoforms. Although IEF seems to provide better resolution than CZE in the cases shown thus far in the literature, it seems somewhat premature to draw any general conclusions. At this stage of development, neither technique is capable of a complete resolution of the many glycoforms encountered. This, however, does not diminish the value of these methods as a rapid "fingerprinting" technique for assessing the glycosylation variants.

The choice of appropriate buffer systems and column types is likely to be important in the optimization of glycoprotein profile runs. This has been demonstrated through the examples of ovalbumin and pepsin glycoforms by Landers et al.[8] These authors found that complexation with borate dramatically enhanced component resolution. Three different HPCE runs (with UV detection at 200 nm) on ovalbumin (Fig. 9) demonstrate this resolving capability. In addition, the role of glycan moieties in borate complexation has been indirectly proven in their paper, as the enzymatic de-

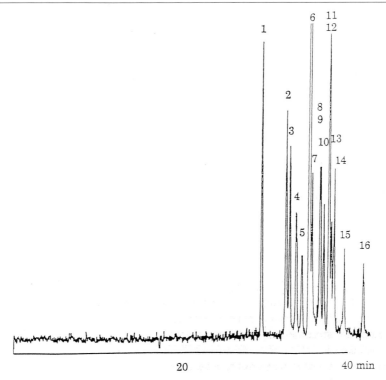

Fig. 7. High-efficiency separation of a complex sugar enantiomeric mixture (fluorescent derivatives). Peak assignment: 1, D-ribose; 2, D-xylose; 3, L-arabinose; 4, D-fucose; 5, D-glucose; 6, L-xylose and reagent (ANA); 7, L-ribose; 8, D-galactose; 9, L-mannose; 10, D-lyxose; 11, L-xylose; 12, D-mannose; 13, L-glucose; 14, D-arabinose; 15, L-fucose; 16, L-galactose. (Reproduced from M. Stefansson and M. Novotny, *J. Am. Chem. Soc.*, **115,** 11573. Copyright 1993 American Chemical Society.)

phosphorylation changed the overall migration rate, but not the basic pattern of peaks. The borate complexation was also found to be beneficial for assessing the glycoform population of native ribonuclease B by Rudd and co-workers.[54]

Glycoprotein Compositional Analysis and Oligosaccharide Mapping

The most essential pieces of information concerning a purified glycoprotein pertain to the knowledge of its monosaccharide composition, variation of the individual glycan moieties, and their incorporation within the polypeptide backbone. At this stage of development, HPCE appears capable

[54] P. M. Rudd, I. G. Scragg, E. Coghill, and R. A. Dwek, *Glycoconjugate J.* **9**(2), 86 (1992).

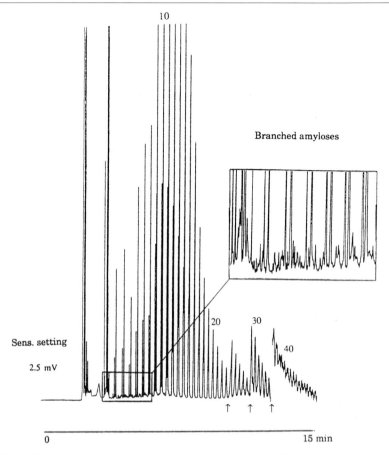

FIG. 8. High-resolution electropherogram of corn amylopectin. The arrows indicate changes in sensitivity setting from 2.5 to 1.0 mV, 250 μV, and 100 μV, respectively. (Modified from *Carbohydr. Res.*, **258**, M. Stefansson and M. Novotny, p 1, Copyright 1994 with kind permission of Elsevier Science–NL, Sara Burgerhartstraat 25, 1055 KV Amsterdam, The Netherlands.)

of delivering such data at very high sensitivity. However, just as with other analytical methodologies, the complete analysis of glycoproteins may involve a number of sample manipulation steps (hydrolysis, chromatographic preconcentration, treatment with enzymes, etc.) that will influence the overall sensitivity possible.

Controlled degradation of glycoproteins at the protein sites may involve the action of different proteases, followed by HPCE as a peptide-mapping procedure; the use of trypsin, chymotrypsin, or a pronase preparation is applicable to various cases in which one observes different glycopeptide maps. Different glycosylation patterns within a glycoprotein molecule can

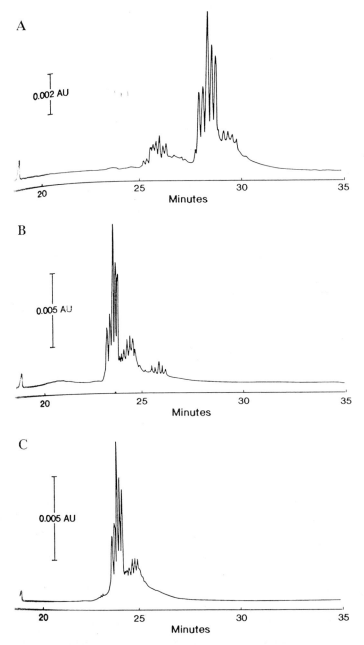

Fɪɢ. 9. Electropherograms demonstrating the enzymatic dephosphorylation of ovalbumin: (A) original ovalbumin sample; (B) ovalbumin treated with calf intestinal alkaline phosphatase; and (C) potato acid phosphatase. (Reproduced from J. P. Landers, B. J. Madden, R. P. Oda, and T. C. Spelsberg, *Anal. Biochem.* **205,** 115 (1992), with permission.)

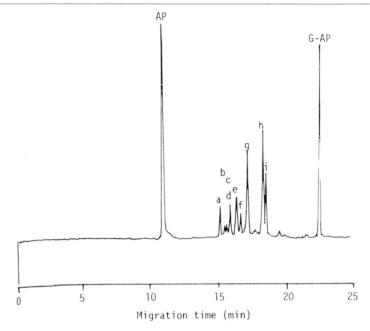

Fig. 10. HPCE of derivatized oligosaccharides originated from ovalbumin. The sample was pyridylaminated, and detected after separation by conventional fluorometry (excitation, 316 nm; emission, 395 nm). (Reproduced from S. Honda, A. Makino, S. Suzuki, and K. Kakehi, *Anal. Biochem.* **191,** 228 (1990), with permission.)

often be spotted on the basis of this approach. Alternatively, sites of glycosylation may be discernible when the glycopeptides are isolated and subjected to sugar analysis.

Several examples of the potential of HPCE in glycoprotein studies have been demonstrated. Using HPCE with UV detection, Nashabeh and El Rassi[55] investigated both tryptic and endoglycosidase cleavage products of α_1-acid glycoproteins of human and bovine origin. The tryptic peptides obtained were further divided into glycosylated and nonglycosylated fractions and subsequently profiled by capillary electrophoresis. The enzymatically released oligosaccharides were converted to 2-aminopyridine derivatives and detected at 240 nm by UV absorbance.

Ovalbumin, a glycoprotein isolated from chicken eggs, has been analyzed through HPCE by Honda *et al.*[36] To determine its main oligosaccharides, the authors subjected this glycoprotein to treatment with pronase, followed by gel filtration, hydrazinolysis, and re-N-acetylation. Aminopyri-

[55] W. Nashabeh and Z. El Rassi, *J. Chromatogr.* **536,** 31 (1991).

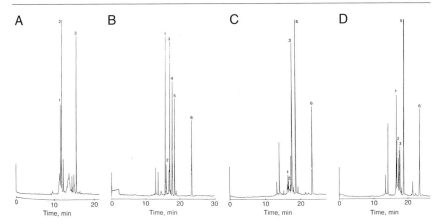

FIG. 11. HPCE analysis of monosaccharides released from bovine fetuin by acid hydrolysis: (A) 4 M HCl cleavage of amino sugars at 100°. Peak assignment: 1, glucosamine; 2, galactosamine; 3, unknown. (B) Standard mixture. Peak assignment: 1, N-acetylglucosamine; 2, N-acetylgalactosamine; 3, mannose; 4, fucose; 5, galactose; 6, galacturonic acid (internal standard). (C) Hydrolysis by 2.0 M trifluoroacetic acid. (D) Total hydrolysis with 4.0 M HCl. (Reproduced from Ref. 31.)

dyl derivatives of the oligosaccharides were subsequently made and measured electrophoretically. Because these derivatives fluoresce on illumination at 316 nm, the oligosaccharide map was measured fluorimetrically at 395 nm (Fig. 10).

Liu *et al.* utilized the fluorogenic reagent CBQCA, and the advantages of laser-induced fluorescence, to profile N-linked oligosaccharides and monosaccharides after degradation of bovine fetuin[31] and soybean trypsin inhibitor.[35] Using bovine fetuin as a model glycoprotein, they compared several nonenzymatic cleavage methods, which yield various monosaccharides (Fig. 11).[31] Oligosaccharides prepared through hydrazinolysis were also developed into a map (Fig. 12). N-Acetylneuraminic acid, originating from the different tryptic fragments of a glycoprotein, can be determined separately after reaction with an aromatic diamine.[49] Following the implementation of suitable sample treatment procedures, it is expected that HPCE with laser-induced fluorescence will become a useful, highly sensitive tool for structural studies of glycoproteins other than the model systems thus far investigated. Availability of appropriate oligosaccharide standards[56] can significantly aid the structural elucidation tasks.

[56] A. Klockow, H. M. Widmer, R. Amado, and A. Paulus, *Fresenius J. Anal. Chem.* **350,** 415 (1994).

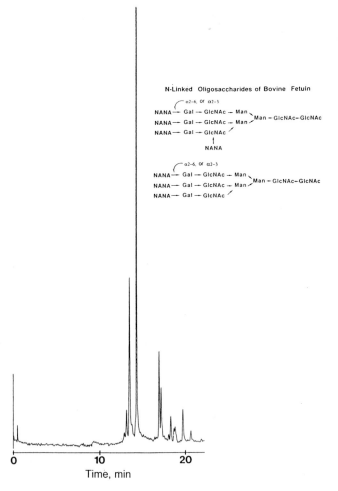

Fig. 12. HPCE oligosaccharide map of N-linked oligosaccharides released from bovine fetuin by hydrazinolysis. (Reproduced from Ref. 31.)

Analysis of Polysaccharides

The importance of polysaccharides to biochemistry, medicine, food industry, nutrition, etc., has been steadily increasing. Unfortunately, the analytical methodologies in this area have not been adequately developed to permit the needed structure–property investigations. The most pressing problem has been the limited separation capabilities and sensitivity of detection. The combination of HPCE with laser-induced fluorescence offers unique possibilites to this long-neglected area.

Important lessons for the HPCE separations of polysaccharides and large proteoglycan molecules can be derived from certain studies of their degradation products in slab gels.[57–59] Intact polysaccharides and proteoglycans are biomolecules of enormous complexity, existing furthermore in various aggregated forms. It seems that comprehensive studies need both the capabilities of separating various intact molecules, as well as the availability of reproducible techniques for their controlled degradation and further separation. Potentially, HPCE can play a role in both directions. However, at this stage, studies of polymer degradation products appear methodologically simpler than investigations of polysaccharide polyelectrolytes, which often exhibit peculiar physical and chemical properties.

Sulfated residues of chondroitins, dermatans, keratans, heparans, and heparins, as well as hyaluronic acid and its derivatives, provide the natural charged moieties for electrophoretic separations. Various glycosaminoglycans can be depolymerized to a convenient molecular size through the use of various polysaccharide lyases. The hydrolysis products featuring unsaturation can be detected by UV absorption at 232 nm, as demonstrated by Linhardt and co-workers[60,61] and Carney and Osborne.[62] This general approach is exemplified with an HPCE separation of chondroitin sulfate disaccharides hydrolyzed from a beagle cartilage sample[62] and its comparison with electromigration of standard disaccharides (Fig. 13). In the low molecular weight range, separations of this kind can be accomplished in the open-tubular analytical mode.

Examples of oligosaccharide separations performed in a free buffer medium include those of maltodextrin,[37] branched oligosaccharides of the xyloglucan type,[63] and others. As the number of sugar units in oligosaccharide molecules increases, separations become more difficult. Yet fairly complex mixtures of various oligosaccharides have been shown at high resolution.[39,46,64]

To increase the scope of HPCE applications to larger polysaccharides, it is essential to resolve effectively the molecules of interest according to their molecular size because of the possibilities of sample polydispersity,

[57] M. K. Cowman, M. F. Slahetka, D. M. Hittner, J. Kim, M. Forino, and G. Gadelrab, *Biochem. J.* **221,** 707 (1984).
[58] J. E. Turnbull and J. T. Gallagher, *Biochem. J.* **251,** 597 (1988).
[59] K. G. Rice, M. D. Rottink, and R. J. Linhardt, *Biochem. J.* **244,** 515 (1987).
[60] A. Al-Hakim and R. J. Linhardt, *Anal. Biochem.* **195,** 68 (1991).
[61] S. A. Ampofo, H. M. Wang, and R. J. Linhardt, *Anal. Biochem.* **199,** 249 (1991).
[62] S. L. Carney and D. J. Osborne, *Anal. Biochem.* **195,** 132 (1991).
[63] W. Nashabeh and Z. El Rassi, *J. Chromatogr.* **600,** 279 (1992).
[64] C. Chiesa and C. Horvath, *J. Chromatogr.* **645,** 337 (1993).

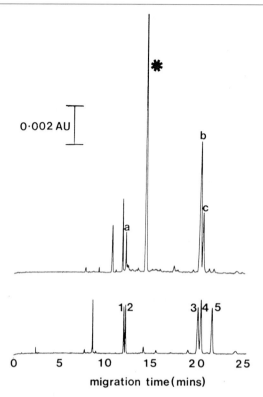

Fɪɢ. 13. HPCE separation of chondroitin sulfate disaccharides hydrolyzed from a
beagle cartilage sample (top) and compared to disaccharide standards (bottom). Peak designation: a, b, and c correspond to tentatively assigned structures of Δdi-HA, Δdi-6S, and
Δdi-4S, respectively. The major peak marked with an asterisk has not been identified.
Peaks 1–5 correspond to Δdi-HA, Δdi-0S, Δdi-6S, Δdi-4S, and Δdi-UA 2S, respectively.
(Reproduced from S. L. Carney and D. J. Osborne, *Anal. Biochem.* **195,** 132 (1991), with
permission.)

while other structural types due to branching, modification of some sugar
units, etc., can further complicate the overall resolution. Fluorescent tagging
provides the best opportunity for detection. However, the fact that there
may be only a limited number of derivatization sites in a polymer molecule
will challenge even the best detection technologies. In our preliminary work
in the area, we observed various fluorescently labeled polysaccharides,
such as chitosan, dextran, and various cellulose derivatives, all charged
additionally through complexation with borate, to migrate readily through
open tubes, however, with little tendency to separate. In concentrated,
immobilized polyacrylamide gels, the polymer network did not permit pene-

tration of such large molecules. Consequently, these polysaccharides and often even their enzymatic hydrolysis products merely concentrated at the inlet of our gel separation capillaries. It, however, appears that various fluorescently labeled polysaccharides can be made to migrate through different solutions of entangled polymers, such as linear polyacrylamide, galactomannan, and hydroxypropylcellulose. Unlike chemically immobilized and cross-linked gels, the entangled matrices "yield" readily toward the large macroions as they migrate toward the respective electrode.

Separation of large oligosaccharides in entangled polymer solutions is exemplified in Fig. 14[20] with a sample of carboxymethylcellulose treated briefly with a cellulase preparation. The same sample could not be success-

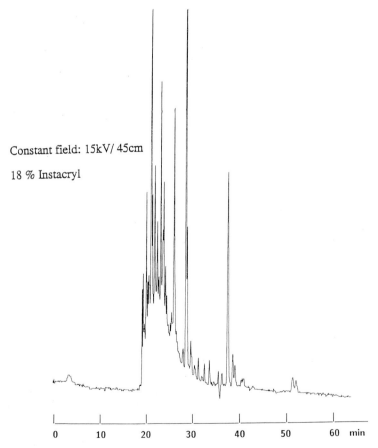

Constant field: 15kV/ 45cm

18 % Instacryl

FIG. 14. Separation of oligosaccharides released from a sample of carboxymethylcellulose by enzymatic cleavage. (Reproduced from Ref. 20, with permission.)

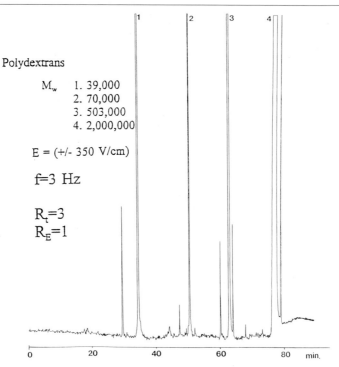

FIG. 15. Separation of the polydextran standards by pulsed-field capillary electrophoresis. Peak assignments: 1, 39,000 Da; 2, 70,000 Da; 3, 503,000 Da; 4, 2,000,000 Da. (Reproduced from Ref. 20, with permission.)

fully run through a gel-filled capillary. The size of "dynamic pores" and viscosity of entangled polymer solutions at constant concentration can be varied by altering the length of polymer chains. In addition, temperature may also become an effective variable of the method.[65]

The extent of separation according to the molecular weight of polysaccharides appears sometimes complicated by molecular elongations, under conditions of high electric field. The corresponding "reptation effect,"[66] a condition analogous to the phenomena observed with electromigration of large DNA molecules in constant fields,[67] was observed in our work with model polydextrans.[20] The polydextrans of different molecular weight migrated at the same speed under constant voltage (50–300 V/cm), but when a potential gradient along the separation capillary was periodically inverted

[65] A. Guttman, J. Horvath, and N. Cooke, *Anal. Chem.* **65,** 199 (1993).
[66] P. G. de Gennes, *J. Chem. Phys.* **55,** 572 (1971).
[67] J. Noolandi, *Annu. Rev. Phys. Chem.* **43,** 237 (1992).

at a 180° angle, the polysaccharide molecules underwent shape transitions that appeared to favor separation according to molecular size (Fig. 15).[20] This form of pulsed electrophoresis appears highly promising in additional separations of large glycoconjugates.

The HPCE of fluorescently labeled polysaccharides is an unexplored area of separation science with significant potential. Future studies must concentrate on exploring various separation matrices, conditions of pulsing experiments, and the relations between electrophoretic migration and other physical or structural parameters of polysaccharides. Moreover, it will be important to find ways of molecular weight determination. Laser desorption mass spectrometry is probably the best candidate in this direction.

Section III

Mass Spectrometry

[15] Protein Structure Analysis by Mass Spectrometry

By JOHN R. YATES

Introduction

Innovations in mass spectrometry have greatly advanced the ability to sequence and study the structures of proteins. In particular, protein analysis strategies have greatly benefited from the introduction of electrospray ionization. On-line mass spectral analyses of peptides and proteins are now routine. This chapter provides an overview of peptide and protein analysis by mass spectrometry and instrumentation and strategies employing tandem mass spectrometers are described.

Structure of Peptides and Proteins

Proteins consist of ~20 different amino acids covalently linked by amide bonds to form primarily linear heteropolymers. The polypeptide backbone has a repeating mass of 56 atomic mass units (u) with the side chains of each amino acid contributing masses from 1 to 130 u. A variety of molecular entities constitute the side-chain groups and each contributes to the rich chemical diversity inherent in protein structure. The linear sequence of amino acids within the heteropolymer must be elucidated to understand fully the functional and mechanistic properties of the protein. Two important components of protein sequence analysis are the accurate determination of the protein molecular weight and determination of the linear order of amino acid residues. In addition to amino acids, proteins may also contain posttranslational modifications such as glycosylation, phosphorylation, acetylation, amidation, formylation, myristoylation, and prenylation. This is by no means an exhaustive list.[1] Mass spectrometry is frequently the best method for elucidation of posttranslational modifications.

Mass Spectrometry

Mass spectrometers are capable of measuring the masses of molecules. To achieve these measurements two physical events are required. The first involves the creation of gas-phase ions. Solution-bound species such as peptides are generally polar and charged and to be moved into the gas phase must overcome formidable energetic barriers. Once gas-phase ions

---

[1] R. G. Krishna and F. Wold, *Adv. Enzymol. Relat. Areas Mol. Biol.* **67,** 265 (1993).

are formed, they must be focused into a device that separates the ions by their mass-to-charge (m/z) ratio and quantitates the ions present at each value. If the charge states of the ions are known then the mass of each ion can be calculated. In the simplest case the charge is 1 and the mass is simply the m/z value. Peptides and proteins are frequently ionized as the protonated molecule, $(M + H_n)^{n+}$, but may also be cationized, $(M + Na_n)^{n+}, (M + K_n)^{n+}$. Atmospheric pressure ionization (API) produces multiple charging and the mass of the peptide or protein can be calculated by considering the m/z values of two peaks separated by one charge or by using a transforming function to deconvolute the m/z envelope. Several chapters describing mass spectrometry are available for those who wish to explore the topic further[2,3] (see also [20]–[23] Volume 270[3a]).

Quadrupole Mass Spectrometry

One of the most common types of mass spectrometer is based on the quadrupole mass filter. Mass separation is achieved by establishing an electric field in which ions of a certain m/z have stable trajectories through the field. The electric fields are created by placing a direct current (dc) voltage and an oscillating voltage [ac voltage at rf (radio frequencies)] on four metal rods, the quadrupoles. Adjacent rods have opposite polarity. Ions pass through the center of the rods inscribing circles as they pass the length of the rods. By increasing the magnitude of the dc and rf voltages while maintaining the appropriate dc-to-rf ratio, stable trajectories are created, allowing ions of different m/z to pass through the quadrupoles and exit to the detector. Mass resolution is dependent on the number of rf cycles an ion spends in the field. However, the more cycles an ion undergoes the lower the ion transmission and the greater the loss of ions at the selected m/z.

The mass filtering effect of quadrupoles can be viewed as a separation process. By coupling quadrupole mass filters together, a powerful approach for structural analysis can be created. Placing a reaction region such as a collision cell between the two quadrupoles allows ions to be dissociated to obtain structural information. Typically, a collision cell is constructed from a quadrupole mass filter operated without a dc voltage on the rods to

[2] R. A. Yost and R. K. Boyd, *Methods Enzymol.* **193**, 154 (1990).
[3] K. L. Busch, G. L. Glish, and S. A. McLuckey, *in* "Mass Spectrometry/Mass Spectrometry: Techniques and Applications of Tandem Mass Spectrometry." VCH, New York, 1988.
[3a] W. E. Seifert, Jr. and R. M. Capnoli, *Methods Enzymol.* **270**, Chap. 20, 1996; C. M. Whitehouse and F. Banks, Jr., *Methods Enzymol.* **270**, Chap. 21, 1996; R. C. Beavis and B. T. Chait, *Methods Enzymol.* **270**, Chap. 22, 1996; J. C. Schwartz and I. Jardine, *Methods Enzymol.* **270**, Chap. 23, 1996.

function as a high-pass filter. All ions above a set mass value are focused through the quadrupole. In addition, enclosing the quadrupole in a cell allows the pressure to be raised to a level that permits multiple, low-energy collisions in the E_{lab} range of 10–40 eV. Ions undergoing multiple, low-energy collisions in a short time frame will become sufficiently activated to fragment. The principal benefit of a quadrupole collision cell is the ability to refocus ions scattered from collisions with the neutral gases.

Peptide and Protein Standards

To calibrate and evaluate the performance of the mass spectrometer using API, peptides and proteins suitable for use as standards are commercially available. A commonly used protein standard for mass calibration is the protein horse heart apomyoglobin (Fig. 1). Under electrospray ioniza-

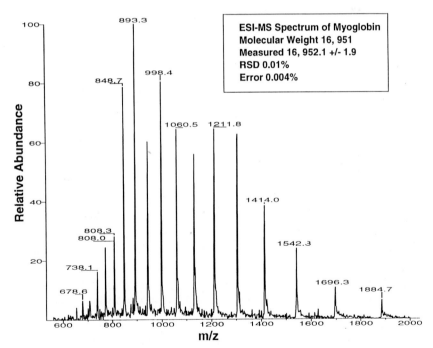

ESI-MS Spectrum of Myoglobin
Molecular Weight 16, 951
Measured 16, 952.1 +/- 1.9
RSD 0.01%
Error 0.004%

Fig. 1. A single-scan ESI–MS spectrum of horse heart apomyoglobin. A 5-pmol/μl solution was infused at 5 μl/min for tuing purposes. The reported molecular mass of this protein is 16,951 Da [R. D. Smith, J. A. Loo, C. G. Edmonds, C. J. Barinaga, and H. R. Udseth, *Anal. Chem.* **62,** 882 (1990)]. The measured molecular weight is 16,952.1 ± 1.9. This represents a mass measurement error of 0.004%.

tion (ESI) conditions a series of highly charged ions whose m/z values differ by one charge are produced and the expected m/z values can be used to calibrate the mass scale of the mass spectrometer. This standard must be prepared fresh from lyophilized material just prior to use. Mixtures of polypropylene glycols (425, 1000, and 2000) have also been used for mass calibration. A good standard for verifying MS/MS performance is the peptide [Glu[1]]fibrinopeptide B (human) (Fig. 2) above the tandem mass spectrum. Under low-energy collision conditions (~17 eV) the predominant ion type observed is the y-ion series. The b- and y-type ions expected from the amion acid sequence of [Glu[1]]fibrinopeptide B are indicated in Fig. 2. Using the expected fragment ions the collision conditions can be modified to maximize fragment ion abundance and signal-to-noise ratios.

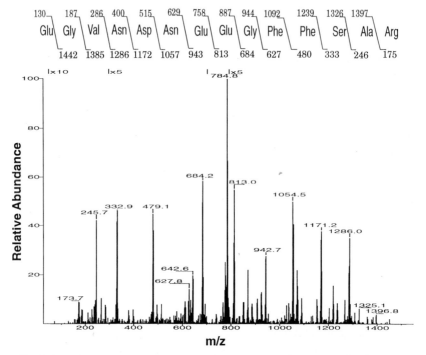

Fig. 2. A tandem mass spectrum of the double-charged ion (m/z 786) of [Glu[1]]fibrinopeptide B (human) is displayed. The spectrum was obtained by flow injection of 1 pmol of the peptide at a flow rate of 150 μl/min. This spectrum represents the sum of the five spectra acquired. The precursor ion was selected with a 3-m/z (FWHH) wide window in the first mass analyzer. The resulting fragment ions were recorded with a peak width of 1.5 amu.

Materials

Solvents

All solvents should be high-quality high-performance liquid chromatography (HPLC) grade and water should be deionized and distilled to minimize the formation of sodium or potassium adducts during ionization.

Methods

Protein Digestion

Proteins are digested by dissolving the material in 50–100 mM ammonium bicarbonate or 50–100 mM Tris-HCl, pH 8.6. Proteolysis may require the addition of denaturants to promote complete protein digestion. Typically, sample can be dissolved in a small amount of buffer containing 8 M urea or guanidine hydrochloride and slowly diluted to 2 M because many proteases remain active under these conditions. Proteolytic enzyme is added in a ratio of 1 : 50 or 1 : 100 (w/w). Sequencing-grade enzymes or modified forms of sequencing enzymes are available that reduce nonspecific cleavage or autolysis of the enzyme. References for manipulation of proteins prior to sequencing are available.[4]

Reversed-Phase High-Performance Liquid Chromatography

Analytical High-Performance Liquid Chromatography–Mass Spectrometry Analysis of Peptides. Most API sources can utilize the high flow rates produced by standard chromatographic analysis. The principal advantage is the ability to connect the HPLC effluent directly to the mass spectrometer. Thus, procedures optimized for HPLC separations using columns 1.0, 2.1, and 4.6 mm in diameter can be conveniently connected to the API source. In addition, effluent can first be monitored by ultraviolet absorbance and then transferred to the mass spectrometer. The effluent from the column can also be readily split to send part of the sample to the mass spectrometer and part to a fraction collector for additional studies. Most solvent systems employing volatile buffers or acids, including trifluoroacetic acid, can be used in the API source and small amounts of nonvolatile salts that may elute in the breakthrough peak can be tolerated. However, system down time can be minimized, and the necessity of frequent source maintenance,

---

[4] P. T. Matsudaira (ed.), "A Practical Guide to Protein and Peptide Purification for Microsequencing." Academic Press, San Diego, CA, 1989.

by diverting the breakthrough peak from the mass spectrometer inlet. An alternate procedure involves activation of the voltage applied to the electrospray needle after the breakthrough peak has eluted.

Microscale High-Performance Liquid Chromatography–Mass Spectrometry Analysis of Peptides. The analysis of small quantities of sample is better suited to microscale separation techniques. Reduction of the inner diameter of packed columns can yield improved efficiency by significantly increasing the theoretical plates for a separation. The construction of columns from small-diameter fused silica tubing also improves sample recoveries by minimization of metallic surfaces and by reducing the amount of reversed-phase packing material to which the sample is exposed. The dependence of flow rate and column diameter for detection of peptides by ultraviolet (UV) absorbance has been illustrated by Simpson and co-workers.[5] In addition, an API source can act as a concentration-dependent detector that can yield improved signal-to-noise measurements at lower flow rates. Procedures have been described for the construction of micro-columns as well as their use for separations.[6,7]

Microcapillary columns can be constructed following the procedures of Kennedy and Jorgenson.[6] A brief description of the procedure is presented here. A 30-cm piece of fused silica capillary (236-μm o.d. × 100-μm i.d.) is rinsed with 2-propanol and dried. A frit is created by tapping the end of the column into a vial of underivatized silica, 5 μm in diameter, and sintering the silica in the end of the capillary with an open flame. A polypropylene Eppendorf centrifuge tube (1.5 ml) is filled with ~100 μg of packing material, 1 ml of 2-propanol, and sonicated briefly to suspend the material and minimize aggregation. The solution is placed in a high-pressure packing device (Fig. 3) and the column is inserted with the inlet of the column placed in the solution. Helium gas at a pressure of ~500 psi is used to drive the packing material into the column while following the progress of the procedure under a microscope. Packing is continued until the material fills a length of the capillary corresponding to 10–20 cm. The pressure is then allowed to drop slowly to zero. The column bed is packed by rinsing with 2-propanol at 1000 psi for a few minutes. The column is conditioned by rinsing with 100% solvent B and slowly reducing the percentage of solvent B until it reaches initial HPLC conditions (100% solvent A). A linear gradient of 100% solvent A to 20% solvent A over 30 min is used to finish conditioning the column. The exit column of the microcapillary HPLC

[5] R. J. Simpson, R. L. Moritz, G. S. Begg, M. R. Rubira, and E. C. Nice, *Anal. Biochem.* **177,** 221 (1989).
[6] R. T. Kennedy and J. W. Jorgenson, *Anal. Chem.* **61,** 1128 (1989).
[7] D. C. Shelley, J. C. Gluckman, and M. V. Novotny, *Anal. Chem.* **56,** 2990 (1984).

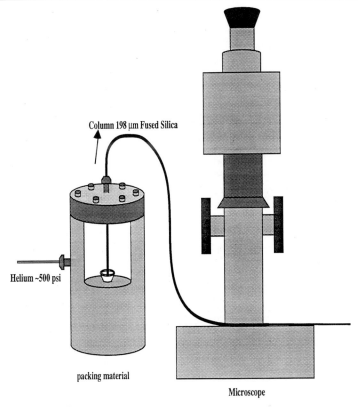

Column 198 μm Fused Silica

Helium ~500 psi

packing material

Microscope

FIG. 3. Setup of equipment for packing microcolumns for liquid chromatography.

column is inserted into the 26-gauge stainless steel electrospray needle until the fused silica capillary exits the tube. The fused silica is then drawn 1–2 mm back into the needle. A configuration for microcolumn HPLC is shown in Fig. 4. Solutions of peptides can be injected or applied to the column using pneumatic injection. The solvent stream is split precolumn to create a final flow rate of 0.5–2 μl/min.

Capillary Electrophoresis–Mass Spectrometry of Peptides

The formation of an electrospray is dependent on a steady flow of liquid, thus it is ideally suited as a method to interface liquid-based separation techniques to a mass spectrometer (see [21] Volume 270[7a]). One of the first

[7a] C. M. Whitehouse and F. Banks, Jr., *Methods Enzymol.* **270**, Chap. 21, 1996.

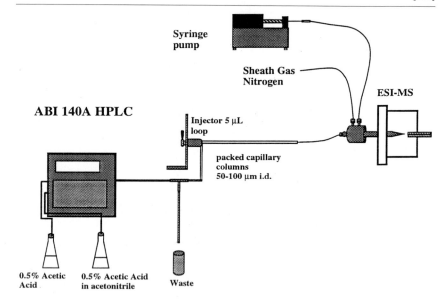

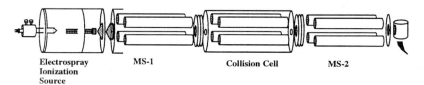

FIG. 4. Configuration of the HPLC and mass spectrometer for analysis of peptide mixtures.

demonstrations of a separation technique interfaced to a mass spectrometer through an API interface was by Smith and Udseth.[8] The potential for high-resolution separations and sensitivities that can be achieved by capillary electrophresis (CE) created a fair amount of interest in the use of this technique for the analysis of peptides in conjunction with mass spectrometry. Ionization of peptides by API performs best at acidic pH values, but achieving good separations and minimizing adsorption of peptides and proteins to the walls of the fused silica are difficult with underivatized columns at low pH. Better separations and sensitivities at low pH for proteins and peptides for CE–mass spectrometry (MS) can be achieved using the Applied Biosystems (Foster City, CA) Microcoat system or by

[8] R. D. Smith and H. R. Udseth, *Nature* (*London*) **331,** 639 (1988).

covalently modifying the surface of the capillary with aminopropylsiloxane to create a cationic layer.[9,10] The electroosmotic flow in acidic solutions is reversed while achieving good separations and femtomole-level sensitivities. Small-bore column (5 to 30-μm i.d.) produce extremely high sensitivities by lowering the flow rate and taking advantage of the concentration dependence of API. Attomole-level sensitivities have been achieved for the analysis of peptides with smaller bore capillaries. A fundamental problem associated with CE has been the ability to preconcentrate samples onto the column, precluding the analysis of dilute samples. Samples can be preconcentrated onto a capillary column using isotachophoresis.[11,12] The sample is injected behind a solution of lower electrical conductivity that forms a slow-moving band. Background electrolyte is injected after the sample is loaded. The sample is focused to a concentrated band by compression of the sample up against the slow-moving band by the fast-moving background electrolyte. This method improves the ability to inject dilute samples and should become a useful technique for CE–MS.

Peptide Mapping with Liquid Chromatography–Atmospheric Pressure Ionization Mass Spectrometry

The earliest applications of liquid chromatography–atmospheric pressure ionization mass spectrometry (LC–APIMS) extended the peptide-mapping techniques of Morris *et al.* and Gibson and Biemann to verify amino acid sequence and identify posttranslational modifications in recombinant proteins.[13,14] The LC–MS analysis of digested proteins realized greatly improved efficiency and mass range over the original fast atom bombardment (FAB)-based peptide-mapping strategies. A natural extension of this procedure is the mapping of recombinant proteins to verify structure or to confirm protein identity. Coupling peptide mapping to commonly used protein separation techniques such as gel electrophoresis allows determination of the enzymatically produced peptide masses derived from the separated proteins. Thus, a complex mixture of proteins can be separated and the peptide maps determined for each protein to identify or partially characterize a protein of interest. Proteins can be enzymatically

[9] P. Thibault, C. Paris, and S. Pleasance, *Rapid Commun. Mass Spectrom.* **5**, 484 (1991).

[10] M. A. Moseley, J. W. Jorgenson, J. Shabanowitz, D. F. Hunt, and K. B. Tomer, *J. Am. Soc. Mass Spectrom.* **3**, 289 (1992).

[11] R. D. Smith, S. M. Fields, J. A. Loo, C. J. Barinaga, H. R. Udseth, and C. G. Edmonds, *Electrophoresis* **11**, 709 (1990).

[12] F. Foret, E. Szoko, and B. L. Karger, *Electrophoresis* **14**, 417 (1993).

[13] H. R. Morris, M. Panico, and G. W. Taylor, *Biochem. Biophys. Res. Commun.* **117**, 299 (1983).

[14] B. W. Gibson and K. Biemann, *Proc. Natl. Acad. Sci. U.S.A.* **81**, 1956 (1984).

digested in the polyacrylamide gel slices or, after electroblotting, on the surface of membranes such as nitrocellulose, polyvinylidene difluoride (PVDF), or CD-Immobilon and analyzed by LC–MS. Splitting the effluent from LC–MS allows material to be collected for Edman sequencing. Peptide m/z values can then be correlated to Edman sequence data.

Peptide Mapping and Database Analysis. Peptide mapping can provide information beyond the fidelity of translation and sites of modification. A peptide map created by predictable proteolytic cleavage produces a highly informative fingerprint for a protein sequence. Mass values obtained from a digested "unknown" protein can be used to find other protein sequences that would produce the same set of masses under the same digestion conditions. Matching the peptide map from a protein of "unknown" amino acid sequence to a known protein sequence can indicate a high probability of protein identification. Computer programs designed to perform searches of databases using peptide mass data generated by API–MS and matrix-associated laser desorption ionization (MALDI)–MS have been described.[15–19] Accurate protein identifications can be made using mass tolerances as large as 5 amu and as little as 7% of the total protein mass. The accuracy of this technique is sufficient to distinguish among members of highly similar protein families, such as cytochrome c, where small differences in amino acid sequence exist. The peptide $(M + H)^+$ values predicted for trypsin-digested cytochrome c proteins derived from different species are shown in Table I. Inspection of the values shows the diversity of these values even though these proteins have highly similar sequences. In Table II results from searches of the database with values obtained by LC–MS peptide mapping of six different proteins are shown. The presence of post-translational modifications changes the mass of only those peptides containing modifications. For example, the sex steroid-binding glycoprotein, containing 13% carbohydrate by weight, can still be identified because the subset of peptides used was not glycosylated. Peptide mapping and database searching can be combined to identify proteins isolated from two-dimensional (2D) gel electrophoresis.[16] If no match or a poor match is found additional sequencing experiments must be performed to characterize the peptides more fully. This approach is restricted to the digestion products of a homogeneous protein sample and cannot be readily used with the mass

[15] J. Yates, P. Griffin, S. Speicher, and T. Hunkapiller, *Anal. Biochem.* **214,** 397 (1993).

[16] W. Henzel, T. Billeci, J. Stults, S. Wond, C. Grimley, and C. Watanabe, *Proc. Natl. Acad. Sci. U.S.A.* **90,** 5011 (1993).

[17] P. James, M. Qaudroni, E. Carafoli, and G. Gonnet, *Biochem. Biophys. Res. Commun.* **195,** 58 (1993).

[18] D. Pappin, P. Hojrup, and A. Bleasby, *Curr. Biol.* **3,** 327 (1993).

[19] M. Mann, P. Hojrup, and P. Roepstorff, *Biol. Mass Spectrom.* **22,** 338 (1993).

TABLE I

$(M + H)^+$ Values Created by Predicted Digestion of Proteins in Cytochrome c
Family with Protease Trypsin[a]

Species	Region					
	9–13	14–22	28–38	40–53	56–72	92–99
Arabian camel	634	1019	1169	1459	2011	907
Chimpanzee	651	1035	1169	1429	2009	808
Dog	634	1019	1169	1457	2011	907
Gray whale	634	1019	1169	1459	2011	907
Hippopotamus	634	1019	1169	1473	2011	907
Honeybee	634	1234	1195	1473	1383	1322
Pacific lamprey	620	1035	1119	1457	2051	836
House mouse	634	1019	1169	1431	1997	907
Ostrich	634	1035	1147	1489	1997	907
Domestic rabbit	634	1019	1169	1459	1997	907
Spider monkey	651	1035	1169	1475	2009	907
Skipjack tuna	622	1247	1216	1505	2051	950

[a] The species is listed in the first column and the regions of the proteins the $(M + H)^+$
values represent are listed at the head of each column.

TABLE II

Searches of PIR Database with Peptide Mass Maps Created by Liquid
Chromatography–Mass Spectrometry of Six Different Proteins

Protein designation	Source	Molecular mass (Da)	Number of peptides	Peptides matched	Number of proteins matched
Cytochrome c	Dog	13,000	4	4	3 (all cyt. c, dog, bat, elephant seal)
Cytochrome c	Pigeon	13,000	4	4	3 (all cyt. c, pigeon, duck, penguin)
GTP-binding protein	Human	26,000	7	7	2 (human and bovine)
Uteroferrin	Porcine	38,000	5	5	1
Sex steroid-binding protein	Human	45,000	8	8	1
DNA polymerase	*Thermus flabus*	100,000	4	4	2 (DNA-directed polymerase I and *SEC7* gene sequence)

TABLE III
MODIFIED AMINO ACIDS AND THEIR MASSES[a]

Modified amino acid	Monoisotopic mass, u^b	Average chemical mass, u^b
N-Methylglycine	71.037	71.079
N-Methyl-L-alanine	85.053	85.106
N-Acetylglycine	100.039	100.097
4-HYDROXY-L-proline	113.047	113.116
N-Acetyl-L-alanine	114.055	114.124
N,N,N-Trimethyl-L-alanine	114.092	114.167
N,N-Dimethyl-L-proline	127.099	127.186
N^4-Methyl-L-asparagine	128.058	128.131
L-erythro-β-Hydroxyasparagine	130.037	130.103
N-Acetyl-L-serine	130.050	130.123
L-erythro-β-Hydroxyaspartic acid	131.021	131.088
N^5-Methyl-L-glutamine	142.074	142.158
N-Acetyl-L-valine	142.086	142.178
N^6-Methyl-L-lysine	142.110	142.201
L-Glutamic acid 5-methyl ester	143.058	143.142
N-Acetyl-L-threonine	144.066	144.150
5-Hydroxy-L-lysine	144.089	144.174
N-Methyl-L-methionine	145.056	145.219
N-Acetyl-L-cysteine	146.027	146.184
L-3-Phenyllacetic acid	149.060	149.169
L-Selenocysteine	149.997	150.039
3-Methyl-L-histidine	150.067	150.160
N^6,N^6-Dimethyllysine	156.126	156.228
N-Acetyl-L-aspartic acid	158.045	158.133
N-Formyl-L-methionine	160.043	160.211
N-Methyl-L-phenylalanine	161.084	161.203
O-Phospho-L-serine	166.998	167.058

generated by a single peptide without additional sequence or structural information.

Liquid Chromatography–Mass Spectrometry Analysis of Posttranslational Modifications

As the genome initiatives proceed and the sequences for genes are determined, identification of posttranslational modifications and their biological functions will become increasingly important. Through posttranslational modifications organisms can increase chemical diversity in a protein, control enzymatic activity, or conduct signals. As many as 200 types of modifications may exist. A partial list of modified amino acids and their

TABLE III (*continued*)

Modified amino acid	Monoisotopic mass, u[b]	Average chemical mass, u[b]
N^6-Acetyl-L-lysine	170.105	170.211
ω-N-Methyl-L-arginine	170.117	170.214
N-Acetyl-L-glutamine	171.077	171.176
N^2-Acetyl-L-lysine	171.113	171.219
N^6,N^6,N^6-Trimethyl-L-lysine	171.149	171.263
N-Acetyl-L-glutamic acid	172.061	172.161
N^6-Carboxy-L-lysine	172.085	172.184
L-γ-Carboxyglutamic acid	173.032	173.125
N-Acetyl-L-methionine	174.059	174.238
O-Phospho-L-threonine	181.014	181.085
ω-N,ω-N'-dimethyl-L-arginine	184.132	184.241
3-Methyl-L-lanthionine	186.046	186.229
L-Aspartic 4-phosphoric anhydride	194.993	195.069
S-(L-Isoglutamyl)-L-cysteine	214.041	214.239
N^6-(4-Amino-2-hydroxybutyl)-L-lysine	215.163	215.296
L-3-Phosphohistidine	217.025	217.121
O-Sulfo-L-tyrosine	243.020	243.234
O-Phospho-L-tyrosine	243.029	243.156
2-[3-Carboxamido-3-(trimethylammonio)propyl]-L-histidine	280.177	280.349
N^6-Lipoyl-L-lysine	316.128	316.477
N^6-Biotinyl-L-lysine	354.173	354.467
N^6-Pyridoxal phosphate-L-lysine	357.109	357.303
N^6-Retinal-L-lysine	394.298	394.601
3,3',5-Triiodo-L-thyronine	632.766	632.949
L-thyroxine	758.658	758.841

[a] The masses represented are the residue masses. Masses used for selenium and iodine are 78.96 and 126.9, respectively.

[b] u = atomic mass units.

Adapted from: J. S. Garavelli, *Protein Science* **2**, 133 (1993) suppl. 1.

masses is given in Table III. In addition, protein engineering frequently uses host organisms to manipulate and then to produce large quantities of a protein. These organisms may lack the necessary enzymatic machinery to duplicate the site and type of posttranslational modification desired or observed on the human protein. Thus, there are compelling reasons to verify the structural uniformity of engineered proteins including posttranslational modifications. Mass spectrometry has been an ideal method for the study and characterization of posttranslational modifications. Strategies employing LC–MS for the analysis of two common forms of posttranslational modification are described.

Glycosylation. Characterization of the extent and sites of glycosylation in proteins is an important component of protein analysis studies. The glycosylation sites of recombinant proteins must be identified to verify proper assembly, an issue of great importance for therapeutics based on recombinant proteins. Guzzetta *et al.* employed LC–MS to verify the proper assembly of recombinant tissue plasminogen activator (rtPA).[20] Glycopeptides were identified using contour plots that display characteristic patterns produced by glycopeptides. Negatively sloping patterns of ions are observed from the mass heterogeneity of the glycopeptides and are a useful diagnostic feature by which to identify glycopeptide ions. Digestion of large glycosylated proteins can yield complex mixtures of peptides and glycopeptides complicating the process of identifying sites of glycosylation. Carr *et al.* developed a procedure to identify N- and O-linked glycopeptides selectively.[21,22] By incorporating precursor ion MS/MS scans into LC analysis, carbohydrate-specific fragment ions could be created to identify the location and m/z values of glycopeptide ions (Figs. 5 and 6). An ion at m/z 204, representing a fragment ion of N-acetylhexose, was monitored throughout the LC analysis to identify the elution positions of the glycopeptides. N- and O-linked sugars are differentiated by treatment of the digested mixture with peptide : N-glycanase to remove the N-linked sugars from the peptides. The analysis is repeated to identify the O-linked glycopeptides. This procedure simplifies the identification of carbohydrate attachment sites in glycoproteins.

Phosphorylation. The phosphorylation–dephosphorylation reaction is an important control mechanism in signal transduction and enzyme activation. Understanding the interactions of proteins in biological processes requires identifying the sites of phosphorylation in the protein. Affinity chromatography methods have been employed to isolate phosphopeptides specifically on the basis of their affinity for iron.[23] This approach has been used to isolate phosphopeptides selectively for subsequent analysis by mass spectrometry.[24] Nuwaysir and Stults adapted this technique for direct, on-line analysis of phosphopeptides.[25] The metal-chelating matrix is charged with iron and the mixture of peptides is applied to the column. Unbound peptides pass through the column and into the mass spectrometer. Phosphopeptides bound to the column are eluted with a basic buffer into the mass spectrometer for mass measurement. Peptides containing acidic domains also have a strong affinity for the iron column, leading to less than

[20] A. W. Guzzetta, L. J. Basa, W. S. Hancock, B. A. Keyt, and W. F. Bennett, *Anal. Chem.* **65**, 2953 (1993).
[21] S. A. Carr, M. J. Huddleston, and M. F. Bean, *Protein Sci.* **2**, 183 (1993).
[22] M. J. Huddleston, M. F. Bean, and S. A. Carr, *Anal. Chem.* **65**, 877 (1993).
[23] L. Andersson and J. Porath, *Anal. Biochem.* **154**, 250 (1986).
[24] H. Michel, D. F. Hunt, J. Shabanowitz, and J. Bennett, *J. Biol. Chem.* **263**, 1123 (1988).
[25] L. M. Nuwaysir and J. T. Stults, *J. Am. Soc. Mass Spectrom.* **4**, 662 (1993).

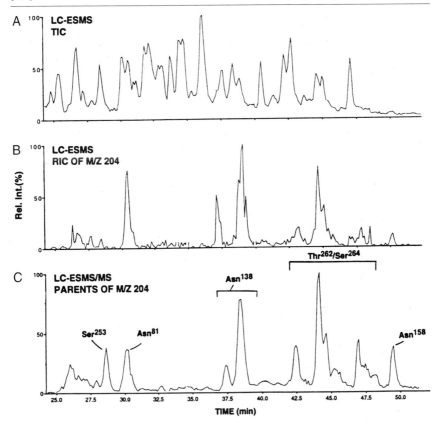

FIG. 5. (A) LC–ESMS total ion current (TIC trace with the orifice voltage ramped from 120 V at mz 150 to 65 V at m/z 500; from m/z 500 to m/z 2000 the orifice voltage was held constant at 65 V). (B) Reconstructed ion chromatogram (RIC) of m/z 204 (HexNAc$^+$) from the ramped orifice LC–ESMS data. (C) LC–ESMS, parent ion scan of m/z 204 (HexNAc$^+$), TIC trace. [From S. A. Carr, M. J. Huddleston, and M. F. Bean, Selective identification and differentiation of N- and O-linked oligosaccharides in glycoproteins by liquid chromatography–mass spectrometry. *Protein Sci.* **2**(2), 183 (1993). Copyright 1993 by Protein Society. Reprinted with permission of Cambridge University Press.]

specific analyses.[26] An alternate approach is to generate phosphopeptide specific marker ions during the LC–MS analysis to locate the phosphopeptides in the chromatographic analysis.[27] The marker ions, m/z 63 (PO$_2^-$) and m/z 79 (PO$_3^-$), can be created with highly energetic collisions in the

[26] G. Muszyska, G. Dobrowolska, A. Medin, P. Ekman, and J. O. Porath, *J. Chromatogr.* **604,** 1 (1992).
[27] M. J. Huddleston, R. S. Annan, M. F. Bean, and S. A. Carr, *J. Am. Soc. Mass Spectrom.* **4,** 710 (1993).

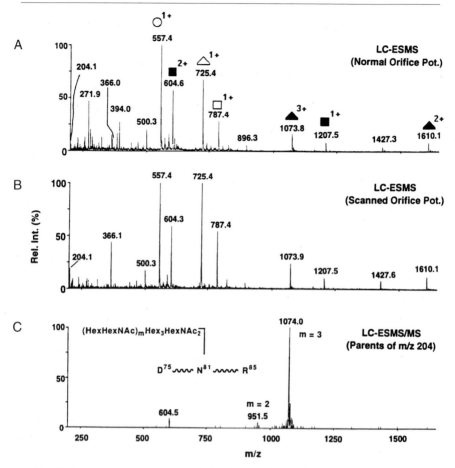

FIG. 6. Electrospray mass spectra of the glycopeptide eluting at ca. 30 min in the TIC traces shown in Fig. 5. (A) LC–ESMS using normal orifice voltage (65 V). Symbols identify a number of the parent ions present. (B) LC–ESMS performed with the voltage ramped from 120 V at m/z 150 to 65 V at m/z 500; from m/z 500 to m/z 2000 the orifice voltage was held constant at 65 V. (C) LC–ESME/MS, parent ion scan of m/z 204 (HexNAc$^+$). [From S. A. Carr, M. J. Huddleston, and M. F. Bean, Selective identification and differentiation of N- and O-linked oligosaccharides in glycoproteins by liquid chromatography–mass spectrometry. *Protein Sci.* **2**(2), 183 (1993). Copyright 1993 by Protein Society. Reprinted with permission of Cambridge University Press.]

high-pressure region of the API source. The orifice potential is raised to -350 V while the mass spectrometer is scanning low m/z values and then lowered to -65 V as scans reach higher m/z values. By scanning in this manner, highly energetic collisions occur only during the low-m/z section

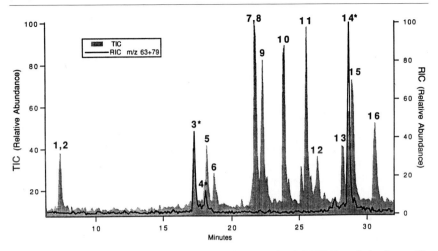

FIG. 7. Negative ion, stepped collisional excitation, scanning LC-ESMS analysis of a peptide mixture. TIC, Total ion chromatogram; RIC, reconstructed ion chromatogram for m/z 63 (PO_2^-) and m/z 79 (PO_3^-); phosphopeptides are denoted by asterisks (*). (Reprinted by permission of Elsevier Science, Inc., from "Selective detection of phosphopeptides in complex mixtures by electrospray liquid chromatography/mass spectrometry," by M. J. Huddleston, R. S. Annan, M. F. Bean, and S. A. Carr, JOURNAL OF THE AMERICAN SOCIETY FOR MASS SPECTROMETRY, Vol. 4, No. 9, p. 716. Copyright 1993 by the American Society for Mass Spectrometry.)

of the scan. At the higher m/z values the collision energies are considerably lower, producing no dissociation of the precursor ions, and they are recorded as intact ions. A plot of the reconstructed ion current for ions at m/z 63 and 79 indicates the positions of phosphopeptides in the analysis (Fig. 7). This strategy represents a useful approach for the identification of phosphopeptides in a complex mixture of peptides and will become an important tool in studies of biological systems.

Matrix-Assisted Laser Desorption Ionization–Tandem Mass Spectrometry of Peptides

A second ionization technique useful for the analysis of mixtures of peptide and proteins is matrix-assisted laser desorption ionization (MALDI) (see [22], Volume 270[27a]). This ionization process produces a pulsed beam of ions and requires a mass spectrometer capable of ion storage or very fast mass analysis. Time-of-flight mass spectrometers measure the m/z values of ions by determining the time required to traverse a region

[27a] R. C. Beavis and B. T. Chait, *Methods Enzymol.* **270,** Chap. 22, 1996.

free of magnetic or electric fields. The pulsed and rapid nature of the mass analysis is ideal for an MALDI source. The capability to ionize mixtures of peptides, frequently in the presence of buffers and salts, is attractive to combine with a mass spectrometry technique for creating sequence information. One approach for sequence analysis of peptides utilizes MALDI interfaced to a reflectron time-of-flight mass spectrometer (RETROFMS). A reflectron lens is placed at the end of the flight path to reverse the direction of the ions, lengthening the flight path of the ions and correcting spatial distributions occurring during the ionization event (Fig. 8). Ions dissociating in the field-free region (postsource decay, PSD) will assume nearly the same velocity as the precursor ion, but have a different kinetic energy. The depth of penetration into the reflectron is dependent on the kinetic energy of the ion. Thus, fragment ions created as a result of acceleration (10–20 keV) through the plume of neutrals ejected during the desorption/ionization event or from collisions in the field-free region will exit the reflectron earlier than their precursor ions.

Dissociation of the ions is sufficient to obtain nearly complete sequence information for small linear peptides. The predominant ions observed are the a-, b-, d-, and y-type ions (Fig. 9) with relatively abundant losses of ammonia observed from a- and b-type ions. The mass accuracy recorded for the fragment ions appears to be excellent, frequently within 0.2 amu.[28] The pioneering application of MALDI-RETOFMS to the sequence analysis of a peptide was performed by Poppeshriemer et al. on a peptide of unknown structure distributed to the participants of a sequencing workshop at the 1990 meeting of the American Society for Mass Spectrometry.[29] The peptide was 17 residues in length and was sequenced using a combination of sequence ions produced by MALDI-RETOF and enzymatic digestion. The principal deterrent to this approach is currently the inability to select a precursor ion with sufficient resolution to prevent other ions from entering the flight path and striking the detector. Consequently, the nondiscriminating ionization capability of MALDI cannot, at this time, be fully exploited. Once methods are improved for precursor ion selection, this will be an important technique for the analysis of mixtures of peptides.

Liquid Chromatography–Tandem Mass Spectrometry

Combining reversed-phase chromatography with tandem mass spectrometry creates a powerful approach for sequencing peptides. The

[28] R. Kaufmann, B. Spengler, and F. Lutzenkirchen, Rapid Commun. Mass Spectrom. 7, 902 (1993).
[29] N. Poppeshriemer, D. R. Binding, W. Ens, F. Mayer, K. G. Standing, X. Tang, and J. B. Westmore, Int. J. Mass Spectrom. Ion Process. 111, 301 (1991).

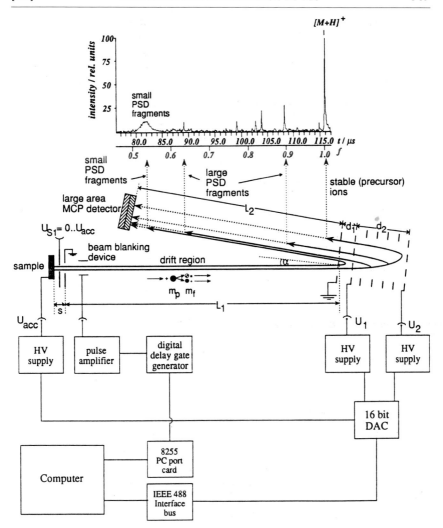

Fig. 8. Schematic diagram of RETOF mass spectrometer used for PSD fragment ion analysis. [From R. Kaufmann, B. Spengler, and F. Lutzenkirchen, Mass spectrometric sequencing of linear peptides by product-ion analysis in a reflectron time-of-flight mass spectrometer using matrix-assisted laser desorption ionization. *Rapid Commun. Mass Spectrom.* **7**(10), 904 (1993). Copyright 1993 by John Wiley and Sons, Ltd. Reprinted by permission of John Wiley & Sons, Ltd.]

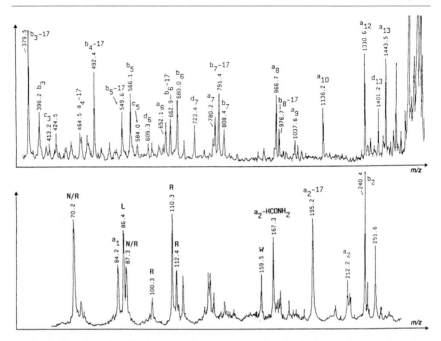

FIG. 9. Complete the PSD fragment ion spectrum of bombesin (p-Glu-Gln-Arg-Leu-Gly-Asn-Aln-Trp-Ala-Val-Gly-His-Leu-Met-NH$_2$, M + H = 1620.8 u) in a 2,5-dihydroxybenzoic acid matrix reconstructed from 12 spectral segments. [From R. Kaufmann, B. Spengler, and F. Lutzenkirchen, Mass spectrometric sequencing of linear peptides by product-ion analysis in a reflectron time-of-flight mass spectrometer using matrix-assisted laser desorption ionization. *Rapid Commun. Mass Spectrom.* **7**(10), 904 (1993). Copyright 1993 by John Wiley and Sons, Ltd. Reprinted by permission of John Wiley & Sons, Ltd.]

strengths of this approach are speed, sensitivity, the ability to obtain sequence data on mixtures of peptides that have not been resolved by the separation technique, and the ability to sequence peptides with blocked N termini. Tandem mass spectrometry has been successfully applied to the sequence analysis of peptides and proteins and to the characterization of posttranslational modifications. This ability to obtain sequence information is not limited to homogeneous proteins, but can be extended to the analysis of peptides contained in a mixture and derived from many different proteins. An excellent illustration is the application of LC–MS/MS to the analysis of peptides presented by the major histocompatibility molecules of antigen-processing cells.[30] These peptides are derived from intracellular

[30] D. F. Hunt, R. A. Henderson, J. Shabanowitz, K. Sakaguchi, H. Michel, N. Sevilir, A. L. Cox, E. Apella, and V. N. Engelhard, *Science* **255**, 1261 (1992).

proteins or from phagocytized extracellular proteins. These proteins are proteolyzed and displayed on the surface of the cell for inspection by circulating cytotoxic T cells (CTLs). Sequence analysis of the presented peptides has revealed information about amino acid length of the bound peptides, antigen processing, and peptide diversity.

The display of peptides on the surface of each cell allows the immune system to survey the health of the cells throughout the body. Viral attack or transformation of cells results in the presentation of peptides obtained by proteolysis of the foreign or newly synthesized proteins. The immune system should then detect cells displaying these antigens and destroy them. Some malignant cells display tumor-specific antigens that are recognized as foreign by the immune system. In cases in which the malignant cells multiply it is presumed the immune system has not mounted a normal response to these antigens, even though a small population of CTLs reactive against the tumor cells can be found. The identification of peptides recognized by the tumor-specific CTLs would allow the use of immunotherapy to boost the level of the immune response. The large number of class I-related antigens displayed on the surface of the cells required the development of methods to correlate immunoreactivity to the m/z values observed by mass spectrometry and kill the tumor cells. By incorporating a novel low flow rate stream splitter, Cox et al. separated peptides derived from HLA-A2.1 molecules on the melanoma cell line DM6 into two fractions.[31] One-sixth of the eluent from the microcolumn reversed-phase HPLC was sent to a microtiter plate for chromium release assays with melanoma-specific CTLs and the remaining material was sent to the mass spectrometer for mass analysis. Specific cell lysis was used to identify the peptide m/z values for sequencing. A subsequent tandem mass spectrometry experiment, similar to the one described below, was used to generate sequence data to identify the amino acid sequence of the peptide stimulating cell lysis. Both the amino acid sequence and activity of the peptide were verified by synthesis and chromium release assay, respectively. Interestingly, the peptide exhibiting the activity was derived from melanin, a pigmentation protein.

Sequencing experiments on peptides are performed by selecting a peptide ion with a 2- to 3-u (full-width half-height, FWHH) wide window in the first quadrupole mass analyzer.[32] Transmitting ions through the first mass analyzer at this lower mass resolution improves sensitivity. These ions pass into a quadrupole collision cell filled with argon to a pressure of

[31] A. L. Cox, J. Skipper, Y. Chen, R. A. Henderson, T. L. Darrow, J. Shabanowitz, V. H. Engelhard, D. F. Hunt, and T. L. Slingluff, Jr., Science **264,** 716 (1994).

[32] D. F. Hunt, J. R. Yates III, J. Shabanowitz, S. Winston, and C. R. Hauer, Proc. Natl. Acad. Sci. U.S.A. **84,** 620 (1986).

~3–5 mtorr and suffer collisions in the 20- to 50-eV (E_{lab}) range. Multiple collisions with the neutral gas vibrationally excite the peptides, inducing dissociation of the ions. The fragment ions produced in the collision cell are transmitted to the second mass analyzer, which is scanned over a mass range from m/z 50 to the molecular weight of the peptide to record the tandem mass spectrum.

As an illustration of the high specificity of the LC–MS/MS sequencing approach, all of the proteins from 10^8 *Saccharomyces cerevisiae* cells were collected and digested with the enzyme trypsin. A 2-μl aliquot of this complex mixture of peptides representing ~1/5000 of the total material was injected onto a microcolumn packed with 10 cm of C_{18} reversed-phase packing material. A 40-min linear gradient of 0–80% acetonitrile containing 20% of 0.5% aqueous acetic acid was used to elute the peptides from the column and into the mass spectrometer (Fig. 10). Tandem mass spectrometry was performed on a doubly charged ion of m/z 933. The tandem mass spectrum is shown in Fig. 11. Peptide ions fragment primarily at the amide bonds, resulting in a set of fragment ions indicative of the amino acid sequence of the peptides (Fig. 12). The amino acid sequence determined from the fragmentation pattern is shown above the spectrum. In the collision

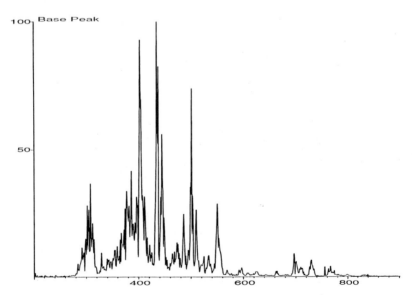

FIG. 10. Reconstructed total ion chromatogram showing the ion current intensity of the most abundant ion at each scan for peptides derived from digestion of the total proteins from 10^8 *S. cerevisiae* cells. An aliquot of the digestion mixture representing approximately 1/5000 of the total solution was injected onto the column for analysis.

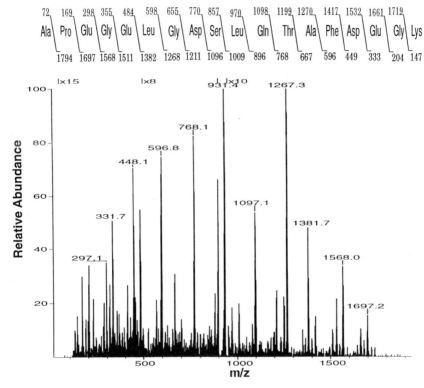

FIG. 11. The tandem mass spectrum obtained from a peptide ion at m/z 933 from a complex mixture of *S. cerevisiae* peptides. The amino acid sequence for the peptide is displayed above the mass spectrum. The average mass for this peptide calculated from the doubly and singly charged ions is 1864 Da. Type b ion fragment ions are listed above the amino acid and type y ions are listed below the sequence. The tandem mass spectrum was acquired during a 40-min linear gradient of 0–70% acetonitrile containing 0.5% acetic acid on a 30 cm × 100 μm packed capillary column with a 12-cm C_{18} bed. The amino acid sequence of the peptide corresponds to residues 114–131 of the initiation factor 5A-2 protein.

dissociation process a ladder of sequence ions is produced, in which the difference between consecutive ions indicates the residue mass of the amino acid at that position in the sequence (Table IV). However, a variety of ion types can be produced in the fragmentation process such that ions that retain the charge on the C terminus represent a different portion of the amino acid sequence than do ions that retain the charge of the N terminus. Neutral losses from fragment ions result in lower intensity ions that accompany the major fragment ions. Interpretation requires determining whether an ion or ions in the spectrum originate from the N or C terminus. Once

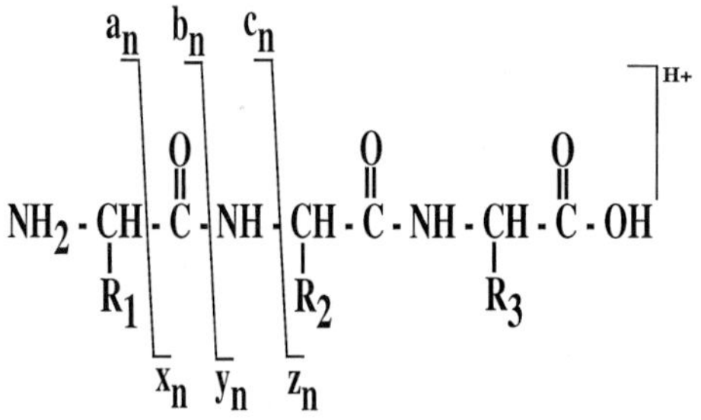

FIG. 12. Designations for the fragment ions obtained by collision-induced dissociation of peptides. Fragment ions representing charge retention on the N terminus are designated type a_n, b_n, or c_n ions and fragment ions representing charge retention on the C terminus are designated type x_n, y_n, or z_n ions. Ions of type b and y are the principal fragment ions observed under multiple, low-energy conditions.

TABLE IV

SINGLE- AND THREE-LETTER CODES FOR AMINO ACIDS AND RESIDUE MASSES

Amino acid	Single-letter code	Three-letter code	Monoisotopic mass, u[a]	Average chemical mass, u[a]
Glycine	G	Gly	57.021	57.059
Alanine	A	Ala	71.037	71.078
Serine	S	Ser	87.032	87.078
Proline	P	Pro	97.052	97.116
Valine	V	Val	99.068	99.132
Threonine	T	Thr	101.047	101.105
Cysteine	C	Cys	103.009	103.138
Leucine	L	Leu	113.084	113.159
Isoleucine	I	Ile	113.084	113.159
Asparagine	N	Asn	114.042	114.103
Aspartic acid	D	Asp	115.026	115.088
Glutamine	Q	Gln	128.058	128.130
Lysine	K	Lys	128.094	128.174
Glutamic acid	E	Glu	129.042	129.115
Methionine	M	Met	131.040	131.192
Histidine	H	His	137.058	137.141
Phenylalanine	F	Phe	147.068	147.176
Arginine	R	Arg	156.101	156.187
Cmc Cystine	C	Cmc	161.014	161.175
Tyrosine	Y	Tyr	163.063	163.176
Tryptophan	W	Trp	186.079	186.213

[a] u = atomic mass units.

ion types have been determined, subtraction of mass differences between consecutive ions identifies the amino acids. Thus, a set of y-type ions, from low mass to high mass, defines the amino acid sequence from the C terminus to the N terminus. The final sequence assignment must agree with the measured molecular weight of the peptide. A limitation to low-energy fragmentation processes is the inability to differentiate leucine and isoleucine or glutamine and lysine. The latter pair of amino acids can be differentiated by specificity of enzyme cleavage or through chemical modification. Computer algorithms designed to aid in the interpretation of the complex spectra have been developed, but the utility of these programs is limited.

Liquid Chromatography–Tandem Mass Spectrometry with Database Analysis. Sequencing of the human genome and the genomes of various organisms is proceeding at a rapid rate.[33] The current collection of nucleotide sequence information is already providing a valuable tool for the study of biological systems. As the collection of data grows it is increasingly important to screen new data through the database to prevent duplication of sequence analysis efforts. In addition, sequence information may have been previously determined, but the experimental context in which the information is rediscovered may be relevant to the biological process under study.[31] Typically, peptide sequence data from tandem mass spectrometry must be interpreted prior to performing a search of the database. Interpretation can be a time-consuming process. Eng *et al.* have developed a database-searching approach that utilizes uninterpreted tandem mass spectrometry data.[34] This approach utilizes the protein database as pseudo-mass spectral library to determine similarities to the experimental data. The amino acid sequences are prefiltered using a scoring procedure to calculate the number of sequence ions predicted from the database sequence that match the experimental data. A list of the 500 best matches is produced. Each of these sequences is compared to the experimental data, using a cross-correlation analysis. The output from the analysis of the tandem mass spectrum shown in Fig. 11 is shown in Table V. The large difference between the normalized cross-correlation value of the first and second answer indicates a high probability of a successful match to the highest ranking result from the database search. A search of the Genpept dasebase (~78,000 proteins translated from nucleotide sequences) requires 10–15 min of computer time, while search of a species-specific subset of the Genpept database requires 3–5 min. Peptide sequences identified using this program for the tandem mass spectra acquired from the digested *S. cerevisiae* proteins are shown in Table VI.

[33] M. Olson, *Proc. Natl. Acad. Sci. U.S.A.* **90,** 4338 (1993).
[34] J. K. Eng, A. L. McCormack, and J. R. Yates, *J. Am. Soc. Mass Spectrom.* **5,** 976 (1994).

TABLE V

SEARCH OF YEAST SEQUENCE-SPECIFIC DATABASE USING TANDEM MASS SPECTRUM[a]

No.	$(M + H)^+$	C_n	S_p	Ions	Ion %	I_m	Ascension No.	Amino acid sequence
1	1865.0	1.0000	3531.3	27/34	79.4	1678.1	X56236	APEGELGDSLQTAFDEGK
2	1867.2	0.2845	157.8	8/32	25.0	549.0	X65124	ELDEIAPVPDAFVPIIK
3	1864.1	0.2616	140.2	8/32	25.0	457.6	X15003	TGVIVGEDVHNLFTYAK
4	1865.8	0.2193	115.0	8/34	23.5	454.8	X16385	GLDDESGPTHGNDSGNHR
5	1862.1	0.2028	42.9	5/34	14.7	253.4	X56084	AIPSLSSSIPYSVPNSNK
6	1865.2	0.2010	217.7	9/32	28.1	631.7	M69017	GFFSVQTTSSKPLPIVR
7	1868.0	0.1973	3.2	1/28	3.6	89.0	M60416	EDINFEFVYFTDISK
8	1867.1	0.1902	45.3	5/30	16.7	252.7	L08070	GNSLPINYPETPHLWK
9	1868.1	0.1882	45.9	5/34	14.7	311.9	X69964	SDGMGNVLLNATVVDSFK
10	1867.2	0.1771	199/0	9/32	26.5	613.6	A33622	TVDAVLNATPIVLGLITR

[a] Spectrum shown in Fig. 11. Column designations: No., ranking of the sequences from the search; $(M + H)^+$, mass of the peptide plus a proton; C_n, the normalized cross-correlation value; S_p, the preliminary scoring procedure; Ions, the number of ions matched versus the number predicted for the sequence; Ion %, the percentage of matched ions observed; I_m, the summed intensities for the matched ions; Ascension No., the database designation for the sequences matching the tandem mass spectrum; Amino acid sequence, the sequence identified in the search.

TABLE VI

SEARCHES OF GENPEPT DATABASE WITH TANDEM MASS SPECTROMETRY SPECTRA OF PEPTIDES[a]

Ion m/z spectrum obtained with $[(M + 2H)^{2+}]$	Amino acid sequence identified	Protein source
801	IPAGWQGLDNGPESR	Phosphoglycerate kinase
728	LPGTDVDLPALSEK	Pyruvate kinase
1093	IEDDPFVFLEDTDDIFQK	Hexokinase PI
666	EEALDLIVDAIK	Enolase
681	NPTVEVELTTEK	Enolase
933	APEGELGDSLQTAFDEGK	Initiation factor 5A-2
618	TGGGASLELLEGK	Phosphoglycerate kinase
1075	QAFDDAIAELDTLSEESYK	GMH1
803	DPFAEDDWEAWSH	Enolase

[a] Obtained by microcolumn LC–MS/MS analysis of peptides derived from trypsin digestion of the total proteins from 10^8 S. cerevisiae cells. Spectra were obtained on doubly charged peptide ions.

Future Developments

Significant instrumental improvements in both cost and performance are to be expected as several types of mass spectrometers reach a level of refinement that permits commercial versions to be introduced. Ion trap mass spectrometers are increasingly used for the analysis of biomolecules with greater sensitivity and mass resolution than quadrupole mass spectrometers.[35–37] Sequence information for peptides can now be obtained on reflectron time-of-flight mass spectrometers and improvements in the ability to scan the reflectron voltages are expected.[28] Improvements in surface-induced dissociation (SID) have been achieved by coating metal surfaces with self-assembled monolayers,[38,39] greatly increasing the efficiency of fragment ion recovery from the surface. In addition, fragment ions have been observed in SID studies of peptides that allow the differentiation of leucine and isoleucine on quadrupole-type mass spectrometers.[40] These developments will ensure that mass spectrometry continues to have a significant role in biological research.

[35] K. A. Cox, J. D. Williams, R. G. Cooks, and R. E. Kaiser, Jr., *Biol. Mass Spectrom.* **21,** 226, 1992.
[36] J. A. Castoro and C. L. Wilkins, *Anal. Chem.* **65,** 2621 (1993).
[37] M. W. Senko, S. C. Beu, and F. W. McLafferty, *Anal. Chem.* **66,** 415 (1994).
[38] B. E. Winger, R. K. Julian, R. G. Cooks, and C. D. Chidsey, *J. Am. Chem. Soc.* **113,** 8967 (1991).
[39] V. H. Wysocki, J. L. Jones, and J. M. Ding, *J. Am. Chem. Soc.* **113,** 8969 (1991).
[40] A. L. McCormack, A. Somogyi, A. R. Dongre, and V. H. Wysocki, *Anal. Chem.* **65,** 2859 (1993).

[16] Carbohydrate Sequence Analysis by Electrospray Ionization–Mass Spectrometry

By VERNON N. REINHOLD, BRUCE B. REINHOLD, and STEPHEN CHAN

Introduction

Carbohydrate materials with a diversity of structural types and acid lability pose unique analytical problems. A treatise in this series has been totally devoted to these endeavors.[1] This chapter may serve as a bridge between this volume and an earlier volume on mass spectrometry.[2] This

[1] G. W. Hart and W. J. Lennarz (eds.), *Methods Enzymol.,* **230,** (1993).
[2] J. A. McClosky (ed.), Mass spectrometry. In *Methods Enzymol.* **193,** (1992).

strong focus to understand carbohydrate structure becomes most germane with the growing interest in the functional roles attributed to these moieties. As researchers sharpen their focus on the biochemical basis of cell–cell interaction, carbohydrates and their glycoconjugates are often found to be crucial participants. The evidence for these interactions has been growing and is strongly supported by the use of carbohydrase inhibitors, deglycosylating enzymes, and site-directed mutagenicity of participating glycans. From these studies carbohydrates appear to serve as the "Velcro" of adhesion, the decoys against bacterial invasion, the modulators of protein structure, and epitopes for molecular targeting and trafficking. Numerous human parasites, through molecular mimicry, have capitalized on these intricate functional roles to subvert immune surveillance, invade, and thrive within human cells. In nitrogen fixation, bacterial oligosaccharides play major roles in their relationships with plants. Starting from recognition and attachment, and concluding with a new plant organ, the root nodule, carbohydrate involvement in these symbiotic events is most specific and pronounced.[3,4] Cell-specific carbohydrate structures are known to regulate events during differentiation,[5] and the same sites serve as microbe attachment and invasion sites.[6] Receptors on lymphocytes that mediate adherence have been found to be a small family of glycosylated molecules, the selectins.[7] Wherever confining membrane surfaces occur, and signal transduction is carried out, the diversity and specificity of the carbohydrate molecule are likely to be utilized as the conduit for this communication. Although a precise understanding of function is frequently compromised by structural complexity, the overall conclusions remain that these residues are major participants in the fine tuning of cellular processes. As in many areas of science, revelation will come with an understanding of detail.

From a structural standpoint, oligosaccharides are ideally suited for events requiring molecular specificity, e.g., their polyhedric character provides a platform with numerous functional groups for modification or interaction; monomers within a chain can participate in multiple linkage and branching patterns, creating arrays of unique possibilities. It is for these very reasons that methodology for carbohydrate sequencing remains undeveloped. With the absence of optical properties for detection or specific functional groups for modification, and their copious structural isomerism,

[3] L. J. Halverson and G. Stacey, *Microbiol. Rev.* **53,** 193 (1986).
[4] H. Spaink, D. Sheeley, A. van Brussel, J. Glusha, W. York, T. Tak, O. Geiger, E. Kennedy, V. N. Reinhold, and B. Lugtenberg, *Nature* (*London*) **354,** 125 (1991).
[5] Z. Zhu, L. Cheng, Z. Tsui, S. Hakomori, and B. A. Fenderson, *J. Reprod. Fertil.* **95,** 813 (1992).
[6] K.-A. Karlsson, *Annu. Rev. Biochem.* **58,** 309 (1989).
[7] L. A. Lasky, *Science* **258,** 964 (1992).

carbohydrates remain the analytical challenge of contemporary biology. Pursuit of linkage and branching detail is a major undertaking and rarely are characterizations complete. Methylation analysis, and the identification of methylated alditol acetates, fails in principle to identify all entities uniquely, and this information cannot be integrated into a sequential array. Moreover, methylation analysis requires large quantities of material compared with the typical operational range of a mass spectrometer.

Sequence has a somewhat more entangled meaning for a variably linked and multiply branched structure than for the linear biopolymers of amino and nucleic acids. Microheterogeneity, a consequence of variable chain termination, adds further complication, and these features are compounded by the stereochemistry of glycosidic bonds (anomers) and isobaric residues. We can take some satisfaction in the knowledge that sequence determination reduces to defining the type of motif (glycotype), its distribution (glycoform), and a precise linkage and branching pattern (glycomer; the term *glycomer* is an extension of earlier notation, *glycoform* and *glycotype*,[7a] defining the detailed structure of a glycan). This diversity of function, the fidelity of biosynthesis at any one site or from any one cell, represent a small part of the many intriguing questions in glycobiology.

Electrospray Ionization–Mass Spectrometry

The major theme of this chapter relates to a detailed characterization of carbohydrates with component applications focused on glycoprotein glycans. A previous volume covered the broad area of mass spectrometry with selected chapters on electrospray ionization.[2] This chapter introduces and discusses electropray ionization–mass spectrometry (ES–MS) as applied to the sequence analysis of glycoprotein oligosaccharides. Here, we describe the instrumental power of mass separation and collision-induced dissociation (CID) to arrive at detailed carbohydrate structure. Developments in generating ions from a condensed phase have brought new dimensions to our understanding of biopolymers. Undertakings that were inconceivable a few short years ago are now routine, and much of that credit goes to the desolvation technique of electrospray[8] (ES) ionization. As applied to biopolymers, the first established success of ES was realized with polar molecules ionized from aqueous aerosols, a development highly suited to peptide sequencing. In contrast to this important focus, this chapter emphasizes the value of ES when applied to lipophilic samples evaporated

[7a] T. W. Rademacher, R. B. Parekh, and R. A. Dwek, *Ann. Rev. Biochem.* **57,** 785 (1988).

[8] J. B. Fenn, M. Mann, C. K. Meng, S. F. Wong, and C. M. Whitehouse, *Science* **246,** 64 (1989).

from nonpolar solvents, a strategy concordant with the detailed analysis of oligosaccharide linkage and branching.

Mass spectrometry of carbohydrates has been largely based on ballistic ionization strategies, e.g., fast atom bombardment (FAB) MS, which exhibits poor ionization efficiency and allows considerable ion source fragmentation. For the purposes of carbohydrate analysis, ES is the most effective and generally applicable technique currently available for transforming these molecules into gas-phase ions. The technique makes use of high electric fields to aerosolize solutions, creating a fine mist with each droplet carrying an excess surface charge. A heated, countercurrent bath gas at ambient pressure evaporates neutral solvent from the droplet surface, reducing its size and hence increasing the surface charge density. Electrostatic repulsion disrupts the surface, leading to the ejection of smaller charged droplets with the eventual development of gas-phase ions.

In operation, solutions (usually between 1 and 20 μl/min) are directly infused through a stainless steel needle maintained at a few kilovolts relative to the chamber walls, which induces the surface charging of the emerging liquid and aerosol formation. The droplets migrate toward a capillary inlet through the countercurrent bath gas, which hastens solvent evaporation and serves to isolate the high-vacuum analyzer from atmospheric pressures. Ions that enter the glass capillary emerge as a supersonic free jet and pass through skimmers into the analyzer (Scheme I).

Electrospray provides two features important for biopolymer characterization: efficient ionization and multiple charging, and both factors contribute to successful CID spectra. The former characteristic produces intense ion beams while the latter allows higher molecular weight determinations. Electrospray spectra of a single protein provides an envelope of ions with the degree and distribution of charge a property of the protein and its milieu. For oligosaccharides, additional ion envelopes ensue as a consequence of microheterogeneity, where any one envelope specifies a glycoform distribu-

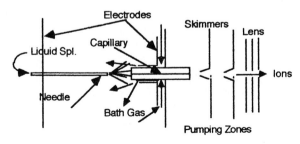

SCHEME I

tion. Although the combination of multiple charge states and microhetero-geneity produces a complex spectrum, in practice they can be easily decon-voluted to yield data from which molecular weights are extracted. Treatment of such data in specific ways suppresses artifacts and improves spectral dynamic range and confidence in peak detection.[9] We illustrate several applications in an effort to arrive at a global strategy for oligosaccha-ride sequencing.

Carbohydrate Glycan Profiling

Introduction

The structural characterization of an oligosaccharide involves gathering molecular detail on three levels: the compositional makeup, the biopolymer array or branching pattern, and an understanding of interresidue linkage. In general, similar elements of structure comprise the glycans of glycoproteins, which have been referred to as *glycotype, glycoform,* and *glycomer.* Carbo-hydrate chains on glycoproteins generally are either O-linked (to threonine or serine) or N-linked (to asparagine). The N-linked glycans extend from the peptide through a constant core region that comprises two 2-acetamido-2-deoxyglucosamine (GlcNAc) residues and a branched mannose (Man) trisaccharide. Except for reducing-terminus fucosylation, this pentasaccha-ride core ($Man_3GlcNAc_2$) remains unmodified during biosynthesis for all N-linked glycans. Extensions from this core ($Man_3GlcNAc_2$-peptide) gives rise to antenna in three basic motifs (glycotypes); high mannose (see Table I), hybrid, and complex (see Table II). Generally, high-mannose glycans possess only Man, the complex glycans, GlcNAc, and galactose (Gal), while the hybrid antennae include residues of both glycotypes. Galactose residues frequently terminate with neuraminic acid (Neu5Ac), a feature termed *capping.* These basic components make up most glycoproteins, with annota-tions to these structures by appropriately positioned fucose, sulfate, phos-phate, and acyl groups giving rise to numerous specific biological activities. Glycan composition provides an indication of glycotype and ES of an intact glycoprotein details the glycoform distribution. These data provide a basis for directing additional structural inquiry. Although not always experimen-tally possible, intact profiles would best represent glycan heterogeneity, uncompromised by selective enzymatic and isolation techniques. Deglyco-sylation may be discriminating or incomplete and chromatography could enrich or exclude particular subforms, thus a constant challenge during

---

[9] B. B. Reinhold and V. N. Reinhold, *J. Am. Soc. Mass Spectrom.* **3,** 207 (1992).

TABLE I
List of Theoretical Mannosidase Trimming Products That May Be Anticipated from High-Mannose Glycotype within Range Man$_{5-9}$GlcNAc^a

Product	Structure

M$_9$ — 9-26-ABC
```
M²-M⁶
        M⁶
M²-M³      M⁴-GlcNAc
        M⁴
M²-M²-M³
```

M$_8$

8-24-BC
```
  M²-M⁶
         M⁶
  M²-M³     M⁴-GlcNAc
  M²-M³
```

8-24-AC
```
  M²-M⁶
         M⁶
  M³      M⁴-GlcNAc
  M²-M²-M³
```

8-24-AB
```
     M⁶
         M⁶
  M²-M³     M⁴-GlcNAc
  M²-M²-M³
```

M$_7$

7-22-A
```
   M⁶
  M³ M⁶
      M⁴-GlcNAc
  M²-M²-M³
```

7-22-B
```
       M⁶
  M²-M³ M⁶
        M⁴-GlcNAc
  M²-M³
```

7-22-C
```
  M²-M⁶
  M³ M⁶
      M⁴-GlcNAc
  M²-M³
```

7-18-AB
```
       M⁶
  M²-M³  M⁶
         M⁴-GlcNAc
  M²-M²-M³
```

7-21-AC
```
  M²-M⁶
       M⁶
       M⁴-GlcNAc
  M²-M²-M³
```

7-22-BC
```
  M²-M⁶
  M²-M³ M⁶
        M⁴-GlcNAc
  M³
```

M$_6$

6-16-A
```
  M³ M⁶
      M⁴-GlcNAc
  M²-M²-M³
```

6-19-A
```
  M⁶
     M⁶
     M⁴-GlcNAc
  M²-M²-M³
```

6-20
```
  M⁶
  M³ M⁶
      M⁴-GlcNAc
  M²-M³
```

6-20-C
```
  M²-M⁶
  M³ M⁶
      M⁴-GlcNAc
  M³
```

6-20-B
```
       M⁶
  M²-M³ M⁶
        M⁴-GlcNAc
  M³
```

6-19-C
```
  M²-M⁶
       M⁶
       M⁴-GlcNAc
  M²-M³
```

6-16-B
```
       M⁶
  M²-M³ M⁶
        M⁴-GlcNAc
  M²-M³
```

M$_5$

5-18
```
  M⁶
  M³ M⁶
      M⁴-GlcNAc
  M³
```

5-17-C
```
  M²-M⁶
       M⁶
       M⁴-GlcNAc
  M³
```

5-14-B
```
  M²-M³ M⁶
         M⁴-GlcNAc
  M³
```

5-13-A
```
  M⁶
     M⁴-GlcNAc
  M²-M²-M³
```

5-17
```
  M⁶
     M⁶
     M⁴-GlcNAc
  M²-M³
```

5-14
```
  M³ M⁶
      M⁴-GlcNAc
  M²-M³
```

^{a} Three-digit notation (e.g., 9-26-ABC) represents mannose residues, sum of linkages, and antenna identification; A, bottom; B, middle; and C, top.

TABLE II
PARTIAL LIST OF COMPLEX GLYCOTYPE GLYCANS COMPILED WITH MOLECULAR WEIGHT
AND PARENT IONS IN +2, +3, AND +4 CHARGE STATE AS SODIUM ADDUCTS[a]

Glycoform	M_r	$[MNa_2]/2$	$[MNa_3]/3$	$[MNa_4]/4$
BiNA$_1$	2583.8	1314.9^{+2}	0884.3^{+3}	0669.0^{+4}
BiNA$_2$	2945.2	1495.6^{+2}	1004.7^{+3}	0759.0^{+4}
BiLac$_1$NA$_1$	3033.3	1539.7^{+2}	1034.1^{+3}	0781.3^{+4}
BiLac$_2$NA$_1$	3482.8	1764.4^{+2}	1183.9^{+3}	0893.7^{+4}
BiLac$_1$NA$_2$	3394.7	1720.4^{+2}	1154.6^{+3}	0871.7^{+4}
BiLac$_2$NA$_2$	3844.2	1945.1^{+2}	1304.4^{+3}	0984.1^{+4}
TriNA$_1$	3033.3	1539.7^{+2}	1034.4^{+3}	0781.3^{+4}
TriNA$_2$	3394.7	1720.4^{+2}	1154.6^{+3}	0871.7^{+4}
TriNA$_3$	3756.1	1901.0^{+2}	1275.0^{+3}	0962.0^{+4}
TriLac$_1$NA$_1$	3482.8	1764.4^{+2}	1183.9^{+3}	0893.7^{+4}
TriLac$_2$NA$_1$	3932.3	1989.1^{+2}	1333.8^{+3}	1006.1^{+4}
TriLac$_3$NA$_1$	4381.8	2213.9^{+2}	1483.6^{+3}	1118.4^{+4}
TriLac$_1$NA$_2$	3844.2	1945.1^{+2}	1304.4^{+3}	0984.1^{+4}
TriLac$_2$NA$_2$	4293.7	2169.8^{+2}	1454.2^{+3}	1096.4^{+4}
TriLac$_3$NA$_2$	4743.2	2394.6^{+2}	1604.1^{+3}	1208.8^{+4}
TriLac$_1$NA$_3$	4205.6	2125.8^{+2}	1424.9^{+3}	1074.4^{+4}
TriLac$_2$NA$_3$	4655.1	2350.5^{+2}	1574.7^{+3}	1186.8^{+4}
TriLac$_3$NA$_3$	5104.6	2575.3^{+2}	1724.5^{+3}	1299.1^{+4}
TetraNA$_1$	3482.8	1764.4^{+2}	1183.9^{+3}	0893.7^{+4}
TetraNA$_2$	3844.2	1945.1^{+2}	1304.4^{+3}	0984.1^{+4}
TetraNA$_3$	4205.6	2125.8^{+2}	1424.9^{+3}	1074.4^{+4}
TetraNA$_4$	4567.0	2306.5^{+2}	1545.3^{+3}	1164.7^{+4}
TetraLac$_1$NA$_3$	4655.1	2350.5^{+2}	1574.7^{+3}	1186.8^{+4}
TetraLac$_1$NA$_4$	5016.5	2531.2^{+2}	1695.2^{+3}	1277.1^{+4}
TetraLac$_2$NA$_3$	5104.6	2575.3^{+2}	1724.5^{+3}	1299.1^{+4}
TetraLac$_2$NA$_4$	5466.0	2756.0^{+2}	1845.0^{+3}	1389.5^{+4}
TetraLac$_3$NA$_3$	5554.1	2800.1^{+2}	1874.4^{+3}	1411.5^{+4}
TetraLac$_3$NA$_4$	5915.5	2980.8^{+2}	1994.8^{+3}	1501.9^{+4}

[a] All listed molecular weights include fucosylated and permethylated structures with single and multiple charging a consequence of sodium cation adduction (23 amu); $[M_r + z(23)]/Z$ is the isotopically measured mass.

isolation is to maintain the fidelity of structures inherent with the starting material. Structural details of attachment site, linkage, and branching are not available in a profile analysis. These features require glycopeptide maps, deglycosylation, and alternative approaches. In this brief chapter we specifically focus on the analysis of isolated glycans and oligosaccharides that are chemically modified to impart greater molecular specificity. Electrospray of these modified analytes provides marked improvements in sensitivity and the products are suitable to CID for an identification of linkage and branching.[10,11]

Oligosaccharide Methylation and Sensitivity

Methylation can be effectively carried out by dissolving the samples in a suspension of dimethyl sulfoxide (DMSO)–NaOH, (prepared by vortexing DMSO and powdered sodium hydroxide) followed by the addition of methyl iodide.[12] The methylated glycans are partitioned into chloroform and extraneous materials back extracted with dilute acetic acid. Methylated oligosaccharides protonate poorly and these samples must be infused into the MS with dilute salt (sodium acetate) solutions, the adducting ions of which provide the vehicle for focusing following desolvation.[11,13] Analyzed in this way, several general features are brought together: (1) easy cleanup by organic solvent extraction, (2) a single monodispersed adducting species appropriate for quantitative profiling, (3) following CID and glycosidic cleavage, a measure of sequence and branching, and from secondary cross-ring cleavages an identification of interresidue linkage, and (4) compatibility with periodate chemistry to augment linkage and branching detail.

An indication of detecting sensitivity can be seen in Fig. 1A–C for a commercially available heptasaccharide prepared by methylation. The plots were prepared from scans over a 10-mass unit interval that included the parent ion for a series of sample dilutions. Although all plots were normalized, an indication of sensitivity can be seen with the change in signal-to-noise ratios on dilution. Importantly, these values exhibit a linear response, (Fig. 1D), a feature not observed with matrix desorption techniques, where surface activity and matrix partitioning are complications of fundamental significance. As discussed below, this dynamic range of detection provides for a wide cross-section of structural features to be measured within a single experiment, from the facile glycosidic linkages to the less abundant cross-ring cleavages important for interresidue linkage assignment.

As noted above, carbohydrate samples and their conjugates are heterogeneous. A comparative profile of this distribution is a critical piece of

[10] V. N. Reinhold, B. B. Reinhold, and S. Chan, *in* "Biological Mass Spectrometry: Present and Future" (T. Matsuo, Y. Seyama, R. C. Caprioli, and M. L. Gross, eds.), pp. 403–434. John Wiley & Sons, New York, 1994.

[11] B. B. Reinhold, S.-Y. Chan, L. Reuber, G. C. Walker, and V. N. Reinhold, *J. Bacteriol.* **176**(7), 1997 (1994).

[12] I. Ciucanu and F. Kerek, *Carbohydr. Res.* **131**, 209 (1984).

[13] M. A. Recny, M. A. Luther, M. H. Knoppers, E. A. Neidhardt, S. S. Khandekar, M. F. Concino, P. A. Schimke, U. Moebius, B. B. Reinhold, V. N. Reinhold, and E. L. Reinherz, *J. Biol. Chem.* **267**, 22428 (1992).

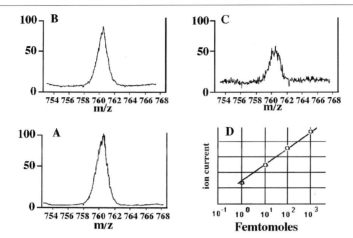

FIG. 1. Detecting sensitivity of methylated maltoheptulose using ES–MS. Standard sample weighed, methylated, and analyzed in serial dilutions by direct infusion; (A) 1.0 pmol; (B) 100 fmol; (C) 10 fmol. Scanned over molecular weight-related ion (m/z 759–762) for parent ion, m/z 760.7, $[(Glu)_7] \cdot 2Na^{2+}$; (D) plot of ion current vs detecting sensitivity.

analytical information that mirrors three fundamental characteristics: the complement of glycosyltransferases expressed during biosynthesis, the *in vivo* physiological milieu of the cell, and the time frame of biosynthesis. Production and marketing of recombinant glycoprotein pharmaceuticals have reinforced the need for reliable procedures for product evaluation, and ES–MS can provide a powerful method by which to fingerprint oligomer distributions and monitor glycosylated products with exceptional fidelity.

The combination of methylation, sodium adduction, and ES–MS yields a quantitative oligosaccharide profile at excellent sensitivity. These attributes also provide for multiple charging and high mass analysis. As seen in Fig. 2, ES–MS at the upper mass limit of a maltodextran sample indicates a degree of polymerization extending beyond 40 residues. Three ion envelopes, centered at 1108.8^{+7}, 1289.9^{+6}, and 1543.3^{+5}, represent the oligosaccharides $(Glu)_n$ ($n = 36$–43). The number of charges on each ion can be determined by deconvolution or from the mass intervals between each ion, (e.g., 29.2, 34.0, and 40.8 Da). Thus, 8-kDa oligosaccharides appear to electrospray as expected, and assuming a comparable charge density one may expect 18- to 20-kDa samples to be easily measured.

N-Linked Glycans

High-Mannose Glycotype. Deglycosylation techniques are well established either by enzymatic or chemical means,[1] and as discussed above,

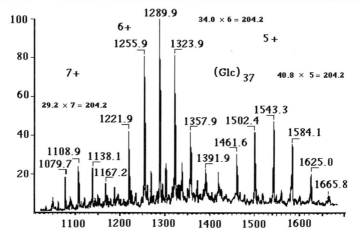

FIG. 2. ES–MS of methylated maltodextran sample [degree of polymerization (DP), 2–43] analyzed by direct infusion; spectrum taken at the high mass end, m/z 1543.3 = [(Glc)$_{40}$] · 5Na^{5+}. Mass intervals determine degree of charging (e.g., 204.2/34 = 6).

glycan methylation and ES–MS provide sensitivity into the low-picomole range. As an example of ES–MS profiling, glycans were obtained from ribonuclease B (single-site N-linked high-mannose glycotype) following enzymatic deglycosylation, methylation, extraction, and ES–MS (Fig. 3). Mass intervals of 162.2/2 indicate hexose residues, a feature characteristic

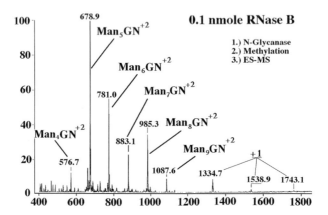

FIG. 3. ES–MS profile analysis of high-mannose glycans obtained from ribonuclease B by endoglycosidase digestion, direct methylation, extraction, and ES–MS. Spectrum showing single charge state for glycan series Man$_n$GlcNAc (n = 4–9). See Scheme VII for detailed structure of Man$_8$GlcNAc.

of a high-mannose glycotype. Glycomers were detected in the two charge states, m/z 576.7, 678.9, 781.0, 883.1, 985.3, and 1087.6 representing the distribution, $(Man)_n GlcNAc$ (n = 4–9). When compared with capillary electrophoresis of the intact glycoprotein (inset, Fig. 3), these spectra showed an identical profile, suggesting that deglycosylation, methylation, and ES–MS exhibited no disproportionate influence. Some carbohydrate glycans, however, are isomeric and molecular weight profiles fail to resolve individual glycomers. For this information, CID or further chemical modification must be considered (see below).

Complex Glycotype. For glycans of the complex type mass profiling provides an assessment of neuraminyl capping and fucosylation, and partial insight into antenna extension and branching, heterogeneity frequently encountered in this glycotype. Presented in Fig. 4 is the glycan profile obtained from a single N-linked site following deglycosylation and methylation. Ion mass intervals equal to Neu5Ac and lactosylamine (Gal1 → 4GlcNAc) indicate the glycans to be of the complex type. The most abundant ion, m/z 1544.6, can be accounted for as a fucosylated TetraNA$_4$ glycan in the three-charged state (Table II, m/z 1545.3^{+3}), with higher homologs of one and two additional lactosamine residues at m/z 1694.8^{+3} and 1844.4^{+3}. A comparable series can be observed at m/z 1424.3^{+3}, 1574.3^{+3},

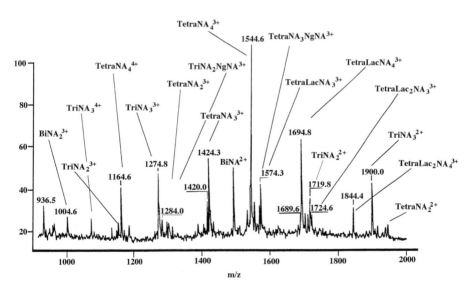

FIG. 4. ES–MS profile analysis of complex glycans obtained from erythropoietin glycopeptide (N-83) following *N*-glycanase deglycosylation, methylation, and ES–MS. Spectrum showing both +2, +3, and +4 charge states. Bi-, Tri-, and Tetra-, number of antennae; NA, neuraminic acid; Lac, Gal(1 → 4)GlcNAc. See Scheme X for detailed structure of TriNA$_3$.

and 1724.6^{+3}, corresponding to the trineuraminyl capped glycans. From this single N-linked site, 11 major structures were detected, all fully fucosylated and never showing more than 2 uncapped antenna (TetraNA$_2$), or 3 additional polylactosamine residues. From ion abundance this profile provides a direct measure of glycan concentration and characterizes its composite glycoform distribution. The task of understanding ion profiles of this complexity is greatly simplified by use of composition–mass tables compiled for each homologous series, (e.g., Neu5Ac$_1$, Neu5Ac$_2$, Neu5Ac$_3$, Lac$_1$, Lac$_2$, Lac$_3$), at +2, +3, and +4 charge states (natural abundance) (Table II).

Linkage and Branching Information

Collision-Induced Dissociation (ES–MS/CID/MS)

In addition to enhanced ionization efficiency and easier product isolation, methylation stabilizes monomers from noninformative, small mass losses following CID. Probably the most significant attributes are a series of glycosidic and cross-ring cleavages that combine to allow an understanding of sequence, branching and linkage.

Three factors make this possible: first, the adducting ion(s) show a statistical distribution for all residues in the polymer, and thus, for any glycosidic bond ruptured, the probability is high that both product ions will appear charged and mass measurable. Second, the glycosidic cleavage is unsymmetrical, e.g., nonreducing fragments eliminate across the C-1–C-2 bond to become unsaturated, while the reducing end fragment retains the glycosidic oxygen and accepts a hydrogen (Scheme II). Thus, reducing end and nonreducing fragments show a respective 14- and 32-Da decrement from their methylated residue weights, providing a direct indication of

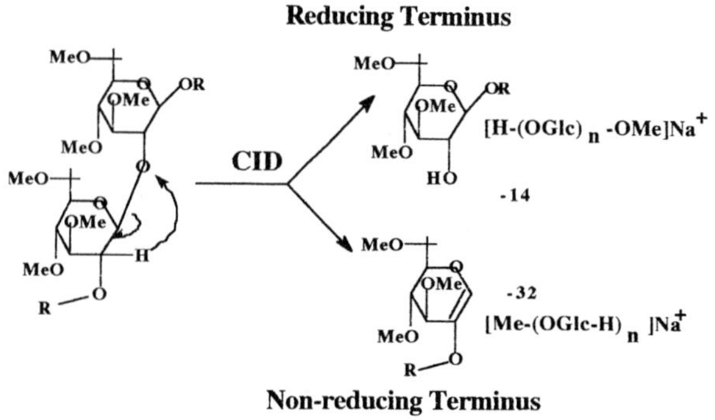

SCHEME II

origin. This feature is not only helpful in defining each fragment, but is enormously important in the structural interpretation of branched molecules, (see below). The mass disparity between these fragments could not be assessed in the absence of methylation. Third, the mass of residual cross-ring fragments pendent to nonreducing termini allows a direct assignment of linkage type.

Collision-Induced Dissociation of Linear Oligosaccharide. The significance of these features can be best illustrated by considering the CID spectrum of a methylated, linear homopolymer represented by the notation $CH_3(OGlc)_7-OCH_3$ (Fig. 5). Collision of the parent ion, m/z 1498, $[CH_3(OGlc)_7-OCH_3] \cdot Na^+$, provided six abundant glycosidic fragments starting at m/z 259 and ending with m/z 1280. The mass of these ions is consistent with reducing end fragments, $[H-(OGlc)_n-OCH_3] \cdot Na^+$ ($n = 1-6$), exhibiting the unsymmetrical rupture with increments of one hexosyl residue. An additional six nonreducing fragments can be observed starting at m/z 241 and ending with m/z 1262, $[CH_3(OGlc)_n-(OGlc-H)] \cdot Na^+$ ($n = 0-5$) (see inset, Fig. 5). The two major ion series define sequence with unique masses from both the reducing and nonreducing termini. It is interesting to note from their abundance that metal ion adduction shows greater tenacity

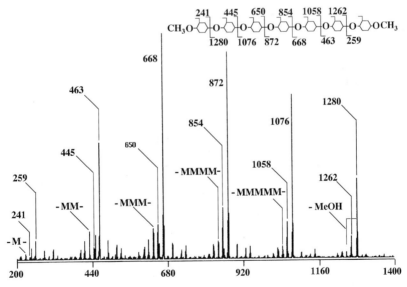

FIG. 5. Tandem MS; collision-induced dissociation (CID) spectrum of ES ionized methylated maltoheptulose, m/z 760.7, $[(Glc)_7] \cdot 2Na^{2+}$. Major fragments as a result of glycosidic cleavage from reducing and nonreducing ends.

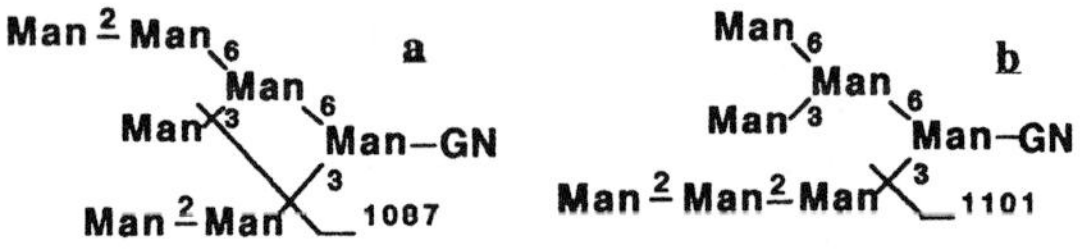

Man $\overset{2}{-}$ Man$_6$ a   Man$_6$ b
      Man$_6$ Man$_6$
Man$\overset{3}{\diagdown}$ Man–GN   Man$\diagup$^{3} Man–GN
      $\diagup$^{3}
Man $\overset{2}{-}$ Man $\underline{}$1087   Man $\overset{2}{-}$ Man$\overset{2}{-}$Man$\overset{3}{\diagdown}$$\underline{}$1101

SCHEME III

to reducing end fragments than nonreducing termini (cf. Fig. 5, m/z 872 and 650).

Collision-Induced Dissociation of Branched Structures. Fragments derived from a glycosidic rupture can undergo additional cleavage(s) at alternate sites that define branching. Such fragments are diminished in abundance, but are easily detected from the low ES background. With methylated samples each glycosidic rupture provides a mass shift that identifies branching frequency. This important difference allows a unique measure of branching not available with underivatized samples and allows selected structural elements to be readily detected, even within structural isomers.

As an example of this, a Man$_7$GlcNAc structure responsible for CD2–LFA-3 cell adhesion was isolated and provided a CID spectrum indicating a trimannosyl loss fragment, m/z 1087. This was characterized as the double-cleavage ion and was accounted for by considering the loss of three mannosyl residues from two loci (i.e., two glycosidic cleavages), because it was 28 Da (2 × 14) lower than the fully methylated ion (Scheme IIIa).[13,14] By contrast, a Man$_7$GlcNAc glycan isolated from ribonuclease B showed a different CID spectrum, also with a unique trimannosyl loss fragment 14 Da higher, m/z 1101. This indicated a single glycosidic rupture that could only arise from a completed lower antenna (Scheme IIIb). Thus, the CD2 glycomer was either 7-22-B or 7-22-C (Table I), which was resolved by further examination of cross-ring cleavages (see below) to be the latter structure.

Collision-Induced Dissociation and Interresidue Linkage. Several laboratories have presented data indicating CID spectra could yield linkage information. Reports have utilized differential ion abundance of glycosidic fragments,[15,16] unique ions generated by periodate oxidation,[17–19] lithium

[14] B. B. Reinhold, M. A. Recny, M. H. Knoppers, E. L. Reinherz, and V. N. Reinhold, *in* "Techniques in Protein Chemistry III" (R. H. Angeletti, ed.), p. 287. Academic Press, New York, 1992.

[15] R. A. Laine, K. M. Pamidimukkala, A. D. French, R. W. Hall, S. A. Abbas, R. K. Jain, and K. L. Matta, *J. Am. Chem. Soc.* **110,** 6931 (1988).

[16] R. A. Laine, E. Yoon, T. J. Mahier, S. A. Abbas, B. de Iappe, R. K. Jain, and K. L. Matta, *Biol. Mass Spectrom.* **20,** 505 (1991).

[17] A. S. Angel, F. Lindh, and B. Nilsson, *Carbohydr. Res.* **168,** 15 (1987).

[18] A. S. Angel and B. Nilsson, *Biomed. Environ. Mass Spectrom.* **19,** 721 (1990).

[19] A. S. Angel, P. Lipniunas, K. Erlansson, and B. Nilsson, *Carbohydr. Res.* **221,** 17 (1991).

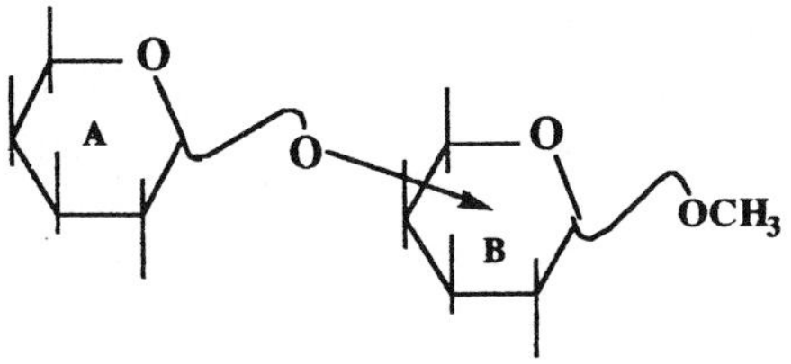

SCHEME IV

ion chelation,[20] or high-energy CID of methylated samples.[21] This latter report continued studies, initiated three decades earlier, showing specific fragments for each of the three linkage types $(1 \rightarrow 6, 1 \rightarrow 4,$ and $1 \rightarrow 2)$ using methylated disaccharides fragmented by electron impact.[22] In this study, two sets of ions carried linkage information and both formed as a result of proximal ring rupture with residual fragments attached through the intervening glycosidic bond (Scheme IV).

As expected, the two bond cross-ring ruptures of the A ring were found to be common among all linkage types, while rupture of the B ring provided ions reflecting linkage position (Scheme IV). Because only high-energy electron ionization was available, larger saccharides could not be evaluated for a higher oligomer sequence strategy.

With the development of "soft" ionization techniques, high molecular weight polymers could be ionized and much attention was directed to native (unmodified) oligosaccharides, where minor cross-ring cleavages were not detected. The differential instrumental approach of low energy for ionization with the higher energies needed for detailed fragmentation was not available. This was partly resolved when two groups[23,24] independently demonstrated that ions could be individually fragmented by introducing a collision gas between the magnet and electrostatic analyzer, providing a high-energy collision-induced dissociation (CID). Most importantly, this new technique provided spectra qualitatively similar to those obtained by electron ionization. Thus, instruments combining "soft" ionization and CID exposed for the first time the possibility of low-energy polymer ionization with the higher energy needs of sequence. As mentioned above, high-energy CID of methylated oligosaccharides ionized by FAB provided cross-ring fragments that identify linkage position in oligomers of 5–10 residues,

[20] Z. Zhou, O. Ogden, and J. Leary, *J. Org. Chem.* **55,** 5444 (1990).

[21] J. Lemoine, B. Fournet, D. Despeyroux, K. R. Jennings, R. Rosenberg, and E. de Hoffmann, *J. Am. Soc. Mass Spectrom.* **4,** 197 (1993).

[22] O. S. Chizhov, L. A. Polyakova, and N. K. Kochetkov, *Doklady Akademii Nauk SSSR* **158,** 685 (1964).

[23] K. R. Jennings, *Int. J. Mass Spectrom. Ion Phys.* **1,** 227 (1968).

[24] W. F. Hadden and F. W. McLafferty, *J. Am. Chem. Soc.* **90,** 4745 (1968).

while collisions at lower energies were reported to be unsuccessful.[21] The ballistic technique of FAB requires a desorption matrix that contributes large backgrounds. This is not a complication for the detection of labile peptide or glycosidic bonds, but does constrain detailed investigation of minor fragments needed to understand oligosaccharide linkage and branching. Electrospray ionization,[8] by contrast, is a desolvation technique not compromised by a matrix and provides a better opportunity to evaluate minor fragments. We reevaluated the earlier work of Chizhov et al.,[22] and applied it to several structural problems using ES and the lower CID energies available in the triple quadrupole.[25,26] The spectral differences in CID between the high-energy sector and the low-energy quadrupole instruments suggest that fragmentation pathways in the former instrument are less selective. In high-energy collisions, energy is deposited on the ion in a single collisional event, in a short time frame. In the triple quadrupole, by contrast, collisions occur over much longer time scales, allowing multiple events. In these circumstances dissociation paths having low activation energies will predominate even if their associated phase space volume or entropy is low. Thus, the direct correlation of abundance to collision energy may not follow, and this could account for the qualitative differences observed in high- and low-energy CID spectra.

As indicated above, a linkage between any two residues can be determined by ions defining an intact glycosidic bond linked to fragments from the reducing side ring, thus only nonreducing terminal fragments will carry linkage information. Fortunately, each linkage type gives rise to a different pendant fragment that defines interresidue linkage, e.g., 2-O linkages, a 74-Da increment; 3-O linkages, no related ions; 4-O linkages, an 88-Da increment; and 6-O linkages, a combination of two ions, 60- and 88-Da increments (Scheme V). Thus, for the simple $(1 \rightarrow 4)$-linked maltoheptulose (Fig. 5), each nonreducing glycosidic fragment can be identified at m/z 241, 445, 650, 854, 1058, and 1262 (inset, Fig. 5). Because linkage-defining ions represent mass increments, they are found as satellite peaks to the right of these glycosidic fragments. To appreciate these structural details, an expansion of the disaccharide region, m/z 380–580, is presented in Fig. 6. In this spectral expansion, three cross-ring cleavage ions can be identified, m/z 491, 519, and 533, but only the latter two ions provide linkage information. The absence of the m/z 505 ion (+60), with the detection of m/z 533

[25] M. A. Velardo, R. A. Bretthauer, A. Boutaud, B. B. Reinhold, V. N. Reinhold, and F. J. Castellino, J. Biol. Chem. **256**, 17902 (1993).
[26] M. A. Deeg, D. R. Humphrey, S. H. Yang, T. R. Ferguson, V. N. Reinhold, and T. R. Rosenberry, J. Biol. Chem. **267**, 18573 (1992).

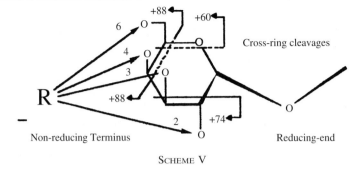

SCHEME V

and 519 (methyl loss fragment), indicates the linkage to be 4-linked. The former ion, m/z 491, represents a reducing end fragment that carries no linkage information. Because this sample was a homopolymer, all fragment intervals are identically shifted only by 204 Da, the residue mass of each monomer.

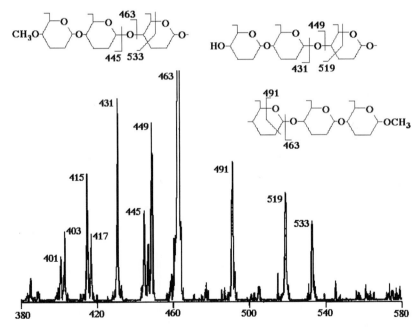

FIG. 6. Tandem MS; collision-induced dissociation (CID) spectrum of ES ionized methylated maltoheptulose, m/z 760.7, $[(Glc_7)] \cdot 2Na^{2+}$. Expanded mass range from Fig. 5 (m/z 380–580), detailing cross-ring cleavages.

In addition to an unsymmetrical cleavage at the glycosidic bond, nonreducing and reducing fragments from N-linked high-mannose samples are easy to define by their additional lack of molecular symmetry. With internal or double-cleavage fragments, however, where this symmetry is lost, the glycosidic rupture assists in defining polarity. Establishing the nonreducing polarity of an ion is important, for it defines the area to search for linkage fragments. Linkage discrimination between the high-mannose glycans often requires an examination of the cross-ring cleavages associated with 6- and 3-linked antenna. In theory, when provided any individual structure within the high-mannose motif, a combination of glycosidic fragments and cross-ring cleavages will provide adequate linkage, sequence, and branching information to assign correctly each of the 26 high-mannose glycomers, $Man_n GlcNAc$ ($n = 4$–9) (Table I).

As an example of how these principles can define structural detail, the analysis of a single $Man_8 GlcNAc$ glycomer is described below. The composition, $Man_8 GlcNAc$, could fit any of three possible structures of the high-mannose glycotype (8-24-AB, 8-24-AC, and 8-24-BC; Table I). The doubly charged molecular weight-related ion was selected, m/z 978.2^{+2} [$(Man)_8 GlcNAc] \cdot 2Na^{2+}$, for CID, which provided the spectrum in Fig. 7. Single glycosidic cleavage fragments dominate the spectrum, with a smaller

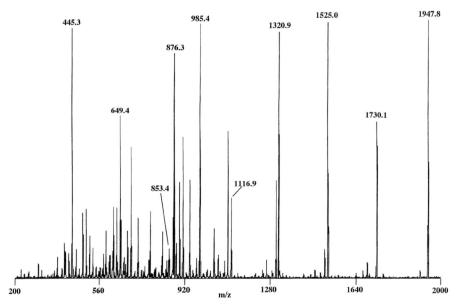

FIG. 7. Tandem MS; collision-induced dissociation (CID) spectrum of ES ionized [$(Man)_8$-GlcNAc] $\cdot 2Na^{2+}$, m/z 978.2^{+2}.

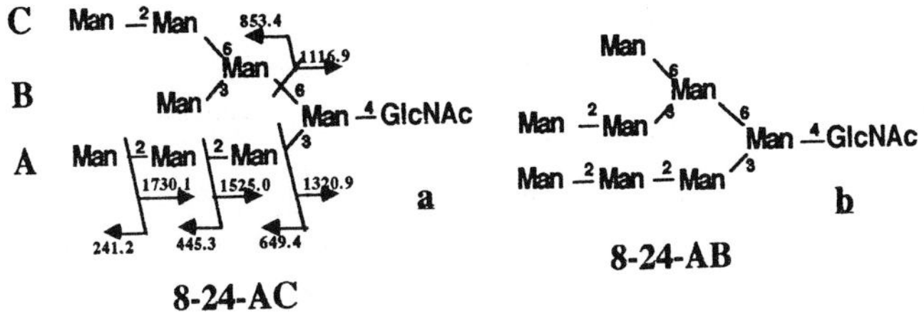

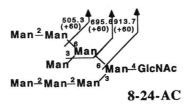

C Man —^{2}Man 853.4

B Man—^{3}

A Man $|^2$ Man $|^2$ Man—^{3}

Man ^{6} Man

1116.9

Man—^{4}GlcNAc

1730.1 1525.0 1320.9

241.2 445.3 649.4

8-24-AC

Man

^{6}Man

Man —^{2}Man—^{6} Man—^{4}GlcNAc

Man —^{2}Man —^{2}Man—^{3}

b

8-24-AB

SCHEME VI

set of double glycosidic and cross-ring cleavages. The highest mass ion can be accounted for by sodium loss, m/z 1947.8 [(Man)$_8$GlcNAc] · Na$^+$, followed by three reducing end fragments, all indicating a single glycosidic cleavage (e.g., 14 Da lower than their methylated mass composition), m/z 1730.1, [(Man)$_7$GlcNAc] · Na$^+$; m/z 1525.0, [(Man)$_6$GlcNAc] · Na$^+$; and m/z 1320.9, [(Man)$_5$GlcNAc] · Na$^+$ (Scheme VI). Only the latter fragment defines a specific segment, loss of the completed "A" antenna. This is supported by the corresponding nonreducing fragments, m/z 445.3, [CH$_3$(OGlc)-(OGlc-H)] · Na$^+$, and m/z 649.4, [CH$_3$(OGlc)$_2$-(OGlc-H)] · Na$^+$, with the latter ion again defining a completed 3-O-linked "A" antenna. These data eliminate the 8-24-BC glycomer, leaving the two structures in Scheme VI as possible candidates.

 Defining which of these two structures are present (8-24-AC, 8-24-AB), requires locating the 2-linked terminal mannose to either the "B" or "C" antenna, a problem resolved with cross-ring cleavage fragments. Three nonreducing terminal fragments, all showing 60-Da increments, define a completed "C" antenna, m/z 505.3, 695.6, and 913.7 (Scheme VII). Expansion of Fig. 7 between m/z 640–750 (Fig. 8) and m/z 840–980 (Fig. 9) shows two of these latter fragments and their associated 88-Da increment. Equally supportive are the nonreducing fragments, m/z 853.4, [(Man)$_4$] · Na$^+$ (which must include components of the "B" and "C" antenna) and the missing ions within each sequence, notably [(Man)$_3$GlcNAc] · Na$^+$, [(Man)$_6$GlcNAc] · Na$^+$, and (Man)$_5$ · Na$^+$.

505.3 695.6 913.7

(+60) (+60) (+60)

Man—^{2}Man

^{6}

$_3$ Man

Man—^{7}

^{6} Man—^{4}GlcNAc

Man—^{2}Man—^{2}Man—^{3}

8-24-AC

SCHEME VII

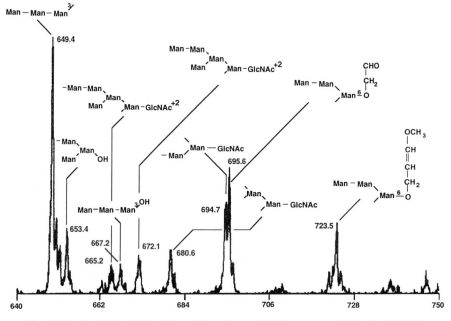

FIG. 8. Tandem MS; collision-induced dissociation (CID) spectrum of ES ionized [(Man)$_8$-GlcNAc] · 2Na^{2+}, m/z 985.4^{+2}. Expanded mass range (m/z 640–750) from Fig. 7, showing collision product ion detail for obtaining linkage and branching detail.

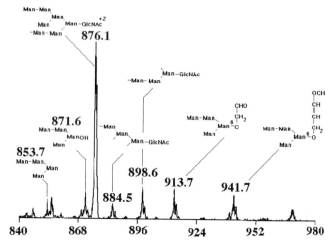

FIG. 9. Tandem MS; collision-induced dissociation (CID) spectrum of ES ionized [(Man)$_8$-GlcNAc] · 2Na^{2+}, m/z 985.4^{+2}. Expanded mass range (m/z 840–980) from Fig. 7, showing product ion detail for obtaining linkage and branching detail.

Identifying components within isomeric mixtures can be particularly difficult where the power of ion selection for CID is lost. However, as demonstrated above, any one component may give rise to a unique fragment and these ions can be utilized to unravel mixtures. The assignment of ion structures in the above data has been consistent with CD_3 permethylation and deuterium reduction prior to permethylation.

Oxidation, Reduction, and Methylation of N-Linked Glycans

Introduction. Although CID analysis provides considerable insight into the complexities of glycan structure, difficulties may still arise with isomeric mixtures. Under these circumstances a quite different approach for understanding structural detail can be utilized. This relies on chemical selectivity to differentiate isomeric structures by modifying residue weights in a manner unique to linkage type (Scheme VIII). This chemical method exploits the presence of adjacent hydroxyl groups and their susceptibility to cleavage by oxidation.[17-19] The shift in molecular weight on oxidation, reduction, and methylation (ORM) is a sum of residue weights that change with each linkage type (Scheme VIII).

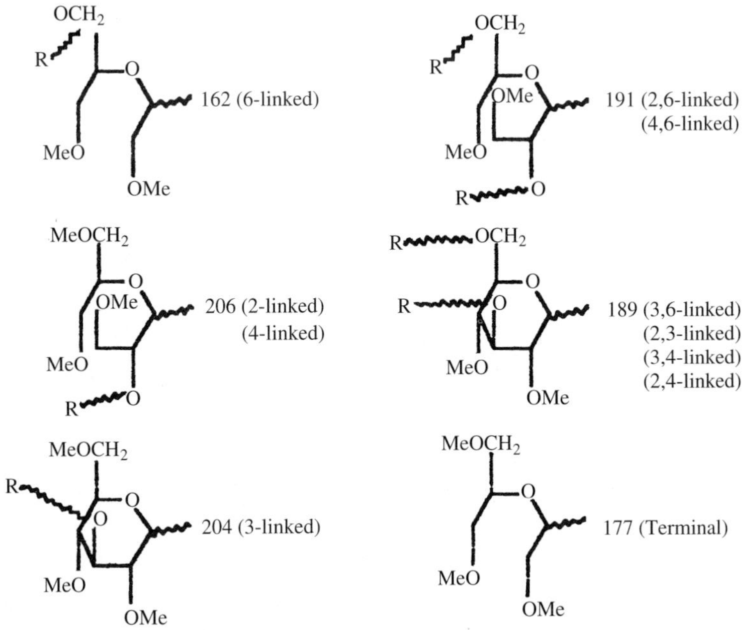

SCHEME VIII

In the information acquired, ORM parallels methylation analysis (methylated alditol acetates), but the amounts of sample required are considerably less. Moreover, the products can be subjected to further collision analysis for sequence, both linkage and branching. Where tandem MS instrumentation is unavailable, glycan profiling following ORM limits the number of structures for consideration, and in some cases may define specific glycomers.

Oxidation, Reduction, and Methylation of High-Mannose Glycans. The following example illustrates the advantages of ORM chemistry with a high-mannose glycan mixture isolated from a plant glycoprotein. The fraction was divided, one fraction methylated with CD_3I and the other prepared by ORM chemistry. The CD_3-methylated glycans showed four major ions, $[(Man)_{5-8}GlcNAc] \cdot 2Na^{2+}$, when analyzed by ES–MS (Fig. 10a). From Table I, this combination of 4 ions indicates 22 possible candidates should be considered. The second fraction showed the appropriate mass shift for four

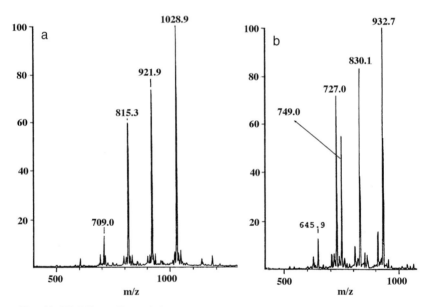

FIG. 10. ES–MS profile analysis of methylated glycans obtained from a soybean lectin prepared by endoglycosidase treatment, methylation, extraction, and direct MS infusion. (a) Parent ions, m/z 709.0, $[(Man)_5GlcNAc] \cdot 2Na^{2+}$; m/z 815.3, $[(Man)_6GlcNAc] \cdot 2Na^{2+}$; m/z 921.9, $[(Man)_7GlcNAc] \cdot 2Na^{2+}$; m/z 1028.9, $[(Man)_8GlcNAc] \cdot 2Na^{2+}$, respectively. (b) Same glycan mixture analysis preceded by periodate oxidation and reduction before methylation; parent ions, m/z 645.9 $[(Man)_5GlcNAc] \cdot 2Na^{2+}$; m/z 727.0 $[(Man)_6GlcNAc] \cdot 2Na^{2+}$; m/z 749.0 $[(Man)_6GlcNAc] \cdot 2Na^{2+}$; m/z 830.1 $[(Man)_7GlcNAc] \cdot 2Na^{2+}$; and m/z 932.7, $[(Man)_8GlcNAc] \cdot 2Na^{2+}$, respectively.

of the ions, with an additional ion for $[(Man)_6GlcNAc] \cdot 2Na^{2+}$, m/z 727.0^{+2} and 749.0^{+2} (Fig. 10b). The ion intervals for all fragments indicate a high-mannose glycotype and the mass shifts due to ORM chemistry limit the isomeric possibilities to four for the Man$_7$GlcNAc (7-22-A, 7-22-B, 7-22-C, and 7-22-BC), two for the Man$_6$GlcNAc at m/z 748.9^{+2} (6-16-A and 6-16-B), and three for the Man$_6$GlcNAc at m/z 727.0^{+2} (6-20-B, 6-20-C, and 6-20), decreasing the glycomer possibilities by 10. The chemistry of ORM does not limit the Man$_8$GlcNAc possibilities and they remain at three (8-24-AC, 8-24-AB, and 8-24-BC). Presented in Fig. 11 is the CID spectrum of the Man$_8$GlcNAc peak along with an inset of the determined structure compiled from the residue masses. A series of fragments in both the +1 and +2 charge states define the 3-branch (m/z 1666.5, 1461.0, and 1254.5), with this latter ion a unique signature for completion, again limiting the glycomer structure to either 8-24-AB or 8-24-AC. Further identification of this gly-comer would require the characterization of cross-ring cleavage fragments as discussed above (Scheme V).

Oxidation, Reduction, and Methylation of Complex Glycans. Profiling glycoprotein glycans by ES–MS provides an important first level of structural inquiry relating to glycotype and glycoform distribution. In complex glycans, the more subtle details of structure (e.g., location of polylactosa-mine groups, neuraminyl capping, linkage, and branching) are positional

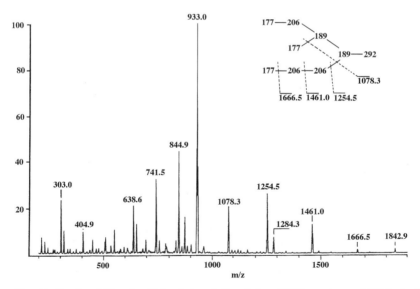

FIG. 11. Tandem MS and collision-induced dissociation (CID) spectrum of ES ionized $[(Man)_8GlcNAc] \cdot 2Na^{2+}$, m/z 933.0^{+2}, selected from spectrum presented in Fig. 10b.

isomers transparent to profiling (Fig. 4). The chemistry of periodate oxidation provides a selective strategy to unravel these features. Also, the positional location of lactosylamine groups relative to the 4 or 4′ core mannose in triantennary structures yields unique products by ORM (Scheme IX). With these structural entities, each neuraminyl residue will decrease the glycan mass by 88 Da, and the 4′-mannosyl moiety will increment by 2 Da (4 Da if a 4′-linked triantenna). Furthermore, a terminal fucosyl group will cause a decrease in mass by 42 Da (adjacent glycols), as will all uncapped galactosyl termini. Not only does the new pattern of ions unravel difficult isomeric problems, it also provides an opportunity to reexamine and verify structures previously profiled following methylation.

The mass differences imparted to selected structures following ORM chemistry can be as little as 2 Da (Scheme X), and multiple charging diminishes the measured mass even further, (e.g., $2/z$). This can be offset by reducing with $NaBD_4$ during the reduction step. An example of this approach is presented in Fig. 12, which profiles the glycans obtained from one glycopeptide. As an example, the most abundant glycan (m/z 1544.5^{+3}, TetraNA$_4$), would be expected to undergo a loss of 352 Da from the four neuraminyl groups (C-8 and C-9 loss), 42 Da from the fucosyl residue (C-3 loss), two mass units would be added due to the single cis-glycol group on

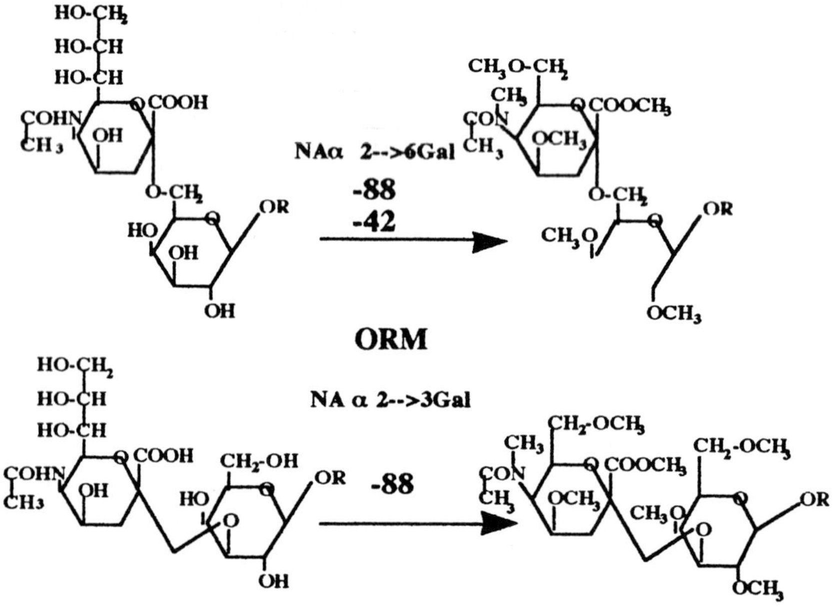

SCHEME IX

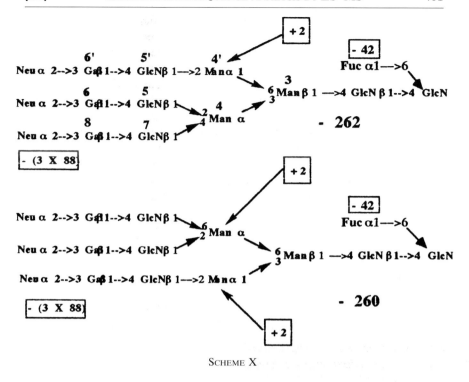

SCHEME X

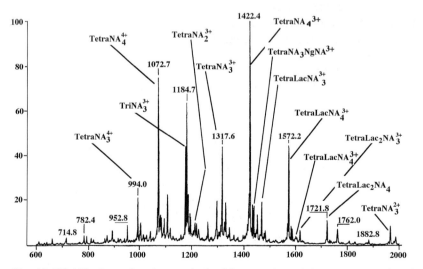

FIG. 12. ES–MS of complex glycans obtained from erythropoietin glycopeptide (N-83) following *N*-glycanase treatment, periodate oxidation, reduction, and methylation.

the 4'-mannose, and the reducing terminus would increment the mass by 16 Da (glucosylaminitol formation during reduction) for a summed loss of 376 Da, m/z 1419.7^{+3}, when using NaBH$_4$. Because this sample was reduced with NaBD$_4$, the expected product ion would shift to m/z 1422.9^{+3}. The experimental data provided an abundant product ion at m/z 1422.2^{+3} (Fig. 12), accounting for the modifications discussed. The rest of the ions can be rationalized using the same considerations.

The significance of this approach was realized in detailing the glycan structures on two adjacent glycopeptides from erythropoietin, N-38 and N-83. The product ions indicated a decided difference in branching (4- vs 4'-) at the two sites with masses at m/z 1181.2^{+3} and 1182.5^{+3}, respectively. This was confirmed by reanalysis after mixing the samples and showing resolution of the ion pair.

A series of suppositions is frequently made by mass spectroscopists during carbohydrate analysis, that core hexose and aminosugar mass increments, and all anomeric configurations, are as previously assigned. Most of these assumptions are appropriate when assessing modifications to established motifs, but if adequate amounts of material are available these features should be confirmed by nuclear magnetic resonance (NMR).

Summary

This chapter summarizes several strategies for a more complete understanding of carbohydrate structure with a focus on N-linked glycans. The techniques include periodate oxidation to impart greater molecular specificity, functional group blocking by methylation, electrospray for "soft" and efficient ionization, collision-induced dissociation to obtain detailed fragmentation, and tandem mass spectrometry for mass separation and analysis. The lipophilic products following derivatization contribute to product cleanup by solvent extraction; they also improve sensitivity during ES and, when combined with CID, yield detailed sequence, linkage, and branching information.

[17] Reversed-Phase Peptide Mapping of Glycoproteins Using Liquid Chromatography/Electrospray Ionization–Mass Spectrometry

By William S. Hancock, A. Apffel, J. Chakel, C. Souders,
T. M'Timkulu, E. Pungor, Jr., and A. W. Guzzetta

Introduction

Carbohydrate chemistry, while highly complex, is of great importance in the study of biological processes. The nature of the diversity of monosaccharides and the variety of possible linkages give rise to the complexity. Unlike amino acids, which are linked through an amide bond, monosaccharides are joined through a variety of hydroxyl groups present on the sugar to form glycosidic linkages.[1] Hellerqvist and Sweetman give an extraordinary example of the linkage possibilities of a hexasaccharide yielding a possible 4.76×10^9 structures.[2] This number, however, is arrived at without using the constraints and conventions of mammalian carbohydrates that exhibit common core structures. Even with such constraints, we are left with a large degree of compositional and linkage variety at any one glycosylation site on the protein. Determining the carbohydrate composition, type, and branching pattern is an important step in understanding the biological function of these molecules as well as in the development of a recombinant DNA-derived glycoprotein as a pharmaceutical.[3–7] For a broad scope of the role of oligosaccharides in biology the reader is referred to a review.[8]

The complexity of carbohydrate structures mandates that a variety of analytical methods be used for the study of these forms. The reader is directed to the chapters on carbohydrate analysis by high-performance

[1] C. G. Hellerqvist, *Methods Enzymol.* **193,** 554 (1990).

[2] C. G. Hellerqvist and B. J. Sweetman, "Biomedical Applications of Mass Spectrometry," Vol. 34: Methods in Biochemical Analysis, pp. 97–143. John Wiley & Sons, New York, 1990.

[3] D. T. Liu, Glycoprotein pharmaceuticals: Scientific and regulatory considerations, and the US orphan drug act. *Trends Biotechnol.* **10,** 114 (1992).

[4] J. A. Chakel, J. A. Apffel, and W. S. Hancock, *LC-GC* 1995 (in press).

[5] R. S. Rush, P. L. Derby, P. L. Smith, D. M. Smith, C. Merry, G. Rogers, M. F. Rhode, and V. Katta, *Anal. Chem.* **67,** 1442 (1995).

[6] P. L. Weber, T. Kornfelt, N. K. Klausen, and S. M. Lunte, *Anal. Biochem.* **225,** 135 (1995).

[7] A. P. Hunter and D. E. Games, *Rapid Commun. Mass Spectrom.* **9,** 42 (1995).

[8] A. Varki, *Glycobiology* **3**(2), 97 (1993).

liquid chromatography (HPLC) (see [6] in this volume[8a]), capillary electrophoresis (see [14] in this volume[8b]), and mass spectrometry (see [16] in this volume[8c]) for further details. A promising new procedure is described in this chapter that uses HPLC in conjunction with electrospray ionization mass spectrometry (ESI/MS) as a tool to identify the sites of glycosylation and the general nature of the glycosylation. ESI/MS can detect whether an oligosaccharide is O-linked or N-linked. It can also differentiate between complex, high-mannose, or hybrid forms. This chapter also examines the use of this approach to gain limited linkage order information using collision-induced dissociation (CID) with both a single- and triple-quadrupole mass spectrometer.

While the approach of on-line MS mapping is a useful addition to the battery of techniques used to characterize carbohydrates it supplements but does not replace other important methods, such as nuclear magnetic resonance spectroscopy (NMR), used to determine linkage positions or anomeric configurations. The reader is also referred to the chapter on matrix-assisted laser desorption/ionization–time of flight (MALDI-TOF) mass spectrometry, as the new analytical method allows the analysis of both the glycoprotein (peptide) as well as the oligosaccharides that can be released by chemical or enzymatic cleavages. A key to the ability of MALDI-TOF to allow the analysis of such a wide range of solutes is in the choice of a suitable matrix for the laser desorption process, and better selections have been reported for oligosaccharide analysis.[9]

Application of Mass Spectrometry to Glycoprotein Analysis

Until recently mass spectrometry was not widely available to the biology researcher, owing to the high cost of the technology and the mass range limitations. Most were limited to gas chromatography–mass spectrometry (GC–MS) systems and the low mass range capabilities of this instrument. The introduction of electrospray ion sources that are coupled with the quadrupole mass filters has produced a mass spectrometer that is easily compatible with HPLC and capillary electrophoresis (CE) analyses and measurement of biological samples and at a reasonable expense (see other chapters in this volume). The design of the ion source is described elsewhere

[8a] R. R. Townsend, L. J. Basa, and M. W. Spellman, *Methods Enzymol.* **271**, Chap. 6, 1996 (this volume).

[8b] M. V. Novotny, *Methods Enzymol.* **271**, Chap. 14, 1996 (this volume).

[8c] V. N. Reinhold, B. B. Reinhold, and S. Chan, *Methods Enzymol.* **271**, Chap. 16, 1996 (this volume).

[9] M. Mohr, K. O. Borosen, and M. H. Widmer, *Rapid Commun. Mass Spectrom.* **9**, 809 (1995).

in this volume and produces multiply charged ions of large biomolecules, making these samples fit within the mass range of most quadrupoles.[10–15] The advantage of the multiple charging phenomena is twofold. Each related ion in a series can be thought of as an independent measurement of the parent mass. These multiple assessments of the parent mass are used to obtain highly accurate mass information, typically <0.02%.[16] For example, the m/z ratio of N-glycanase-treated, reduced, and S-carboxymethylated recombinant tissue plasminogen activator (rtPA) (61221.6 Da), with 40 proton adducts, is 1531.5. This mass-to-charge value is well within the mass range of the analyzer (our unpublished result, 1995), because a mass scan is usually within the mass range of 100 to 2000 Da. Usual scan times can range from 1 to 8 sec, depending on how the instrument parameters are set.

High-Performance Liquid Chromatography

The liquid chromatograph, in this case an HPLC, has been the standard in establishing protein pharmaceutical integrity and is relied on by the quality control departments within the biotechnology industry.[3] The HPLC is used to establish purity for the intact protein and in a more detailed manner peptide maps are used to look for microheterogeneity of the sample as well as degradation products. Because most MS ion sources require flow rates at less than common chromatographic levels, a split of the HPLC eluent stream must be accomplished. For example, most microbore columns with an internal diameter of 1 to 2 mm utilize flow rates of 40 and 200 μl/ min, respectively). Most often this split is made before the ultraviolet (UV) detector to guard against possible dilution effects caused by the flow cell.[17] The flow rate capabilities of the mass spectrometer are presently less of an issue with the increasing popularity of capillary HPLC, which can generate flow rates of a few microliters per minute and has the added advantage of allowing increased sensitivity of detection (typically 10-fold more than traditional analytical systems). The reader is referred to Novotny ([14] in this volume[8b]) for further details. Capillary packed-HPLC columns have an internal diameter ranging from 500 μm to a present lower limit of 50

[10] R. J. Anderegg, "Biomedical Applications of Mass Spectrometry," Vol. 34: Methods in Biochemical Analysis, pp. 1–89. John Wiley & Sons, New York, 1990.

[11] C. K. Meng, M. Mann, and J. B. Fenn, Z. Phys. C. Atoms Molecules Clusters **10,** 361 (1988).

[12] J. A. Loo, H. R. Udseth, and R. D. Smith, Biomed. Environ. Mass Spectrom. **17,** 411 (1988).

[13] A. P. Bruins, T. R. Covey, and J. D. Henion, Anal. Chem. **59,** 2624 (1987).

[14] U. A. Mirza, S. L. Cohen, and B. T. Chait, Anal. Chem. **65,** 1 (1993).

[15] B. T. Chait and B. H. Kent, Science **257,** 1885 (1992).

[16] J. D. Henion, B. A. Thompson, and P. H. Dawson, Anal. Chem. **54,** 451 (1980).

[17] A. Apffel, S. Fisher, G. Goldberg, P. Goodley, and F. E. Kuhlman, J. Chromatogr, **712,** 177 (1996).

μm. In general the diameter value at which "micro" changes to "capillary" is somewhat arbitrary. The 300-μm capillary column requires a flow rate of 3 to 5 μl/min, which makes it compatible with most electrospray ion sources.[13] A high-flow nebulizer that allows LC/MS to be performed at flow rates of up to 1 ml/min has been developed, and thus the effluent from an analytical column can be introduced into the MS without stream splitting. A typical configuration uses N_2 as the nebulizing gas (1.5 liters/min) and with drying at 300°.

Figure 1 shows the result of a reversed-phase separation of an rtPA

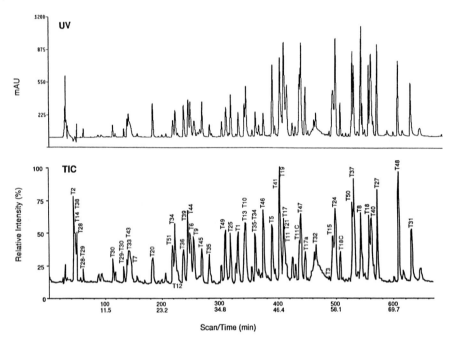

FIG. 1. Total ion current (TIC) map generated with 30 pmol (ca. 1.8 mg), of a tryptic digest of rtPA separated using a 0.32 × 250 mm Nucleosil C_{18}, 5-μm capillary reversed-phase column. The separation conditions were 0 to 60% B in 90 min. Solvent A consisted of water with 0.05% TFA. Solvent B was acetonitrile with 0.05% TFA. The gradient was developed with an HP1090 HPLC at 150 μl/min and split down to 5 μl/min using a microflow processor from LC-Packing, Inc. The flow from the column was fed directly into the spray tip of an API III mass spectrometer. Quadrupole three was scanned from 300 to 2000 Da using a 0.5-Da step size. The dwell time per step was set at 1.2 msec, for a scan duration of 4.39 sec. The peptides were identified in the map and designated with a tryptic map number. The tryptic peptides are numbered sequentially from the N terminus as T1 through T51, with T51 being the C-terminal peptide. The peptides were identified in the map by their expected ions and the LC/MS map was labeled with the tryptic map numbers.

tryptic digest on a 0.32 × 250 mm packed capillary column. The analysis required 3 μg of digest, ca. 50 pmol, to generate a total ion current chromatogram on an electrospray triple-quadrupole mass spectrometer. From this LC/MS map it was possible to assign the peptides and glycopeptides in the separation. This capillary column technique gives equal if not higher sensitivity response for the peptides and glycopeptides than a conventional reversed-phase column, owing to the low flow rates involved, and reduced dilution effects. The low flow rate allows a direct feed of the column eluent to the electrospray tip without a connecting junction.

The low-pH solvent systems used in reversed-phase chromatography help to charge biomolecules positively in electrospray. Some of the more common organic modifiers such as trifluoroacetic acid (TFA) and formic acid work moderately well. It is important to use the purest forms of these acids and solvents to eliminate signal background. Trifluoroacetic acid can be purchased in 1-g ampoules for improved purity, and sequencer grade is often of superior purity. Several researchers have reported increased sensitivities using acetic acid or other organic acids as modifiers.[16,17] It has been reported that the signal-suppressing effects of trifluoroacetic acid on the electrospray process could be largely alleviated by a previously reported method, referred to as the "TFA Fix," which uses a postcolumn addition of 75% (v/v) propionic acid, 25% (v/v) 2-propanol to the effluent immediately before the electrospray process. The modifier was delivered using an HPLC pump and was added into the column effluent after the detector and after the column switching valve. In this manner the HPLC effluent could be monitored without significant degradation in separation efficiency.

Experimental

High-Performance Liquid Chromatography

The HPLC separation is performed on a Hewlett-Packard 1090 liquid chromatography system equipped with a DR5 ternary solvent delivery system, diode array UV–Vis detector, autosampler, and heated column compartment (Hewlett-Packard Co., Wilmington, DE). All HPLC separations are done using a Vydac C_{18} (Separations Group, Hesperia, CA) (5-μm particle, 300-Å pore size) reversed-phase column. A standard solvent system of H_2O (solvent A) and acetonitrile (solvent B), both with 0.1% TFA, is used with a flow rate of 0.2 ml/min. The gradient for the separation is constructed as 0–60% B in 90 min. The column temperature is maintained at 45° throughout the separation.

Electrospray Liquid Chromatography/Mass Spectrometry

Mass spectrometry is done on a Hewlett-Packard (HP) 5989B quadrupole mass spectrometer equipped with extended mass range, high-energy dynode detector (HED) and a Hewlett-Packard 59987A API-electrospray source with high-flow nebulizer option. Both the HPLC and MS are controlled by the HP Chemstation software allowing simultaneous instrument control, data acquisition, and data analysis. The high-flow nebulizer is operated in a standard manner with N_2 as nebulizing (1.5 liters/min) and drying (15 liters/min at 300°) gases.

To counteract the signal-suppressing effects on electrospray LC/MS of trifluoroacetic acid, a previously reported method,[17] referred to as the "TFA Fix," is employed. The TFA Fix consists of postcolumn addition of 75% propionic acid, 25% 2-propanol at a flow rate of 100 μL/min. The TFA Fix is delivered using an HP 1050 HPLC pump and is teed into the column effluent after the diode array detector (DAD) and after the column switching valve. Column effluent is diverted from the MS for the first 5 min of the chromatogram, during which time excess reagents and unretained components elute.

For peptide mapping, MS data are acquired in scan mode, scanning from 200 to 1600 Da at an acquisition rate of 1.35 Hz at 0.15-Da step size. Unit resolution is maintained for all experiments. Data are filtered in the mass domain with a 0.03-Da Gaussian mass filter and in the time domain with a 0.05-min Gaussian time filter. The fragment identification is done with the aid of HP G1048A Protein and Peptide Analysis software, a software utility that assigns predicted fragments from a given sequence and digest with peak spectral characteristics.

For the in-source collison-induced dissociation (CID) method for detecting fragments indicative of glycopeptides, the electrospray capillary exit (CapEx) voltage is set to 200 V instead of the standard 100 V. Data acquisition is done in selected ion monitoring (SIM) mode, monitoring ions at m/z 147, 204, 292, and 366, each with a 1-Da window and a dwell time of 150 msec resulting in an acquisition rate of 1.5 Hz. Data are filtered in the time domain with a 0.05-min Gaussian time filter.

Chemicals and Reagents

HPLC-grade acetonitrile and trifluoroacetic acid (TFA), as well as EDTA, are obtained from J. T. Baker (Phillipsburg, NJ). Distilled, deionized Milli-Q water (Millipore, Bedford, MA) is used. Urea is obtained from GIBCO (Gaithersburg, MD) and DL-dithiothreitol (DTT), iodoacetic acid, glycine, bicine, NaOH, and $CaCl_2$ are obtained from Sigma (Sigma, St. Louis, MO). The enzyme endoproteinase Lys-C (Wako BioProducts, Richmond, VA) is used for mapping; recombinant *Desmodus* salivary plasminogen activator [DSPA(α1)] is supplied by Berlex Biosciences (Richmond, CA) and the starting concentration is 0.5 mg/ml.

Methods

Reduction and Alkylation. The molecule is first denatured in urea in order to keep it soluble during manipulations. To facilitate complete proteolytic digestion, the disulfide bonds are broken via reduction with DTT. Alkylation prevents them from reforming. The sample is then passed over a size-exclusion column in order to remove the above reactants from the sample to minimize interference with the protease during digestion.

PROCEDURE

1. Dilute the protein solution to a concentration of 0.5 mg/ml in 200 mM glycine, pH 4.5.
2. Dissolve urea crystals in the sample to a concentration of 6 M.
3. Titrate the sample to slightly alkaline pH using 6 N NaOH.
4. Add 1 M DTT to give a molar excess to protein of 300:1, and incubate in a 55° water bath for 3 hr.
5. Add 1 M iodoacetic acid to give a 2.2 molar excess over protein and incubate a further 30 min in the 55° water bath.
6. Pass the samples over a size-exclusion column eluted in 200 mM glycine, pH 4.5. Ascertain the protein concentration of eluted fractions by checking absorbance at 280 nm. Pool the corresponding fractions.

Lys-C Digestion. Under these conditions, the use of endoproteinase Lys-C has resulted in nearly complete digestion with little nonspecific cleavage. Another advantage of its use is the lack of dependence on cofactors. Arg-C, for example, requires the presence of DTT, EDTA, and CaCl$_2$ for optimal results.

PROCEDURE

1. Titrate the sample up to pH 9 with 6 M NaOH.
2. Add endoproteinase Lys-C to give an enzyme-to-substrate ratio of 1:100 and incubate for 18 hr in a 37° incubator.
3. Quench the reaction by lowering the pH to 1.5.

Enzymatic Deglycosylation. The sample should be desalted by a size-exclusion column or a microdialysis apparatus into water and lyophilized. Owing to possible steric hindrance the glycoprotein is either denatured with sodium dodecyl sulfate (SDS) and 2-mercaptoethanol for 10 min at 100° or reduced and alkylated as described previously. However, because most glycanases are inhibited by SDS, it is essential to have Nonidet P-40

(NP-40) or Triton X-100 present in the reaction. Some native glycoproteins may require longer incubation times than recommended below, as well as the addition of more enzyme after the initial 6 hr of incubation.

PROCEDURE

Hydrolysis of N-linked structures with PNGase F

1. Denature 20 nmol of freeze-dried DSPA(α1) in 50 mM sodium phosphate buffer pH 7.5, with 0.5% SDS and 1% 2-mercaptoethanol at 100° for 10 min.
2. Allow the reaction to cool to room temperature, then add NP-40 to a final concentration of 1%.
3. Add 62,500 units of PNGase F (New England Biolabs, Beverly, MA) and incubate for 2 hr at 37°.
4. The sample is extensively dialyzed into the appropriate buffer to remove the detergents, which might interfere with the analysis of the deglycosylated sample.
5. Samples that have been reduced and alkylated are incubated directly in 50 mM sodium phosphate buffer, pH 7.5, at 37° in the presence of 62,500 units of enzyme for 2 hr.

Hydrolysis of Ser/Thr-linked NeuNAc(α2-3)Gal(β1,3)GalNAc(α1) with O-glycosidase

1. Twenty nanomoles of reduced and alkylated DSPA(α1) is digested in 50 mM sodium phosphate buffer, pH 5.5, at 37° in the presence of 200 mU of neuraminidase III (Glyko, Inc., Novato, CA) for 2 hr.
2. After the initial incubation, add 40 mU of O-glycosidase to the same reaction vial and incubate for a further 6 hr.
3. The sample can now be dialyzed into an appropriate buffer for analysis.
4. It is not always necessary to denature or reduce and alkylate the sample, as sialylated oligosaccharides are found on the outside of folded proteins and, therefore, are readily available for digestion.

Mammalian Protein Carbohydrate Chemistry

Monosaccharides. The most commonly occurring monosaccharides found in oligosaccharide attachments to mammalian proteins are D-mannose (Man or Hex), D-galactose (Gal or Hex), D-glucose (Glc or Hex), L-fucose (Fuc or dHex), N-acetylglucosamine (GlcNAc or HexNAc), N-acetylgalactosamine (GalNAc or HexNAc), and N-acetylneuraminic acid (sialic acid or NANA). Fucose is different from the other hexose (Hex) sugars mentioned because it is missing a hydroxyl group at the C-6 position,

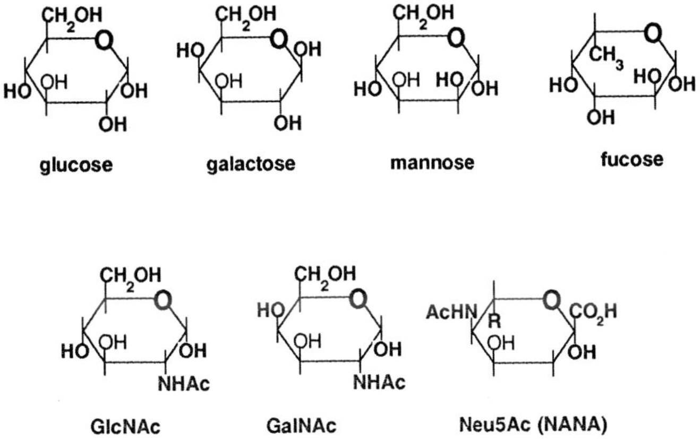

FIG. 2. *Top:* The Haworth formulas of the most commonly encountered pyranose ring forms of monosaccharides found in mammalian glycoprotein pharmaceuticals are shown. The less encountered sugars, such as xylose (pentose, $C_5O_5H_{10}$), were omitted. GlcNAc and GalNAc refer to the N-acetylated derivatives of glucose and galactose, respectively. NANA, Sialic acid or *N*-acetylneuraminic acid.

and is thus referred to as deoxyhexose or simply abbreviated dHex. Because many of the structures have isobaric mass they are often noted with these generic terms; see Fig. 2 for the monosaccharide structures and nomenclature.[18] The coincident masses would appear to simplify and at the same time confound the world of carbohydrate chemistry for the mass spectrometrist, with only four common distinct monosaccharide masses observable. There are a number of approaches that can be made to distinguish between these isobaric forms, such as the assumption that *N*-acetylglucosamine is the monosaccharide that links all structures that are attached to asparagine, or "N-linked." There are further assumptions that can be made when examining N-linked oligosaccharide structures and these are discussed below.

Forms of Mammalian Polysaccharides

N-LINKED STRUCTURES. There are three types of carbohydrate moieties commonly recognized as asparagine-linked, (N-linked) structures: complex, high-mannose, and hybrid. It is helpful to know that all N-linked structures share a common core of two GlcNAc and three mannose sugar residues (see (Fig. 3). This common core structure is noted in Fig. 3 as a triangle around the diantennary complex glycoform. When linked to a peptide, this core adds 893 Da to the peptide mass. The core can have a fucose attached

---

[18] R. J. Sturgeon, *in* "Carbohydrate Chemistry," Chap. 7 (J. F. Kennedy, ed.). Oxford University Press, Oxford, 1988.

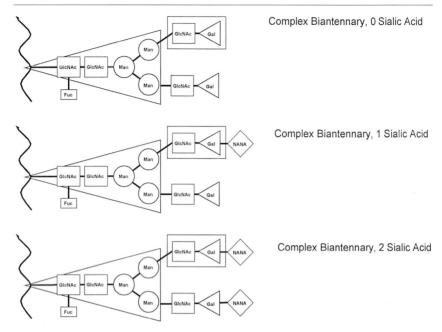

Complex Biantennary, 0 Sialic Acid

Complex Biantennary, 1 Sialic Acid

Complex Biantennary, 2 Sialic Acid

FIG. 3. The basic forms of the common N-linked oligosaccharides are shown. All N-linked oligosaccharides possess a common core of two *N*-acetylglucosamine residues and three mannose residues; this core is highlighted within large triangles on the complex diantennary structures. N-Linked structures have a common motif (termed *complex*) that can have from two to four antennae. The antennae usually terminate in galactose or sialic acid. Sialylation adds the greatest degree of microheterogeneity to a glycoprotein. Other sources of heterogeneity can arise from variable core fucosylation. The common antenna unit is *N*-acetylglucosamine-galactose, and is shown in the large rectangles on the diantennary complex glycoforms.

to the GlcNAc that is linked to the asparagine in the polypeptide, and it can also possess a bisecting GlcNAc residue attached to the central mannose of the core structure. Core fucosylation and bisecting GlcNAc residues are just a few of the many variations that can add to the complexity of the oligosaccharide.

Complex-type structures usually have from two to four branches attached to the two outer core mannose residues. The branches are distributed over the two terminating core mannose residues. The complex structures are termed *diantennary*, *triantennary*, and *tetraantennary*, referring to the number of antenna. The basic branch structures are composed in most instances of one GlcNAc and one galactose residue, highlighted by the rectangle on the diantennary arm in Fig. 3. This branched residue unit mass is 365 Da. If ions are observed that differ by a multiple of this unit it may be an indication of a complex-type glycosylation. To complicate matters further each of these arms can terminate in sialic acid. For example, it is possible to have a tetraantennary complex glycoform with zero to four terminating sialic acid residues, including variable core fucosylation. The second type of N-linked structure is called *high mannose*. It contains the common core of two GlcNAc and three mannose sugar residues and has additional mannose residues emanating from the two terminating mannose residues of the core. The different forms are termed 4 Man, 5 Man, 6 Man, etc., with the three core mannose residues being counted in this terminology. A series of high-mannose glycoforms in a mass spectra will differ by multiples of 162 Da, the residue mass for a hexose, such as mannose.

O-LINKED STRUCTURES. O-Linked structures are more difficult to define. They are linked through the oxygen present on the side-chain hydroxyl of serine and threonine, but a consensus sequence for their attachment has not been reported. O-linked structures in general appear to be less complex than N-linkages in the number of antenna and monosaccharides. However, they can be fucosylated and sialylated.

Examples of O-linked structures include a report in which a single fucose was attached to Thr-61 in recombinant tissue plasminogen activator.[19] Also reported have been single O-linked GlcNAc residues on various proteins.[20] *O*-Glycosylproteins having an inner core of Gal(α1–3)GalNAc(α1–3)Ser or Thr are often referred to as *mucin-like* structures even though they are present in biological fluids and cell membranes. The linkage is alkali labile. Glycosaminoglycans, for example serylchondroitin 4-sulfate, have the following inner core: Gal(β1–3)Gal(β1–3)Xyl(β1–3)Ser. They are sometimes referred to as *acid mucopolysaccharides*, which are a homogeneous family of glycans. This linkage is also alkali labile. Collagen-type linkages between D-galactose and 5-hydroxy-D-lysine and extensin-type linkages between L-arabinofuranose and 4-hydroxy-L-proline are found in plant glycoproteins and are alkali stable.[21] Also, *N*-acetylglucosamine moieties that are O-glycosidically linked by an alkali-labile linkage to serine or threonine residues have been described; all are phosphorylated and often form multimeric complexes depending on their state of phosphorylation.[22]

[19] R. J. Harris, C. K. Leonard, A. W. Guzzetta, and M. W. Spellman, *Biochemistry* **30**, 2311 (1991).

[20] J. R. Turner, A. M. Tartakoff, and N. S. Greenspan, *Proc. Natl. Acad. Sci. U.S.A.* **87**, 5608 (1990).

[21] S. Montreuil, H. Bouquelet, Debray, B. Foumet, G. Spik, and G. Strecker, *in* "Carbohydrate Analysis: A Practical Approach" (M. F. Chaplin and J. F. Kennedy, eds.), p. 143. IRL Press, 1986.

[22] R. S. Haltiwanger, W. G. Kelly, E. P. Roquemore, M. A. Blomberg, L.-Y. D. Dong, L. Kreppel, T.-Y. Chou, and G. W. Hart, *Biochem. Soc. Trans.* **20**, 264 (1992).

Choosing Method of Glycoprotein Cleavage

Proteins. The primary objective in choosing a cleavage scheme for a glycoprotein is to isolate the glycosylation sites onto separate peptides. The second priority should be that the glycopeptides do not coelute in the reversed-phase separation. In the rtPA reversed-phase tryptic map two of the glycopeptides have nearly coincident elution.[15,23,24] While a coelution is difficult to predict and difficult to remedy within the limitation of mobile phases compatible with the mass spectrometer, it can often be helped by trying a different protein cleavage method. An example is reviewed in the next section. These suggestions will hold down the heterogeneity of the spectra to a manageable level, and make the data easier to interpret. The ES spectra of an intact glycoprotein will often appear as a broad hump of ions, and in most cases interpretation is nearly impossible. The large number of glycoforms divides the ion current and thus reduces the ion intensity of any single form. Lower resolution techniques such as matrix-assisted laser desorption, time of flight mass spectrometry (MALDI),[15] or even sodium dodecyl sulfate-polyacrylamide gel electrophoresis (SDS–PAGE),[25] will often yield better general mass information about such heterogeneous samples.

Primary Amino Acid Sequence and Peptide Mapping. As described elsewhere (see [2], [3], [11], and [15] in this volume[26]) there are a number of methods for protein fragmentation, including cyanogen bromide cleavage (chemical) and trypsin (enzymatic).[26a] For MS/MS sequence analysis, for example when examining the C- or N-terminal peptide of a protein,[27] or when isolating an oxidized methionine on a peptide, one should aim for a peptide of reasonable length (e.g., 20 residues is the maximum); in many cases a smaller peptide can give complex fragmentation data that are difficult to interpret.[28]

Another factor in the choice of a suitable reagent for cleavage of the

[23] V. Ling, A. W. Guzzetta, E. Canova-Davis, J. T. Stults, T. R. Covey, B. I. Shushan, and W. S. Hancock, *Anal. Chem.* **63,** 2909 (1991).

[24] A. W. Guzzetta, L. J. Basa, W. S. Hancock, B. A. Keyt, W. F. Bennett, *Anal. Chem.* **65,** 2953 (1993).

[25] W. Chen and O. P. Bahl, *J. Biol. Chem.* **266**(10), 6246 (1991).

[26] E. R. Hoff and R. C. Chloupek, *Methods Enzymol.* **271,** Chap. 2, 1996 (this volume); K. M. Swiderek, T. D. Lee, and J. E. Shively, *Methods Enzymol.* **271,** Chap. 3, 1996 (this volume); E. C. Rickard and J. K. Towns, *Methods Enzymol.* **271,** Chap. 11, 1996 (this volume); J. R. Yates, *Methods Enzymol.* **271,** Chap. 15, 1996 (this volume).

[26a] T. D. Lee and J. E. Shively, *Methods Enzymol.* **193,** 361 (1990).

[27] D. A. Lewis, A. W. Guzzetta, and W. S. Hancock, *Anal. Chem.* **66**(5), 585 (1994).

[28] D. F. Hunt, J. Shabanowitz, J. R. Yates, N.-Z. Zhu, D. H. Russell, and M. E. Castro, *Proc. Natl. Acad. Sci. U.S.A.* **84,** 620 (1987).

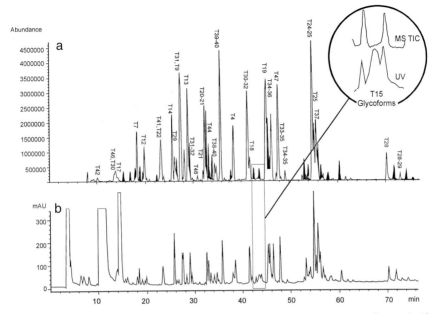

FIG. 4. LC/MS of enzyme digest of *Desmodus* salivary plasminogen activator [DSPA(α1)] reduced, alkylated, and digested with endoproteinase Arg-C: 78 pmol total protein injected. (a) Mass spectrometry total ion current; (b) detection at 214 nm.

glycoprotein is the complexity of the resulting map. In the analysis of a complex glycoprotein recombinant *Desmodus* salivary plasminogen activator [DSPA(α1)][29] it was found that cleavage with a trypsin-like protease gave a complex map in which the glycopeptides were largely obscured by the more abundant nonglycosylated peptides (some 50 in total). Digestion of the protein with a more specific protease, such as Lys-C or Arg-C, gave fewer fragments (approximately 20–25) and simplified detection of the glycopeptides (see Fig. 4 for the result of an Arg-C digest). Table I lists the predicted and found peptides for a Lys-C digest in a format typical for the output of an LC/MS analysis. A further advantage of this approach is that many of the peptides contained multiple internal basic residues, which had the effect of increasing the charge state of the peptide and in this case allowed the detection of a high molecular weight glycopeptide that had been undetected with the alternative procedure. In fact, a combination of maps produced by proteases of different specificities is often required to detect all of the glycosylation sites.

[29] A. Apffel, J. A. Chakel, S. Udiavar, W. S. Hancock, C. Souders, and E. Pungor, *J. Chromatogr.* **717,** 41 (1995).

TABLE I
PREDICTED AND FOUND FRAGMENTS IN LIQUID CHROMATOGRAPHY/MASS SPECTROMETRY
ANALYSIS OF DSPA(α1)[a]

Fragment	Residues	Retention time (min)	Molecular mass (Da)	Sequence
K1	1–7	24.1	768.4	AYGVA CK
K2	8–29	NF[b]	2808.4	DEITQ MTYRR QESWL RPEVR SK
K3	30–82	NF	6367.6	RVEHC QCDRG QARCH TVPVN SCSEP RCFNG GTCWQ A VYFS DFVCQ CPAGY TGK
K4	83–152	40.5	8162.7	RCEVD TRATC YEGQG VTYRG TWSTA ESRVE CINWN S SLLT RRTYN GRMPD AFNLG LGNHN YCRNP NGAPK
K5	153–159	36.79	965.5	PWCYV IK
K6	160–162	NA[b]	274.2	AGK
K7	163–175	29.3	1488.6	FTSES CSVPV CSK
K8	176–182	21.94	805.4	ATCGL RK
K9	183–184	NA[b]	309.2	YK
K10	185–248	53.8	7295.5	EPQLH STGGL FTDIT SHPWQ AAIFA QNRRS SGERF LCGG I LISSC WVLTA AHCFQ ESYLP DQLK
K11	249–258	29.48	1189.7	VVLGR TYRVK
K12	259–267	24.15	1063.5	PGEEE QTFK
K13	268–269	NA[b]	245.2	VK
K14	270–270	NA[b]	146.1	K
K15	271–275	23.40	658.4	YIVHK
K16	276–292	41.33	2026.9	EFDDD TYNND IALLQ LK
K17	293–330	44.4	4314.9	SDSPQ CAQES DSVRA ICLPE ANLQL PDWTE CELSG YGK
K18	331–332	NA[b]	283.2	HK
K19	333–343	32.1	1271.6	SSSPF YSEQL K
K20	344–358	28.51	1756.9	EGHVR LYPSS RCAPK
K21	359–363	34.98	667.4	FLFNK
K22	364–419	NF[b]	6084.6	TVTNN MLCAG DTRSG EIYPN VHDAC QGDSG GPLVC M NDNH MTLLG IISWG VGCGE K
K23	420–427	27.31	877.5	DVPGV YTK
K24	428–441	33.0	1730.9	VTNYL GWIRD NMHL

[a] Map produced by digestion of the protein with Lys-C.
[b] NA, Fragment below mass range; NF, fragment not detected.

Specifically, Fig. 4 shows the ES LC/MS analysis of the Arg-C digest of DSPA(α1) together with the UV signal at 220 nm for comparison. Note that the sensitivities of the two detectors (UV and MS) are similar and that most peaks show up in both traces although the individual intensities may differ. The UV detector is useful because of the presence of the peptide backbone, which has significant absorption at 200–205 nm. In the region of the chromatographic profile denoted by the inset in Fig. 4, the elution of the oligosaccharides N-linked at asparagine at residue 117 is shown in fragment T15(111–123). The abundance of these forms is low as can be seen in the inset in Fig. 4, in which the total ion current (TIC) is essentially flat in this region of the map. In the same region, a single broad peak is observed by UV detection presumably because each of the glycoforms has a similar spectrum but the significant mass degeneracy defeats the TIC monitoring. This example illustrates the power of combined UV and MS detection for peptide-mapping separations. This type of analysis is particularly useful for confirming the bulk of the amino acid sequence of a rDNA protein [see Table I for an example with DSPA(α1)].

Predicting N-Linked Glycosylation Sites and Choosing Method of Protein Cleavage. Possible N-linked glycosylation sites can be predicted if the primary amino acid sequence is known. This postulation can be made on the basis of the N-linked consensus sequence where the first amino acid is asparagine, the second amino acid is not defined, but is most likely not proline, and the third amino acid is threonine or serine.[3] The most widespread method of cleavage is with the enzyme trypsin.[11] Trypsin enjoys this popularity for a number of reasons: it cleaves the protein reproducibly, it is inexpensive, and it gives peptides of reasonable length (often ca. 10 amino acids). Tryptic digestions are carried out with enzyme-to-protein ratios ranging from 1 : 25 to 1 : 100 and at slightly above pH 8, either at ambient temperature for 24 hr, or at 37° for 2 to 4 hr, and often involve addition of enzyme at two time points to ensure complete digestion. With an N-terminal amino group and a C-terminal arginine or lysine residue (due to the specificity of trypsin) the resultant peptides will invariably protonate at the two sites at low pH, except for C-terminal peptides, which may not have a basic residue at their C termini. Tryptic peptides are almost certainly an advantage when MS/MS sequence analysis is performed, because CID fragmentation is enhanced near chargeable residues.[27]

Locating Glycopeptides in Liquid Chromatography/Mass Spectrometry Analysis of Peptide Maps of Proteins

Generating LC/MS data from peptide maps can be a simple task. Interpreting the data from an LC/MS run and being confident of the conclusions are not as simple. An LC/MS run can consist of 1000 single mass scans and the interpretation of a data set can be a time-consuming task. However, there are patterns that can be observed and techniques that can be used to identify sites of glycosylation in this complex data set.

Determining whether a protein is glycosylated is achieved by other approaches. For example, glycoproteins appear as broad, fuzzy bands on SDS-polyacrylamide gels that will collapse to a tighter band when treated with glycosidases.[28] This is an approach that can be used for determination of N- or O-linkage with the help of glycosidases.[29] In LC/ES/MS there are also glycopeptide signatures that can be used to note possible sites of glycosylation. These include the observation of clusters of ions in the contour plots of m/z vs scan data, and the observation of oxonium ions in CID data.[29]

Locating Glycopeptides by Peptide Assignments. Often the question of glycosylation characterization is secondary in the characterization of an LC/MS tryptic map. The researcher will try to identify as many of the predicted tryptic peptides as possible and will invariably find some masses missing. Sometimes the missing peptides can be accounted for by an unusual cleavage or a posttranslational modification of the protein unrelated to glycosylation.[26] (Yates, this volume). In some cases the sequence of the missing peptide will contain an N-linked consensus sequence, and in other cases the possibility of an O-linked attachment.

As was seen in the LC/MS analysis of the DSPA(α1) enzyme map the application of a software interpretation utility has allowed detection of essentially all of the peptides predicted from the cDNA sequences. Interestingly, in the case of DSPA(α1) those digest fragments that are expected to be glycosylated are either missing or are found at very low abundance relative to other fragments, and with prediction of likely masses can be used as a rough predictor of glycosylation sites and subsequent ion extractions (see Table II as an example).

Collision-Induced Dissociation Analysis. The first reports of peptide and protein fragment ions generated in the high-pressure region of electrospray ion sources were made between 1988 and 1991.[26] (Yates, this volume). The analysis of oligosaccharides and glycopeptides has been carried out in a similar manner.[30] This advance allowed the collection of primary sequence information using a single-quadrupole electrospray mass analyzer. It has been noted that peptides tend to fragment near charged residues, and prolines in a sequence appear to enhance fragmentation.

The studies described previously involved peptides of low molecular

---

[30] S. A. Carr, M. J. Huddleston, and M. F. Bean, *Protein Sci.* **2**, 183 (1993).

TABLE II
GLYCOFORMS OF DSPA(α1) FRAGMENT (K21) IDENTIFIED

Type	Sialic acid	Fucose	Retention time	Charge states
High mannose				
Man$_3$GlcNAc$_2$	0	0	30.40	+2, +3, +4
Man$_4$GlcNAc$_2$	0	0	29.8	+2
Man$_5$GlcNAc$_2$	0	0	27.2	+2, +3
Man$_6$GlcNAc$_2$	0	0	26.8	+3, +4
Man$_7$GlcNAc$_2$	0	0	26.6	+3, +4
Man$_8$GlcNAc$_2$	0	0	26.1	+2, +3, +4, +5
Complex				
Biantennary	0	0	32.8	+5
	1	0	32.4	+4
	2	0	32.3	+3
	0	1	31.3	+2, +3
	1	1	31.15	+2, +3
	2	1	30.98	+2, +3
	0	2	32.3	+3, +4
	1	2	32.3	+3, +4, +5
	2	2	31.8	+3, +4, +5, +6
Triantennary	0	0	33.65	+2, +3, +4
	0	1	30.55	+2
	1	1	30.68	+2, +3
	2	1	30.88	+2, +3
	3	1	31.0	+3
	2	2	33.2	+5
	3	2	32.2	+3, +4, +5, +6
Tetraantennary	0	0	31.0	+2, +3
	1	0	33.6	+3, +4, +5
	0	1	30.7	+2, +3
	1	1	30.52	+3
	2	1	30.38	+3
	3	1	30.25	+3
	0	2	33.65	+3, +4, +5
	2	2	33.1	+4

mass (ca. 1000–2000 Da). What might not appear to be obvious is that this technique is quite amenable to analyzing glycopeptides that may have molecular masses ranging up to 8000 Da and possibly larger. Reference 29 shows that oxonium ion fragments could be obtained through collisions produced in the declustering region, pre-Q1, and demonstrated that they could be used as markers for glycopeptides in LC/ES/MS maps. Others showed that fragmentations resulting from glycopeptide CID could be

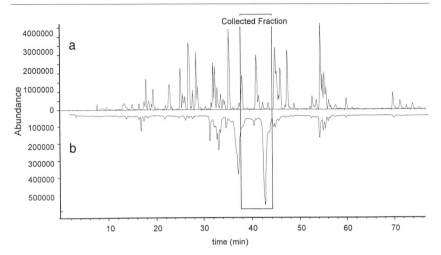

Fɪɢ. 5. CID detection of glycopeptides. A total of 78 pmol of protein was injected. (a) Total ion chromatogram from scan acquisition (CapEx = 100 V). (b) For the in-source collision-induced dissociation (CID) method for detecting fragments indicative of glycopeptides, the CapEx voltage was set to 200 V instead of the standard 100 V. Data acquisition was done in selected ion monitoring (SIM) mode, monitoring ions at m/z 147, 204, 292, and 366.

generated in the high-pressure region of the electrospray ion source[30,31] These authors also demonstrated that the oxonium ion fragments produced in these experiments could be used as markers for glycosylations of both N- and O-linked glycoforms in LC/MS runs of glycoprotein digests. The method involves increasing the orifice potential to induce the fragmentation during the low mass range of the 150- to 500-Da mass scan to observe the carbohydrate oxonium ion fragments [at the following m/z values: Hex-HexNAc$^+$, m/z 366; NeuAc$^+$, m/z 292; HexNAc$^+$, m/z 204; Hex$^+$, m/z 163; dHex$^+$ (fucose or xylose), m/z 147] and then changing the orifice potential back to a setting that does not cause fragmentation for the mass range 500–2000 Da. This method allows for both the observation of the carbohydrate oxonium ion marker fragments, and in the later portion of the scan the parent glycopeptides.

An example of the use of in-source collision-induced dissociation in the analysis of DSPA(α1) is illustrated in Fig. 5. The CID mass spectral acquisition was carried out in selected ion monitoring (SIM) mode to increase the sensitivity. The reproducibility of the HPLC separation is sufficiently good that areas identified in the SIM analysis can be accurately compared with separate full spectral acquisitions conducted at lower frag-

[31] M. J. Huddleston, M. F. Bean, and S. A. Carr, *Anal. Chem.* **65,** 877 (1993).

mentation energy. This allows the parent glycopeptide to be identified. In cases in which there is a large degree of microheterogeneity due to glycosylation, there may not be a sufficient amount of an individual glyco-peptide present to allow detection in either the UV or TIC signal. However, the sensitivity from the SIM CID data is sufficient to detect forms of low abundance, because all of the different chromatographically unresolved glycoforms will contribute fragment ions to the electrospray signal. How-ever, the summation of the signals due to the marker ions results in relatively broad-looking chromatographic peaks. On the basis of the CID data, it is possible to infer some glycostructure differentiation. For example, all N-linked structures will generate an m/z 204 ion corresponding to the HexNAc in the backbone structure. All complex N-linked glycopeptides will generate an m/z 366 ion corresponding to the HexNAc + Hex structure indicative of the branching in multiantennary structures. Only fucosylated glycopep-tides will generate an m/z 147 ion and only sialylated glycopeptides will generate an m/z 292 ion.

Locating Glycopeptides in Contour Plot. The contour plot allows one to visualize all of the ions in an entire LC/MS run. The first impression is that of a two-dimensional (2D) gel, with spots in almost every scan. One of the most obvious signatures of a glycopeptide in an LC/MS reversed-phase map is a cluster or cloud of ions that appears to have a negative slope in the plot of mass to charge vs. time and scan. This cloud may have 20 glycoforms, with ions at several charge states, all within a 1-min time frame in the LC/MS run. Figure 6 shows the comparison of the patterns obtained from the contour mapping of two complex glycoproteins; a tryptic digest of recombinant tissue plasminogen activator (rtPA) (three sites of glycosylation) and the Arg-C digest of DSPA(α1) (six potential sites of glycosylation). The circled regions correspond to the elution position of some of the glycopeptides. Although the conventional chromatographic view of the peptide map shown in the upper part would suggest that the rtPA map has more components than the DSPA(α1) map, the 2D map shows that the latter digest is much more complex, as would be expected from the greater number of glycosylation sites in DSPA(α1). The more heterogeneous is the sample, as in the case of DSPA(α1), the more glyco-forms over which the ion current will be distributed. Unlike the UV chro-matogram in which the different unresolved glycoforms contribute to a moderately intense, relatively broad chromatographic peak, the contour map has a single point for each mass state present for the glycoform.

Figure 7 shows the contour plot for a section of a tryptic map of DSPA(α1) from 40 to 50 min, with the corresponding integrated signal (TIC) shown above. The highlighted areas are due to peptide T15, which contains residues 111–123 and a glycosylation site comprising complex,

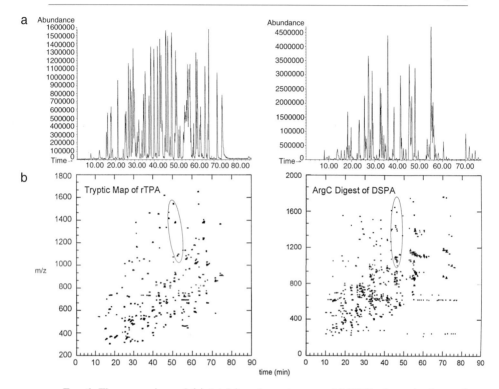

FIG. 6. The comparison of (a) total ion chromatogram of LC/MS of tryptic digest of recombinant tissue plasminogen activator (rt-PA) (three sites of glycosylation) and the trypsin-like digest of DSPA(α1) (six sites of glycosylation) and (b) Contour plot obtained from 2D mapping of the LC/MS data.

sialylated biantennary structures. The circled areas contain signals due to three major charge states (+1, +2, and +3). The approximate slopes of the pattern of glycosylation heterogeneity are also plotted to demonstrate that characteristic patterns are present in these complex data sets. We are investigating the application of contour plots and multivariate statistical techniques to the examination of different lots of the drug substance because such an approach may have value in demonstrating consistency in glycosylation.

With a closer inspection of the contour plot other intriguing patterns often begin to emerge. If one examines the duration of ions (peptides) it can become apparent that some signals will last 10 scans (which corresponds to a 1-min peak width) while others will endure for only 3 or fewer (see Fig. 7 for examples). To some degree this is related to their concentration

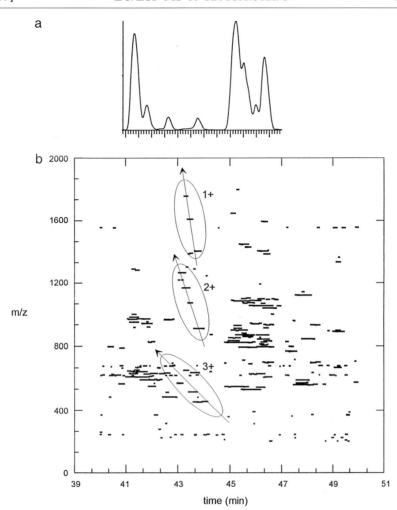

Fig. 7. The plot from the contour plot for a section of the tryptic-like map of DSPA(α1). (a) TIC of the elution profile from 40 to 50 min. (b) Contour plot obtained from the 2D mapping of the LC/MS data for this region of the analysis. The circled areas show the signal due to fragment T15[111–123] complex sialylated biantennary structures, with three different charge states (+1, +2, and +3) being shown as well as the approximate slope of the pattern of glycosylation heterogeneity.

or the characteristics of a particular reversed-phase column or to the elution behavior of the peptide. The population of ions in the plot shows a gradual positive slope over the course of the reversed-phase run. This is attributed to the fact that as the organic gradient develops the more hydrophobic (and generally larger) peptides elute near the end of the run, while the hydrophilic (and generally smaller) peptides elute earlier. Often what upsets this pattern is the appearance of a glycopeptide. Because a glycosylation can make a peptide more hydrophilic while greatly increasing its mass, the glycopeptides tend to not follow the general pattern of elution.

This observation led us to the identification of the glycopeptides in the analysis of a monoclonal antibody.[27] The antibody had only two major glycoforms, but they were on a peptide that eluted early in the gradient (see Fig. 8, labeled glycopeptides at a retention time of 30 min). In general, low MW peptides are polar and elute early in the map, whereas glycopeptides are polar but of a much higher molecular weight. In this situation, the 2D plot of the data showed a series of ions that do not fit the characteristic trend of an increase in mass with an increase in retention time. Thus, despite the heterogeneity of the map owing to the complexity of a monoclonal antibody, this simple visual location allowed detection of mass differ-

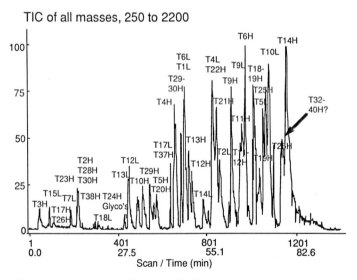

Fig. 8. Total ion current recording of an LC/MS map generated on a tryptic digest of antibody, humanized anti-TAC separated using a 2.1 × 250 mm Nucleosil C_{18}, 5-μm capillary reversed-phase column. The separation conditions were 0 to 50% solvent B in 90 min. Solvent A consisted of water with 0.05% TFA. Solvent B was acetonitrile with 0.05% TFA. H and L, Fragments of the heavy and light chain, respectively.

ences in the individual scans that were indicative of carbohydrate forms. This information was then used to propose both the site of attachment and glyco structures.[27]

Analyzing Complex Series of Ions. Once the glycopeptides are identified in an LC/MS run either through the observation of the glycosylation heterogeneity observed in an *m/z*-vs-time contour plot or by the observation of oxonium ions produced by CID the next step is characterization of the ions, as is described below.

Using Sugar Residue Differences and Calculated Compositions to Propose Carbohydrate Structures. After location of the glycopeptide, the next step is to average the scans defined by these glycopeptide signatures. If heterogeneity is present, the glycopeptide can be characterized initially by the monosaccharide mass differences. For example, a high-mannose glycopeptide may exhibit a series of ions that differ by the residue mass of mannose at a particular charged state. It may then be possible to go to different charge state in this same average of scans and observe another cluster of ions that differ by mannose residues. With this evidence one can be reasonably confident that one has found a high mannose-type glycopeptide. The next step is to link this observation with a potential candidate peptide. With the charge state established through the relationship of the ions to one another, the parent mass can be calculated, even if another charge state cannot be found on which to base the calculation. With the assumption that this is an N-linked high-mannose glycopeptide, the core mass of two GlcNAc and three mannose residues can be subtracted from the established observed parent mass. Mannose residues can then be sequentially subtracted until the mass of one of the candidate peptides is matched. This method will give carbohydrate compositional data and general structural information.

An example of this method is demonstrated in Fig. 9 for the analysis of the glycoforms of fragment K21(359–419) of *Desmodus* salivary plasminogen activator [DSPA(α1)]. As an example, the upper profile shows the extracted ion profiles of the biantennary complex glycoforms that contain one fucose and variable sialic acid (*m/z* 1219). Note that in the total ion chromatogram, this area consists of a broad unresolved zone, which contains the overlapping family of K21 glycopeptides as well as other unrelated fragments. The lower part of Fig. 9 shows the extracted ion chromatograms for this zone for the *m/z* 292 (NeuAc) and *m/z* 366 (Hex + HexNAc) marker ions. Thus, having predicted the presence of the glycoforms, ion chromatograms could be extracted and used to confirm the proposed structures (the glycomarker ions *m/z* 292 and 366 agree with the presence of sialylated N-linked glycopeptides and the elution profiles in both parts of Fig. 9 are consistent). Other pieces of information can also be used; for

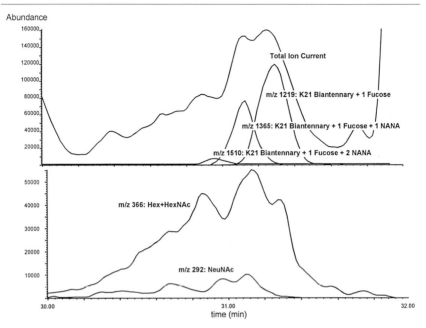

FIG. 9. Identification of specific K21 glycopeptides isolated from *Desmodus* salivary plasminogen activator [DSPA(α1)] reduced, alkylated, and digested with endoproteinase Lys-C. *Top:* Total ion current and three extracted ion chromatograms. *Bottom:* m/z 292 and 366 CID glycomarkers.

example, for each of these peaks there should be at least two charge states present in the spectra. Another important observation is the elution position of members of the glycopeptide family. The closely related glycopeptides should elute adjacent to each other and as the degree of sialylation increases, the retention should decrease, as is seen in Fig. 9 (upper trace). It must be remembered that true linkage information can be obtained only through traditional carbohydrate chemistry methods and MS/MS or NMR studies.[32–34]

Conclusion

This chapter has shown that there is considerable potential in the analysis of complex glycoprotein samples by hyphenated liquid-phase separations

[32] R. R. Townsend, M. R. Hardy, O. Hindsgaul, and Y. C. Lee, *Anal. Biochem.* **174,** 459 (1988).
[33] F. Maley, R. B. Trimble, A. L. Tarentino, and T. H. Plummer, *Anal. Biochem.* **180,** 195 (1989).
[34] J. F. G. Vliegenthart, L. Dorland, and H. Van Halbeek, *Adv. Carbohydr. Chem.* **41,** 209 (1983).

and mass spectrometry. Such information should prove invaluable in determining the role in biology of the carbohydrate moiety in glycoproteins as well as reducing the approval barriers for the pharmaceutical use of glycosylated proteins produced by mammalian fermentation systems. For example, this new analytical method can be used to deduce possible carbohydrate structures by determining both the mass and elution position of individual glycopeptides. In a complex map the region of the map that contains the glycopeptides can be deduced by looking for characteristic patterns in the 2D plot or by the observation of oxonium ions produced by collision-induced dissociation (CID).

Electrospray mass spectrometry on-line with reversed-phase HPLC has greatly expanded the power of peptide mapping to identify carbohydrate structures that are attached to asparagine, serine, or threonine residues. One can use in-source collision-induced dissociation to scan the map for regions with a high concentration of glycopeptides.

The use of contour maps (m/z vs retention time) as a facile approach to rapid 2D mapping of complex samples has some similarity to the popular 2D techniques currently used in biochemistry, such as a combination of isoelectric focusing and SDS–PAGE, but offers a different combination of orthogonal separation methods. Such maps are readily available from the data generated by an LC/MS analysis and can give valuable information about glycosylation patterns and product consistency.

[18] Peptide Characterization by Mass Spectrometry

By BETH L. GILLECE-CASTRO and JOHN T. STULTS

Introduction

There are three distinct goals in peptide characterization: (1) confirmation of putative sequence, (2) identification and localization of amino acid modifications, and (3) sequence determination of unknown peptides. Owing to its speed, sensitivity, and versatility, mass spectrometry, and more specifically tandem mass spectrometry utilizing high- or low-energy collision-induced dissociation, plays a pivotal role in accomplishing each of these goals. Other widespread methods of peptide characterization such as automated Edman degradation and amino acid analysis are not discussed in this chapter; however, their absence here does not belittle their importance. On the contrary, these techniques can, in many cases, efficiently provide the data necessary to solve a particular problem. The usefulness of these more traditional techniques, along with tandem mass spectrometry, in providing an integrated approach to peptide characterization cannot be overstated.

The first major foray of mass spectrometry into routine peptide characterization began with the discovery of fast atom bombardment (FAB)–mass spectrometry,[1,1a] and the related liquid secondary ionization mass spectrometry (LSIMS) method,[2] which were recognized for their ability to produce protonated peptide molecular ions from peptides of up to 10 kDa. With the development of electrospray ionization (ESI) the upper detectable mass limit for biomacromolecules was extended to over 100 kDa and at the same time detection limits were decreased to the low-picomole level.[3–9a] Importantly, ESI uses liquid sample introduction, and therefore mass spectrometers having this type of ion source are readily interfaced with high-performance liquid chromatography (see [3] in this volume[9b]) or with capillary electrophoresis systems (see [19] in this volume[9c]). Such systems are highly advantageous in that they facilitate simultaneous separation of complex mixtures along with ultraviolet absorption and mass detection. Electrospray ionization spectra are characterized by molecular ions in multiple charge states with little or no fragmentation. Traditionally, electrospray ion sources are used in conjunction with quadrupole mass analyzers. However, the higher mass resolving power of sector[10,11] and Fourier transform mass spectrometers[12] have more recently been utilized with electrospray ionization.

[1] M. Barber, R. S. Bordoli, R. D. Sedgwick, and A. N. Tyler, *Nature* **293**, 270 (1981).

[1a] W. E. Seifert, Jr., and R. M. Caprioli, *Methods Enzymol.* **270**, Chap. 20, 1996.

[2] W. Aberth, K. M. Straub, and A. L. Burlingame, *Anal. Chem.* **54**, 2029 (1982).

[3] C. M. Whitehouse, R. N. Dreyer, M. Yamashita, and J. B. Fenn, *Anal. Chem.* **57**, 675 (1985).

[4] R. D. Smith, C. J. Berinaga, and H. R. Udseth, *Anal. Chem.* **60**, 1948 (1988).

[5] T. R. Covey, R. Bonner, B. Shushan, and J. D. Henion, *Rapid Commun. Mass Spectrom.* **2**, 249 (1988).

[6] J. B. Fenn, M. Mann, C. K. Meng, S. F. Wong, and C. M. Whitehouse. *Science* **246**, 64 (1989).

[7] M. Mann, *Org. Mass Spectrom.* **25**, 575 (1990).

[8] I. Jardine *Nature* (*London*) **345**, 747 (1990).

[9] S. K. Chowdhury, V. Katta, and B. T. Chait, *Biochem. Biophys. Res. Commun.* **167**, 686 (1990).

[9a] C. M. Whitehouse and F. Banks, Jr., *Methods Enzymol.* **270**, Chap. 21, 1996.

[9b] K. M. Swiderek, T. D. Lee, J. E. Shively, *Methods Enzymol.* **271**, Chap. 3, 1996 (this volume).

[9c] R. D. Smith, H. R. Udseth, J. H. Wahl, D. R. Goodlet, and S. A. Hofstadler, *Methods Enzymol.* **271**, Chap. 19, 1996 (this volume).

[10] B. S. Larsen and C. N. McEwen, *J. Am. Soc. Mass Spectrom.* **2**, 205 (1991).

[11] R. B. Cody, J. Tamura, and B. D. Musselman, *Anal. Chem.* **64**, 1561 (1992).

[12] K. D. Henry and F. W. McLafferty, *Org. Mass Spectrom.* **25**, 490 (1990).

Peptide molecular masses can also be readily obtained with high sensitivity by matrix-assisted laser desorption/ionization (MALDI).[12a] First introduced in 1988 by Hillenkamp and co-workers,[13,14] the method allows for rapid data acquisition using only low-femtomole quantities of peptides. Laser desorption/ionization-derived molecular weight determinations have been combined with Edman microsequencing data for the analysis of peptides.[15] An extension of this combination, known as ladder peptide sequencing, has been used to characterize peptides including those with modifications on selected amino acid residues.[16] The high sensitivity of MALDI mass spectrometry has opened new approaches to protein analysis[17] such as the analysis of electroblotted proteins from two-dimensional (2D) gel electrophoresis by *in situ* proteolytic digestion and molecular mass determination of the resultant peptides.[18] The observed peptide masses are matched with lists of expected peptide masses that are generated by calculation of the digest fragment molecular weights for each protein in the database.[18]

Information used to determine the sequence of peptides and the ability to isolate the molecular species of interest from a mixture, can be provided by tandem mass spectrometry. As shown in Fig. 1a, single molecular species are isolated in the first mass analyzer for transmission into a collision cell. There, sufficient energy to break peptide bonds can be introduced by collision of peptide ions with inert gas molecules causing them to undergo collision-induced dissociation (CID). Typically, the peptide fragment molecular masses are determined in a second mass analyzer, and this combination is referred to as tandem mass spectrometry or MS–MS. The types of fragment ions are shown in Fig. 1b, and are classified in two types: the backbone cleavages (a, b, c, x, y, and z ions) and the side-chain cleavages (d, w, and v ions). Collision-induced dissociation may be performed with a variety of mass spectrometers including two-, three-, and four-sector, single- and triple-quadrupole, hybrid sector-quadrupole, Fourier transform, and quadrupole ion trap configurations. Each technique has particular ad-

[12a] R. C. Beavis and B. T. Chait, *Methods Enzymol.* **270,** Chap. 22, 1996.

[13] M. Karas and F. Hillenkamp, *Anal. Chem.* **60,** 2299 (1988).

[14] F. Hillenkamp, M. Karas, R. Beavis, and B. T. Chait, *Anal. Chem.* **63,** 1793 (1991).

[15] S. Geromanos, P. Casteels, C. Elicone, M. Powell, and P. Tempst, *in* "Techniques in Protein Chemistry V" (J. W. Crabb and W. A. Jones, eds.), p. 143. Academic Press, San Diego, CA, 1994.

[16] R. Wang, B. T. Chait, and S. B. H. Kent, *in* "Techniques in Protein Chemistry IV" (R. H. Angeletti, ed.), p. 471. Academic Press, San Diego, CA, 1993.

[17] T. M. Billeci and J. T. Stults, *Anal. Chem.* **65,** 1709 (1993).

[18] W. J. Henzel, T. M. Billeci, J. T. Stults, S. C. Wong, C. Grimley, and C. Watanabe, *Proc. Nat. Acad. Sci. U. S. A.* **90,** 5011 (1993).

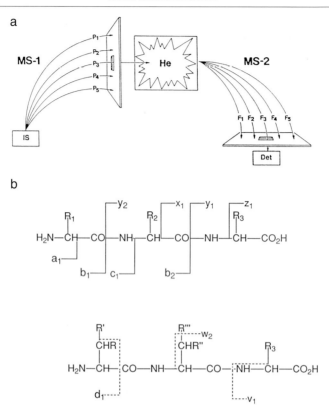

FIG. 1. Tandem mass spectrometry (a) involves selection of a single precursor ion mass for collision with neutral gas molecules (He or Ar) followed by mass analysis of the fragment product ions. (Reprinted with permission from Biemann and Scoble, 1987.[33]) Nomenclature for peptide CID product ions (b) follows that of Johnson and co-workers.[61] The a, b, and y ion types are formed by low- and high-energy CID; the others are formed only by high-energy CID. Under low-energy CID conditions these ions may also lose ammonia (17 Da) or water (18 Da) molecules. The y ion formation requires the addition of two protons to the fragment shown. Likewise, fragments formed by side-chain cleavage under high-energy CID conditions (d, w, and v ions) require proton transfers.[61]

vantages, and descriptions of the common tandem mass spectrometers are given elsewhere.[19–21] The availability and performance of instruments with various configurations are changing at a rapid pace, making many options available for mass spectrometric characterization of peptides. Applications

[19] K. R. Jennings and G. G. Dolnikowski, *Methods Enzymol.* **193,** 37 (1990).
[20] R. A. Yost and R. K. Boyd, *Methods Enzymol.* **193,** 154 (1990).
[21] R. N. Hayes and M. L. Gross, *Methods Enzymol.* **193,** 237 (1990).

of tandem mass spectrometry using electrospray ionization on a triple-quadrupole instrument and FAB ionization on a four-sector instrument are described in this chapter. Both of these instrumental configurations have been used to solve many problems related to peptide characterization.

Experimental

Sample Preparation

Peptides are analyzed from low ionic strength (<10 mM) volatile buffers, and in the absence of surfactants and chaotropic agents. If these requirements are not met, high-performance liquid chromatography (HPLC) with a volatile buffer [such as trifluoroacetic acid (TFA) or ammonium acetate] must be used as the final step of the purification, or on-line liquid chromatography–mass spectrometry (LC–MS) may be used (details may be found in [17] in this volume[21a]). A rapid or stepped gradient provides a convenient desalting step prior to ionization as an alternative to traditional dialysis or size-exclusion chromatography. Concentrations of the peptide for electrospray ionization are typically >0.5 pmol/μl for routine measurement with direct infusion of the peptide, or >1 pmol total peptide for injection on packed-capillary HPLC. For direct infusion, the sample is dissolved in 10–20 μl of 50% (v/v) aqueous acetonitrile or aqueous methanol with 1% (v/v) acetic acid. If the sample is a fraction collected directly from reversed-phase (RP)-HPLC, that fraction can be infused directly in the typical water–acetonitrile–TFA solution.

Peptides analyzed by FAB ionization are in a matrix of thioglycerol. For static FAB 1 μl of thioglycerol and 1 μl of the peptide solution (100 pmol/μl) are mixed and analyzed. For continuous-flow FAB (frit-FAB), 1 μl of the peptide solution is diluted with 50 μl of aqueous methanol solution (50:50,v/v) containing 2.5% (v/v) monothioglycerol to a final concentration of 10–100 fmol/μl. The solutions are infused into the frit-FAB ion source at 3 μl/min.

Mass Measurement

The instrument should be properly calibrated, and the calibration and tuning checked before sample analysis. Resolving power should be adjusted to provide unit mass resolution at m/z 1000 while analyzing a mixture of polypropylene glycols or peptides. This requirement ensures that one is able

---

[21a] W. S. Hancock, A. Apffel, J. Chakel, T. M'Timkulu, and A. W. Guzetta, *Methods Enzymol.* **271,** Chap. 17, 1996 (this volume).

to distinguish singly charged from more highly charged ions in electrospray ionization. Resolving power on the sector instrument is set to provide unit mass resolution for the peptide of interest or up to m/z 3000. In cases in which very high sensitivity is necessary, the resolution may be reduced with a concomitant increase in signal but decrease in selectivity.

Collision-Induced Dissociation Fragmentation

The molecular masses of the peptides are first determined by direct analysis or LC–MS. Individual fractions or peptides are then analyzed by MS–MS. If the peptides are separated by on-line LC–MS, two chromatograms are normally required, and the data system is programmed to acquire fragmentation spectra during the second chromatogram for the appropriate precursor masses that appear at particular retention times during the first chromatogram. Interpretation of the CID spectra is done primarily without the aid of computer software; however, programs have been written that can suggest candidate sequences.[22]

Instrumentation

Collision-Induced Dissociation on Four-Sector Mass Spectrometer. High-energy collision-induced dissociation (CID) analyses are performed on a JEOL (Tokyo, Japan) JMS HX110/HX110 four-sector tandem mass spectrometer with fast atom bombardment (FAB) ionization.[23] The static FAB spectra are obtained with an accelerating voltage of 10 kV and a mass resolution of 3000. Collision-induced dissociation spectra are obtained with an accelerating voltage of 10 kV, a mass resolution of 1000, and a collision cell voltage of 6 kV. Helium gas pressure is adjusted to decrease the protonated molecule peak to 30% of its original intensity. The CID product ions are collected by array detection on a 3-in. microchannel plate held at 1.5 kV while the associated phosphor is held at 4.0 kV. The CID scans are collected every 2 min typically scanning from m/z 70 to the molecular weight of the peptide.

Collision-Induced Dissociation on Triple-Quadrupole Mass Spectrometer. Low-energy collision-induced dissociation (CID) analyses[24] are performed on a Sciex (Thornhill, Canada) API III triple-quadrupole mass spectrometer with electrospray ionization. The intact molecular weights

[22] R. S. Johnson and K. Biemann, *Biol. Environ, Mass Spectrom.* **18,** 945 (1989).

[23] K. Sato, T. Asada, M. Ishihara, F. Kunihiro, Y. Kammei, E. Kubota, C. E. Costello, S. A. Martin, H. A. Scoble, and K. Biemann, *Anal. Chem.* **59,** 1652 (1987).

[24] D. F. Hunt, J. R. Yates, III, J. Shabanowitz, S. Winston, and C. R. Hauer, *Proc. Natl. Acad. Sci. U.S.A.* **83,** 6233 (1986).

are measured by infusion of micromolar solutions (1–10 pmol/μl) of the peptides in 50% aqueous acetonitrile or aqueous methanol with 1% acetic acid. The solutions are infused at 1.5–3 μl/min. The mass spectrometer is operated with the Sciex Ionspray articulated nebulizer at 4600 V, the interface plate at 600 V, and the orifice at 80–120 V. Data are collected at 0.1- to 0.25-m/z increments with the quadrupole typically scanning from 600 to 2200 m/z for molecular weight analysis. Two types of CID experiments are performed on this instrument. The first is MS–MS, or tandem mass spectrometry, with a single m/z from the first quadrupole passed into the second, radio frequency (rf) only, quadrupole containing argon. The peptide daughter ions formed by the collision are separated by mass in the third quadrupole. The second type of CID experiment involves collisions induced by raising the orifice (skimmer) potential to 160 V in the high-pressure region of the ion source. The orifice potential collisions are used to observe glycopeptide fragment ions.[25,26] The CID data are collected every 0.25 m/z with the quadrupole typically scanning from m/z 60 to the molecular weight of the peptide in 6–10 sec.

High-Performance Liquid Chromatography. The HPLC system consists of a gradient HPLC pump, flow splitter, injector, capillary column, detector, and a capillary transfer line to the mass spectrometer.[27] The flow splitter [LC Packings (San Francisco, CA) splitter or Valco Instruments (Houston, TX) tee] decreases the flow rate produced by reciprocating or syringe pumps (typically 100 μl/min) to 3–4 μl/min. This flow rate is compatible with the packed-capillary HPLC columns (5-μm C$_{18}$ particles in 320 μm × 15 cm columns). A capillary flow cell (LC Packings) is used to modify the ultraviolet absorbance detector.

On-line High-Performance Liquid Chromatography–Tandem Mass Spectrometry

Peptides are separated on packed-capillary reversed-phase C$_{18}$ columns. The low flow rates of these columns can be accommodated easily by mass spectrometer ion sources and afford high sensitivity. From 5 to 100 pmol of proteolytic peptides has been separated with the packed-capillary C$_{18}$ column (320 μm × 150 mm). Columns with 75-μm (and narrower) diameters yield detection limits in the low-femtomole range. Gradient elution in aqueous acetonitrile with 0.1% TFA from 98 to 40% water is accomplished in 60 min. The flow rate is programmed to load the sample at 20 μl/min with isocratic 2% acetonitrile. The gradients are run at a flow rate of 3–4 μl/min. Peptides isolated from major histocompatibility complex (MHC)

[25] J. J. Conboy and J. D. Henion, *J. Am. Soc. Mass Spectrom.* **3,** 804 (1992).
[26] M. J. Huddleston, M. F. Bean, and S. A. Carr, *Anal. Chem.* **65,** 877 (1993).
[27] W. J. Henzel, J. H. Bourell, and J. T. Stults, *Anal. Biochem.* **187,** 228 (1990).

class I molecules are eluted from the reversed-phase column with a shallower gradient from 85 to 60% water in 50 min. To interpret the LC–MS data files, eliminate background peaks, and reduce broad peaks to a single spectrum by computer subtraction, routines such as PeptideMap or Enhance (PE/SCIEX, Thornhill, Ontario, Canada) are used. Once masses of interest are identified with their retention times, a second chromatogram is run under CID conditions set to obtain fragment spectra for individual components. It should be noted that overlapping peaks cannot be analyzed in this way, but if enough material is available to collect fractions of >1 pmol/μl, these fractions can be used to obtain fragment ion data on multiple peaks in a mixture.

Results and Discussion

Molecular Weight Analysis

Quite often the sequences of known proteins are confirmed by matching the molecular weights of peptides generated by proteolytic cleavage against the expected values (see [15] and [17] in this volume[27a]). Similarly, peptides produced by posttranslational processing within cells can be identified by comparing the observed molecular weights with the genetic sequence of the intact protein. For example, an investigation of cell-based production of relaxin, a mammalian hormone similar in size to insulin, utilized mass analysis to identify enzymatic cleavage sites.[28] Relaxin is involved in remodeling the birth canal during parturition, and was synthesized as a preprohormone precursor. Following signal sequence removal, the endoproteolytic enzyme prohormone convertase 1 and carboxypeptidase enzymes removed a central portion of the protein, leaving the mature, active disulfide-linked species secreted by the cell (see Fig. 2).[28] Molecular weight analysis using electrospray ionization was performed on several mutants of preprorelaxin. Immunopurified relaxins were analyzed by reversed-phase HPLC on a 2×100 mm C_4 column using a linear aqueous (0.1% trifluoroacetic acid) acetonitrile gradient from 10 to 60%. Fractions were collected at 1-min intervals and major peaks of ultraviolet (UV) absorbance were mass analyzed after drying and dissolution with aqueous acetonitrile (50:50, v/v) with 1% acetic acid. Mass spectra taken from two HPLC fractions from a single mutant $K_{136}R/R_{137}A$ (Fig. 2) showed a number of peptides with masses 6051.3, 6207.3, and 6336.1 Da in fraction 4, and masses 7378.0 and 7393.6 Da in fraction 5. All of these arose from different sites of processing.

[27a] J. R. Yates, *Methods Enzymol.* **271,** Chap. 15, 1996 (this volume); W. S. Hancock, A. Apffel, J. Chakel, C. Sounders, T. M'Timkulu, and A. W. Guzetta, Chap. 17, 1996 (this volume).
[28] D. Marriott, B. Gillece-Castro, and C. M. Gorman, *Mol. Endocrinol.* 1441 (1992).

The small peak at 7393.6 Da reflected minor methionine oxidation. The wild-type molecular weight (5963.1 Da) reflected a pyroglutamic acid residue on the A chain.[28] These spectra demonstrated the power of mass measurements to identify mixtures of components, including many types of modifications.

Collision-Induced Dissociation Fragment Spectra from Tandem Mass Spectrometry

The first analyses of peptides by FAB in the early 1980s were soon followed by CID spectra of peptides to obtain sequence information.[29–32] Soon CID spectra obtained on four-sector and triple-quadrupole mass spectrometers were being used to identify unknown or modified peptide sequences.[33–37] All of the following examples utilize CID spectra in the characterization of peptides, disulfide-linked peptides, phosphopeptides, glycopeptides, and unknown peptides.

Disulfide Bond Assignment. Protein and peptide secondary structure has been elucidated by coupling mass spectrometry with enzymatic digestion to map disulfide bond positions.[38,39] FAB–mass spectrometry was used to determine the molecular weights of disulfide-bonded peptides. The molecular weights of the component peptides were then determined following on-probe reduction of the disulfide bonds.[40] A thiol [e.g., monothioglycerol or dithiothreitol (DTT)] was added to the probe tip with a volatile buffer such

[29] I. J. Amster, M. A. Baldwin, M. T. Cheng, C. J. Proctor, and F. W. McLafferty, *J. Am. Chem. Soc.* **105,** 1654 (1983).

[30] W. Heerma, J. P. Kamerling, A. J. Slotboom, G. J. M. van Scharrenburg, B. N. Green, and I. A. S. Lewis, *Biol. Mass Spectrom.* **10,** 13 (1983).

[31] T. Matsuo, H. Matsuda, S. Aimoto, Y. Shimonishi, T. Higuchi, and Y. Maruyama, *Int. J. Mass Spectrom. Ion Process.* **46,** 423 (1983).

[32] K. Eckart, H. Schwartz, K. B. Tomer, and M. L. Gross, *J. Am. Chem. Soc.* **107,** 6765 (1985).

[33] K. Biemann and H. A. Scoble, *Science* **237,** 992 (1987).

[34] E. A. Robinson, T. Yoshimura, E. J. Leonard, S. Tanaka, P. R. Griffin, J. Shabanowitz, D. F. Hunt, and E. Appella, *Proc. Natl. Acad. Sci. U.S.A.* **86,** 1850 (1989).

[35] S. A. Carr, B. N. Green, M. E. Heming, G. D. Roberts, R. J. Anderegg, and R. Vickers, *in* "Proceedings of 35th ASMS Conference on Mass Spectrometry and Allied Topics, Denver, CO, 1987."

[36] S. Akashi, K. Hirayama, A. Murai, M. Arai, S. Murao, R. Takahashi, K.-i. Fukuhara, N. Oouchi, K. Tanaka, and I. Nojima, *Biochemistry* **26,** 6483 (1987).

[37] D. M. Desiderio, *Int. J. Mass Spectrom. Ion Process.* **74,** 217 (1986).

[38] H. R. Morris and P. Pucci. *Biochem. Biophys. Res. Commun.* **126,** 1122 (1985).

[39] A. M. Buko and B. A. Fraser, *Biol. Mass Spectrom.* **12,** 577 (1985).

[40] J. T. Stults, J. H. Bourell, E. Canova-Davis, V. T. Ling, G. R. Laramee, J. E. Winslow, P. R. Griffin, E. Rinderknecht, and R. L. Vandlen, *Biomed. Environ. Mass Spectrom.* **19,** 655 (1990).

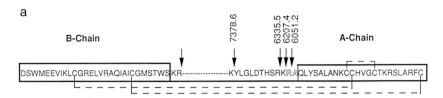

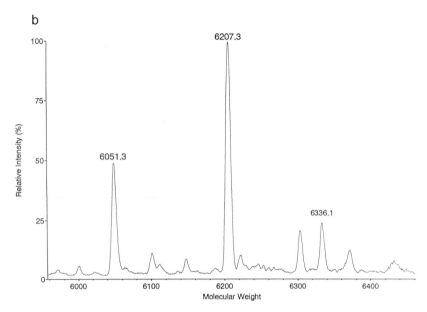

FIG. 2. Molecular weight determination of the double-mutant prorelaxin, K136R/R137A (highlighted residues).[28] (a) Sequence and observed processing sites of mutant prorelaxin coexpressed with the enzyme prohormone convertase 1 (PC1). The A and B chains are held together by two disulfide bonds in the mature protein after removal of the connecting peptide. Following cleavage by PC1 (arrows) the B chain is shortened by carboxypeptidase removal of lysine and arginine residues to end at a serine residue. (b) Reconstructed electrospray ionization mass spectrum of one isolated HPLC fraction of the processed relaxin. (c) Reconstructed electrospray ionization mass spectrum of a second isolated HPLC fraction of the processed relaxin.

as ammonium bicarbonate or ammonium hydroxide to provide the required basic pH. Subsequent addition of an acid (oxalic) provided optimal ionization. Alternatively, reduction or reduction/alkylation can be performed prior to LC–MS by electrospray ionization. In this method two LC–MS analyses are often performed, one prior to and one following reduction. In the latter experiment the disulfide-bonded peptides are each separated into two peaks with new retention times and masses. Subdigestion may be

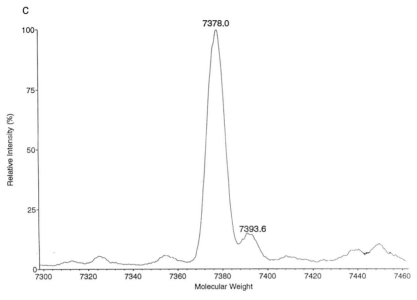

FIG. 2. (*continued*)

required for peptides containing more than one cysteine residue, or MS–MS
may be required when the cysteine residues are adjacent.

As an example, native human relaxin contains adjacent cysteine residues
at positions 10 and 11 in the A chain (see Fig. 2a). Reduction of native
human relaxin (data not shown) showed that the A-chain molecular mass
of 2656 Da corresponded to residues 138–161 with pyroglutamic acid at
the N terminus. The B-chain molecular mass of 3313 Da corresponded to
residues 1–29, four residues shorter than predicted by homology to other
species. The A chain contains four cysteine residues, including the adjacent
residues at positions Cys-A10 and Cys-A11. Tryptic mapping of relaxin
gave one fragment that contained one peptide each from the A and B
chains, including two disulfide bonds (see Fig. 3c).[40] The CID spectra from
a four-sector mass spectrometer shown in Fig. 3 were used to identify the
disulfide linkage pattern for relaxin. The spectrum from the recombinant
relaxin tryptic fragment (Fig. 3a) showed abundant cleavage of the disulfide
between the two peptide chains. Unfortunately, fragmentation was lacking
for residues in the cyclized region formed by the intrachain disulfide. Peptides containing the possible disulfide linkages were chemically synthesized.
The CID spectrum shown in Fig. 3b (corresponding to the peptide shown
in Fig. 3c) matched the native relaxin spectrum identically. In this case

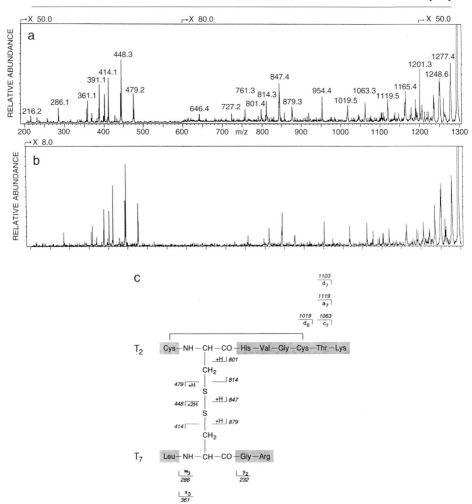

Fig. 3. Collision-induced dissociation spectra obtained by high-energy CID on a four-sector mass spectrometer. (a) CID spectrum of a disulfide-bonded trypsin digest fragment from recombinant human relaxin containing two disulfide bonds. (b) CID spectrum of a synthetic disulfide-bonded peptide. (c) Fragments observed in the CID spectra. (Results printed with permission from Stults et al., 1990.[40])

tandem mass spectrometry successfully identified adjacent cysteine disulfide bonds, a challenging peptide structural element.

   Chemical Derivatives for Collision-Induced Dissociation Spectra. Because CID spectra contain only the molecular masses of fragment ions, the

N- or C-terminal origin of a fragment is not always immediately obvious. Several strategies have been used to modify the amino terminus or carboxyl terminus of peptides to label them for identification in CID spectra. Carboxylic acids have been exchanged with ^{18}O-labeled water for a 2-Da shift in mass or the formation of a mass doublet with 50% ^{18}O-labeled water. Esterification has also been used extensively, with methyl and hexyl esters preferred. The methyl esters can be prepared using commercial methanolic hydrochloric acid (1 M) and are the most common.[41] The hexyl esters are advantageous for FAB/LSIMS ionization, because the hydrophobicity of the hexyl group improves sensitivity for hydrophilic peptides.[42] Amino-terminal derivatives include acetylation of amines (N-α-amino and lysine ε-amino groups), which gives information on the number of amino groups, and they shift the fragment ions in CID spectrum. Because protonation or a fixed positive charge can direct fragmentation in high-energy CID spectra, derivatives to add a charge at the amino terminus of peptides have been developed.[43,44] The fragment ion spectra shown in Fig. 4 demonstrate the shift in charge retention from the carboxy-terminal lysine (w, x, y, and z ions) to the amino-terminal derivative (a, b, c, and d ions). The 2-(dimethyl-hexylamino)acetyl group was coupled to a trypsin digest fragment from thymosin (SDAAVDTSSEITTK). All of the product ions of the CID experiment have charge retained on the amino terminus of the peptide, simplifying the sequence analysis of this 14-amino acid peptide. It should be noted that owing to the high-energy collisions many of the fragment ions are w-, v-, and d-type cleavages, which involve the side chains of amino acid residues. In particular, the isoleucine residue gives rise to ions w_{4a} and w_{4b} (Fig. 4a) and d_{11a} and d_{11b} (Fig. 4b), distinguishing it from leucine. Thus, these spectra demonstrate the utility of high-energy CID and derivatization for sequencing peptides. Fragmentation in low-energy CID occurs predominantly at sites of proton attachment, so fixed charge derivatives actually reduce the fragmentation in these spectra.

Phosphopeptide Characterization. The phosphorylation state of proteins is controlled by the activity of the enzymes that transfer phosphate to them (kinases) or remove phosphate from them (phosphorylases). To determine sites of attachment for phosphate, isolated phosphopeptides have been analyzed by CID. The phosphoryl ester bonds formed with hydroxyl groups

[41] T. Krishnamurthy, L. Szafraniec, D. F. Hunt, J. Shabanowitz, J. R. Yates, III, C. R. Hauer, W. W. Carmichael, O. Skulberg, G. A. Codd, and S. Missler, *Proc. Natl. Acad. Sci. U.S.A.* **86,** 770 (1989).

[42] A. M. Falick and D. A. Maltby, *Anal. Biochem.* **182,** 165 (1989).

[43] D. S. Wagner, A. Salari, D. A. Gage, J. Leykam, J. Fetter, R. Hollingsworth, and J. T. Watson, *Biol. Mass Spectrom.* **20,** 419 (1991).

[44] J. T. Stults, J. Lai, S. McCune, and R. Wetzel, *Anal. Chem.* **65,** 1703 (1993).

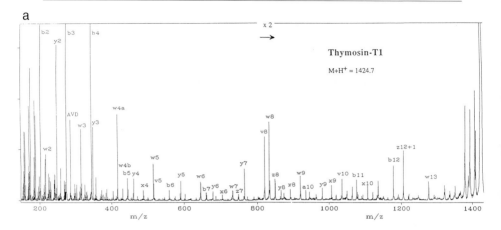

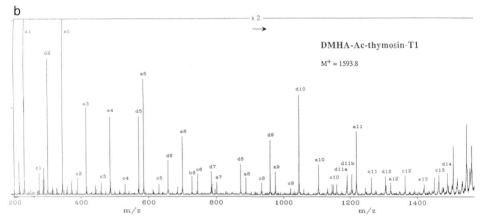

Fig. 4. Collision-induced dissociation (CID) spectrum of (a) thymosin-T1 (SDAAVDTS-SEITTK). CID spectrum of an amino-terminal derivative (dimethylhexylaminoacetyl) of thymosin-T1 (b). These spectra were obtained by high-energy CID on a four-sector mass spectrometer. (Results printed with permission from Stults *et al.*, 1993.[44])

on serine, threonine, and to a lesser extent tyrosine, are quite labile under high-energy CID conditions and in negative-ion mode form strong PO_3^- and HPO_4^- ions in addition to peptide bond cleavage.[45] A particularly innovative approach to characterize phosphoproteins makes use of fragmentation that is generated by electrospray source excitation. By raising the electric field in the desolvation region (typically the orifice or nozzle–skimmer voltage), the ions with the surrounding gas molecules are acceler-

[45] B. W. Gibson and P. Cohen, *Methods Enzymol.* **193,** 480 (1990).

ated through and undergo extensive collisions that cause fragmentation. Huddleston and co-workers have utilized electrospray source fragmentation to detect phosphopeptides in LC–MS experiments. In negative-ion mode the appearance of the PO_3^- and HPO_4^- fragment ions from phosphopeptides were induced by collisions in this way. The orifice voltage is then lowered to its normal value for the remainder of the mass scan to identify peptide and phosphopeptide molecular weights.[46] A separate method for the identification of phosphorylated peptides utilizes neutral loss scans for the loss of phosphoric acid from the peptide by collisional activation.[47]

Once phosphopeptides are identified in a digest mixture, CID spectra can be used to identify the phosphorylated residue(s). In the positive-ion mode, the primary fragments formed in a triple-quadrupole MS–MS experiment are the result of peptide bond cleavages. For example, the CID spectrum from a synthetic phosphopeptide, TFLPVPEpYINQSV (primary autophosphorylation site of epidermal growth factor receptor kinase, molecular weight 1586.7), gave sufficient fragments to localize the phosphate moiety to the Tyr-Ile dipeptide (see Fig. 5). Complementary N-terminal and C-terminal product ions (b_9 at m/z 1141.5 and y_6 at m/z 803.5, respectively) were 80 Da higher in mass than predicted for the nonphosphorylated peptide. These mass shifts are used to identify the sites of modification on peptides of known sequence. The previously described procedures apply equally well to phosphorylated serine, threonine, and tyrosine residues; however, the less common phosphorylated histidine and aspartic acid residues are chemically unstable. Sites of phosphorylation on aspartic acid residues have been identified by chemical reduction to the stable homoserine residue, which can be sequenced by tandem mass spectrometry.[48,49]

Glycopeptide Characterization. Glycosylation of proteins is the most complex of the posttranslational modifications. The heterogeneity of protein-linked oligosaccharides and their relatively large molecular masses (typically 600–3000 Da) have made molecular mass determinations of glycopeptides an important tool in glycopeptide characterization. Collision-in-

[46] M. J. Huddleston, R. S. Annan, M. F. Bean, and S. A. Carr, *J. Am. Soc. Mass Spectrom.* **4,** 710 (1993).

[47] T. Covey, B. Shushan, R. Bonner, W. Schroder, and F. Hucho, *in* "Methods in Protein Sequence Analysis" (H. Jörnvall, J. O. Höög, and A. M. Gustavsson, eds.), p. 249. Birkhauser, Basel, Switzerland, 1991.

[48] D. A. Sanders, B. L. Gillece-Castro, A. M. Stock, A. L. Burlingame, and J. D. E. Koshland, *J. Biol. Chem.* **264,** 21770 (1989).

[49] D. A. Sanders, B. L. Gillece-Castro, A. L. Burlingame, and J. D. E. Koshland, *J. Bacteriol.* **174,** 5117 (1992).

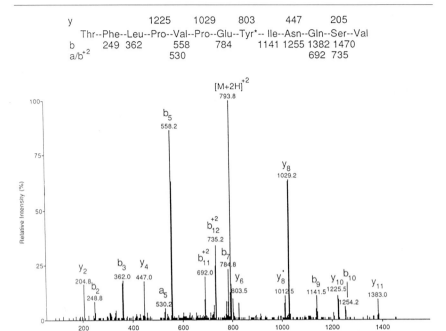

FIG. 5. Low-energy collision-induced dissociation spectrum of a phosphotyrosine-containing peptide.[62] The fragments show that the phosphate group is located on the Tyr-Ile dipeptide. (The phosphopeptide was a gift from G. Tregear of the Florey Institute, Melbourne, Australia.)

duced dissociation spectra of glycopeptides have been used to identify oligosaccharide structures and sites of glycosylation.[25,50,51] For example, analyses of glycopeptides utilizing high-energy CID on a four-sector mass spectrometer have identified sites of O-linked glycosylation, including the specific serine or threonine residue modified in each peptide.[50,51] The innovative approach of fragmentation generated by electrospray source excitation discussed above as a method for phosphopeptide characterization is also an excellent means of detecting glycosylated peptides.[25,26] In the positive-ion mode, the N-linked and O-linked carbohydrates yield oxonium fragment ions that are indicative of their parent saccharide subunits (N-

[50] K. F. Medzihradszky, B. L. Gillece-Castro, C. A. Settineri, R. R. Townsend, F. R. Masiarz, and A. L. Burlingame, *Biomed. Environ. Mass Spectrom.* **19,** 777 (1990).

[51] C. A. Settineri, K. F. Medzihradszky, F. R. Masiarz, A. L. Burlingame, C. Chu, and C. George-Nascimento, *Biomed. Environ. Mass Spectrom.* **19,** 665 (1990).

acetylhexosamine, m/z 204; N-acetylneuraminic acid, m/z 292; and the disaccharide hexose + N-acetylhexosamine, m/z 366). During each scan a high orifice voltage (160 V) is used to induce the fragmentation when scanning the mass region of the low mass fragment ions. The orifice voltage is then lowered to its normal value for the remainder of the mass scan to identify peptide and glycopeptide molecular weights. Figure 6 shows the tryptic peptide map of recombinant human tissue plasminogen activator (rtPA), a fibrin-specific serine protease important in clot lysis, which contains three N-linked glycosylation sites. Electrospray source fragmentation produced carbohydrate fragment ions (204, 292, and 366 Da) that identified three regions of the chromatogram containing glycopeptides. Within these regions were found four to eight glycoforms per peptide. Identification of these glycoform mixtures with overlapping elution patterns is rapidly accomplished by on-line LC–MS. From these data the compositions of hexose, N-acetylhexosamine, deoxyhexose, and sialic acid residues on each peptide were deduced.

Unknown Sequence Analysis. Tandem mass spectrometry has been used to sequence unknown peptides and proteins in relatively few labora-

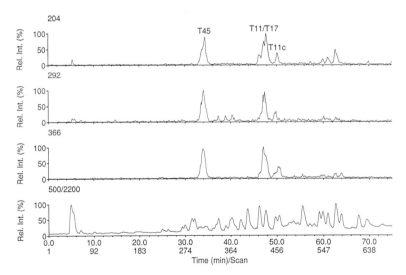

FIG. 6. Glycopeptide identification by in-source collisions with electrospray ionization. A peptide map was made of the trypsin digest of recombinant human tissue plasminogen activator (rtPA) by LC–MS.[63] The oligosaccharide-containing peptides have fragment ions at m/z 204, 292, and 366. Profiles for each ion show the retention times of glycopeptides and spectra in these regions give the molecular ions for many glycoforms on each peptide. The total ion current (TIC) shown below these profiles indicates peaks for all peptides.

tories.[34,52-56] However, these groups have demonstrated the value of this technique for projects such as sequencing proteins from 2D gel spots[54] and complex mixtures of peptides bound to MHC molecules.[55,56] Furthermore, these sequences have been obtained using low-femtomole to low-picomole amounts of peptide.

In our laboratory, capillary LC–MS was used to analyze the complex mixture of peptides bound by one MHC molecule, HLA-B0702. The MHC class I molecules are glycoproteins that present peptides to cytotoxic T cells, thus providing a defense against viruses and other intracellular parasites, or malignant cells. To identify the most abundant peptide molecular ions data were acquired from a capillary LC–MS experiment. Subsequently, the low-energy CID data for a number of the peptides were acquired during a second LC–MS–MS experiment on a triple-quadrupole mass spectrometer. In the spectrum shown in Fig. 7 the product ions were formed by collision of a doubly protonated molecule, m/z 520.3, with argon atoms at 70-eV energy. Sequence ions were identified starting with the highest masses observed, the first pair of which differed by 28 Da (m/z 908–880). This combination typically corresponds to an a/b ion pair, which differ by a carbon monoxide molecule (see Fig. 2). Because the putative b ion was 131 Da lower in mass than the MH$^+$ ion (m/z 1039), a leucine or isoleucine residue (indistinguishable in this experiment) must be at the C terminus of the peptide. Note that the highest mass b fragment ion is formed by loss of the C-terminal residue, including the hydroxyl group and one additional proton, or 18 Da more than all other b- or y-ion differences. The next pair of intense peaks (m/z 795 and m/z 767) also differed by 28 Da and therefore fell in the a/b series. They were 113 Da lower in mass than the first pair, corresponding to another leucine or isoleucine residue. The following a/b pairs of ions (m/z 696/668, 583/555, and 469/441) were 99, 113, and 113 Da lower in mass, corresponding to valine, leucine or isoleucine, and asparagine residues. Therefore the C-terminal portion of this peptide reads NXV$X$$X$, where X is leucine or isoleucine. The low mass product ions reflect the amino acid composi-

[52] R. S. Johnson and K. Biemann, *Biochemistry* **26,** 1209 (1987).

[53] K. F. Medzihradszky, B. F. Gibson, S. Kaur, Z. Yu, D. Medzihradszky, A. L. Burlingame, and N. M. Bass, *Eur. J. Biochem.* **327,** 203 (1992).

[54] S. C. Hall, D. M. Smith, F. R. Masiarz, V. W. Soo, H. M. Tran, L. B. Epstein, and A. L. Burlingame, *Proc. Natl. Acad. Sci. U.S.A.* **90,** 1927 (1993).

[55] D. F. Hunt, R. A. Henderson, J. Shabanowitz, K. Sakaguchi, H. Michel, N. Sevilir, A. L. Cox, E. Appella, and V. H. Engelhard, *Science* **255,** 1261 (1992).

[56] E. L. Huczko, W. M. Bodnar, D. Benjamin, K. Sakaguchi, N. Z. Zhu, J. Shabanowitz, R. A. Henderson, E. Appella, D. F. Hunt, and V. H. Engelhard, *J. Immunol.* **151,** 2572 (1993).

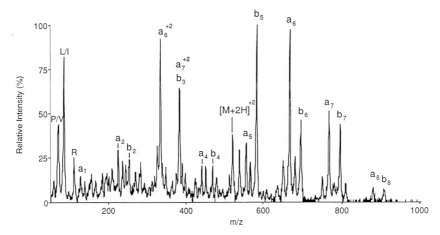

FIG. 7. Capillary LC–MS–MS spectrum of a peptide eluted from an MHC class I molecule, HLA-B0702. The low-energy CID data for a number of peptides were acquired at different times during an LC–MS–MS run, and the spectrum shown is for the ion at *m/z* 520.5, which eluted at 42 min. The ion was made to collide with argon atoms at 70-eV energy. The masses of the products (immonium ions) reflect an amino acid composition including proline, valine, arginine, and leucine or isoleucine (leucine and isoleucine are indistinguishable in this experiment). Charge was retained on the N-terminal basic amino acids, giving rise to a and b fragments (shown in nominal mass) reflecting a sequence that matches a portion of CD20, RPKSNIVLL, found in the sequence database.

tion[57] including proline (*m/z* 70), valine (*m/z* 72), arginine (*m/z* 112/129), and leucine or isoleucine (*m/z* 86). The a/b pair at *m/z* 226/254 corresponds to a dipeptide consisting of proline and arginine. The a_1 ion at *m/z* 129 corresponds to the N-terminal arginine, but could arise as an immonium ion. Edman microsequencing of an HPLC fraction (still a complex mixture) containing this peptide gave strong signals for N-terminal Arg-Pro-Lys residues. The sequence then identified was RPK?NXVXX, where the residue in question had a mass of 87 Da, which corresponds to a serine residue. The final sequence, RPKSNIVLL, matches a portion of CD20 found in the sequence database, and a similar spectrum was reported by Huczko and co-workers.[56]

[57] A. M. Falick, W. M. Hines, K. F. Medzihradszky, M. A. Baldwin, and B. W. Gibson, *J. Am. Soc. Mass Spectrom.* **4,** 882 (1993).

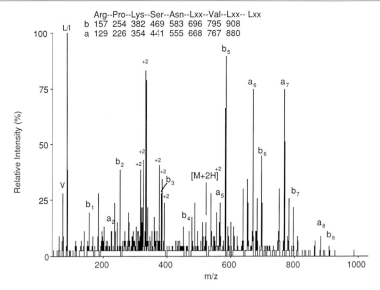

Fig. 8. Low-energy CID spectrum of synthetic RPKSNIVLL using an upgraded, enclosed collision cell that allows multiple, very low-energy collisions. The precursor ions at m/z 520.5 were made to collide with argon atoms at 5-eV energy. The product ions reflect much improved resolution (compare with Fig. 7), which allowed the assignment of the fragment at m/z 334.5 as the doubly charged a_6 ion, and allowed assignment of the b_3 ion (m/z 382), now resolved from the doubly charged a_7 ion (m/z 384). Approximately 400 fmol was used to obtain this spectrum.

The peptide RPKSNIVLL was synthesized and an identical spectrum was produced by CID, using the same conditions. The low-energy CID spectrum of RPKSNIVLL shown in Fig. 8 was produced using an upgraded collision cell, whereby the precursor ion at m/z 520.5 was made to collide with argon atoms at 5-eV energy. The product ions reflect much improved mass resolution, which allowed the assignment of the abundant fragment at m/z 334.5 as the doubly charged a_6 ion. Importantly, the b_3 ion (m/z 382) was resolved from the doubly charged a_7 ion (m/z 384), which allowed clear assignment of the b_3 ion, identifying the lysine residue in position 3 from the mass spectrum.

The high-energy CID spectrum of synthetic RPKSNIVLL shown in Fig. 9 was obtained on a four-sector mass spectrometer. Approximately 250 fmol was used to obtain this spectrum, using continuous-flow FAB ionization and tandem mass spectrometry. The singly protonated peptide at m/z 1040 was made to collide with argon atoms at 4000 eV. The resulting spectrum is more complex than the low-energy CID counterparts shown in Figs. 7 and 8, but the side-chain cleavages induced in the process are indicative of

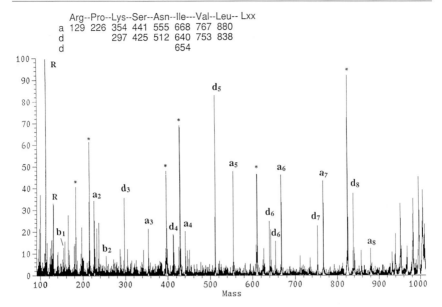

FIG. 9. High-energy CID spectrum of synthetic RPKSNIVLL. Approximately 250 fmol was used to obtain this spectrum using continuous-flow FAB ionization and MS–MS. Asterisks indicate matrix-related background ions. The singly protonated peptide at m/z 1039.7 was made to collide with argon atoms at 4000 eV. The resulting spectrum is more complex than the low-energy CID counterparts shown in Figs. 7 and 8, but is still readily interpreted. The full series of a ions (shown in nominal mass) defines the sequence. The d_6 and d_8 fragments allow identification of isoleucine and leucine at positions 6 and 8, respectively.

each amino acid residue. In particular, the isoleucine side chain can be differentiated from leucine by the presence of the d_6 ions 14 and 28 Da lower in mass than the a_6 ion (m/z 668). Thus, triple-quadrupole and four-sector mass spectrometers are capable of sequencing unknown peptides. Because on-line HPLC–electrospray ionization with mass spectrometry is slightly more tolerant of background buffers and detergents than FAB ionization, and is available on triple-quadrupole instruments, peptide characterization by LC–ESI–CID is the commoner method.

Summary and Projections

Characterization of peptides using mass spectrometric techniques, especially when coupled to high performance separations, can accomplish much. In particular, posttranslational modifications are readily identified and the modified amino acid residue(s) are pinpointed. Continuing efforts to improve sensitivity and spectral information, e.g., Fourier transform,[12,58] time of flight,[58a] and ion trap mass spectrometers[59,60] (see also [23] in this volume[60a]), should further open the field of peptide characterization to the low levels of peptides and proteins found in nature.

[58] S. Guan, M. C. Wahl, and A. G. Marshall, *Anal. Chem.* **65,** 3647 (1993).
[58a] R. C. Beavis and B. T. Chait, *Methods Enzymol.* **270,** Chap. 22, 1996.
[59] K. A. Cox, J. D. Williams, R. G. Cooks, and R. E. Kaiser Jr., *Biol. Mass Spectrom.* **21,** 226 (1992).
[60] V. M. Doroshenko and R. J. Cotter **7,** 822 (1993).
[60a] J. C. Schwartz and I. Jardine, *Methods Enzymol.* **270,** Chap. 23, 1996.
[61] R. S. Johnson, S. A. Martin, and K. Biemann, *Int. J. Mass Spectrom. Ion Process.* **86,** 137 (1988).
[62] L. M. Nuwaysir and T. Sakuma, unpublished results (1993).
[63] V. Ling, A. W. Guzzetta, E. Canova-Davis, J. T. Stults, and W. S. Hancock, *Anal. Chem.* **63,** 2909 (1991).

[19] Capillary Electrophoresis–Mass Spectrometry

By Richard D. Smith, Harold R. Udseth, Jon H. Wahl, David R. Goodlett, and Steven A. Hofstadler

Introduction

In Volume 270 of this series the techniques of capillary electrophoresis (CE) are described in considerable detail, covering the range of methods and electrophoretically driven separation processes that can be conducted in capillaries. Broad areas of CE application relevant to biological macromolecules are considered elsewhere in this volume. As with any separation technique, the power of CE in macromolecular characterization often depends on the nature of the detector. The focus of this chapter is on the instrumentation, methods, and advantages of the combination of this high resolution separation technique with equally powerful mass spectrometric (MS) methods.

The mass spectrometer is potentially the most powerful of all CE detectors. The mass spectrometer provides the equivalent of up to several thousand discrete and selective "detectors," functioning in parallel and capable of providing both molecular weight information from the intact "molecular ion," as well as structurally related information from its dissociation in the mass spectrometer.[1] The CE–MS combination is, in one sense, nearly ideal:

[1] J. A. McCloskey (ed.), *Methods Enzymol.* **193** (1990).

CE is based on the differential migration of ions in solution, while MS analyzes ions by their mass-to-charge ratio (m/z) in the gas phase. On the other hand, these two highly orthogonal analytical methods exploit ion motion in two quite different environments; moderately conductive liquid buffers and high vacuum, respectively. The CE–MS combination places significantly different demands on an interface than does liquid chromatography (LC)–MS[2] because CE flow rates can be quite low or negligible, the buffer is moderately conductive, and, most important, electrical contact must be maintained with both ends of the capillary so as to define the CE field gradient. Moreover, if the high separation efficiencies possible with CE are to be realized, any extra band broadening due to laminar flow, which may arise from pressure differences between the capillary termini, or dead volume associated with detection, needs to be avoided or minimized.

Ongoing developments in MS instrumentation continue to provide improvements in the sensitivity, speed, and level of structural detail obtainable from this powerful detector. Some developments in MS have been highlighted in Volume 270 of this series. In addition, an earlier volume in this series is dedicated to MS.[1] Developments involving tandem MS methods (i.e., MS–MS) and higher order MS methods (i.e., MS^n, where $n \geq 3$) promise future instrumentation offering even greater selectivity and structural information. Although MS is still one of the most complex and expensive CE detectors, costs continue to decrease and user "friendliness" increases, and the density of information that can be gained from this combination increasingly mitigates these disadvantages.

In this chapter we review methods, experimental considerations, and selected applications for CE–MS interfacing methods based on electrospray ionization (ESI). We emphasize CE separations by free-solution electrophoresis, i.e., capillary zone electrophoresis (CZE) and, to a lesser extent, capillary isotachophoresis (CITP). This chapter reflects the explosive growth in ESI applications, the increasing availability of such instrumentation, the compatibility with MS methods capable of providing extensive structural information, and the belief of the authors that the ESI interface is currently the method of choice for CE–MS owing to its sensitivity, versatility, and ease of implementation. Finally, we wish to describe clearly the current status of CE–MS, a highly promising combination, but still somewhat limited by the sensitivity and scan-speed constraints of conventional quadrupole mass spectrometers. We show some of the steps that can be taken to mitigate the limitations of conventional mass spectrometers, and describe new instrumental approaches that should allow the full potential of the CE–MS combination to be more fully realized.

[2] P. Arpino and J. Fresenius, *Anal. Chem.* **337,** 667 (1990).

Electrospray Interface for Capillary
Electrophoresis–Mass Spectrometry

Although this chapter is primarily focused on CE–MS using ESI interfaces,[2a,2b] other approaches have been examined. The earlier reports of an off-line CE–MS combination[3,4] should be noted, as well as the more promising reports based on plasma desorption mass spectrometry[5] and the exciting initial report of Castoro and co-workers based on laser desorption methods.[6] A considerable number of reports also describe initial results based on either "liquid-junction" fast atom bombardment (FAB) interfaces,[7–15] or "coaxial" continuous-flow FAB interfaces.[16–22]

[2a] R. D. Smith, J. H. Wahl, D. R. Goodlett, and S. A. Hofstadler, *Anal. Chem.* **65,** A574 (1993).

[2b] R. D. Smith, D. R. Goodlett, and J. H. Wahl *in* "Handbook of Capillary Electrophoresis," p. 185. CRC Press, Boca Raton, FL, 1994.

[3] E. Kenndler and D. Kaniansky, *J. Chromatogr.* **209,** 306 (1981).

[4] E. Kenndler, E. Haidl, and J. Fresenius, *Anal. Chem.* **322,** 391 (1985).

[5] R. Takigiku, T. Keough, M. P. Lacey, and R. E. Schneider, *Rapid Commun. Mass Spectrom.* **4,** 24 (1990).

[6] J. A. Castoro, R. W. Chiu, C. A. Monnig, and C. L. Wilkins, *J. Am. Chem. Soc.* **114,** 7571 (1992).

[7] R. M. Caprioli, W. T. Moore, M. Martin, and B. B. DaGue, *J. Chromatogr.* **480,** 247 (1989).

[8] N. J. Reinhoud, E. Schroder, U. R. Tjaden, W. M. A. Niessen, M. C. Ten Noever de Brauw, and J. van der Greef, *J. Chromatogr.* **516,** 147 (1990).

[9] M. J.-F. Suter, B. B. DaGue, W. T. Moore, S.-N. Lin, and R. M. Caprioli, *J. Chromatogr.* **553,** 101 (1991).

[10] W. T. Moore and R. M. Caprioli, *in* "Techniques in Protein Chemistry II," p. 511. Academic Press, New York, 1991.

[11] M. J.-F. Suter and R. M. Caprioli, *J. Am. Soc. Mass Spectrom.* **3,** 198 (1992).

[12] R. D. Minard, D. Luckenbill, R. Curry, Jr., and A. G. Ewing, *in* "Proceedings of the 37th ASMS Conference on Mass Spectrometry and Allied Topics," p. 950. Miami Beach, FL, 1989.

[13] S. M. Wolf, P. Vouros, C. Norwood, and E. Jackim, *J. Am. Soc. Mass Spectrom.* **3,** 757 (1992).

[14] N. J. Reinhoud, W. M. A. Niessen, and U. R. Tjaden, *Rapid Commun. Mass Spectrom.* **3,** 348 (1989).

[15] E. R. Verheij, U. R. Tjaden, W. M. A. Niessen, and J. van der Greef, *J. Chromatogr.* **554,** 339 (1991).

[16] J. S. M. deWit, L. J. Deterding, M. A. Moseley, K. B. Tomer, and J. W. Jorgenson, *Rapid Commun. Mass Spectrom.* **2,** 100 (1988).

[17] M. A. Moseley, L. J. Deterding, K. B. Tomer, and J. W. Jorgenson, *J. Chromatogr.* **480,** 197 (1989).

[18] M. A. Moseley, L. J. Deterding, K. B. Tomer, and J. W. Jorgenson, *Rapid Commun. Mass Spectrom.* **3,** 87 (1989).

[19] M. A. Moseley, L. J. Deterding, K. B. Tomer, and J. W. Jorgenson, *Anal. Chem.* **63,** 109 (1991).

[20] L. J. Deterding, M. A. Moseley, K. B. Tomer, and J. W. Jorgenson, *J. Chromatogr.* **554,** 73 (1991).

[21] L. J. Deterding, J. R. Perkins, and K. B. Tomer, *in* "Proceedings of the 40th ASMS Conference on Mass Spectrometry and Allied Topics," p. 392. Washington, DC, 1992.

[22] L. J. Deterding, C. E. Parker, J. R. Perkins, M. A. Moseley, J. W. Jorgenson, and K. B. Tomer, *J. Chromatogr.* **554,** 329 (1991).

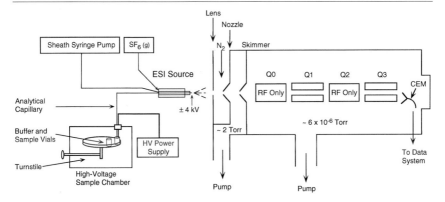

FⅠG. 1. Schematic illustration of the experimental arrangement for CE–MS using an electrospray interface.

Although the choice of interface will often be governed by the available MS instrumentation, the present trend favors ESI methods. Figure 1 shows a schematic illustration of a typical arrangement of instrumentation used for CE–MS based on ESI. The ESI method is exceptionally well suited for CE–MS interfacing because it produces ions directly from liquid solution at atmospheric pressure.[23] Considerations for interfacing generally derive from some limitations on CE buffer composition and the desire to position the ESI source (i.e., the point of charged droplet formation) as close as possible to the analytical capillary terminus, avoiding lengthy transfer lines.

In the first CE–MS interface, the electrical connection was made by silver metal deposition onto the terminus of the bare fused silica capillary (Fig. 2A).[24,25] With this interfacing method, it was necessary to select CE conditions that gave rise to a net electroosmotic flow in the direction of the MS. The metallized capillary terminus allowed contact with the CE eluent under conditions of high electroosmotic flow for conventional 50- to 100-μm i.d. capillaries, but exhibited unstable behavior for some CE conditions and limited lifetime of the metallized terminus. The ease of interfacing was dramatically improved by introducing the sheath flow design

[23] R. D. Smith, J. A. Loo, C. G. Edmonds, C. J. Barinaga, and H. R. Udseth, *Anal. Chem.* **62,** 882 (1990).

[24] J. A. Olivares, N. T. Nguyen, C. R. Yonker, and R. D. Smith, *Anal. Chem.* **59,** 1230 (1987).

[25] R. D. Smith, J. A. Olivares, N. T. Nguyen, and H. R. Udseth, *Anal. Chem.* **60,** 436 (1988).

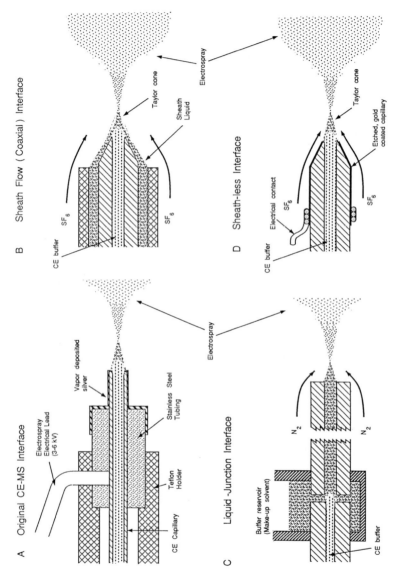

FIG. 2. Schematic illustration of CE–electrospray interfaces: (A) the original design utilizing a metallized capillary terminus, (B) sheath flow (coaxial) interfaces used for CE–MS, (C) electrospray ionization based on a liquid junction, and (D) a new sheathless interface design being explored at the laboratory of the authors.

in 1988.[26,27] In this design, a small coaxial flow (about 1–5 μl/min) of liquid serves to establish the electrical contact with the CE effluent[28] and facilitate the electrospraying of buffers that could not be directly electrosprayed with earlier interfaces. Figure 2B shows a schematic illustration of a more recent implementation of the sheath flow design that is distinguished by an etched conical tip at the terminus of the CE capillary. This design serves to enhance the electric field gradient at the capillary terminus, minimize the effective mixing volume between the sheath liquid and the CE effluent, and increase the stability of the electrospray process.[29] These developments allowed the demonstration of related techniques, such as the combination of capillary isotachophoresis (CITP) with MS,[30,31] and CE–MS/MS.[31,32] Other researchers have reported similar methods and interfaces.[33–63]

[26] R. D. Smith, H. R. Udseth, J. A. Loo, B. W. Wright, and G. A. Ross, *Talanta* **36,** 161 (1989).

[27] R. D. Smith, C. J. Barinaga, and H. R. Udseth, *Anal. Chem.* **60,** 1948 (1988).

[28] A. P. Bruins, T. R. Covey, and J. D. Henion, *Anal. Chem.* **59,** 2642 (1987).

[29] L. L. Mack, P. Kralik, A. Rheude, and M. Dole, *J. Chem. Phys.* **52,** 4977 (1970).

[30] H. R. Udseth, J. A. Loo, and R. D. Smith, *Anal. Chem.* **61,** 228 (1989).

[31] R. D. Smith, S. M. Fields, J. A. Loo, C. J. Barinaga, H. R. Udseth, and C. G. Edmonds, *Electrophoresis* **11,** 709 (1990).

[32] C. G. Edmonds, J. A. Loo, S. M. Fields, C. J. Barinaga, H. R. Udseth, and R. D. Smith, *in* "Biological Mass Spectrometry" (A. L. Burlingame and J. A. McCloskey, eds.), p. 77. Elsevier Science Publishers, Amsterdam, 1990.

[33] M. Hail, J. Schwartz, I. Mylchreest, K. Seta, S. Lewis, J. Zhou, I. Jardine, J. Liu, and M. Novotny, *in* "Proceedings of the 38th ASMS Conference on Mass Spectrometry and Allied Topics," p. 353. Tucson, AZ, 1990.

[34] S. Pleasance, P. Thibault, and J. Kelly, *J. Chromatogr.* **591,** 325 (1992).

[35] C. E. Parker, J. R. Perkins, and K. B. Tomer, *J. Am. Soc. Mass Spectrom.* **3,** 563 (1992).

[36] J. R. Perkins, C. E. Parker, and K. B. Tomer, *J. Am. Soc. Mass Spectrom.* **3,** 139 (1992).

[37] D. F. Hunt, J. E. Alexander, A. L. McCormack, P. A. Martino, H. Michel, J. Shabanowitz, M. A. Moseley, J. W. Jorgenson, and K. B. Tomer, *in* "Techniques in Protein Chemistry II" (J. J. Villafranca, ed.), p. 441. Academic Press, San Diego, CA, 1991.

[38] D. F. Hunt, J. E. Alexander, A. L. McCormack, P. A. Martino, H. Michel, J. Shabanowitz, N. Sherman, M. A. Moseley, J. W. Jorgenson, and K. B. Tomer, *in* "Techniques in Protein Chemistry II" (J. J. Villafranca, ed.), p. 441. Academic Press, San Diego, CA, 1991.

[39] P. Thibault, C. Paris, and S. Pleasance, *Rapid Commun. Mass Spectrom.* **5,** 484 (1991).

[40] P. Thibault, S. Pleasance, and M. V. Laycock, *J. Chromatogr.* **542,** 483 (1991).

[41] P. Thibault, S. Pleasance, and M. V. Laycock, *in* "Proceedings of the 39th ASMS Conference on Mass Spectrometry and Allied Topics," p. 593. Nashville, TN, 1991.

[42] S. Pleasance, S. W. Ager, M. V. Laycock, and P. Thibault, *Rapid Commun. Mass Spectrom.* **6,** 14 (1992).

[43] M. A. Moseley, J. Shabanowitz, D. F. Hunt, K. B. Tomer, and J. W. Jorgenson, *J. Am. Soc. Mass Spectrom.* **3,** 289 (1992).

[44] I. Mylchreest and M. Hail, *in* "Proceedings of the 39th ASMS Conference on Mass Spectrometry and Allied Topics," p. 316. Nashville, TN, 1991.

[45] Y. Shida, C. E. Parker, J. R. Perkins, K. O. O'Hara, and K. B. Tomer, "Proceedings of the 39th ASMS Conference on Mass Spectrometry and Allied Topics," p. 587. Nashville, TN, 1991.

In the sheath flow electrospray interface an organic liquid (typically pure methanol, methoxyethanol, or acetonitrile, but frequently augmented by as much as 10 to 20% formic acid, acetic acid, water, or other reagents) flows through the annular space between the $\sim$200-μm o.d. CE capillary and a fused silica or stainless steel capillary (generally >250-μm i.d.). Potential problems from the formation of bubbles in the connecting lines are minimized by the inclusion of a gas trapping volume. Enhanced stability can be obtained by degassing the organic solvents used in the sheath and by minimizing heating (e.g., from the ion source or a "countercurrent" gas flow). Cooling of the ESI source or the use of less volatile sheath solvents such as methoxyethanol has also been useful in some situations.[44] In addition to those noted earlier, a number of reports from our laboratory on the use of the sheath flow approach are now in the literature.[64-70]

[46] K. A. Halm, S. E. Unger, R. L. St. Claire, R. F. Arendale, and M. A. Moseley, *in* "Proceedings of the 39th ASMS Conference on Mass Spectrometry and Allied Topics," p. 591. Nashville, TN, 1991.

[47] E. D. Lee, W. Mück, J. D. Henion, and T. R. Covey, *Biomed. Environ. Mass Spectrom.* **18,** 253 (1989).

[48] E. D. Lee, W. Mück, J. D. Henion, and T. R. Covey, *Biomed. Environ. Mass Spectrom.* **18,** 844 (1989).

[49] I. M. Johansson, E. C. Huang, J. D. Henion, and J. Zweigenbaum, *J. Chromatogr.* **554,** 311 (1991).

[50] I. M. Johansson, R. Pavelka, and J. D. Henion, *J. Chromatogr.* **559,** 515 (1991).

[51] W. M. Mück and J. D. Henion, *J. Chromatogr.* **495,** 41 (1989).

[52] F. Garcia and J. D. Henion, *in* "Proceedings of the 39th ASMS Conference on Mass Spectrometry and Allied Topics," p. 312. Nashville, TN, 1991.

[53] E. D. Lee, W. Mück, J. D. Henion, and T. R. Covey, *J. Chromatogr.* **458,** 313 (1988).

[54] T. Wachs, J. C. Conboy, F. Garcia, and J. D. Henion, *J. Chromatogr. Sci.* **59,** 357 (1991).

[55] F. Garcia and J. Henion, *J. Chromatogr.* **606,** 237 (1992).

[56] A. P. Tinke, N. J. Reinhoud, W. M. A. Niessen, U. R. Tjaden, and J. van der Greef, *Rapid Commun. Mass Spectrom.* **6,** 560 (1992).

[57] T. J. Thompson, F. Foret, P. Vouros, and B. L. Karger, *Anal. Chem.* **65,** 900 (1993).

[58] A. J. Tomlinson, L. M. Benson, G. F. Scanian, K. L. Johnson, K. A. Veverka, and S. Naylor, *in* "Proceedings of the 41st ASMS Conference on Mass Spectrometry and Allied Topics," p. 1046. San Francisco, CA, 1993.

[59] J. F. Kelly, P. Thibault, H. Masoud, M. B. Perry, and J. C. Richards, *in* "Proceedings of the 41st ASMS Conference on Mass Spectrometry and Allied Topics," p. 1079. San Francisco, CA, 1993.

[60] K. Tsuji, L. Baczynskj, and G. E. Bronson, *Anal. Chem.* **64,** 1864 (1992).

[61] J. R. Perkins, C. E. Parker, and K. B. Tomer, *Electrophoresis* **14,** 458 (1993).

[62] J. Varghese and R. B. Cole, *in* "Proceedings of the 41st ASMS Conference on Mass Spectrometry and Allied Topics," p. 1041. San Francisco, CA, 1993.

[63] C. E. Parker, J. R. Perkins, K. B. Tomer, Y. Shida, K. O'Hara, and M. Kono, *J. Amer. Soc. Mass Spectrom.* **3,** 563 (1992).

[64] R. D. Smith, H. R. Udseth, C. J. Barinaga, and C. G. Edmonds, *J. Chromatogr.* **559,** 197 (1991).

Another variation on the CE–MS interface was introduced by Lee and co-workers, in which a liquid-junction interface was used to establish electrical contact with the analytical capillary and provided an additional makeup flow of buffer (Fig. 2C).[47,48,53] For the liquid-junction interface variation, electrical contact with the capillary terminus is established through a liquid reservoir that surrounds the junction of the analytical capillary and a transfer capillary. The gap between the two capillaries is typically adjusted to 10–20 μm, a compromise resulting from the need for sufficient makeup liquid being drawn into the transfer capillary while avoiding analyte loss by diffusion into the reservoir. The flow of makeup liquid arises from a combination of gravity-driven flow, due to the height of the makeup reservoir, and flow induced in the transfer capillary owing to a mild vacuum generated by the Venturi effect of the nebulizing gas used at the ESI source.[49–55]

It appears that most of the distinctions between the sheath flow and liquid-junction ESI interfaces are mechanical; the performance in terms of spectral quality is similar and most other MS-related considerations are similar. Some ESI sources allow the liquid effluent to be at ground potential, and others require the capillary terminus to be operated at 3–5 kV relative to ground potential. The ESI source requires production of a high electric field, which causes charge to accumulate on the liquid surface at the capillary terminus so as to disrupt the liquid surface. The ESI liquid nebulization process can be pneumatically assisted using a high-velocity annular flow of gas at the capillary terminus, and it sometimes is referred to as "ion spray,"[28] an approach originally described for ESI by Mack and co-workers.[29] However, some differences have been noted. Thibault *et al.* have compared the liquid-junction variation of Henion and co-workers with the sheath flow interface using a pneumatically assisted ESI interface, and have noted that the latter provided a "more robust and reproducible interface with the added advantage of offering an independent means of calibration and quantitation ... through the sheath."[39] In a comparison for a separation of marine toxins, where care was taken to obtain comparable conditions,[34] significantly improved signal-to-noise ratios and slightly improved separa-

[65] J. H. Wahl, D. R. Goodlett, H. R. Udseth, and R. D. Smith, *J. Amer. Chem. Soc.* **115**, 803 (1993).

[66] C. G. Edmonds, J. A. Loo, C. J. Barinaga, H. R. Udseth, and R. D. Smith, *J. Chromatogr.* **474**, 21 (1989).

[67] J. A. Loo, C. G. Edmonds, and R. D. Smith, *Anal. Chem.* **63**, 2488 (1991).

[68] J. A. Loo, H. K. Jones, H. R. Udseth, and R. D. Smith, *J. Microcol. Sep.* **1**, 223 (1989).

[69] R. D. Smith, J. A. Loo, C. J. Barinaga, C. G. Edmonds, and H. R. Udseth, *J. Chromatogr.* **480**, 211 (1989).

[70] R. D. Smith, J. A. Loo, C. G. Edmonds, C. J. Barinaga, and H. R. Udseth, *J. Chromatogr.* **516**, 157 (1990).

tions were obtained with the sheath flow interface. In addition, a difference in migration times was evident and attributed to modification of the liquid-junction buffer due to flushing of the CE capillary between analyses. A disadvantage of both approaches, however, is that they depend on an additional flow of liquid incorporating charge-carrying species (buffer, solvent impurities, etc.) that invariably degrades detection sensitivity to some extent. Consequently, interest remains in the development of a more versatile and reliable design that does not depend on an additional liquid flow. A sheathless design for this purpose, such as that illustrated in Fig. 2D, is one type that we are currently investigating.

Electrospray ionization is an extremely "soft" ionization technique, and will yield, under appropriate conditions, intact molecular ions. Molecular weight measurements for large biopolymers that exceed the MS "mass range" can be obtained because ESI mass spectra generally consist of a distribution of molecular ion charge states, without contributions due to dissociation (unless induced during transport into the MS vacuum).[23,71,72] The envelope of charge states for proteins, arising generally from protonation for positive-ion ESI, yields a distinctive pattern of peaks owing to the discrete nature of the electronic charge; i.e., adjacent peaks vary by addition or subtraction of one charge. Although mass spectra of noncovalently bonded species, such as multimeric proteins, typically show only the individual subunits, the detection of such labile species is feasible.[71] The growing experience with ESI and its broad range of applications promises to expand further the utility of the CE–MS combination.

The dependence of ESI ion current on solution conductivity is relatively weak, generally $0.1–0.3$ μA at atmospheric pressure, and about $10–100$ pA integrated ion current (i.e., the sum of all ions) is actually focused into the MS and transmitted for detection. Electrospray ionization currents for a typical water–methanol–5% (v/v) acetic acid solution are in the range of $0.1–0.3$ μA, only a small portion of which ($<10^{-4}$) is generally transmitted for MS analysis. An electron scavenger is often used to inhibit electrical discharge at the capillary terminus, particularly for ESI of aqueous solutions. Sulfur hexafluoride has proven particularly useful for suppressing corona discharges and improving the stability of negative-ion ESI. Introduction of this gas is most effectively accomplished using a gas flow ($\sim$100–250 ml/min) through an annular volume surrounding the sheath liquid.

Ion formation by ESI requires conditions affecting solvent evaporation

[71] R. D. Smith, J. A. Loo, R. R. Ogorzalek Loo, M. Busman, and H. R. Udseth, *Mass Spectrom. Rev.* **10**, 359 (1991).
[72] J. B. Fenn, M. Mann, C. K. Meng, S. F. Wong, and C. M. Whitehouse, *Mass Spectrom. Rev.* **9**, 37 (1990).

from the initial droplet population. Droplets must shrink to the point at which repulsive Coulombic forces approach the level of droplet cohesive forces (e.g., surface tension). This evaporation can be accomplished at atmospheric pressure by a countercurrent flow of dry gas at moderate temperatures (~80°), by heating during ion transport through the sampling region, and (particularly in the case of ion-trapping MS methods) by energetic collisions at relatively low pressure.[71]

An alternative and increasingly popular approach to droplet desolvation for ESI relies solely on heating during droplet transport through a heated metal or glass capillary.[68,69] A countercurrent gas flow is not essential and, in fact, can decrease obtainable ion currents. The electrospray source can be closely positioned to the sampling capillary orifice (typically 0.3 to 1.0 cm) because the spacing is not used for desolvation, resulting in more efficient charge transport. The charged droplets from the ESI source are swept into the heated capillary, which heats the gas sufficiently to provide effective ion desolvation, particularly when augmented by a voltage gradient in the capillary–skimmer region.

An important attribute of a CE–MS interface base on ESI is the efficiency of sampling and transport of ions into the MS. The simplest commercially available instrument utilizes a 100- to 130-μm pinhole sampling orifice to a vacuum region maintained by a single stage of high-speed cryopumping (Sciex, Thornhill, Ontario, Canada). Charged droplets formed by ESI drift against a countercurrent flow of dry nitrogen, which serves to speed desolvation and exclude high-m/z residual particles and solvent vapor. As the ions pass through the orifice into the vacuum region, further desolvation is accomplished as the gas density decreases owing to collisions as the ions are accelerated by the ion optics of the mass spectrometer. Also widely used is the interface developed by Fenn and co-workers,[72] which is available in versions compatible with many mass spectrometers (Analytica, Branford, CT). A countercurrent flow of nitrogen bath gas assists solvent evaporation similar to the approach described above. Ions migrate toward the sampling orifice, where some small fraction is entrained in the gas flow entering a glass capillary (metallized at both ends to establish well-defined electric fields). Ions emerge from the capillary, in the first differentially pumped stage of the MS (~1 torr), as a component of a free-jet expansion. A fraction of the ions is then transmitted through a "skimmer" and additional ion optics into the MS. The electrically insulating nature of the glass capillary provides considerable flexibility because ion transmission does not depend strongly on the voltage gradient between the conducting ends of the capillary. This approach has the advantage that the CE capillary terminus can be at ground potential, simplifying current measurements and voltage manipulation during injection.

Experimental Methods and Considerations

Separation Conditions

In general, all the concerns in CE relevant to sample injection, buffer composition, capillary surface interactions, and separation efficiency apply to CE–MS, and the reader is referred to other chapters in this volume. Most CE–MS has been conducted with 50- to 100-μm i.d. capillaries, although it is now apparent that smaller diameters have advantages in optimizing sensitivity when sample size is extremely limited.[73–75] Both the liquid junction and sheath flow ESI interfaces allow a wide range of CE buffers to be successfully electrosprayed, owing to the effective dilution of the low CE elution flow by a much larger volume of liquid, providing considerable flexibility despite some constraints (e.g., high salt and surfactant concentrations are problematic). Buffer concentrations of at least 0.1 M can be used for CE–MS owing to dilution by the sheath liquid. However, owing to practical sensitivity constraints, CE buffer concentrations are generally minimized, and it has also been found that sensitivity can vary significantly with buffer composition. In general, the best sensitivity is obtained with the use of volatile buffer components, such as acetic acid at the lowest practical concentration, and by minimizing nonvolatile components. In addition, buffer components that interact strongly with the sample (e.g., denaturants) degrade sensitivity. In particular, the use of either ionic or nonionic surfactants, sodium dodecyl sulfate (SDS) for example, can give rise to intense background signals in both positive- and negative-ion ESI, presenting a major obstacle when practical surfactant concentrations are required for capillary micellar electrokinetic chromatography–MS applications.

The MS sensitivity, which is ultimately limited by the ability to analyze efficiently ions produced at atmospheric pressure by ESI, is probably the most important factor related to MS detection. Sensitivity does not appear to depend greatly on the mechanical details of the electrospray interface, although such factors can influence ESI stability and ease of operation. Increasingly, CE–MS instrumentation benefits from the features of the automated CE instruments, including precise electrokinetic or hydrostatic injection, on-capillary spectrophotometric detection, and capillary temperature control.[64] The advantages of computer-controlled injection are significant for both accuracy and precision, as well as freedom from artifacts arising from less reliable manual injection methods.

[73] J. H. Wahl, D. R. Goodlett, H. R. Udseth, and R. D. Smith, *Anal. Chem.* **64,** 3194 (1992).
[74] J. H. Wahl, D. C. Gale, and R. D. Smith, *J. Chromatogr.* **A659,** 217 (1994).
[75] J. H. Wahl, D. R. Goodlett, H. R. Udseth, and R. D. Smith, *Electrophoresis* **14,** 448 (1993).

Proper CE capillary surface coatings and conditioning are essential for many applications. As an example, our initial CE–MS of proteins in acidic buffers (pH 3–5) with uncoated capillaries produced poor separations for proteins such as myoglobin and cytochrome c owing to interactions with capillary surfaces.[68] However, excellent separations of myoglobin mixtures have been obtained at higher pH (>8) owing to the net negative charge of the protein, and reasonable detection limits (<100 fmol) obtained using multiple ion monitoring in 10 mM Tris.[64] In this case, methanol–water sheath liquid containing 5 to 10% acetic acid was used to obtain the acidic conditions for ESI–MS. However, the Tris buffer component, and perhaps limitations on mixing of the CE and sheath flows, resulted in reduced sensitivity compared to that obtained for the same separations conducted in volatile acidic buffers.[64] Thibault et al.[39] have demonstrated that protein separations conducted in acidic solution (e.g., 10 mM acetic acid, pH 3.4), using reversible amino-based coated capillaries, result in substantially improved sensitivity. Similar CE–MS results have more recently been demonstrated by others.[43,44] In general, the separation systems mentioned have an electroosmotic flow component with the separation. It is possible, however, through capillary surface modification to effectively eliminate this flow component. Yet, this approach may be more problematic because the sheath fluid, which is necessary to maintain electrical contact with the CE terminus itself, may adversely effect the separation process.[76]

Considerations for Mass Spectrometry Detection

Probably the most widely used buffers to date for CE–ESI/MS are acetic acid, ammonium acetate, and formic acid systems, choices made primarily on the basis of their volatility. Nonvolatile buffer systems may be used for CE–MS; however, different ESI interfaces may be required for their routine use. For example, phosphate buffers produce cluster ions extending to at least m/z 1000 with sufficient abundance to limit CE–MS applicability substantially. Conversely, acetic acid buffers often produce a broad range of background ions, presumably due to solution impurities, which illustrates the need to use high-purity buffer systems in CE–MS. For example, Fig. 3 shows mass spectra obtained during elution of a substituted [Ala^2]Met-enkephalin from a CE–MS separation in a 20-μm i.d. capillary using a 10 mM acetic acid buffer. Figure 3 (top) shows the "raw" mass spectrum (above m/z 300) for which a major peak arises at m/z 589 due to the protonated molecule from an 8-fmol injection in a binary mixture,

---

[76] T. J. Thompson, F. Foret, D. P. Kirby, P. Vouros, and B. L. Karger, in "Proceedings of the 41st ASMS Conference on Mass Spectrometry and Allied Topics," p. 1057. San Francisco, CA, 1993.

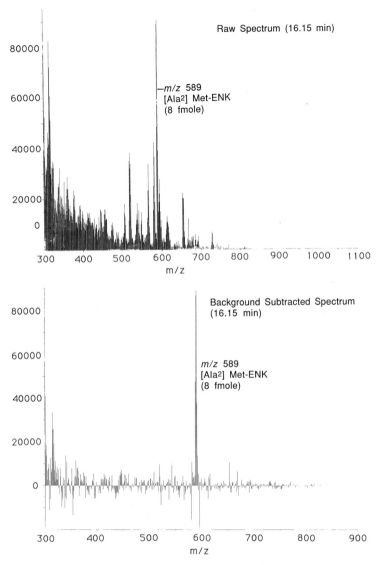

Fig. 3. CE–mass spectrum obtained during elution of [Ala²]Met-enkephalin from an ∼8-fmol injection using a 1 m × 20-μm i.d. capillary with a 10 mM acetic acid buffer. *Top:* Raw spectrum with large "background" contributions arising from the buffer. *Bottom:* "Background subtracted" spectrum.

with Leu-enkephalin. Clearly, a large "background" exists below $m/z \sim 700$, which is likely due to impurities and degradation products in the acetic acid and "clustering" involving solvent association with trace ionic impurities. A subtraction of the "background" using spectra obtained just before peak elution gives the mass spectrum shown in Fig. 3 (bottom), which is dominated by the eluting analyte.

The major considerations relevant to MS detection generally arise in response to the nature and complexity of the particular sample and pragmatic constraints due to MS detection sensitivity, resolution, and the related scan-speed compromises. For quadrupole mass spectrometers, single or multiple ion monitoring, sometimes referred to as selected ion monitoring (SIM), leads to significantly enhanced detection limits compared to scanning operation owing to the greater "dwell time" at specific m/z values. This approach can be problematic, however, because relative ionization efficiencies can be different for direct infusion of mixtures compared to the separated components. Thus, important components may be overlooked during direct introduction. One of the advantages of CE–MS is that mixture components strongly discriminated against in such direct infusion experiments often show much more uniform response after CE separation. For samples in which analyte molecular weights are known, and m/z values can be predicted, SIM detection is an obvious choice. If sufficient sample is available, direct infusion can be used to produce a mass spectrum of the unseparated mixture and the results used to guide selection of a suite of m/z values for SIM detection.

The advantages in terms of sensitivity gain with SIM detection vs scanning (e.g., quadrupole) instruments can be large. For example, Fig. 4 shows a SIM profile for the simple pentapeptide mixture from which the (scanning) mass spectrum shown in Fig. 3 was obtained. Figure 4 shows three SIM plots (from a total of seven m/z values selected) for injection of samples (from serial dilution) covering more than two orders of magnitude lower concentration than shown in Fig. 3. For the 5- to 10-fmol injection (Fig. 4, top), comparable to Fig. 3, SIM detection provides excellent signal-to-noise ratios, and the CE performance and peak shape are captured by the MS detector in near-ideal fashion. For injection sizes corresponding to 0.5 to 1 fmol (Fig. 4, middle), the excellent peak shape is retained, but increased baseline "noise" due to buffer impurities is evident. A shift in elution times was also noted that became more dramatic as the capillary aged, owing to a decrease in electroosmotic flow. For injections of 40 to 75 amol (Fig. 4, bottom) good peak shape was retained, but a noisy and elevated baseline suggests detection limits of ~ 10 amol, unless a reduction of background "chemical noise" can be obtained. Such a reduction of background may provide a basis for an extension to subattomole detection limits.

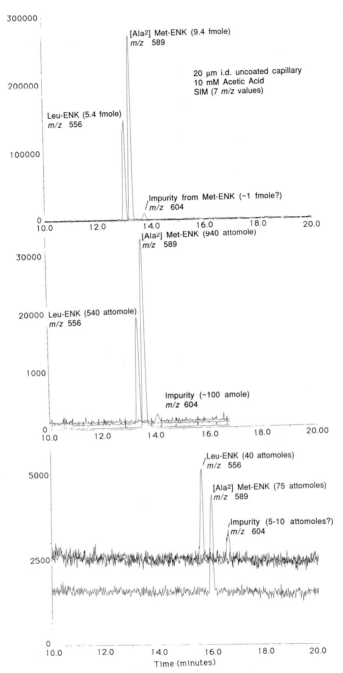

FIG. 4. Selected ion electropherograms (of seven ions monitored) for injections of a pentapeptide mixture of Leu-enkephalin and [Ala²]Met-enkephalin for sample dilutions covering approximately two orders of magnitude in sample size, using the same conditions for the mass spectrum shown for Fig. 3.

An advantage of MS relative to other detectors is its high specificity. As shown in Fig. 4, the migration times for the analytes vary somewhat between runs. These differences are due to experimental variations, such as capillary surface modification, which in some cases may be unavoidable. Consequently, substantial care and effort are necessary to obtain reproducible migration times to identify eluents. In contrast, highly reproducible elution times are far less important for MS detection because most components give distinctive mass spectra. Thus, relative migration times are usually sufficient, and one is generally more concerned with either MS data quality to determine molecular weights with good precision or signal intensity so that dissociation of molecular ions can be used to provide structural information. Implicitly, one is also concerned with separation quality (i.e., resolution) or separation time because these directly impact these goals. Obtaining the maximum number of theoretical plates possible with combined CE–MS is rarely required unless components are of nearly similar molecular weight.

Sensitivity for larger biomolecules (e.g., proteins) is substantially less with ESI than for smaller peptides, and even greater needs arise for the use of tandem MS methods for obtaining structural information, as in polypeptide sequencing, owing to the much lower signal intensities obtained for dissociation products with conventional tandem instruments. These limitations are a major driving force for the implementation of improved MS instrumentation.

There are several possible approaches to resolving these sensitivity limitations. One is to increase the efficiency of ion transport from the ESI source into the mass spectrometer, now generally only ~0.01% in overall efficiency.[71] A second approach is to analyze all the ions that enter the mass spectrometer, using either ion-trapping methods or array detection. Indeed, a number or workers are beginning to examine the quadrupole ion trap mass spectrometer (ITMS) for this purpose.[77,78] Because the ITMS can trap and accumulate ions over a wide m/z range, followed by a rapid swept ejection/detection, the fraction of total ions detected compared to that entering the ITMS is thus much greater, allowing the utilization of a much larger fraction of the total ESI current (assuming efficient ion injection and trapping).[79]

[77] G. J. Van Berkel, G. L. Glish, and S. A. McLuckey, *Anal. Chem.* **62,** 1287 (1990).

[78] R. G. Cooks, G. L. Glish, S. A. McLuckey, and R. E. Kaiser, *Chem. Eng. News* March 25, 26 (1991).

[79] A. V. Mordehai, G. Hopfgartner, T. G. Huggins, and J. D. Henion, *Rapid Commun. Mass Spectrom.* **6,** 508 (1992).

The best MS performance would, in principle, be obtained with full-range array detection, but practical instrumentation for this purpose is not yet available, and would likely be much more expensive than either the ITMS or existing instrumentation. Alternatively, the "orthogonal" (i.e., perpendicular ion extraction) time-of-flight mass spectrometer has the potential of obtaining high ion utilization efficiency, and initial results with an electrospray ion source has demonstrated efficiency of $\sim$2.5% for ions transmitted into the orthogonal ion drawout region[80] (but overall ion utilization efficiency was likely at least two orders of magnitude lower). Other developments are also promising. Henry and co-workers have pioneered the combination of ESI with Fourier transform-ion cyclotron resonance mass spectrometry (FTICR),[81] and shown that very high resolution is obtainable.[82] Hofstadler and Laude have demonstrated that substantial gains in sensitivity are obtainable by having the ESI interface in the high magnetic field of an FTICR superconducting magnet.[83] The potential for higher order mass spectrometry, MS^n ($n \geq 3$), using FTICR or ITMS instruments also appears promising.

Approaches to Improved Capillary Electrophoresis–Mass Spectrometry Sensitivity

The small solute quantities in CE require highly sensitive detection methods. The low signal intensities generally produced by ESI–MS, typically resulting in maximum analyte ion detection rates of no greater than 10^5 to 10^6 ion counts per second, effectively limit the maximum practical scan speeds with quadrupole mass spectrometers. Thus, depending on the desired m/z range, solute concentration, and other factors related to the nature of the solute and buffer species, maximum m/z scan speeds are often insufficient to exploit the high-quality separations feasible with CE when coupled to quadrupole or other scanning mass spectrometers.

For fundamental reasons the maximum electrospray ion current is a weak function of solution conductivity. When the amount of charge-carrying solute entering the ESI source exceeds the capability of the electrospray process, then the efficiency of solute ionization decreases. Thus, at higher analyte concentrations or flow rates, ESI–MS signal strengths become relatively insensitive to analyte mass flow rate. At very low flow rates or concentrations the ESI signal strength becomes limited by the number of

[80] J. G. Boyle and C. M. Whitehouse, *Anal. Chem.* **64**, 2084 (1992).
[81] K. D. Henry, E. R. Williams, B. H. Wang, F. W. McLafferty, J. Shabanowitz, and D. F. Hunt, *Proc. Natl. Acad. Sci. U.S.A.* **86**, 9075 (1989).
[82] K. T. Henry, J. P. Quinn, and F. W. McLafferty, *J. Am. Chem. Soc.* **113**, 5447 (1991).
[83] S. A. Hofstadler and D. A. Laude, *Anal. Chem.* **64**, 569 (1992).

charge-carrying species in solution. In this regime, optimum sensitivity will be obtained. However, most ESI work, and nearly all CE–MS, has been conducted in the former regime where the efficiency of solute ionization is substantially limited. Capillary electrophoresis separations generally incur higher currents (5 to 50 μA) than typical ESI currents (0.1 to 0.1 μA), and therefore deliver charged species to the ESI source at a rate at which ionization is necessarily inefficient. Any contaminant present may also compete with the analyte for available charge, thus decreasing the solute ionization efficiency. Among other characteristics, ideal buffer components facilitate CE separation, are volatile, can be discriminated against during ESI, and form minimal gas phase contributions to the mass spectra arising from ionized clusters of solvent and buffer constituents.

The most obvious approach to improving sensitivity is to modify the injection step to load more sample onto the capillary. Such methods are of general interest to CE techniques and the reader is referred to other chapters on CE in this volume. Such techniques have already been used for CE–MS; a particularly useful approach involving isotachophoretic sample preconcentration has been reported by Tinke *et al.,*[84] Thompson and co-workers,[57] and Lamoree *et al.*[85] Our discussion in the remainder of this section is from the viewpoint of subsequent steps that can be taken to enhance CE–MS sensitivity.

One approach developed at our laboratory is based on reduced elution speed (RES) detection, which aids in alleviating both the sensitivity and scan-speed limitations of CE–MS with scanning mass spectrometers.[86] Involving only step changes in the CE electric field strength, the technique is simple and readily implemented. Prior to elution of the first analyte of interest into the ESI source, the electrophoretic voltage is decreased and elution of solutes is slowed. This potential decrease allows more scans to be recorded without a significant loss in ion intensity when the amount of solute entering the ESI source per unit of time exceeds its ionization capacity. As a result, a greater fraction of the analyte ions can be sampled.[86] The prolonged analyte elution into the electrospray source can be exploited by (1) increasing the m/z range scanned, (2) increasing the number of scans recorded during migration of a given solute, and (3) enhancement of sensitivity, for a given solute. The method does not increase solute consumption and incurs very little loss in ion intensity, which is particularly

[84] A. P. Tinke, N. J. Reinhoud, W. M. A. Niessen, U. R. Tjaden, and J. van der Greef, *Rapid Commun. Mass Spectrom.* **6,** 560 (1992).

[85] M. H. Lamoree, N. J. Reinhoud, U. R. Tjaden, W. M. A. Niessen, and J. van der Greef, *in* "Proceedings of the 41st ASMS Conference on Mass Spectrometry and Allied Topics," p. 1046. San Francisco, CA, 1993.

[86] D. R. Goodlett, J. H. Wahl, H. R. Udseth, and R. D. Smith, *J. Microcol. Sep.* **5,** 57 (1993).

important for tandem MS methods and their potential application to peptide sequencing.

Comparison of constant electric field strength and RES CE–MS for a 40-fmol injection of peptides produced by digestion of bovine serum albumin (BSA) with trypsin is shown in Fig. 5A and results in only a small decrease (20%) in ion intensity when the electrophoretic voltage was decreased (Fig. 5B). More than 100 tryptic fragments can be readily extracted from the separation.[86] The method results in a reduction in the complexity of individual scans and facilitates data interpretation of complex mixtures owing to the greater number of scans obtained during elution of a given component and the reduced likelihood of other components eluting during the same scan.

The utility of RES CE–MS is particularly relevant for larger polypeptides and proteins, where broad m/z range spectra can be useful in molecular weight determination. Figure 6A shows the total ion electropherogram (TIE) for the separation of 60 fmol (injected) each of carbonic anhydrase II, aprotinin, myoglobin, and cytochrome c using a constant electric field strength of 300 V/cm during CE–ESI/MS. In comparison, Fig. 6B shows the TIE for the same mixture where the electric field strength was reduced to 60 V/cm 1 min prior to migration of the first protein into the ESI source. The inset in Fig. 6B shows the same TIE after the abscissa has been

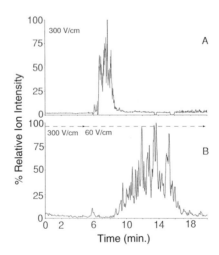

FIG. 5. Comparison of constant field strength (A) and reduced elution speed (B) CE–MS analysis of bovine serum albumin after digestion by trypsin. The constant field strength CE–MS analysis was conducted at 300 V/cm. The reduced elution speed CE–MS analysis was conducted at 300 V/cm until 1 min prior to elution of the first peptide, when the electric field strength was reduced to 60 V/cm.

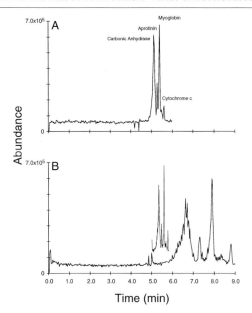

FIG. 6. Comparison of constant field strength (A) and reduced elution speed (B) CE–MS analysis of a mixture containing carbonic anhydrase II, aprotinin, myoglobin, and cytochrome c. The constant field strength CE–MS analysis was conducted at 300 V/cm. The reduced elution speed CE–MS analysis was conducted at 300 V/cm until 1 min prior to elution of the first protein, when the electric field strength was reduced to 60 V/cm.

compressed by a factor of five. Only a 14% reduction in the ion intensity is observed for the strongest response (myoglobin), without observable loss in separation quality. The quality of mass spectral data for a scan from 600 to 1200 m/z is sufficient to allow molecular weights to be calculated to 0.02–0.1%. The additional scans collected by RES CE–MS provide better mass spectral quality and more precise molecular weight measurements, as well as modest improvements in signal-to-noise ratios.[86]

An alternative approach described by Wahl et al.[73–75] to obtaining greater sensitivity involves the use of smaller diameter capillaries than conventionally used for CE. As noted above, optimum ESI/MS sensitivity is generally obtained at low CE currents at which the rate of delivery of charged species to the ESI source is minimized. An optimum CE capillary diameter ideally meets several criteria: it should (1) be available commercially or readily prepared, (2) be amenable to alternative detection methods, and (3) provide the necessary detector sensitivity. For the last criterion, we would expect that optimum sensitivity would be obtained for CE currents approximately equal to or less than the ESI current. Even for this relatively low-conductivity acetic acid buffer system, CE capillary diameters greater than 40 μm will have currents that exceed that of the ESI source.[74,75]

The effect of capillary diameter on sensitivity and separation quality is illustrated in Fig. 7 for a mixture of standard proteins (aprotinin, 6.5 kDa; cytochrome c, 12 kDa; myoglobin, 17 kDa; and carbonic anhydrase II, 29 kDa; each at 30 μM). The separations obtained for this sample mixture using capillary inner diameters of 50, 20, 10, and 5 μm are shown on the same absolute intensity scale. The samples were injected electrokinetically at −50 V/cm for approximately 3 sec, and the separations are performed at −300 V/cm in capillaries chemically modified with aminopropylsilane.[73–75] Comparable separations were achieved for all four capillary inner diameters, where the migration order was carbonic anhydrase, aprotinin, myoglobin, and cytochrome c. The overall analysis time among the four separations

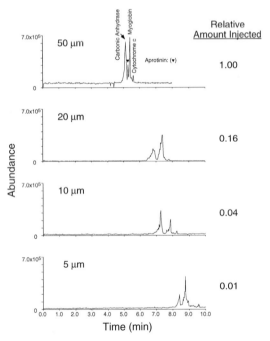

Fig. 7. Total ion current electropherograms obtained for CE–MS analysis of a mixture containing carbonic anhydrase II, aprotinin, myoglobin, and cytochrome c, using 50-, 20-, 10-, and 5-μm i.d. capillaries. Injected amounts were approximately 60 fmol/protein for the 50-μm i.d. capillary to 600 amol/protein for the 5-μm capillary by electrokinetic injection. Experimental conditions: 100-cm capillary lengths; 10 mM acetic acid buffer solution at pH 3.4; CE electric field, −300 V cm^{-1}; MS scanning from m/z 600 to 1200 at 1.5 sec/scan.

gradually increased with decreasing capillary inner diameter, presumably owing to less effective surface coverage of the aminopropyl moiety for the smaller inner diameter capillaries, leading to somewhat poorer separation efficiencies. Most importantly, however, the relative signal intensity decreased by only a factor of two to four between the 50- and 5-μm i.d. capillaries. The amount of each protein injected decreases from approximately 60 fmol to 600 amol for the 50- and 5-μm i.d. capillaries. This represents a 25- to 50-fold gain in sensitivity, which is far outside the day-to-day variation in instrumental response (a factor approximately equal to two). The mass spectra acquired for each solute from the separations obtained in the 50- and 5-μm i.d. capillaries are shown in Fig. 8. From these data, molecular weight determination is generally better than 0.05%. These results show that CE–MS of proteins is feasible at subfemtomole levels, and that sensitivity for SIM mode of MS detection should extend to the low-attomole range for the proteins studied because generally a sensitivity improvement of at least 10-fold is noted relative to scanning detection with quadrupole instruments.

These results suggest the goal of ultrasensitive peptide and protein analysis at low-attomole and even subattomole levels is obtainable. However, a number of methodological problems remain to be addressed. Improved procedures for sample and buffer preparation and handling are also needed to prevent capillary plugging, a more common problem with small inner diameter capillaries. In addition, the ESI interface sensitivity may be improved further because the efficiency of ion transport from the ESI source, which is at atmospheric pressure, to the MS detector remains relatively low. The results indicate that the sheath liquid makes only a modest contribution to the ESI current and that ions delivered to the ESI source from the CE analytical capillary are transferred to the gas phase with very high efficiency for the smaller (≤ 20 μm) diameter capillaries.

Applications of Capillary Electrophoresis–Mass Spectrometry

As yet, routine applications of CE–MS are few, and most reports have aimed at evaluation of its use for specific applications.[58,59] A particular strength, or weakness, depending on one's viewpoint, of CE and CE–MS is the extremely small sample sizes generally utilized. Thus, the emphasis of CE–MS applications would appear to focus most effectively on those situations where sample sizes are inherently limited. Some of the most demanding applications involve the characterization of gene products, and specifically, proteins and polypeptide mixtures generated by enzymatic means.

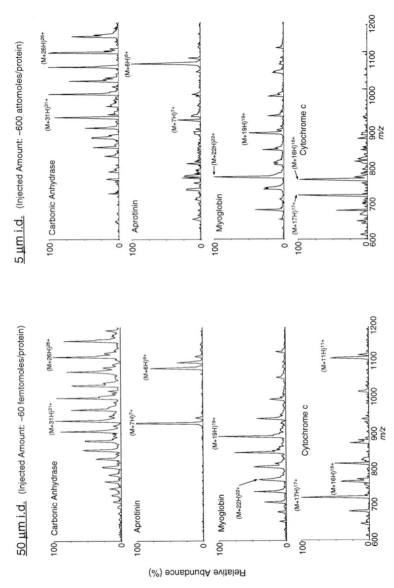

Fig. 8. Mass spectra obtained from the individual solute zones for 60 fmol/protein injected onto a 50-μm capillary (left) and 600 amol/protein injected onto a 5-μm i.d. capillary (right) for the separations and conditions shown in Fig. 7.

Polypeptides and Enzymatic Digests

Application of CE–MS to complex polypeptide mixtures (e.g., tryptic digests) is an extremely important area of application owing to the significance of peptide mapping and sequencing using ever smaller sample sizes. Complex mixtures of peptides generated from tryptic digestion of large proteins present a difficult analytical challenge because the fragments cover a large range of both isoelectric points and hydrophobicities. Because trypsin specifically cleaves peptide bonds on the C-terminal side of lysine and arginine residues, the resulting peptides generally form doubly charged as well as singly charged molecular ions by positive-ion ESI. Such doubly charged tryptic peptides generally fall within the m/z range of modern quadrupole mass spectrometers. The resolving power of these methods is illustrated in Fig. 9, which shows an ultraviolet (UV) electropherogram (top), obtained in conjunction with a CE–MS analysis,[64] for a separation of a tryptic digest of tuna cytochrome c in a 50 mM acetate buffer (at pH 6.1) mixed with an equal volume of acetonitrile. Shown are single ion electropherograms for two charge states of two tryptic fragments. In this work, a commercial CE instrument [Beckman (Fullerton, CA) P/ACE 2000] was modified to allow UV detection of the separation at one-half the MS detection time, providing a "preview" of ESI–MS detection. The individual peaks indicate separation efficiencies of up to $\sim 4 \times 10^5$ theoretical plates. It was noted that the apparent efficiency for MS detection is significantly greater than for UV detection (1.2×10^5 theoretical plates for the same peak), a fact that can be largely attributed to the longer separation time and the very small effective "dead volume" or residence time in the ESI interface.[64]

The total ion electropherograms obtained from tryptic digests of bovine, *Candida krusei,* and horse cytochrome c are shown in Fig. 10.[65] For each separation, a 10 mM ammonium acetate–acetic acid buffer system, pH 4.4, was used. The separation capillary used had an inner diameter of 10 μm, was 50 cm in length, and was chemically modified with 3-aminopropyltrimethoxysilane. The injection size corresponded to approximately 30 fmol of protein before digestion. The CE–MS interface used in these experiments was a variation of the metallized capillary terminus shown in Fig. 2D. In this set of experiments, a silver coating was applied to the capillary terminus; as a result, the effluent from the CE capillary was directly electrosprayed at an approximate flow rate of 10 nl/min. The mass spectrometer was scanned from m/z 600 to 1200 in 2-m/z steps at 0.6 sec/scan. In each case, the separation was complete within 6 min. The MS detection allows each of the tryptic fragments to be identified. For example, shown in Fig. 11 are the extracted electropherograms for the individual tryptic fragments YIPGTK

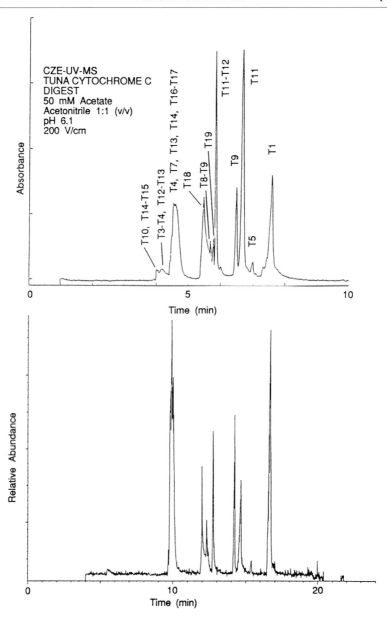

FIG. 9. Comparison of the UV (214 nm) detection (top) and MS reconstructed ion electropherograms (bottom), with time axes adjusted for the corresponding peaks, for a CE–MS separation of a tryptic digest of tuna cytochrome *c*.

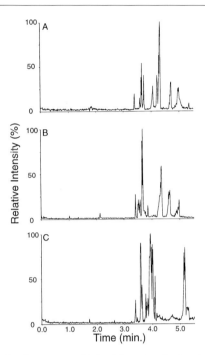

FIG. 10. Total ion electropherograms obtained from tryptic digests of bovine (A), *Candida krusei* (B), and horse cytochrome *c* (C). Separation conditions: 10 m*M* ammonium acetate–acetic acid buffer system, pH 4.4; capillary, 10-μm i.d., 50 cm in length and chemically modified with 3-aminopropyltrimethoxysilane.

(*m/z* 678), which is a common tryptic fragment for all three proteins, EDLIAYLK (*m/z* 964), which is common to bovine and horse cytochrome *c*, and MAFGGLK (*m/z* 723), which is specific to *C. krusei* cytochrome *c*. In addition, the three extracted electropherograms show additional solute zones due to other components producing ions at nearly the same *m/z* as these tryptic fragments. These additional species may arise as a result of incomplete digestion, adduction of ion species, noncovalent complexes, or fragmentation occurring within the interface of the mass spectrometer.

Thibault *et al.* examined a separation of the tryptic fragments of a glucagon digest.[39] The separation was done in a 50-μm i.d. capillary at 390 V/cm, using either a 0.01 *M*, pH 3.4, or 0.0025 *M* acetic acid buffer, pH 2.1. They used a commercial polymeric coating reagent. Two picomoles of digested glucagon was injected. The four expected fragments, which produced both doubly and singly charged ions, were observed when the MS was scanning a 950-*m/z* wide range. In addition, evidence for the partial oxidation of the sample was obtained along with a small amount of CID

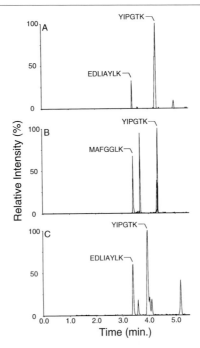

Fig. 11. Extracted ion electropherograms obtained from Fig. 10 for the individual tryptic fragments YIPGTK (m/z 678), which is a common tryptic fragment for all three proteins, EDLIAYLK (m/z 964), which is common to bovine (A) and horse (C) cytochrome c, and MAFGGLK (m/z 723), which is specific to *Candida krusei* (B) cytochrome c. The additional solute zones are probably due to incomplete protein digestion and/or fragmentation occurring within the mass spectrometer interface.

fragmentation that occurred in the ESI/MS interface. Higher efficiency, (~34,000 theoretical plates) but longer analysis times were obtained with the lower pH buffer.

Experience to date suggests that nearly all tryptic fragments can be effectively detected by ESI–MS methods if first separated; infusion of the unseparated digest often results in dramatic discrimination against some components. Fragments observed by the UV detector are almost always detected by ESI–MS. However, detection has sometimes been problematic for very small fragments (i.e., amino acids and dipeptides) owing to difficulties in obtaining optimum ESI–MS conditions over a wide m/z range. Excessive internal excitation or large solvent-related background peaks at low m/z appear to be the origin of some difficulties. There are also indications that some ESI interface designs discriminate against low-m/z ions. However, available evidence indicates that tryptic fragments not detected by CE–MS are generally "lost" prior to the separation, a problem particularly evident with "nanoscale" sample handling. The full realization of more sensitive analytical methods depends on appropriate care in sample handling and preparation.

Proteins

The ESI–MS interface provides the basis for detection of molecules having molecular weights exceeding 100,000. The analysis of proteins by CE–MS is challenging owing to the well-established difficulties associated with protein interactions with capillary surfaces. Often, capillaries must be coated to minimize, or prevent, wall interactions. It is unlikely that one capillary or buffer system will be ideal for all proteins, but rather, as for LC separations, procedures optimized for specific classes of proteins will be developed. Most CE separations require the ionic strength of the buffer to be about 100 times greater than that of the sample to prevent degradation of the separation due to perturbation of the local electric field in the capillary, although much lower relative buffer concentrations are often used in CE–MS to enhance sensitivity. Detection sensitivity for proteins is generally lower than for small peptides, due to the greater number of charges per molecule and the greater number of charge states. Our laboratory was the first to explore CE–MS of proteins,[68] but impressive results have been reported by Thibault *et al.*[39] and Moseley *et al.*[43]

Figure 12 shows a CE–MS separation with SIM detection for three myoglobins obtained on an uncoated 50-μm capillary 80 cm in length.[64] The separation was obtained in 10 mM TRIS buffer at pH 8.3 with an electric field strength of 120 V/cm. Approximately 100 fmol per component was injected onto the capillary. Figure 12 (top) shows the UV trace for the separation and Fig. 12 (bottom) shows two single ion electropherograms from MS detection for the same separation. Whale myoglobin (M_r 17,199) was detected at m/z 861 (due to loss of the heme) and carried 20 positive charges. The resolution of the mass spectrometer was too low to separate the $(M + 19H)^{19+}$ molecular ion of horse myoglobin (M_r 16,950) from sheep myoglobin (M_r 16,923), and both contribute to the m/z 893 ion in the single-ion electropherogram. A single ion at m/z 617 due to the heme group of the myoglobins is also observed and clearly remains associated with the protein during CZE separation.[64]

The use of small-diameter capillaries and optimized buffer systems allows full-scan mass spectrometric data to be obtained for more concentrated protein samples. For example, Fig. 13 shows the separation and mass spectrum obtained for injection of ~30 fmol of the protein carbonic anhydrase I, using a 20-μm i.d. capillary. The high-quality mass spectrum

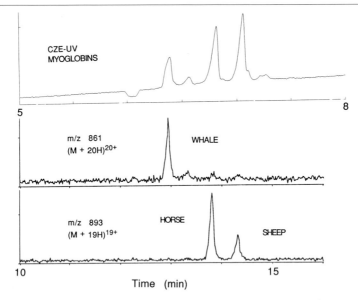

Fig. 12. CE–MS separation of a mixture of whale (M_r 17,199), horse (M_r 17,951), and sheep (M_r 16,923) myoglobins. The on-line UV response and later MS response are shown with adjusted time scales.

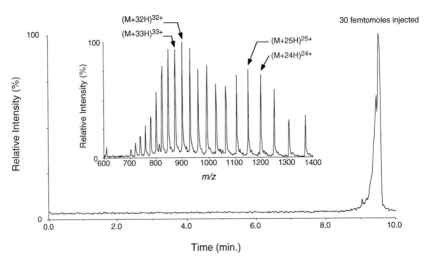

Fig. 13. CE–MS of a 50 μM carbonic anhydrase I sample, showing the separation and mass spectra obtained for a 30-fmol injection with a 20-μM capillary that was chemically modified with 3-aminopropyltrimethoxysilane.

allows molecular weight determination with a precision of approximately 0.02%.

The quality and purity of polypeptides or proteins can be rapidly determined by CE–MS. Figure 14 shows the total ion electropherogram from an injection of approximately 30 fmol of oxidized B-chain insulin onto the same 20-μm i.d. capillary used above. In addition to the major peak due to the oxidized B-chain insulin, three smaller peaks are reproducibly observed. The mass spectra obtained for each of the four peaks from this separation are shown in Fig. 15. The first of the three small peaks has a slightly higher molecular weight, and likely corresponds to a more highly oxidized form of the molecule. Interestingly, the other two peaks give mass spectra that are indistinguishable from the major peak. These peaks are tentatively attributed to other structural forms of the molecule having different electrophoretic mobility. These may correspond to different higher order structures, or perhaps labile multimers that dissociate in the ESI interface.

The potential of CE–MS for protein characterization has only begun to be explored and is expected to grow as more sensitive and powerful MS detection methods become available. Cases in which the combination of CE and MS provides unique information, or provides answers to characterization problems much more rapidly, are expected to become increasingly common. For example, Tsuji *et al.* have demonstrated that CE–MS allows recombinant somatotropins (M_r ~22 kDa) to be characterized as well as detect both mono- and dioxidized homologs that could not be unambigu-

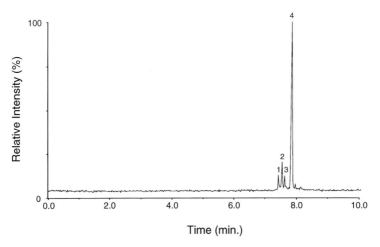

FIG. 14. CE–MS separation obtained for a 50-μM sample of oxidized B-chain insulin (M_r 3496), using a 20-μm i.d. capillary and an ~30-fmol injection. Mass spectra for the four peaks are given in Fig. 15.

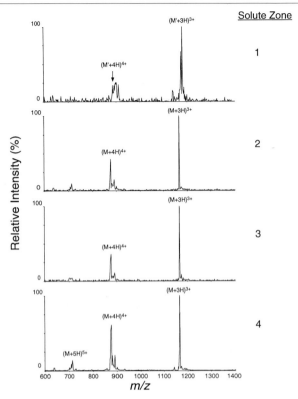

FIG. 15. Mass spectra obtained for the four solute zones shown in Fig. 14. The mass spectra for peaks 2 and 3 are indistinguishable from the major peak, whereas the first peak corresponds to a significantly higher molecular weight, and possibly arises as a result of extensive oxidation.

ously detected by either CE or low-resolution MS.[60] Perkins and co-workers demonstrated the application of CE–MS in the characterization of snake venoms.[61] Here, CE–MS offered a rapid and sensitive technique for molecular determination and provided sequence confirmation for the toxins known to be present. Their results suggest the presence of nearly 70 peptides in the molecular weight range of 6000–9000 Da.

Capillary Isotachophoresis–Mass Spectrometry

Isotachophoresis in a capillary is an alternative electrophoretic method that provides an attractive complement to the conventional free zone electrophoretic technique (i.e., CZE). The instrumentation used for capillary isotachophoresis (CITP) can be nearly identical to that of CZE. In CZE, analyte bands are separated on the basis of the differences in their electrophoretic mobilities in an electric field gradient. In isotachophoresis, the sample is inserted between leading and terminating electrolyte solutions, which have sufficiently high and low electrophoretic mobilities, respectively, to bracket the electrophoretic mobilities of the sample components. For example, for cationic analyses a leading electrolyte solution may contain 5–20 mM potassium acetate/acetic acid while the terminating electrolyte solution may contain 20 mM acetic acid/β-alanine. For anionic analyses, a leading electrolyte solution may contain 1–10 mM HCl/β-alanine while the terminating electrolyte solution may contain 5–10 mM caproic acid, propionic acid, or acetic acid. The sample components are separated into distinct bands on the basis of their electrophoretic mobilities. The electric current in the capillary is determined by the leading electrolyte, which defines the ion concentration in each band. The electric field strength varies in each band, the highest field strength being found in the bands with the lowest mobility (i.e., greatest resistance). The length of a band is then proportional to the ion concentration in the sample. Resolution is ultimately limited by band broadening due to molecular diffusion, Joule heating, and differences in electroosmosis between different bands.

A direct CITP–MS combination is an attractive complement to CZE for a number of reasons.[30] First, the sample size that can be introduced is greater by several orders of magnitude than can be tolerated in CZE, and is limited primarily by the volume of the capillary and the separation time allowed (which is inversely related to the voltage). Second, CITP generally results in concentration of analyte bands (depending on their concentration relative to the leading electrolyte), which is in contrast to the inherent dilution obtained during CZE. These two features can provide a substantial improvement in concentration detection limits compared to CZE. It must be remembered that although 10-amol detection limits have been reported for CZE, the 10-nl injection volume typical of CZE corresponds to a concentration detection limit of about 10^{-7} to 10^{-9} M with SIM detection (depending on ESI and instrumental characteristics). If the ionization efficiency is lower, or full-scan mass spectra are required, the CITP–MS detection limits will be substantially better than those obtainable by CZE–MS. In addition, all the bands have similar ion concentrations, so there should ideally be no large differences in signal intensities between bands, and information on the amount of analyte is also conveyed by the length of the analyte band. The relatively long and stable period of elution of separated bands in CITP would facilitate MS/MS experiments requiring longer integration, signal averaging, or more concentrated samples than provided by CZE. Although CITP–MS has thus far attracted relatively little attention, these characteristics make CITP–MS/MS potentially well suited for characterization of enzymatic digests of proteins and other applications where larger sample volumes exist.

Figure 16 shows selected ion isotachopherograms for the CITP–ESI–MS analysis of a mixture of peptides derived from the tryptic digestion of glucagon (HSQGTFTSDYSKYLDSRRAQDFVQWLMNT). The selected ion isotachopherograms include the tryptic fragments arginine (R, m/z 175), $T1^{2+}$ (HSQGTFTSDYSK, m/z 676), and $T2^+$ (YLDSR, m/z 653) as well as other products. Additional peptide fragments apparently arise from the action of contaminating chymotrypsin on the predicted tryptic fragments (RAQDFVQW, T4-478)$^{2+}$ and (LMNT, T4-1049)$^{2+}$, and on intact glucagon (giving m/z 546 and 366). The ESI mass spectrum of the T1 peptide contains intense 3+ and 2+ peaks, while the singly charged molecular ion is observed to have much lower intensity. The time scale shown on the axis refers to

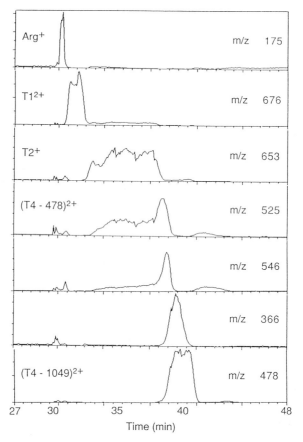

Fig. 16. Selected ion isotachopherograms for the CITP–MS analysis of an enzymatic digest of glucagon with trypsin and contaminating chymotrypsin. T1, HSQGTFTSDYSK; T2, YLDSR; T3, AQDFVQWLMNT; T4, RAQDFVQWLMNT.

the total time from the beginning of the separation. Separation development in CITP may be accomplished before MS detection starts and, as shown here, the actual time for CITP–MS analysis is determined from the time of application of hydrostatic flow due to elevation of the trailing electrolyte reservoir so as to cause elution of the focused bands (here at 27 min). The isotachopherograms show typical bands and some step changes at boundaries between bands, but are far from ideal, with the centers of sample bands showing no obvious structure.

On-Line Capillary Electrophoresis–Fourier Transform Ion Cyclotron Resonance Mass Spectrometry

Generally, CE applications require detectable amounts in the femto-mole range and above because conventional scanning mass spectrometers face a compromise between factors that include m/z scan range, MS resolution, and sensitivity. We have developed ESI–FTICR instrumentation that has the capability of rapidly changing the pressures in the FTICR cell between those that are optimum for ion trapping and cooling ($>10^{-5}$ torr) and those for high-resolution detection ($<10^{-8}$ torr).[87] In this section results are presented for the on-line combination of CE with Fourier transform ion cyclotron resonance (FTICR) mass spectrometry, an approach that is capable of providing both high CE and MS resolution and high sensitivity.[88]

A schematic illustration of the experimental system used for CE–FTICR–MS using an electrospray interface in shown in Fig. 17. The features of this instrumentation include five differentially pumped regions, two high-speed shutters to enhance differential pumping in regions close to the ESI source, and an integral cryopump, which provides effective pumping speeds of $>10^5$ liters/sec in close proximity to the FTICR cell, which extends into the bore of the 7-T superconducting magnet.[87,88]

The analyte mixture chosen for the initial study was composed of so-matostatin (1638 Da), ubiquitin (8565 Da), α-lactalbumin (14,175 Da), lysozyme (14,306 Da), myoglobin (16,951 Da), and carbonic anhydrase I (28,802 Da), each at 50 μM. The analytical capillaries used were approximately 1 m in length and 100 μm $\times$ 20-μm i.d., and the separations were performed at approximately 200 V cm^{-1}. The inner surfaces of the capillaries were treated with aminopropylsilane to reduce interactions of analytes with the capillary inner surface. The CE buffer was a 10 mM acetic acid solution (pH 3.4). The samples were electrokinetically injected at about -35 V cm^{-1}, and amounts injected were determined on the basis of the electromigration rates of the components. The FTICR experimental sequence in-

[87] B. E. Winger, S. A. Hofstadler, J. E. Bruce, H. R. Udseth, and R. D. Smith, *J. Am. Soc. Mass. Spectrom.* **4,** 566 (1993).
[88] S. A. Hofstadler, J. H. Wahl, J. E. Bruce, and R. D. Smith, *J. Am. Chem. Soc.* **115,** 6983 (1993).

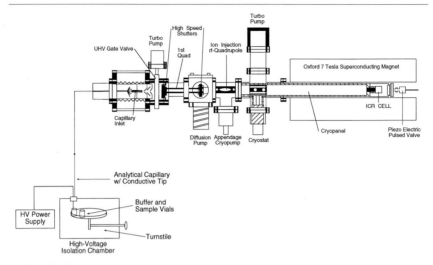

FIG. 17. Schematic illustration of the experimental arrangement for CE–FTICR–MS, using an electrospray interface. The features of this instrumentation include five differentially pumped regions, two high-speed shutters to enhance differential pumping in regions close to the ESI source, and an integral cryopump, which provides effective pumping speeds of $>10^5$ liters/sec in close proximity to the FTICR cell, which extends into the bore of the 7-T superconducting magnet.

volves gas-phase ion injection from the external ESI source using radio frequency (rf) quadrupole ion guides. These ions are collisionally trapped in the FTICR ion cell and are energetically cooled using a nitrogen gas pulse from a pulsed inlet source.[87]

Figure 18 shows a total ion electropherogram (m/z 800 to 1800) obtained using a 100-μm i.d. capillary for the analysis, with the elution rate reduced by one-half just prior to elution of the analytes (see the reduced elution speed discussion in the mass spectrometry sensitivity section), and a CE injection representing approximately 500 fmol/component. Mass spectra were collected every 6 sec. The acquisition rate is a function of ion injection, ion cooling, vacuum pump down in the cell region, and an additional data storage period required by the present data system. Figure 18 shows mass spectra for three of the components from this separation, and insets shows detail for specific peaks demonstrating that MS resolution sufficient to resolve the 1-Da spacing of the isotopic envelope is achieved, even for the largest protein studied, carbonic anhydrase I (28,802 Da). The mass spectra (Fig. 18) yielded an average resolution in excess of 30,000 full-width half-maximum (FWHM), demonstrating the unique combination of sensitivity and high MS resolution afforded by CE–FTICR. This MS resolution is sufficient to resolve the 1-Da isotopic spacing for molecular weights up to approximately 30,000 in the present studies.

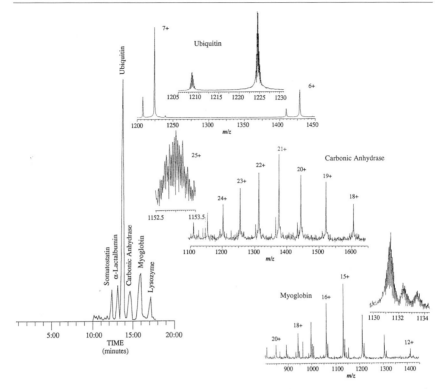

FIG. 18. An example of the application of on-line CE–FTICR mass spectrometry to protein analysis. Shown on the left is the total ion electropherogram for a separation of a polypeptide and five proteins, where the elution rate was reduced by half just prior to elution of the solutes. Mass spectra obtained during this separation are shown on the right for three of the proteins [ubiquitin, 8565 Da; carbonic anhydrase, 28,802 Da; and myoglobin (minus the heme moiety), 16,951 Da]. Insets to the mass spectra show that sufficient resolution [30,000–50,000 (FWHM)] was obtained to resolve the 1-Da spacing due to isotopic contributions for each of the proteins.

In CE-FTICR–MS, even greater MS resolution is obtainable, but this is at the expense of using longer experiment sequences ($\sim$12 sec/spectrum), primarily due to the need for acquisition and storage of larger data sets. As discussed earlier, one of the advantages of CE compared to a liquid chromatographic system is the ability to reduce the migration rate of a solute zone at any period during the separation (e.g., RES). This ability to control the solute migration rate during a separation can be used to exploit the ultrahigh MS resolution capabilities of the FTICR–MS technique. For example, in an experiment comparable to the previous results, where a 100-μm i.d. capillary was used for the analysis of the protein mixture, just prior to elution of somatostatin the CE power supply was set to 0 kV.

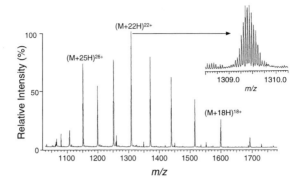

Fig. 19. Mass spectrum demonstrating the ultrahigh resolution capabilities of CE-FTICR-MS. The mass spectrum shown was obtained from a single data acquisition event during the elution of the carbonic anhydrase I (28,802 Da) zone. The inset is an expansion of the $(M + 22H)^{22+}$ molecular ion charge state, illustrating the 1-Da spacing of the isotopic envelope and represents an MS resolution of 164,000 FWHM.

Consequently, only the ESI potential (~4 kV) was used to maintain solute migration to the ESI source. This reduction in the CE electric field strength caused a reduction in migration rate of approximately 20%. As a result, a greater FTICR acquisition period could be afforded, providing ultrahigh MS resolution. This ultrahigh resolution is demonstrated in Fig. 19, where the mass spectrum shown was obtained from a single data acquisition event during the elution of the carbonic anhydrase I (28,802 Da) zone. The inset is an expansion of the $(M + 22H)^{22+}$ molecular ion charge state illustrating the 1-Da spacing of the isotopic envelope and represents a MS resolution of 164,000 FWHM.

Moreover, the results obtained using a 20-μm i.d. capillary and the same protein mixture are shown in Fig. 20, where the amount of sample injected onto the analytical CE capillary is ~6 fmol/component. Because of the small ion injection time used in the pulse sequence of the FTICR instrument, this result represents an average of only 20 amol/component consumed per spectrum. The mass spectrum in Fig. 20 shows that both good resolution and sensitivity are obtained. Preliminary results obtained in our laboratory[89] suggest that direct CE-FTICR-MS analysis of individual cells, or small populations of cells, is feasible with the instrumentation and methodology described here.

These results represented the first on-line combination of CE with FTICR, as well as the first on-line CE with high MS resolution detection. Wilkins and co-workers[6] have demonstrated off-line CE–FTICR using matrix-assisted laser desorption/ionization (MALDI), where minimum CE

[89] S. A. Hofstadler, F. D. Swanek, D. C. Gale, A. G. Ewing, and R. D. Smith, *Anal. Chem.* **67,** 1477 (1995).

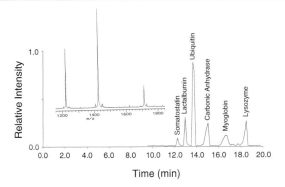

FIG. 20. CE–FTICR ion electropherogram and mass spectrum obtained using a 20-μm i.d. CE capillary, into which approximately 6 fmol/component was injected by electromigration. The mass spectrum for bovine ubiquitin shows that excellent MS sensitivity and resolution are obtained during the separation.

injection sizes were ~5 pmol, and MS resolution for the singly charged molecules with CE was insufficient to resolve the 1-Da spacing of isotopic constituents. Recent studies by Hofstadler and co-workers[90] demonstrated CE-FTICR-MS/MS in which partial amino acid sequence was obtained from a mixture of proteins by dissociation in the trapped ion cell following separation by CZE. The high pumping speed afforded by our cryopumping arrangement is crucial to the relatively rapid mass spectral acquisition rates obtained in this work. Improvements in commercial data stations that allow real-time data acquisition, processing, and storage, in conjunction with FTICR instrumentation presently under development that employs a 12 Tesla magnetic field, are expected to provide significantly higher spectral acquisition rates with improved mass resolution and sensitivity. These results clearly demonstrate that CE–FTICR–MS can be a powerful research tool, which combines the flexibility and high resolution of CE with the high-performance MS characteristics to FTICR.

Summary

In the regime of ultrasensitive analysis, and in particular for biopolymer characterization, the attraction of MS detection is that accurate molecular weight information and component identification can be performed. Moreover, techniques for obtaining structural information based on tandem methods are currently being extended to larger molecules. While only beginning to be realized, this combination of capabilities will ultimately allow the sampling, separation, and structural determination (i.e., sequencing, location of modifications, noncovalent associations, etc.) from attomole

[90] S. A. Hofstadler, J. H. Wahl, R. Bakhtiar, G. A. Anderson, J. E. Bruce, and R. D. Smith, *J. Amer. Soc. Mass Spectrom.* **10,** 894 (1994).

and even subattomole quantities of material. These results, and ongoing developments in mass spectrometric instrumentation, suggest that CE–MS has the potential to move into the realm of ultrasensitive detection.

Acknowledgments

This research was supported by internal exploratory research and the Director, Office of Health and Environmental Research, U.S. Department of Energy. Pacific Northwest National Laboratory is operated by Battelle Memorial Institute for the U.S. Department of Energy, through Contract DE-AC06-76RLO 1830. We thank C. J. Barinaga for helpful discussions and contributions to the work described in this chapter.

[20] Trace Analysis Peptide Mapping by High-Performance Displacement Chromatography

By JOHN FRENZ

Peptide mapping by high-performance liquid chromatography (HPLC) is an indispensable tool for the structural characterization of proteins that is also widely used as a sensitive identity test in the biotechnology industry. The high selectivity of reversed-phase (RP)-HPLC for peptides with even minor sequence differences makes it a useful tool for the discrimination, by peptide mapping, of the identities of closely related proteins and their variants. The keys to the technique are the generation of relatively well-defined and reproducible peptide mixtures by digestion of the protein with sequence-selective proteases and the high resolution separation of the peptide mixture by HPLC.[1] One drawback of peptide mapping, however, is its relatively low sensitivity for detection of variants present at trace levels in a protein preparation. Minor variants of the intact protein may be difficult to resolve by standard separation techniques, and may go undetected in peptide mapping. A technique that enhances the sensitivity of peptide mapping can improve the limit of detection of aberrant proteins in the presence of an excess of the correct molecule. One way of enhancing the capacity and thus sensitivity of a chromatographic column, without sacrificing selectivity, is by operating in the displacement mode.[2–4] Displace-

---

[1] W. S. Hancock, C. A. Bishop, R. L. Partridge, and M. T. W. Hearn, *Anal. Biochem.* **89,** 203 (1978).

[2] A. Tiselius, *Ark. Kemi Mineral Geol.* **16A,** 1 (1943).

[3] C. Horváth, A. Nahum, and J. H. Frenz, *J. Chromatogr.* **218,** 365 (1981).

[4] J. Frenz and C. Horváth, *in* "HPLC—Advances and Perspectives" (C. Horváth, ed.), Vol. 5, p. 212. Academic Press, New York, 1989.

ment chromatography has been investigated as an approach to preparative HPLC,[5-9] where the higher column capacity provides improved throughput, and more recently as a means of improving the detectability of trace components of a mixture.[10-13] The rationale for the latter practice is that high-performance displacement chromatography permits a greater total mass loading of the sample into the column, and maintains concentrated, narrow bands of trace components that are then more easily detected than by conventional elution operating modes. This chapter describes the use of high-performance displacement chromatography (HPDC) for high-sensitivity peptide mapping of a recombinant protein, and the application of the technique to liquid chromatography–mass spectrometry.

Displacement Chromatography

The displacement mode of operating a chromatographic separation was developed by Tiselius 50 years ago[2] primarily as an analytical technique. The feature that distinguishes the displacement mode from other approaches to operating a column is the use of a relatively dilute solution of a strongly adsorbing species, the "displacer," to develop the column and mobilize the components of the sample. The displacer is delivered to the column in a solution at a constant concentration, in a manner that is operationally identical to stepwise elution development. Unlike in elution chromatography, however, the displacer binds more strongly to the column than the sample components and so "displaces" rather than elutes the sample from the column. The eluents commonly used in chromatography contain compounds that have, on a molar basis, weaker affinity for the column than does the sample. Hence, in all interactive types of chromatography, including ion exchange, reversed phase, affinity, etc., the eluting species is present at relatively high concentration and washes the product from the column by mass action. This distinction between the displacement and elution mechanisms is most important when a relatively large sample is loaded onto the column. Column overloading results in a significant degree of

[5] J. Frenz, P. van der Schrieck, and C. Horváth, *J. Chromatogr.* **330,** 1 (1985).

[6] C. Horváth, *in* "The Science of Chromatography" (F. Bruner, ed.), pp. 179–203. Elsevier Science, Amsterdam, 1985.

[7] S. M. Cramer and C. Horváth, *Prep. Chromatogr.* **1,** 29 (1988).

[8] S. M. Cramer, Z. El Rassi, and C. Horváth, *J. Chromatogr.* **394,** 305 (1987).

[9] A. R. Torres, S. C. Edberg, and E. A. Peterson, *J. Chromatogr.* **389,** 177 (1987).

[10] A. M. Katti and G. Guiochon, *Comptes Rendus Acad. Sci. Paris* **309**(II), 1557 (1989).

[11] R. Ramsey, A. M. Katti, and G. Guiochon, *Anal. Chem.* **62,** 2557 (1990).

[12] J. Frenz, J. Bourell, and W. S. Hancock, *J. Chromatogr.* **512,** 299 (1990).

[13] J. Frenz, C. P. Quan, W. S. Hancock, and J. Bourell, *J. Chromatogr.* **557,** 289 (1990).

peak tailing, under conventional elution operating conditions, that impairs resolution and limits column capacity for a given separation. Because the displacer is selected for its efficient displacement of the product from the surface, it effectively prevents tailing of the peak and thereby increases the capacity of the column for high resolution separations. This feature underlies the claims that have been made for the advantages of the displacement mode for large-scale and preparative purification processes[5–9] and for certain analytical applications.[10–13]

A more subtle difference between the displacement and elution modes of operating a column is that the sample moves down the column ahead of the displacer front, rather than commingled with the eluent as in the elution mode. The components of the product, after passing a certain distance down the column, arrange themselves into adjacent bands with, ideally, sharp boundaries between bands, as shown schematically in Fig. 1A. This arrangement of the bands is referred to as the "displacement train." The advancing front of the displacer drives the "train" down the column, and within the sample zone the individual components sort themselves into a series of bands according to relative affinities for the stationary phase, with the component with lowest affinity at the front of the train and that with highest affinity at the rear of the train. The concentration of a

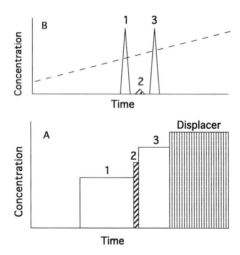

FIG. 1. Comparison of (A) displacement and (B) elution chromatograms of a mixture consisting of two major components (1 and 3) and one trace component (2). The dashed line in (B) illustrates the increase in eluent modifier concentration in the mobile phase in gradient elution chromatography. The larger feed load and focusing effect afforded by operating in the displacement mode are illustrated in (A), and account for the advantages of HPDC for analysis of minor components of a mixture.

particular band within the train depends only on the retention properties of that particular species and the displacer concentration, and is constant across the band, rather than reaching a maximum as in the peaks characteristic of elution chromatography. The characteristic shape and arrangement of the bands in displacement chromatography comes about as a result of displacement processes acting between adjacent species in the displacement train. In practice, band spreading broadens the boundaries between bands, reducing the resolution obtained in the displacement mode. Nevertheless, the displacement mode offers a means to maintain high resolution in a chromatographic separation carried out on a column at high loading.

For the purposes of analytical characterization of a protein, displacement chromatography carried out on columns and equipment developed for HPLC is one way of maintaining high resolution while loading the column with a sufficient amount of the sample to enhance the useful detection of minor components, that is, to increase the sensitivity of detection of minor components. One application of this approach is in peptide mapping of a purified protein, where the displacement mode can permit the detection and isolation of peptides arising from contaminant proteins or of variants of the desired protein that are present at low levels. Two features of displacement chromatography account for the advantages offered by HPDC for analytical characterization. First, the volume and mass of the peptide mixture loaded onto the column can be significantly increased without suffering an untoward decrease in resolution, thus overcoming limits on the sensitivity of the detection and characterization systems employed. The second benefit stems from the feature that the concentration of a species in the displacement train depends on the adsorption properties of that particular component of the product, and, at least in the ideal case, not on the amount of the species present in the displacement train. Hence, the displacement separation process affords the opportunity to focus minor components of the mixture into relatively narrow, concentrated bands that may be more readily detected than the same component separated in a technique without a similar focusing capability, such as most high-resolution elution modes. The contrast is depicted schematically in Fig. 1, which shows the chromatograms of a mixture containing a minor component separated in both a displacement (Fig. 1A) and a gradient elution (Fig. 1B) process. The detectability of the minor component is enhanced both by the increased load onto the column in the displacement separation as well as by the focusing effect that increases the concentration of the component. Figure 1A demonstrates that individual peptides are not resolved in the conventional sense, because the bands in the displacement train are conjoined. Nevertheless, high-performance displacement separations can result in high levels of enrichment of even minor components that constitute less than 0.5% of a mixture and can permit quantification and characterization of these low-level variants. In this way, HPDC mapping can amplify the signal in a peptide map and aid in the identification of all components of a mixture.

Operating Conditions for High-Performance Displacement Chromatography Peptide Mapping

Column Selection

In principle, all of the types of reversed-phase and ion-exchange HPLC systems that have been employed for HPDC[14] can be adapted to peptide-mapping applications. Because the reversed-phase mode generally provides adequate resolution of proteolytically derived peptides, it has become most closely associated with HPLC peptide mapping. In certain applications, charge-based separations involving cation- or anion-exchange HPLC may also be employed. In either case, conventional columns developed for HPLC are also employed for high-performance displacement peptide mapping.

Sample Preparation

Enzymatic digestion of a protein sample for HPDC mapping is essentially identical to that employed for conventional mapping procedures, described elsewhere in this volume. Because a key objective of mapping by displacement chromatography is to enable separation of the peptides in a large amount of feed, the sample preparation should be carried out on the appropriate amount of protein. In the case of a conventional 0.46-cm i.d. column this amount is typically up to 10 mg. Digestion at this scale must be carried out with a sufficiently dilute concentration, or in an appropriate buffer, to ensure that precipitation of hydrophobic fragments of the protein does not interfere with the digestion or sample loading. The digested peptides may need to be filtered through a 0.22-μm pore size membrane prior to loading onto the column. A small aliquot (equivalent to 50 μg of the protein sample) should be analyzed by gradient elution HPLC to ensure

---

[14] J. Frenz, *LC-GC* **10,** 668 (1992).

that digestion is adequate for isolation of the peptide of interest prior to carrying out the HPDC analysis.

High-Performance Displacement Chromatography Tryptic Map

Displacement separations are carried out on standard HPLC instrumentation, modified only to permit relatively large-volume sample injection. In the case of reversed-phase HPDC, column lengths of 25–30 cm are employed; these may be achieved by coupling two or more columns in series. The sample, which may consist of a total volume of 5 ml or more, can be injected as a series of aliquots of smaller volume, e.g., 20 or more injections of 250-μl volumes, if that is the maximum injection volume provided by the available instrumentation. The autosampler, if it is used, should be programmed in such a way that loop overfilling is minimized, to load the entire sample volume most efficiently. After feed loading is complete, the displacer solution is pumped into the column to mobilize the sample and form the displacement train. Note that the displacer is not applied to the column as a gradient, but as a solution at fixed composition until the sample emerges at the column outlet. The column effluent is typically collected as a series of fractions that are subsequently analyzed off-line by HPLC or mass spectrometry[13] to locate the peptide of interest. Peptides within individual fractions may be further purified by gradient elution chromatography in order to permit additional characterization. Alternatively, analysis of the column effluent may be carried out by on-line HPLC, or, by on-line mass spectrometry,[12] as described below. Following the displacement separation, the HPDC column is regenerated by using a strong eluent, such as acidified 2-propanol in the case of a reversed-phase column, or sodium hydroxide for an ion-exchange column, prior to being equilibrated with the carrier for a subsequent analysis.

Displacer Selection

The choice of the type and concentration of the displacer determines the useful range of peptides that are resolved by HPDC mapping. First, only peptides with lower affinity for the stationary phase than the displacer will be displaced. Hence, the displacer should have relatively high affinity in order to ensure that the peptides of interest are contained in the displacement train. Second, the concentrations of individual peptides in the displacement train are determined by the concentration of the displacer, with the peptides at the rear of the train assuming higher concentrations than those preceding them in the train. Thus, in order to achieve a useful peptide concentration the displacer concentration should be adjusted accordingly.

Table I[15-35] lists displacers that have been employed for separations of peptides and proteins in various modes of chromatography. While the displacer is a critical element of the separation, like the eluent in gradient elution chromatography, the few displacers shown in Table I are likely to be appropriate for the majority of HPDC mapping applications. Specifically, cetyltrimethylammonium bromide (CTAB) at a concentration of 2–6 mg/ml has been the "workhorse" displacer in reversed-phase HPDC separations of peptides. This displacer shows relatively high solubility in reversed-phase mobile phases and is more hydrophobic than all but the most strongly retained peptides. The identification and synthesis of novel displacers is currently an active area of research, but prior to the commercial availability of these compounds the displacers shown in Table I are probably the most useful for a broad spectrum of peptide separations.

On-Line High Performance Displacement Chromatography–Mass Spectrometry

The topography of the displacement train is such that peptides in the HPDC map are not "resolved" in the conventional sense familiar in analyti-

[15] G. Guiochon, S. G. Shirazi, and A. M. Katti, "Fundamentals of Preparative and Nonlinear Chromatography," p. 470. Academic Press, San Diego, CA, 1994.

[16] G. Viscomi, S. Lande, and C. Horváth, J. Chromatogr. **440,** 157 (1988).

[17] G. Viscomi, A. Ziggiotti, and A. Verdini, J. Chromatogr. **482,** 99 (1989).

[18] K. Kalghati, I. Fellegvari, and C. Horváth, J. Chromatogr. **604,** 47 (1992).

[19] F. Cardinali, A. Ziggiotti, and G. C. Viscomi, J. Chromatogr. **499,** 37 (1990).

[20] C. Horváth, J. Frenz, and Z. El Rassi, J. Chromatogr. **255,** 273 (1983).

[21] G. Subramanian, M. Phillips, and S. Cramer, J. Chromatogr. **439,** 341 (1988).

[22] G. Subramanian, M. Phillips, G. Jayaraman, and S. Cramer, J. Chromatogr. **484,** 225 (1988).

[23] G. Vigh, G. Farkas, and G. Quintero, J. Chromatogr. **484,** 251 (1989).

[24] A. W. Liao, Z. El Rassi, D. M. LeMaster, and C. Horváth, Chromatographia **24,** 881 (1987).

[25] A. W. Liao and C. Horváth, Ann. N.Y. Acad. Sci. **589,** (1990).

[26] J. Gerstner and S. Cramer, Biotechnol. Progr. **9,** 540 (1992).

[27] A. Torres and E. Peterson, J. Chromatogr. **604,** 39 (1992).

[28] S. Ghose and B. Matthiasson, J. Chromatogr. **547,** 145 (1991).

[29] G. Jayaraman, S. Gadam, and S. Cramer, J. Chromatogr. **630,** 53 (1993).

[30] S. D. Gadam, G. Jayaraman, and S. Cramer, J. Chromatogr. **630,** 37 (1993).

[31] S.-C. Jen and N. G. Pinto, J. Chromatogr. Sci. **29,** 478 (1991).

[32] H. G. Boman, in "Symposium on Protein Strucutre" (A. Neuberger, ed.), p. 100. IUPAC Paris, 1957. Metuthen, London, 1958.

[33] J. Gerstner and S. Cramer, Biopharmacology, November/December, 42 (1992).

[34] S.-C. D. Jen and N. G. Pinto, J. Chromatogr. **519,** 87 (1990).

[35] Y. J. Kim and S. M. Cramer, J. Chromatogr. **549,** 87 (1990).

TABLE I

DISPLACERS EMPLOYED IN SEPARATIONS OF PEPTIDES
AND PROTEINS[a]

Type of chromatography	Ref.
Reversed-phase HPDC	
Benzyldimethyldodecylammonium bromide	16, 17
Benzylhexadecylammonium chloride	18, 19
Benzyltributylammonium chloride	19–22
2-(2-Butoxyethoxy)ethanol	7, 8, 23
Carbowax-400 (polyethylene glycol)	5
Cetyltrimethylammonium bromide	13, 7
Chondroitin sulfate	24, 25
Decyltrimethylammonium bromide	7
Protamine	26
Tetrabutylammonium bromide	20, 5
Ion-exchange HPDC	
Carboxymethyldextran	27, 9
Carboxymethylstarch	28
DEAE dextran	29, 30
Dextran sulfate	29–31
α_2-Globulin	32
Heparin	33
Nalcolyte 7105	22
Polyethyleneimine	21
Poly(vinyl sulfate)	24
Ribonuclease A	35

[a] Adapted from Ref. 15.

cal gradient elution HPLC, at least by absorbance monitoring of the column effluent. Because the peptides occupy adjacent bands in the train, a selective detection method must be employed in order to discern the shape of the boundaries between zones, and to ascertain the degree of resolution of adjacent peptides. The resolution of peptides can be determined by collecting fractions of the effluent of the displacement system and subsequently analyzing them as described above. A more rapid means, however, is by coupling the column outlet to a mass spectrometer through a suitable interface. The mass spectrometer is a nearly universal detector for peptides that provides mass information about the composition of the effluent, and so is a powerful tool for monitoring the separation achieved in HPDC. The coupling of the HPDC instrument to the mass spectrometer involves conventional interfaces developed for HPLC–MS such as electrospray or continuous-flow fast atom bombardment (CF–FAB) devices. No modifications to the hardware are required, except to ensure that the flow rate of the column effluent into the interface is compatible with the requirements of the particular system employed. In general, optimal performance of the mass spectrometer requires relatively low flow rates, typically less than 100 μl/min, as described elsewhere in this volume. To achieve flow rates of this magnitude, either small-diameter columns (320-μm i.d. or less) or a postcolumn stream splitter is employed. Both approaches have been extensively employed for HPLC and in HPDC. The chief advantage of the microcolumn approach is the smaller sample load required. The stream splitter approach has the advantage of allowing fraction collection of that portion of the column effluent that is not diverted to the mass spectrometer, and subsequent further characterization of the collected fractions by other methods. In certain cases, the mass information provided by the mass spectrometer will not be sufficient to identify components of the mixture completely, so the availability of the peptide for further characterization may be essential.

Applications

High-Performance Displacement Chromatography Peptide Mapping

To illustrate the potential of HPDC mapping for separations of large amounts of an enzymatic protein digest, a 9-mg sample of recombinant human growth hormone (rhGH) was digested with trypsin and displaced with CTAB in a reversed-phase system. The resulting displacement train is shown in Fig. 2. The analyses by gradient elution HPLC of fractions collected during the displacement separation are shown in Fig. 3. As expected, the rear of the displacement train contains the most strongly retained peptides and these form the most concentrated bands. This portion of the chromatogram also reveals that essentially all of the peptides through T6–T16 are effectively displaced by CTAB, and the only major peptide absent from the displacement train is the most hydrophobic, T9, which is poorly soluble in aqueous buffers. The abrupt absence of peptides from fraction 35 on indicates that the displacement train components did not tail into the displacer zone, which lends support to the conclusion that CTAB efficiently displaced the peptide mixture under these conditions.

An aliquot of fraction 23 was also analyzed by electrospray mass spectrometry (MS), yielding the spectrum shown in Fig. 4. The mass spectrum shown in Fig. 4 reveals the presence of the T8 peptide and the doubly charged ion of the T15 peptide in fraction 23; these ions have masses of 844 and 745 atomic mass units (amu), respectively. They are the same two peptides shown by HPLC analysis in Fig. 3 to predominate in the fraction. Along with these T8 and T15 ions, the doubly charged T8 ion (m/z 423)

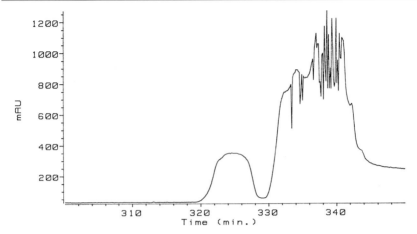

Fig. 2. Displacement chromatogram of 9 mg of rhGH tryptic peptides. The displacer was CTAB at a concentration of 3 mg/ml in a carrier of 0.1% trifluoroacetic acid (TFA) in water. The separation was carried out on two 0.46 mm (i.d.) × 150 mm Nucleosil C$_{18}$ columns connected in series. The sample and displacer were loaded into the column at a flow rate of 1 ml/min, and 1-min fractions were collected. The chromatogram shown was produced by summing the absorbance profiles obtained at 214, 254, and 280 nm. (Reprinted with permission from Ref. 12.)

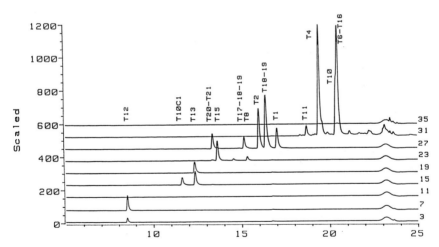

Fig. 3. HPLC analysis of the fractions collected during the displacement separation shown in Fig. 2. Aliquots of the fractions were analyzed by HPLC on a Nucleosil C$_{18}$ column operated at a flow rate of 1 ml/min, using a linear gradient from 0.1% TFA in water to 40% acetonitrile in water. The fraction number is indicated adjacent to each chromatogram. Time, in minutes, is shown on the abscissa (as in Fig. 2). (Reprinted with permission from Ref. 12.)

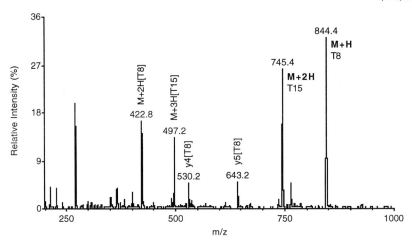

FIG. 4. Mass spectrum of fraction 23 in a solution of 50% acetonitrile–0.1% formic acid. Aliquots of the fractions were infused into the Sciex API-III (Thornhill, ON, Canada) instrument at a rate of 10 μl/min. The orifice voltage was set to 80 V and the ion spray potential to 4500–5000 V. Ions arising from T15 and T8, including fragment ions, are identified. (Reprinted with permission from Ref. 12.)

is apparent along with several ions that are products of collision-induced dissociation in the electrospray interface.[36,37] Aside from the ions derived from the two major peptides in this mixture, there are ions present at much lower abundances that presumably represent components of the fraction present at low concentrations, including the unidentified peptides occurring as minor peaks in the chromatogram of the mixture in Fig. 3. These ions are not readily associated with peptides expected to arise from the tryptic map of rhGH, and require further characterization for definitive assignment.

The unambiguous identification of minor components within a fraction requires further purification and isolation of individual species. Identification of the peptides corresponding with these peaks yields information about the composition of the protein preparation and about the trypsin digestion process. To illustrate the isolation by HPDC of useful amounts of minor components of the peptide digest, individual fractions from the separations shown in Fig. 3 were rechromatographed by gradient elution HPLC and the resulting peaks were manually collected for subsequent

[36] R. D. Smith, J. A. Loo, C. J. Barinaga, C. G. Edmonds, and H. R. Udseth, *J. Am. Soc. Mass Spectrom.* **1**, 53 (1990).
[37] V. Katta, S. K. Chowdhury, and B. T. Chait, *Anal. Chem.* **63**, 174 (1991).

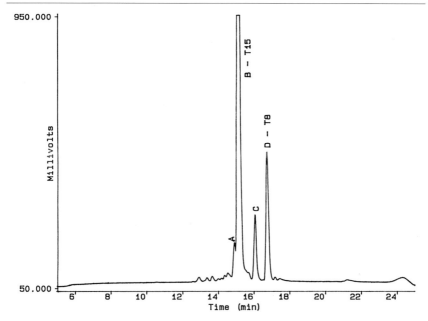

FIG. 5. Reversed-phase HPLC chromatogram of fraction 23 of the HPDC tryptic map. Analyses were carried out as described in Fig. 3. Peptides collected for identification by mass spectrometry are labeled to correspond with the appropriate spectrum in Fig. 11. (Reprinted with permission from Ref. 12.)

analysis by electrospray mass spectrometry. Figure 5 shows the four peaks collected by gradient elution HPLC analysis of fraction 23 of the displacement separation. The mass spectra of the individual peaks are shown in Fig. 6. The identification of the major peaks as the doubly charged T15 (m/z 745) and the singly charged T8 (m/z 844) was confirmed by their mass spectra, as shown in Fig. 6B and D. Two minor peaks were also identified, and corresponded to cleavages that are not commonly seen in trypsin digestion of rhGH. The earlier eluting of these peaks, with a mass of 853 amu, was identified by MS as a doubly charged peptide constituting residues 144–158 of hGH, and represents a chymotryptic-like clip between Tyr-143 and Ser-144 in what would be the T14 peptide. Residue 145 is a lysine, but no cleavage occurred after this residue, suggesting that this amide bond, near the clipped portion of T14, is a relatively poor substrate for trypsin. The fragmentation pattern discerned for this peptide and shown in Fig. 6A confirms the identification of this relatively unusual product of tryptic digestion. Analysis of all the fractions collected indicates that this peptide constitutes approximately 0.9% of the T15 residues in the digest mixture,

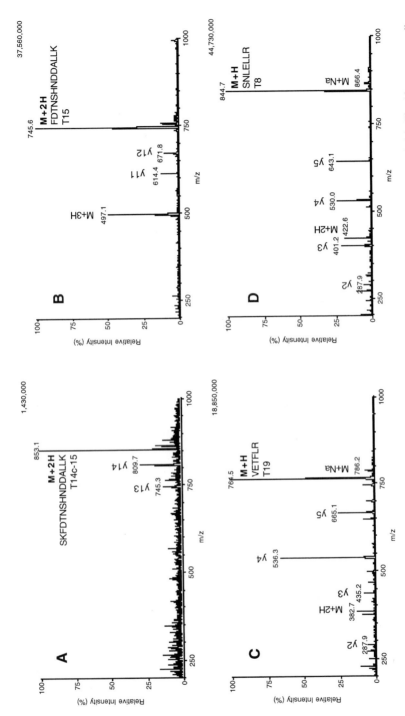

Fig. 6. Mass spectra of peaks collected from the gradient elution chromatogram of fraction 23. The four peaks labeled in Fig. 10 were manually collected and analyzed by electrospray MS, as described in Fig. 4. Principal ions are identified in each of the spectra. (Reprinted with permission from Ref. 12.)

demonstrating the high resolving power of HPDC for minor components of a complex mixture. The other minor peptide identified in fraction 23 is T19, which is usually not cleaved from T18 by trypsin, as indicated in Fig. 3. The fragmentation pattern observed in Fig. 6D confirms that this peptide is present in the digest of hGH. Integration of the chromatograms of fractions containing this peptide shows that it occurs at a level of approximately 1.5% of the T19 residues in the protein, compared with approximately 26.5% present in the T18-19 peptide and the balance in the T17-18-19 peptide.

High-Performance Displacement Chromatography–Mass Spectrometry Peptide Mapping

The HPDC mapping described above was carried out using an analytical, 0.46-cm i.d. column with a sample of 9 mg of a digested protein. In certain applications this amount of material may be unavailable, necessitating a scaling down of the HPDC mapping procedure to narrower columns. Fraction collection from microbore columns may be impractical, thus coupling of such columns to a mass spectrometer may be the simplest means of obtaining an HPDC map. To illustrate the potential of this approach, a displacement separation was implemented in a capillary liquid chromatography (LC) column with an inner diameter of 320 μm,[12] directly coupled through a frit–FAB interface to a magnetic sector mass spectrometer.[38] Figure 7 shows the displacement chromatograms for the separation of 200 μg of digested recombinant methionyl human growth hormone (Met-hGH) in the capillary HPDC–MS system. The chromatogram in Fig. 7A was made by ultraviolet (UV) absorbance detection, while that in Fig. 7B is the total ion current (TIC) chromatogram obtained by mass spectrometric detection. Both chromatograms illustrate the "staircase" pattern expected in displacement-mode separations. The plateau heights measured by UV absorbance of the tryptic peptides increase across the displacement train, as expected, as their size and concentration increase. The TIC chromatogram in Fig. 7B exhibits a converse behavior, with the plateau heights decreasing in size along the displacement train. This behavior reflects the relatively poorer ionization properties of the larger, more hydrophobic peptides that are a feature of the FAB process. The TIC chromatogram shows instability in the ion current at the end of the displacement train where the high peptide concentration may inhibit efficient ionization from the dilute glycerol matrix. As in the previous example, the relatively sharp rear boundary of the UV chromatogram indicates that CTAB, which exhibits little UV absorbance, efficiently displaces peptides under these conditions.

[38] W. J. Henzel, J. H. Bourell, and J. T. Stults, *Anal. Biochem.* **187,** 228 (1990).

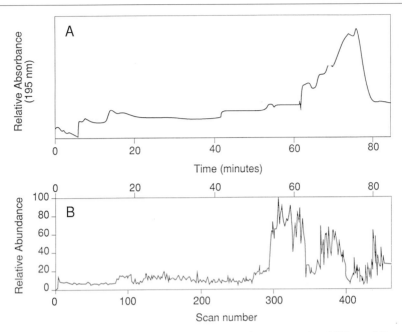

F<small>IG</small>. 7. Column effluent profile during displacement chromatography of 200 μg of digested biosynthetic human growth hormone monitored by (A) absorbance of light at 195 nm and (B) the total ion current in the mass spectrometer. The separation was carried out in a 0.32 mm (i.d.) × 150 mm LC Packings (Zürich, Switzerland) capillary column packed with 3-μm RP-18 silica. The sample was displaced from the column with CTAB (2 mg/ml) in a solution of 1% glycerol–0.1% TFA in water at a flow rate of 5 μl/min. Spectra were acquired in 12-sec scans, and data collection commenced 150 min after the displacer solution started flowing into the column. (Reprinted with permission from Ref. 13.)

The mass spectrometer provides the opportunity to monitor selectively the composition of the displacement train in an on-line fashion and so provide composition information such as obtained previously by fraction collection and subsequent analysis. Figure 8 shows the reconstructed ion current (RIC) profiles for the predominant peptides in the tryptic digest. The RIC chromatograms demonstrate that each peptide occupies a distinct band in the displacement train that in most cases is well separated from adjacent bands. The more strongly retained peptides form extremely narrow, concentrated bands and, in the cases of T11 and T10c2, the band is little more than one scan wide in the RIC chromatograms. The most hydrophobic peptides, as noted above, were not identified in the displacement train, owing to the interruption in ion current in the latter part of the train. The displacer was monitored by the mass at 647.5 (amu), which

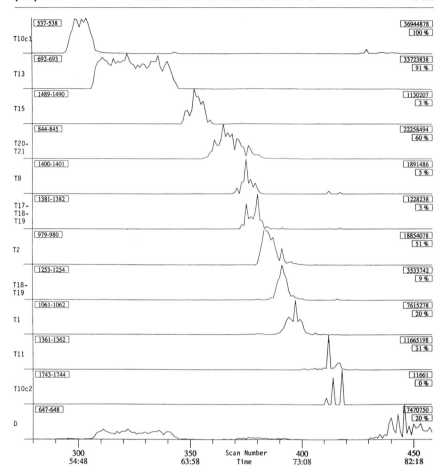

FIG. 8. Reconstructed ion current chromatograms of the ions corresponding to the predominant peptides comprising the tryptic digest of biosynthetic hGH. The RIC chromatograms were produced by monitoring the abundance of ions within a narrow mass window measured by the mass spectrometer during the run. Each RIC thus selectively monitors a single species in the effluent. The mass range employed in the RIC and peptide identity are indicated on the left side of each chromatogram. The abundance of ions relative to that of the T10c1 RIC maximum is given on the right. "D" indicates the RIC of the brominated dimer of cetyltrimethylammonium, used to monitor the displacer front. (Reprinted with permission from Ref. 13.)

corresponds to a brominated CTAB dimer, because the CTAB monomer ($M + H^+ = 285$) was below the mass range scanned in this experiment. The displacer ion appeared on resumption of the ion current, evidently after the disturbance at the probe tip had been removed.

The mass range from 500 to 4000 amu was scanned every 12 sec, providing a "snapshot" of the composition of the column effluent at 12-sec intervals. Subsequent scans can be compared to provide insight into the dynamics of the displacement process, in a manner analogous to the tandem HPDC–HPLC arrangement described previously.[5] Figure 9 shows sequential mass spectra, starting with scan 305, which was started at 61 min. The dominant ion in scan 305 has a molecular mass of 537.4 amu and corresponds to residues 96–99 in Met-hGH. This peptide arises from a chymotryptic-like clip in the T10 peptide, has been observed previously,[2] and is dubbed the "T10c1" peptide. Scan 306 is also dominated by the T10c1 peptide, while scans 307 and 308 represent the boundary between the zone of this peptide and the next in the displacement train, an ion with a mass of 693.2 amu. This peptide corresponds to the T13 fragment of the tryptic digest of Met-hGH. Figure 9 thus shows the small extent of overlap between bands in HPDC, i.e., less than two scans or 24 sec wide. Few on-line or off-line analytical techniques can monitor the effluent composition at higher scan rates than mass spectrometry, demonstrating the usefulness of this approach in studies of the displacement process.

In contrast to the clean transition from T10c1 to T13 observed in Fig. 9, the boundary at the front of the T10c1 band, which marks the transition from T12 to T10c1, contained several small fragment ions, as shown in Fig. 10. The serial spectra shown in Fig. 10 illustrate that the minor components of the digest, such as the 559.4 ion, which can be assigned to a nonspecific cleavage of the T15 peptide, concentrate between the bands of the predomi-

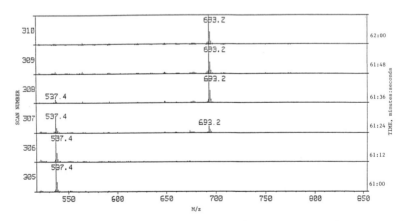

Fig. 9. Mass spectra acquired as scans 305–310 during the displacement separation. The time corresponding to the start of each scan is indicated on the right-hand side and the scan number on the left-hand side. (Reprinted with permission from Ref. 13.)

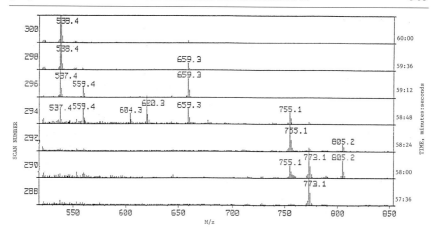

FIG. 10. Mass spectra acquired during the displacement separation by summation of pairs of scans. The spectrum labeled "288" is thus the sum of scans 288 and 289. The time corresponding to the start of the scan is indicated on the right-hand side and the beginning scan number on the left-hand side. (Reprinted with permission from Ref. 13.)

nant peptides and yield intense ions in individual scans that unambiguously indicate their presence and allow precise determination of their mass. This focusing effect illustrates the power of the displacement mode for accumulation and characterization of minor components of a complex mixture. The relative amounts of individual species in the digest mixture could be estimated roughly from the width of the zones occupied in the displacement train, as indicated by Tiselius.[2]

Conclusions

The rationale for adopting high-performance displacement chromatography for mapping the peptides formed by tryptic digestion of a protein is to characterize the minor components of the mixture whose presence conveys information about the composition of the protein preparation digested as well as about the specificity of the enzymatic digestion and side reactions occurring in the digestion. Chief among the objectives of examining the "fine structure" in the enzymatic map is to identify the tell-tale peptides indicating the presence of protein variants in the preparation. These variants can arise from degradative processes acting on the protein, such as deamidation, oxidation, and proteolysis; from incorrect folding and disulfide rearrangement; from errors in translation by the host cell[39]; from chemical

[39] G. Bogosian, B. N. Violand, E. J. Dorward-King, W. E. Workman, P. E. Jung, and J. F. Kane, *J. Biol. Chem.* **264**, 531 (1989).

modifications to the protein caused by processing conditions used, for example, to cleave the target protein from a larger fusion construct; or from the heterogeneity conferred on a glycoprotein by the covalently linked carbohydrate structures. Other peptides occurring at low levels in the tryptic map may arise as artifacts of the digestion process, resulting either from autoproteolysis of trypsin[38] or from nonspecific cleavages of the protein that are observed even in highly purified preparations of proteolytic enzymes. Finally, host cell proteins may also contribute to the low-level contaminants found in the peptide mixture, although in a pharmaceutical-grade protein these contaminants are expected to be present at undetectable levels.[40] Hence, the information content of the minor components of a mixture can be sufficient to warrant the development of methods, such as HPDC, for closer examination of the trace-level components of the mixture.

[40] S. E. Builder and W. S. Hancock, *Chem. Eng. Prog.* August 42 (1988).

[21] Measuring DNA Adducts by Gas Chromatography–Electron Capture–Mass Spectrometry: Trace Organic Analysis

By Roger W. Giese, Manasi Saha, Samy Abdel-Baky, and Kariman Allam

Introduction

This chapter presents our practical experience in method development for the determination of trace amounts of DNA adducts by gas chromatography–electron capture–mass spectrometry (GC–EC–MS). We have detected femtomole amounts of such analytes by optimizing sample preparation (involving extraction, chemical reaction, and purification steps starting with a biological sample) and low-attomole amounts of pure, derivatized standards by GC–EC–MS. Although such methodology is already useful, the concepts and techniques described should extend sample preparation to the attomole level.

In this chapter our work on chemical transformation is emphasized as part of sample preparation. This is a means to broaden the range of compounds that can be detected by GC–EC–MS. Also, our experience in operating a GC–EC–MS to achieve attomole detection limits routinely (for standards) is presented.

New ionization techniques for MS, such as electrospray (see [21] in this volume[1]) and matrix-assisted laser desorption (see [22] in this volume[1a]), are increasing the ability of MS to analyze "nonvolatile" substances present even in aqueous samples. Less new but of continuing importance as a desorption/ionization technique in this respect is fast atom bombardment. In contrast, we are focusing on procedures in which significant chemical treatment of the sample precedes the "old technique" of GC to deliver the analyte into the MS. The desorption approaches are attractive because they can minimize sample preparation. They are also unique in their ability to achieve the direct detection of medium to high molecular weight biopolymers by MS. For trace organic analysis, however, the use of chemical steps to aid in the characterization and purification (including recovery) of smaller analytes by changing their physicochemical properties, coupled with the additional purification provided by GC (including the high purity of GC carrier gases), may be important.

DNA Adducts

DNA *adducts* result from covalent damage incurred *in vivo* by endogenous and exogenous agents. Measuring DNA adducts is of interest largely because carcinogens and mutagens (or their metabolites) tend to react covalently with DNA. Depending on many factors (e.g., adduct structure, location on the DNA, cell type, species) DNA adducts persist, are correctly repaired, or lead to mutations to different degrees. Potentially, the measurement of DNA adducts in human samples can help to assess how much of the human burden of cancer and genetic disease arises from exposure of people to DNA-reactive chemical and physical agents, or at least improve our measurement of individual exposure. High sensitivity is required, because cells contain little DNA (about 1 mg of DNA/g of wet tissue; 45 μg/ml of blood), and DNA contains relatively few adducts (e.g., 1 adduct in 10^7 or more nucleotides), which may nonetheless be biologically significant. Books and reviews are available on the meaning and measurement of DNA adducts, including the use of techniques other than GC–EC–MS for this purpose.[1b–4] These same sources also discuss other biomarkers of

[1] J. F. Banks, Jr., and C. M. Whitehouse, *Methods Enzymol.* **270,** Chap. 21, 1996.
[1a] R. C. Beavis and B. T. Chait, *Methods Enzymol.* **270,** Chap. 22, 1996.
[1b] R. C. Garner, P. B. Farmer, G. T. Steel, and A. S. Wright (eds.), "Human Carcinogen Exposure." IRL Press at Oxford University Press, Oxford, 1991.
[2] H. Bartsch, K. Hemmimki, and J. K. O'Neill (eds.), "Methods for Detecting DNA Damaging Agents in Humans: Applications in Cancer Epidemiology and Prevention." IARC, Lyon, France, 1988.
[3] M. C. Poirier and F. A. Beland, *Chem. Res. Toxicol.* **5,** 749 (1992).
[4] J. W. Groopman and P. W. Skipper, "Molecular Dosimetry and Human Cancer: Analytical, Epidemiological, and Social Considerations." CRC Press, Boca Raton, FL, 1991.

human exposure to chemicals, such as protein adducts and urinary metabolites.

Gas Chromatography–Electron Capture–Mass Spectrometry

Gas chromatography–electron capture–mass spectrometry (GC–EC–MS) is useful for the determination of electrophoric derivatives of DNA adducts. In this sensitive and specific technique, the derivatized sample is dissolved in an organic solvent and injected into a gas chromatograph, where the analyte is separated in the gas phase from many of the impurities that are present. Once the analyte elutes from the GC column, it encounters a cloud of low-energy electrons (and other reactive species including positive ions such as CH_5^+ when CH_4 is the reagent gas) in the ion source of the MS. If the analyte has an ability to capture an electron efficiently under such conditions (few compounds do), it will ionize by electron capture, forming an anionic product. The final, anionic product that is detected may be the initial one that forms (nondissociative electron capture), or a fragment that forms subsequently (dissociative electron capture). Multiple anionic fragments can also arise, depending on the structure of the analyte.[5,6]

A fused silica, bonded phase capillary GC column is employed because of its high performance in terms of chromatographic efficiency (sharp peaks), good analyte recovery (low level of analyte-destroying active sites), low bleed (the stationary phase is both cross-linked and bonded to the capillary wall), and durability. For less polar analytes, columns from different manufacturers give similar results, but a higher quality (usually more expensive) column is a wise choice for more polar analytes. This is because its level of performance will be higher when new and will degrade more slowly with use.

A moderately priced GC–EC–MS instrument can be used. Individual instruments from a given manufacturer may vary in their sensitivity, therefore the manufacturer should be requested to supply appropriate data for the particular instrument before it is shipped, e.g., whether the instrument can detect 50 amol of a representative compound. On the MS instrument in our laboratory [a Hewlett-Packard (HP, Palo Alto, CA) HP 5890A GC coupled to an HP 5988A MS], a reduced-volume, tighter source (narrow inlet hole for the GC column) was selected to enhance the sensitivity,[7] on the basis of a recommendation from the manufacturer.

[5] E. A. Stemmler and R. A. Hites, *Biomed. Environ. Mass Spectrom.* **17,** 311 (1988).
[6] A. G. Harrison, "Chemical Ionization Mass Spectrometry," 2nd Ed. CRC Press, Boca Raton, FL, 1992.
[7] S. Abdel-Baky and R. W. Giese, *Anal. Chem.* **63**(24), 2986 (1991).

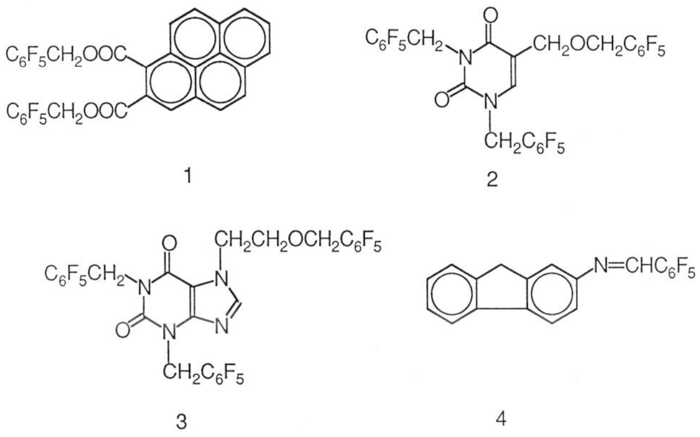

FIG. 1. Structures (**1–4**) of some high-response, single-ion electrophores related to DNA adducts. (Reprinted with permission from S. Abdel-Baky and R. W. Giese, *Anal. Chem.* **63**(24), 2986. Copyright 1991 American Chemical Society.)

High-Response, Single-Ion Electrophores

Although a variety of compounds can be detected by GC–EC–MS, the lowest detection limits result when high-response, single-ion electrophores (compounds that ionize by EC to give a single ion, or nearly so, aside from the isotope peak) are determined with selected ion monitoring. It is important for the ion to be structurally characteristic. The first examples of such compounds were demonstrated in 1978.[8] Some examples related to DNA adducts are shown in Fig. 1 (**1–4**); they all have been detected at the low-attomole level by GC–EC–MS.[7]

Not all high-response, single-ion electrophores have exactly the same sensitivity, for reasons that are not always completely clear. Two- to three-fold differences in response even of analogs are not uncommon. Differences in their susceptibility to losses at active sites in the overall GC–EC–MS system, ease of electron capture, participation in other reactions in the ion source, and inherent stability of the anion of interest (during its brief transit from the ion source to the electron multiplier detector) probably are the main reasons. Thus, one should select a single-ion electrophoric derivative for a given DNA adduct carefully, including testing at the attomole level on an aged GC column. For example, two such compounds may perform equivalently when tested as standards by GC–EC–MS down to the femtomole level, but only one may consistently give a high response at the attomole level, especially as the active sites increase on an aging GC column.

Method Development

For a given adduct, one should first obtain 10 mg or more as a standard to develop a chemical procedure that converts the substance efficiently (yield ≥50%) into a high-response, single-ion electrophore. Once this is achieved, and a stable isotope internal standard is prepared, the procedure next is extended to a trace amount of the adduct as a standard. Down to about the 10-ng level, it is useful to monitor the procedure (yield, formation of side products) by high-performance liquid chromatography (HPLC) with ultraviolet (UV) detection. This reduces the workload on the GC–EC–MS; is rapid and convenient for optimizing the procedure; can reveal nonvolatile starting material, intermediates, and side products; and avoids the introduction of impure samples (early method development) into the GC–EC–MS.

Below the 10-ng level, one switches to GC–EC–MS (GC with electron capture detection can also be used conveniently down to roughly the midpicogram level, but with less convenience due to interference). Near the 10-ng level the reaction conditions (e.g., quantities of reagents and solvents employed in the procedure) are determined, and only dilutions of the analyte are required in proceeding to lower analyte levels. Further reduction in the amounts of the reagents yields concentrations that afford unacceptably long reaction times, or fail to form product efficiently for other reasons. The amount of each reagent and solvent should be minimized because every chemical contains impurities. Impurities increase in chemical reactions; and greater impurity early in a method generally leads to more extraneous peaks later in the chromatogram produced by the GC–EC–MS. Finally, in the method the sequential chemical steps should be compatible, e.g., the sample can simply be evaporated between any two chemical reactions, allowing the reactions to be conducted tandemly in the same vial.

As the amount of standard analyte is progressively reduced to the levels of interest anticipated in real samples, interference and losses will be encountered. Making a dilution of the analyte by only a factor of 10 at lower levels can increase these problems significantly. To address these difficulties, first test the last steps in the procedure, even the final evaporation step, with standards. Conduct each chemical reaction step as a blank (no analyte), and then spike in a known amount of authentic product at the end of the reaction. This approach determines whether it is the yield (from a chemical reaction) or recovery of analyte from a given reaction that is causing a loss. Unfortunately, the "tuning" largely needs to be done at the lower analyte level, because at higher analyte levels "everything works."

The later transition in method development to analyte-spiked, standard DNA, and then to real samples, means that the analyte will no longer be pure, and the impurities may consume some of the reagents being employed.

This can lead to a demand for higher amounts of the reagents than were effective for a pure standard of the analyte. Thus, one may elect to delay completed tuning of the procedure, especially the amounts of the reagents and column capacities, until real samples are tested.

DNA and Adduct Isolation

Genomic DNA is a relatively unique substance (e.g., large size, high negative charge and associated water solubility, resistance to enzymes other than deoxynucleases), so it is relatively easy to isolate from a biological sample, at least in moderate yield and purity. The common procedure involving protein and ribonucleic acid digestion/phenol–chloroform extraction/alcohol precipitation[9] is frequently selected. Many subtle variations of this method can be found in the literature. An automated instrument for this purpose is available from Applied Biosystems Inc. (Foster City, CA). Additional purification by ultracentrifugation in a cesium chloride gradient or by other methods is sometimes also performed.

Commercially available kits relying on extraction of the DNA onto a solid-phase packing (e.g., anion exchanger) are available, but the incomplete disclosure of the kit components by a manufacturer can discourage its use, because troubleshooting and tuning are then more difficult. A general problem with many procedures for DNA extraction is that the recovery of DNA declines with smaller samples.

Both chemical and enzymatic methods have been employed, in conjunction with liquid chromatography, to isolate adducts from the purified DNA. The choice is dictated largely by the special physicochemical properties of the adduct of interest, and the techniques most accessible to the laboratory. Strong acid hydrolysis degrades DNA to nucleobases. Moderate acid hydrolysis primarily depurinates the DNA. The latter depends on protonation at the N-7 position of guanine and adenine, which imposes a positive charge on the purine that makes the associated glycosidic bond susceptible to hydrolysis. Mild acid hydrolysis has been used to liberate benzo[a]pyrene-7,8,9,10-tetrahydrotetrol from its attachment at least to the N-2 position on guanine (see below). So-called "neutral thermal hydrolysis" (heating in buffer near pH 7) releases N^7-guanine and N^7-adenine alkyl adducts (because of the positive charge at the N-7 position) along with a small fraction of the normal purines.[10] Enymatic hydrolysis can be used to convert the DNA to nucleotides or nucleosides.

It is attractive to isolate the adduct from the bulk of normal DNA

[9] D. M. Wallace, *Methods Enzymol.* **152,** 33 (1987).
[10] D. N. Mhaskar, J. M. Raber, and S. M. D'Ambrosio, *Anal. Biochem.* **125,** 74 (1982).

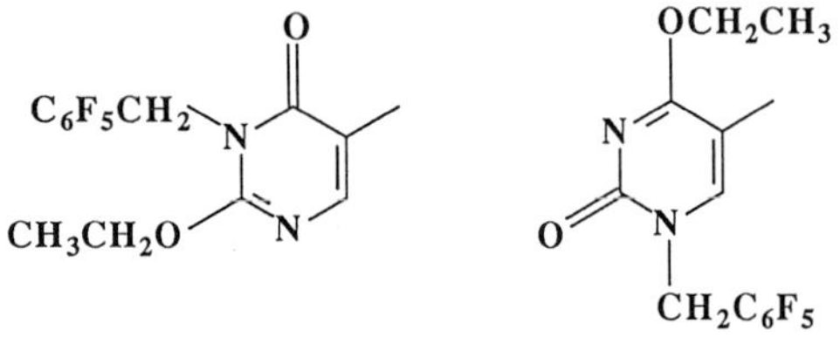

FIG. 2. Structures of pentafluorobenzyl derivatives of O^2-ethylthymine (**5**) and O^4-ethyl-thymine (**6**).

monomers (derived from the hydrolyzed DNA sample) by solid-phase extraction on a short column, assuming that a column that more or less discriminates the adduct can be set up. This isolation procedure is convenient, can keep related adducts together (they can be resolved later by GC–EC–MS), and avoids two potential problems, which may be interrelated when HPLC instead is used at this stage: trace adduct retention times can change with column use, and contaminants can accumulate (including carryover of analyte) in the HPLC system. In one case it was found that 99.9% of the carryover in the HPLC system took place in the injector.[11] These two problems can be tedious to control because the amounts of the adducts tend to be too small for on-line detection. Nevertheless, in spite of these concerns about using HPLC when it has been used early in the overall procedure, it may be necessary to select this approach whenever solid-phase extraction falls short in purifying the adduct. It can be especially attractive to subject the hydrolyzed DNA sample to an antibody affinity column to extract the tiny amount of adduct away from the large excess of normal DNA monomers (nucleotides, nucleosides, or nucleobases). This method will be used increasingly as antibodies become more readily available, and adducts emerge that deserve extensive monitoring.

Chemical Transformation/Derivatization

Overview

The semipurified adduct is next converted into a single-ion electrophore. For some DNA adducts, such as certain alkyl derivatives of the nucleobases, electrophoric derivatization alone can accomplish this. Figure 2 shows such derivatives for O^2-ethylthymine and O^4-ethylthymine (**5** and **6**, respec-

[11] M. Saha and R. W. Giese, *J. Chromatogr.* **631,** 161 (1993).

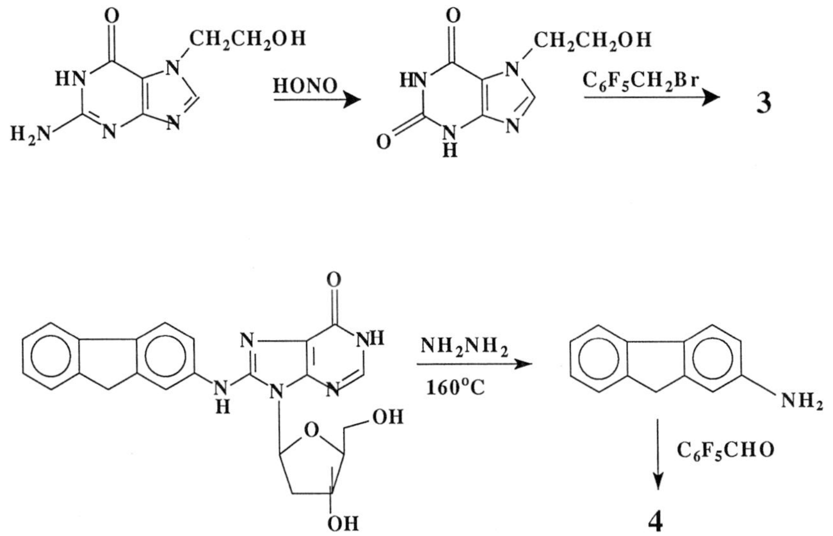

FIG. 3. Chemical transformation and electrophoric derivatization reactions. (Adapted from Allam *et al.*[13] and Bakthavachalam *et al.*[14])

tively).[12] These derivatives, possessing a pentafluorobenzyl moiety on a nucleobase nitrogen, tend to undergo efficient, dissociative electron capture, forming a pentafluorobenzyl radical and a nucleobase anion. Conveniently, the adducting alkyl moiety (the damage of interest on the DNA) reduces the degree of derivatization required for such adducts by inherently masking one of the other ionizable sites on the nucleobase. This type of derivative is obtained by reacting the alkyl nucleobase adduct with pentafluorobenzyl bromide in the presence of a base such as potassium carbonate (e.g., two-phase reaction with solid potassium carbonate in acetonitrile), triethyla-mine, or potassium hydroxide (e.g., phase transfer reaction involving a lower phase of dichloromethane and an upper phase of 1 N aqueous potassium hydroxide; see below).

When derivatization alone cannot convert a DNA adduct into a single-ion electrophore, chemical transformation is employed prior to derivatiza-tion. Two examples are shown in Fig. 3[13,14]: nitrous acid hydrolysis of an N^7-guanine adduct of ethylene oxide to a corresponding xanthine, and hydrazinolysis of an N^7-guanine adduct of 2-aminofluorene to liberate the

[12] M. Saha, G. M. Kresbach, R. W. Giese, R. S. Annan, and P. Vouros, *Biomed. Environ. Mass Spectrom.* **18,** 958 (1989).
[13] K. Allam, M. Saha, and R. W. Giese, *J. Chromatogr.* **499,** 571 (1990).
[14] J. Bakthavachalam, S. Abdel-Baky, and R. W. Giese, *J. Chromatogr.* **538,** 447 (1991).

2-aminofluorene moiety. A third example, mild acid hydrolysis/superoxide oxidation of an N^2-guanine benzo[*a*]pyrenediol epoxide adduct to liberate 2,3-pyrenedicarboxylic acid, is shown as part of an analytical scheme in Fig. 4. Also indicated in Fig. 3 is the subsequent electrophoric derivatization of the chemical transformation products to single-ion electrophores **3** and **4**. Not all single-ion electrophores are easy to obtain and purify in high yield, especially at a trace level, thus alternative chemical transformation/ derivatization procedures sometimes should be explored.

Practical Considerations for Sample Preparation

We have chosen to employ standard, commercially available laboratory ware wherever possible. Conical vials are employed for the reactions, and subjected to extensive cleaning between procedures, including the vigorous use of a test tube brush, at least initially, to dislodge any microscopic particles that may be present on the vial walls. These particles are visible with a handheld microscope and can be present in the vials as received from the manufacturer.[15]

Plastics including polytetrafluoroethylene (PTFE) may need to be avoided. Not only can plastics emit volatile contaminants (e.g., residual plasticizers), but they can cause losses. We have observed analyte losses on PTFE-coated stirring bars. The use of a glass-stoppered vs plastic-capped vial for a trace chemical reaction can reduce the level of extraneous peaks observed by GC–EC–MS.[15]

Dilutions are performed with disposable glass capillary micropipettes (10- to 100-μl size) using a Drummond Captrol III (Drummond Scientific Co., Broomall, PA) to load and dispense aqueous samples, and gravity for samples in organic solvents. For both kinds of sample solutions, it is generally necessary to dispel the residual volume from the micropepette with a pulse of air. For this purpose, we either attach the plastic mouthpiece that comes with the micropipette to the dispensing end of a Pasteur pipette fitted with a rubber bulb, or use residual air in the Captrol III.

Reaction volumes in the 30- to 100-μl range are convenient, permit some evaporation during the reaction, and are adequate to dissolve prior evaporation residues. Reactions that involve mild conditions (no heating and no strong nucleophiles or electrophiles including pH extremes) should be selected, whenever possible, in order to minimize degradation of the reaction components, and thereby reduce interferencing peaks later in the GC–EC–MS chromatogram. Even selection of a more chemically inert solvent such as toluene relative to acetonitrile can reduce interference

[15] S. Abdel-Baky, K. Allam, and R. W. Giese, *Anal. Chem.* **64,** 2882 (1992).

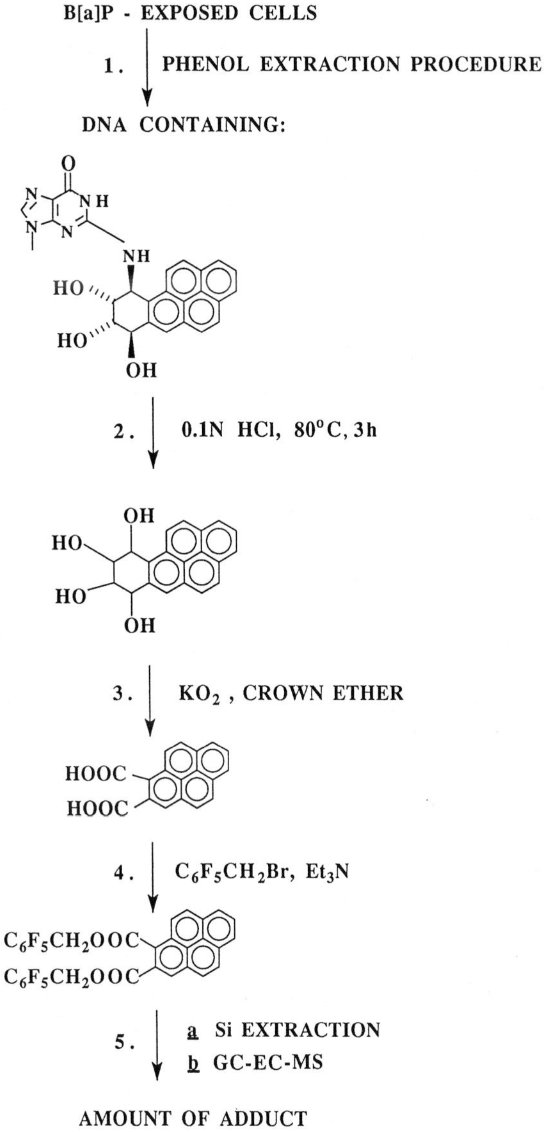
B[a]P - EXPOSED CELLS
1. PHENOL EXTRACTION PROCEDURE
DNA CONTAINING:
O
N
NH
N
N
NH
HO
HO
OH
2. 0.1N HCl, 80°C, 3h
OH
HO
HO
OH
3. KO2 , CROWN ETHER
HOOC
HOOC
4. C6F5CH2Br, Et3N
C6F5CH2OOC
C6F5CH2OOC
5. a Si EXTRACTION
b GC-EC-MS
AMOUNT OF ADDUCT

FIG. 4. Scheme for the detection of an acid-labile benzo[*a*]pyrenediol epoxide DNA adduct in cultured lymphocytes exposed to benzo[*a*]pyrene (B[*a*]P).

significantly.[16] A wider range of solvents can be used for trace vs conventional reactions because the low concentrations of the analyte and reagents reduce the need to select a solvent with good solubility properties toward the reaction components. Reactions should be selected that, at least on workup, yield no solid residues, to minimize adduct losses.

Continuous vortexing is attractive for agitating a trace chemical reaction (thereby avoiding a problematic PTFE or glass-coated stirring bar). The continuous vortexing unit that we employ (Multi-Mixer model 4600; Lab-Line Instruments, Melrose Park, IL) can be operated in an oven up to a temperature of 50°.

For evaporation of the solvent on completion of a chemical reaction, a stream of high-quality nitrogen may be preferred (with mild heating in a heating block, as necessary) rather than a centrifugal vacuum technique. The latter procedure may contaminate the sample (including analyte carryover; we encountered this problem with compound 1) from exposure to plastic, paint, rubber tubing, and vacuum pump oil. However, the vacuum technique is more convenient and thus it may be selected if its performance is acceptable for a given analyte.

Whether it is useful to employ trimethylsilylated vs nontreated glassware for the dilutions, reactions, and evaporations depends on the adduct, reagents, and solvents involved. We resort to trimethylsilylation only when necessary. At the conclusion of an evaporation, it has proven to be better to redissolve the trace analyte immediately rather than store it dry until the GC–EC–MS is available for injection. We have observed that a dry, trace analyte stored in the refrigerator for 1 week can largely disappear relative to an equivalent, evaporated sample that was redissolved and stored as a solution.

Attomole Gas Chromatography–Electron
Capture–Mass Spectrometry

Single-ion electrophores can be detected routinely at the low-attomole level on a GC–EC–MS that is dedicated to the determination of relatively pure, low-concentration compounds. We recommend injecting into the GC–EC–MS only samples that have been highly purified (e.g., by HPLC), form invisible residues on evaporation, and contain <10 pg/μl of analyte when derived from a low-level reaction, or <5 ng/μl of pure compound in the case of scanning detection.

Ultrahigh-purity gases (99.999%; total hydrocarbons < 0.5 ppm, $O_2 <$ 3 ppm; $H_2 < 3$ ppm) are used: helium carrier gas and methane reagent

[16] K. Allam, S. Abdel-Baky, and R. W. Giese, *Anal. Chem.* **64,** 238 (1992).

gas. Each gas is filtered through an Oxysorb 1 in-line system (Med-Tech Gases, Medford, MA) to remove oxygen. The black adsorbent in this glass cartridge turns brown with use, with a lifetime of about 1 year for such ultrahigh-purity gases.

A 10-μl syringe (e.g., Hamilton 710) fitted with a stainless steel plunger (not PTFE tipped at the sample–contact end), and a 7-cm polyimide-coated fused silica needle, are used for 1-μl on-column injections. We have not found it necessary to remove the polyimide coating at the sample–contact end of the needle. Sequentially, 2 μl of air, 1 μl of sample (dissolved in toluene, hexane, or acetonitrile) and 2 μl of air are drawn into the syringe, the needle is blotted between a fold of paper tissue, and a fast (1 sec) on-column injection is made. (Nitrogen instead of air in the syringe improved the response for octafluoronaphthalene, an abnormal model test compound, when a small amount of this compound was injected.[7]) The syringe is then washed by pulling up 10 μl of warm (60°) 10× toluene (from a vial sitting on the top of the warm GC), and placed in a covered box until the next injection. Whenever a more contaminated or higher concentration ($\geq$10-fold) analyte is injected (based on the resulting GC–EC–MS chromatogram), pure solvent is injected prior to the next sample to check for sample carryover. This overall approach has avoided a carryover problem in the GC–EC–MS. About twice a year, the duckbill and spring on the on-column injector (model 114; Hewlett-Packard) needs to be replaced, when needle insertion becomes difficult.

At the end of the day, 1-μl volumes of pure solvent (e.g., toluene) are injected until the baseline reequilibrates to its prior value at the start of the day (e.g., peak height abundance in the range of 1000–2000 units). Typically, a single injection is adequate after a full day of determining femtomole-level samples, whereas about five injections may be necessary when picomole-level samples have been analyzed (scanning conditions). The methane makeup gas then is turned off. (In fact, this should be done whenever a $\geq$1-hr delay is anticipated for the next injection, to minimize buildup of contamination in the system.) The carrier gas head pressure is reduced from 20 to 5 psi, and the column oven temperature is left at 220° until the following morning.

After about 2 weeks, we remove 10 cm from the injection end of the 15- to 30-m (when new) capillary GC column, because at this point the peaks have increased in width about 1.2-fold. This procedure improves the column performance (less so with time) until about 4 months (400 injections) has elapsed, at which point a new column is installed. On an older column, 20 cm is removed every 2 weeks. The peak width for the adduct after 400 injections is about 3 times its initial value on a new column, and no longer improves when an initial segment of the column is removed.

While some manufacturers have claimed that bonded-phase, fused silica capillaries can be restored by solvent washing after their performance degrades, this has not been successful in our work.

Tuning the instrument is accomplished generally every 2 weeks. Friday is a good day, so that the instrument can then "bake out" over the weekend. Tuning is accomplished in four stages, the first three of which [EI (electron impact), PCI (positive chemical ionization), and NCI (negative chemical ionization)] are performed as directed by the manufacturer (except that the source temperature is maintained at 250°), selecting recommended settings for high sensitivity. We then add an additional stage, in which the settings listed in Table I are selected (including, as seen, an elevated head pressure for the GC column). With these settings, the response of single-ion electrophores (at least **1–4**) is increased about 10-fold.[7] Routinely, we keep the emission current at 300 μA, however. Additional sensitivity

TABLE I
INCREASE IN SIGNAL-TO-NOISE RATIO[a]

| | Conditions | | |
Type	Conventional setting	Higher setting	Increase in S/N[b]
Head pressure (top of column), psi	5[c]	20[d]	6.5 ± 0.50 4.8 ± 0.40 (**2**)
Ion source pressure (methane), Torr	1.0[e]	2.0	2.8 ± 0.15 4.5 ± 0.32 (**3**)
Ion source T, °C	150[f]	250	2.1 ± 0.19
Electron energy, eV (ion source)	150	240	2.1 ± 0.15
Emission current, μA	300	450	2.0 ± 0.13

[a] Values of selected conditions in GC/EC–MS are increased one at a time (compound **1** tested except where indicated). (Reprinted with permission from S. Abdel-Baky and R. W. Giese, *Anal. Chem.* **63**(24), 2987. Copyright 1991 American Chemical Society.)

[b] Selected ion monitoring measurements were made. For each measurement, the other settings were conventional, as defined. Attomole amounts injected into 1 μl of toluene as separate solutions were 781 (**1**), 1123 (**2**), and 135 (**3**). The S/N was calculated by the computer for peak height. For each pair of settings (conventional and higher), the measurements were made on the same day. The precision shown is the range for triplicate measurements.

[c] Conventional range is 5–8 psi for a 12 m × 0.2 mm (length × i.d.) column. Linear velocity for the column was 25 cm sec^{-1} at 5.0 psi (air injection).

[d] Linear velocity was 35 cm sec^{-1}.

[e] Conventional range is 0.5–1.0 torr.

[f] Conventional range is 120–180° for EC.

sometimes can be achieved by raising the voltage on the electron multiplier by 600 V (the manufacturer recommends 400 V) above the value set in the EI tuning stage.

After about 6–9 months (when 3000 V is reached for the electron multiplier in the EI tuning stage), the ion source is cleaned and the electron multiplier replaced. (At times the ion source must be cleaned more frequently, e.g., after 4 months.) Individual electron multipliers can vary in sensitivity, so the response of a standard single-ion electrophore should be tested in the GC–EC–MS (with the column and ion source in good condition) immediately on installation of a new electron multiplier. Periodically, a new electron multiplier gives a 10-fold or lower response and needs to be exchanged for another one from the manufacturer. Selecting a sensitive electron multiplier currently is a matter of trial and error. Aside from unusual problems, the sensitivity of the instrument (for a given analyte) ordinarily depends on the condition of the column, ion source, and electron multiplier.

Example I: N^2 Guanine Benzo[a]pyrenediol Epoxide Adduct

Using the scheme shown in Fig. 4, we have detected the acid-labile N^2-guanine adduct of benzo[a]pyrenediol epoxide in cultured human lymphocytes exposed to benzo[a]pyrene (B[a]P). As seen, the method relies on the chemical transformation of the acid-released tetrahydrotetrol of B[a]P with potassium superoxide (step 3). It is attractive, as presented below, that steps 2–4 take place sequentially in the same vial, helping to make the procedure convenient.

After the DNA is isolated from the cell pellet by a conventional phenol extraction procedure, it is redissolved in water in an all-glass vial,[15] (300 μg of DNA in 500 μl of water). Extraction, once each, with 500 μl of water-saturated isoamyl alcohol and ethyl acetate (the intent being to remove nonadducted B[a]P metabolites) is followed by evaporation to 100 μl, dilution to 450 μl with water, addition of 50 μl of 1.2 N HCl, and heating at 90° for 3 hr with occasional vortexing. The sample is evaporated to dryness at 60° under nitrogen, followed by addition of 100 μl of acetonitrile and reevaporation. An internal standard is added (20 μl of methanol containing 106 pg of [1,2,3,4,5,6,11,12-$^2H_8$]9,10-dihydrobenzo[a]pyren-7(8H)-one, synthesized as described[17]) followed by evaporation. Addition of 25 μl of dimethylformamide containing 260 μg each of potassium superoxide and 18-crown-6 (freshly prepared mixture) is performed, followed by continuous vortexing at room temperature for 18 hr. After adding 500 μl of

[17] W. Li, C. Sotiriou-Leventis, S. Abdel-Baky, D. Fisher, and R. W. Giese, *J. Chromatogr.* **588,** 273 (1991).

1% acetic acid, followed by evaporation, 500 μl of 10% triethylamine in toluene is added. The sample is evaporated, 50 μl of toluene containing 2 μl of pentafluorobenzyl bromide and 2.5 μl of triethylamine is added, and the vial is vortexed at 50° for 5 hr and then overnight at room temperature. After evaporation, the sample is treated with 100 μl of hexane, followed by vortexing, evaporation, and addition of 500 μl of hexane. The sample is vortexed and then loaded with a Pasteur pipette onto a silica cartridge column: 100 mg of 40-μm silica gel particles from J. T. Baker (Phillipsburg, NJ) is sandwiched in a Pasteur pipette between two 10-mg plugs of silanized glass wool, followed by gravity washing with 3 ml each of ethyl acetate and hexane before the sample is applied. The column is washed twice with 2 ml of hexane taken through the vial, and the diester product is eluted with two 1-ml washes of ethyl acetate, followed by evaporation under N_2 at 60°. To focus the product at the bottom of the vial, the sample is treated with 100 μl of acetonitrile (the sample can be stored at this point) followed by vortexing and evaporation. Just before injection of 1 μl into the GC–EC–MS, 20 μl of acetonitrile is added and the sample is vortexed.

Representative, selective ion mass chromatograms from B[a]P-exposed and nonexposed cells are shown in Fig. 5.[18] Peak **1** in chromatogram A ("exposed cells" in Fig. 5) corresponds to the injection of 2.5 fmol of diester product. Corresponding peak **1'** in chromatogram A ("nonexposed cells" in Fig. 5) is an interference, as discussed in more detail elsewhere.[15] Detection of the internal standard (peak **2**) in the two kinds of samples is shown in the B chromatograms. On the basis of the analysis of a control sample consisting of calf thymus DNA spiked with a known amount of authentic 7,8,9,10-benzo[a]pyrenetetrahydrotetrol, the overall absolute yield of adduct in the procedure, at least throughout steps 3–5 (Fig. 4), is 22%.

Example II: N^7-(2'-Hydroxyethyl)xanthine

The electrophoric derivatization and detection of N^7-(2'-hydroxyethyl)-xanthine has been achieved. This compound is a chemical transformation product of the N^7-guanine adduct of ethylene oxide, as shown previously in Fig. 3. The final product that is formed and detected, after a two-step derivatization with pentafluorobenzyl bromide, is N^1, N^3-bis(pentafluoro-benzyl)-N^7-[2'-(pentafluorobenzyloxy)ethyl]xanthine (compound **3**).

N^7-(2'-Hydroxyethyl)xanthine (95 pg, 480 fmol) in 10 μl of 1 N HCl is evaporated in a vial under N_2. From a stock solution of tetrabutylammonium hydrogen sulfate (5 mg in 5 ml of 1N KOH), 50 μl (0.15 μmol) is added to the vial, with subsequent addition of 150 μl of CH_2Cl_2 and 10 μl

[18] K. Allam, S. Abdel-Baky, and R. W. Giese, *Anal. Chem.* **65,** 1723 (1993).

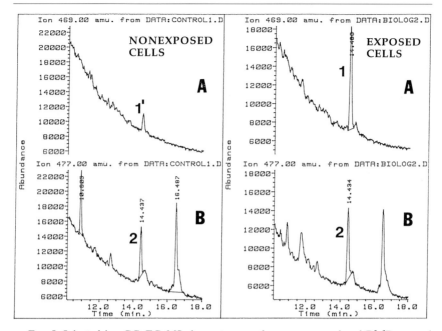

FIG. 5. Selected ion GC–EC–MS chromatograms from nonexposed and B[a]P-exposed cells. (A) m/z 469 to detect a diester product (**1**, which is the final product in Fig. 4) as peak **1**, and an interference (peak **1′**). (B) m/z 477 to detect a corresponding d_8-diester derived from the internal standard. GC–EC–MS: model 5988A mass spectrometer from Hewlett-Packard (Palo Alto, CA) fitted with a chemical ionization detector and connected to a Hewlett-Packard 5890 Series II gas chromatograph via a capillary interface kept at 290°. The capillary GC column (25-m length, 0.22-mm i.d., 0.1-μm film thickness, Ultra 1 from Hewlett-Packard) was temperature programmed from 140° immediately after on-column injection up to 290° at 70° min^{-1} and held for 13 min. Carrier gas, helium (20 psi); CI gas, methane (2 Torr). (Reprinted with permission from K. Allam *et al. Anal. Chem.* **65,** 1723. Copyright 1993 American Chemical Society.)

(0.065 μmol) of pentafluorobenzyl bromide. The reaction mixture is stirred for 20 hr at room temperature and the residual CH_2Cl_2 is slowly evaporated under N_2. Fifty microliters of H_2O and 150 μl of ethyl acetate are added, and the organic layer, after vortexing and centrifugation, is collected. Three more 150-μl ethyl acetate extractions are performed, and the combined organic layer is evaporated under N_2, redissolved in 50 μl of hexane : ethyl acetate, 1 : 1 (v/v), and applied to a Pasteur pipette column containing 200 mg of silica gel retained by silanized glass wool. This column had been prewashed (gravity flow) with 1 ml each of ethyl acetate and hexane. After washing with 4 ml of hexane and 8 ml of hexane : ethyl acetate, 90 : 10 (v/v), the product (**3**) is eluted with 2 ml of ethyl acetate, redissolved in 50

μl of toluene, and 1 μl is injected into the GC–EC–MS. The representative chromatogram shown in Fig. 6b[19] is obtained, corresponding to a yield of 56 ± 15% (n = 4). In Fig. 6a is shown a chromatogram from a blank reaction (no analyte).

Future

One of the attractive features of chemical transformation prior to GC–EC–MS for the detection of DNA adducts is that a given method, in principle, can be applied to an entire class of adducts. For example, the method presented above for the N^2-guanine adduct of benzo[a]pyrenediol epoxide is anticipated to detect diol epoxide polyaromatic hydrocarbon (PAH) DNA adducts in general,[20] including unknowns. Toward this goal, the described method for the tetrahydrotrol of B[a]P was found to detect the model compound chrysene-1,4-quinone.[20] Because the ions from the chemical transformation products of unknown diol epoxide PAH DNA adducts initially would be unknowns (making it impractical to use selected ion monitoring for their first-time detection), it would be attractive to use an MS instrument that simultaneously (essentially) detects many ions with high sensitivity, such as one equipped with an array detector.[21] Fourier transform ion cyclotron resonance and ion trap MS also are of interest in this regard. Once an ion (as a certain m/z value) is revealed with such equipment for an unknown, the compound then can be detected subsequently with high sensitivity by relying on selected ion monitoring on less expensive equipment. The ability to detect the adduct specifically, in turn, makes it possible to begin exploring the biological significance of the adduct even as an unknown, which in turn might trigger an interest in scaling it up for structural characterization.

We believe that GC–EC–MS methodology will be used increasingly in the trace detection of DNA adducts, and also for trace analytes in other areas of biomedical science. Toward this goal, it is important to continue making advances in the understanding and control over losses and interference during trace sample preparation. These problems can be best overcome once their mechanisms are revealed. Also, additional chemical transformation procedures are needed to bring a broader variety of trace analytes into the range of GC–EC–MS. With such advances, the high sensitivity and specificity of GC–EC–MS can be utilized more easily and generally in trace organic analysis.

[19] M. Saha and R. W. Giese, *J. Chromatogr.* **629,** 35 (1993).
[20] C. Sotiriou-Leventis, W. Li, and R. W. Giese, *J. Org. Chem.* **55,** 2159 (1990).
[21] J. A. Hill, J. E. Biller, S. A. Martin, K. Biemann, K. Yoshidome, and K. Sato, *Int. J. Mass Spectrom. Ion Process.* **92,** 211 (1989).

a

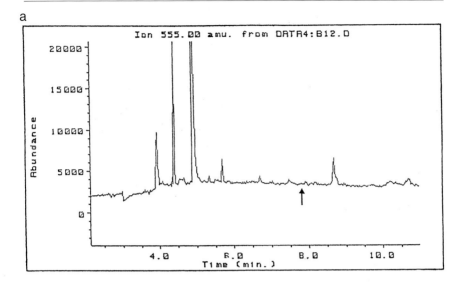

b

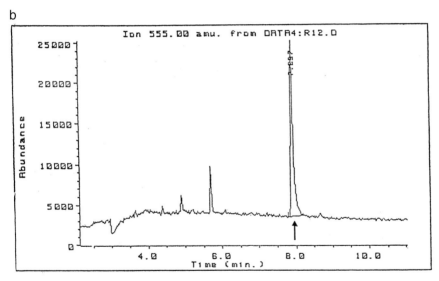

FIG. 6. Selected ion chromatograms obtained by derivatizing 0 pg (a) and 95 pg (b) of N^7-(2-hydroxyethyl)xanthine with pentafluorobenzyl bromide, followed by solid-phase extraction on a short silica column and injection of 1/50 of the final sample volume of 50 μl into a GC–EC–MS. Equipment and conditions were the same as in Fig. 5 except for the following: column, fused silica capillary, Ultra-1, 12 m, 0.2-mm i.d., 0.11-μm film thickness (Hewlett-Packard); GC oven temperatures, 120° start, then 70°/min immediately after injection up to 290° and hold for 6 min. [Reprinted with permission from M. Saha and R. W. Giese, *J. Chromatogr.* **629**, 40 (1993).]

Trace organic analysis by any technique is difficult. The tiny amount of analyte can easily fail to be detected because it is lost or inadequately purified during sample preparation. Additional purification steps tend to cause more losses. The losses and interference sometimes are no reproducible, increasing the difficulty of pinpointing their origin or learning their mechanisms. Highly purified reagents, exacting techniques, and instrumentation in top condition are often essential for success. Fortunately, each success makes it easier to achieve others, speeding up method development for related analytes. The talents and skills of a variety of scientists need to be combined to conquer the challenging frontier of trace organic analysis.

Acknowledgments

This work has been funded by NIH Grants OH02792, CA 65472, and CA 70056, Grant CN-71 from the American Cancer Society, and by a contract to the Health Effects Institute (HEI), an organization jointly funded by the United States Environmental Protection Agency (EPA) (Assistance Agreement X-812059) and automotive manufacturers. The specific grant was HEI Research Agreement 86-82. The contents of this chapter do not necessarily reflect the views of the HEI nor do they necessarily reflect the policies of the EPA or automotive manufacturers. Contribution No. 589 from the Barnett Institute. We thank Ronald Hites at Indiana University for reviewing this manuscript.

Author Index

Numbers in parentheses are footnote reference numbers and indicate that an author's work is referred to although the name is not cited in the text.

A

Abbas, S. A., 390
Abdel-Baky, S., 504, 506–507, 507(7), 511–512, 514, 515(7), 516, 516(7), 517–518, 518(15), 519
Aberth, W., 428
Aebersold, R., 201
Afeyan, N. B., 37, 112
Ager, S. W., 453
Agrawal, S., 152
Ahmed, F., 8, 17(36)
Aimoto, S., 435
Akashi, S., 435
Albin, M., 258
Alexander, J. E., 453
Al-Hakim, A., 344
Allam, K., 504, 511–512, 514, 518, 518(15), 519
Allen, R. O., 312
Alpert, A. J., 9, 17, 44
Amade, R., 341
Amankwa, L. N., 255
Ames, G. F., 181
Ampofo, S. A., 344
Amster, I. J., 435
Anderegg, R. J., 405, 435
Andersen, A. H., 203
Anderson, D. C., 239
Anderson, N. G., 119, 194–195
Anderson, N. L., 194–195
Andersson, L., 364
Andresen, F. H., 93(12), 94, 95(12)
Andrews, P. C., 9, 17
Andrieu, J. M., 313
Andrus, A., 306

Angel, A. S., 390, 397(17–19)
Annan, R. S., 365, 367, 441, 511
Anson, J. F., 179
Ansorge, W., 226, 230
Apella, E., 370, 376(31)
Apffel, A., 403, 405, 407(17), 408(17), 415, 418(29)
Apffel, J. A., 403
Appella, E., 435, 444, 444(34), 445(56)
Arai, M., 435
Arendale, R. F., 453(46), 454
Arold, N., 196
Arpino, P., 449
Arriaga, E., 307
Asada, T., 432
Astephen, N. E., 253, 256(24)
Astrua-Testori, S., 195
Aswad, D. W., 67
Atherton, J. M., 313
Atkins, L. M., 93, 95(10)
Atkinson, D., 315
Avdalovic, N., 147
Axelsson, K., 93(16), 94, 95(16), 96(16)

B

Baba, Y., 306, 311
Baczynskj, L., 453(60), 454, 478(60)
Bahl, O. P., 414
Baine, P., 293
Baker, C. S., 201
Bakthavachalam, J., 511
Baldo, B. A., 200
Baldonado, I. P., 67, 93(28), 94, 96(28)
Baldwin, M. A., 435, 445

Subject Index

H

U

W